Lecture Notes in Artificial Intelligence   12749

Subseries of Lecture Notes in Computer Science

Series Editors

Randy Goebel
  University of Alberta, Edmonton, Canada
Yuzuru Tanaka
  Hokkaido University, Sapporo, Japan
Wolfgang Wahlster
  DFKI and Saarland University, Saarbrücken, Germany

Founding Editor

Jörg Siekmann
  DFKI and Saarland University, Saarbrücken, Germany

Ido Roll · Danielle McNamara ·
Sergey Sosnovsky · Rose Luckin ·
Vania Dimitrova (Eds.)

Artificial Intelligence in Education

22nd International Conference, AIED 2021
Utrecht, The Netherlands, June 14–18, 2021
Proceedings, Part II

 Springer

Editors
Ido Roll 🆔
Technion – Israel Institute of Technology
Haifa, Israel

Sergey Sosnovsky 🆔
Utrecht University
Utrecht, The Netherlands

Vania Dimitrova
University of Leeds
Leeds, UK

Danielle McNamara 🆔
Arizona State University
Tempe, AZ, USA

Rose Luckin
London Knowledge Lab
London, UK

ISSN 0302-9743   ISSN 1611-3349 (electronic)
Lecture Notes in Artificial Intelligence
ISBN 978-3-030-78269-6   ISBN 978-3-030-78270-2 (eBook)
https://doi.org/10.1007/978-3-030-78270-2

LNCS Sublibrary: SL7 – Artificial Intelligence

This Springer imprint is published by the registered company Springer Nature Switzerland AG
The registered company address is: Gewerbestrasse 11, 6330 Cham, Switzerland

Preface

The 22nd International Conference on Artificial Intelligence in Education (AIED 2021), originally planned for Utrecht, the Netherlands, was held virtually during June 2021. AIED 2021 was the latest in a longstanding series of yearly international conferences for the presentation of high-quality research into ways to enhance student learning through applications of artificial intelligence, human computer interaction, and the learning sciences.

The theme for the AIED 2021 conference was "Mind the Gap: AIED for Equity and Inclusion." Over the past decades, racial and other bias-driven inequities have persisted or increased, diversity remains low in many educational and vocational contexts, and educational gaps have widened. Despite efforts to address these issues, biases based on factors such as race and gender persist. These issues have come to the forefront with recent crises around the world. In this conference, we reflected on issues of equity, diversity, and inclusion in regards to the educational tools and algorithms that we build, how we assess the efficacy and impact of our applications, theoretical frameworks, and the AIED society. The use of intelligent educational applications has increased, particularly within the past few years. As a community, development and assessment practices mindful of potential (and likely) inequities are necessary. Likewise, planned diversity, equity, and inclusion practices are necessary within the AIED society and home institutions and companies.

There were 168 submissions as full papers to AIED 2020, of which 40 were accepted as full papers (10 pages) with virtual oral presentation at the conference (an acceptance rate of 23.8%), and 66 were accepted as short papers (4 pages). Of the 41 papers directly submitted as short papers, 12 were accepted. Each submission was reviewed by at least three Program Committee (PC) members. In addition, submissions underwent a discussion period (led by a leading reviewer) to ensure that all reviewers' opinions would be considered and leveraged to generate a group recommendation to the program chairs. The program chairs checked the reviews and meta-reviews for quality and, where necessary, requested that reviewers elaborate their review. Final decisions were made by carefully considering both meta-review scores (weighed more heavily) and the discussions, as well as by rereading many of the papers. Our goal was to conduct a fair process and encourage substantive and constructive reviews without interfering with the reviewers' judgment.

Beyond paper presentations and keynotes, the conference also included the following:

- An Industry and Innovation track, intended to support connections between industry (both for-profit and non-profit) and the research community.
- A series of six workshops across a range of topics, including: empowering education with AI technology, intelligent textbooks, challenges related to education in AI (K-12), and optimizing human learning.

- A Doctoral Consortium track, designed to provide doctoral students with the opportunity to obtain feedback on their doctoral research from the research community.
- A Student Forum, funded by the Schmidt Foundation, that supported undergraduate students in learning about AIED, its past, present, and future challenges, and helped them make connections within the community. Special thanks go to Springer for sponsoring the AIED 2020 Best Paper Award. We also wish to acknowledge the wonderful work of the AIED 2020 Organizing Committee, the PC members, and the reviewers who made this conference possible. This conference was certainly a community effort and a testament to the community's strength.

April 2021

Ido Roll
Danielle McNamara
Sergey Sosnovsky
Rose Luckin
Vania Dimitrova

Organization

General Conference Chairs

Rose Luckin   University College London, UK
Vania Dimitrova   University of Leeds, UK

Program Co-chairs

Ido Roll   Technion - Israel Institute of Technology, Israel
Danielle McNamara   Arizona State University, USA

Industry and Innovation Track Co-chairs

Steve Ritter   Carnegie Learning, USA
Inge Molenaar   Radboud University, The Netherlands

Doctoral Consortium Track Co-chairs

Janice Gobert   Rutgers Graduate School of Education, USA
Tanja Mitrovic   University of Canterbury, New Zealand

Workshop and Tutorials Co-chairs

Mingyu Feng   WestEd, USA
Alexandra Cristea   Durham University, UK
Zitao Liu   TAL Education Group, China

Interactive Events Co-chairs

Mutlu Cukurova   University College London, UK
Carmel Kent   Educate Ventures, UK
Bastiaan Heeren   Open University of the Netherlands, the Netherlands

Local Co-chairs

Sergey Sosnovsky   Utrecht University, the Netherlands
Johan Jeuring   Utrecht University, the Netherlands

Proceedings Chair

Irene-Angelica Chounta   University of Duisburg-Essen, Germany

Publicity Chair

Elle Yuan Wang   Arizona State University, USA

Web Chair

Isaac Alpizar-Chacon   Utrecht University, the Netherlands

Senior Program Committee Members

Ryan Baker   University of Pennsylvania, USA
Tiffany Barnes   North Carolina State University, USA
Emmanuel Blanchard   IDÛ Interactive Inc., Canada
Christopher Brooks   University of Michigan, USA
Min Chi   BeiKaZhouLi, USA
Sidney D'Mello   University of Colorado Boulder, USA
Benedict du Boulay   University of Sussex, UK
Janice Gobert   Rutgers, USA
Peter Hastings   DePaul University, USA
Neil Heffernan   Worcester Polytechnic Institute, USA
Ulrich Hoppe   University of Duisburg-Essen, Germany
Judy Kay   The University of Sydney, Australia
H. Chad Lane   University of Illinois at Urbana-Champaign, USA
James Lester   North Carolina State University, USA
Noboru Matsuda   North Carolina State University, USA
Gordon McCalla   University of Saskatchewan, Canada
Bruce Mclaren   Carnegie Mellon University, USA
Agathe Merceron   Beuth University of Applied Sciences Berlin, Germany
Tanja Mitrovic   University of Canterbury, New Zealand
Inge Molenaar   Radboud University, the Netherlands
Roger Nkambou   Université du Québec à Montréal, Canada
Amy Ogan   Carnegie Mellon University, USA
Andrew Olney   University of Memphis, USA
Luc Paquette   University of Illinois at Urbana-Champaign, USA
Abelardo Pardo   University of South Australia, Australia
Zach Pardos   University of California, Berkeley, USA
Niels Pinkwart   Humboldt-Universität zu Berlin, Germany
Kaska Porayska-Pomsta   University College London, UK
Martina Rau   University of Wisconsin-Madison, USA
Ma. Mercedes T. Rodrigo   Ateneo de Manila University, Philippines
Jonathan Rowe   North Carolina State University, USA
Olga C. Santos   UNED, Spain
Sergey Sosnovsky   Utrecht University, the Netherlands
Erin Walker   Arizona State University, USA
Beverly Park Woolf   University of Massachusetts, USA

Kalina Yacef   The University of Sydney, Australia
Diego Zapata-Rivera   Educational Testing Service, USA

Program Committee Members

Laura Allen   University of New Hampshire, USA
Antonio R. Anaya   Universidad Nacional de Educacion a Distancia, Spain
Roger Azevedo   University of Central Florida, USA
Esma Aïmeur   University of Montreal, Canada
Michelle Banawan   Ateneo de Davao University, Philippines
Michelle Barrett   Edmentum, USA
Ig Bittencourt   Federal University of Alagoas, Brazil
Nigel Bosch   University of Illinois at Urbana-Champaign, USA
Anthony F. Botelho   Worcester Polytechnic Institute, USA
Jesus G. Boticario   UNED, Spain
Kristy Elizabeth Boyer   University of Florida, USA
Bert Bredeweg   University of Amsterdam, the Netherlands
Simon Buckingham Shum   University of Technology Sydney, Australia
Geiser Chalco Challco   ICMC/USP, Brazil
Maiga Chang   Athabasca University, Canada
Pankaj Chavan   IIT Bombay, India
Guanliang Chen   Monash University, Australia
Penghe Chen   Beijing Normal University, China
Heeryung Choi   University of Michigan, USA
Irene-Angelica Chounta   University of Duisburg-Essen, Germany
Andrew Clayphan   The University of Sydney, Australia
Keith Cochran   DePaul University, USA
Mark G. Core   University of Southern California, USA
Alexandra Cristea   Durham University, UK
Veronica Cucuiat   University College London, UK
Mutlu Cukurova   University College London, UK
Rafael D. Araújo   Universidade Federal de Uberlandia, Brazil
Mihai Dascalu   University Politehnica of Bucharest, Romania
Kristin Decerbo   Khan Academy, USA
Anurag Deep   IIT BOMBAY, India
Carrie Demmans Epp   University of Alberta, Canada
Diego Dermeval   Federal University of Alagoas, Brazil
Tejas Dhamecha   IBM, India
Barbara Di Eugenio   University of Illinois at Chicago, USA
Daniele Di Mitri   DIPF - Leibniz Institute for Research and Information
                               in Education, Germany
Vania Dimitrova   University of Leeds, UK
Fabiano Dorça   Universidade Federal de Uberlandia, Brazil
Nia Dowell   University of California, Irvine, USA
Mingyu Feng   WestEd, USA
Rafael Ferreira Mello   Federal Rural University of Pernambuco, Brazil

Carol Forsyth	Educational Testing Service, USA
Reva Freedman	Northern Illinois University, USA
Kobi Gal	Ben Gurion University, Israel and University of Edinburgh, UK
Cristiano Galafassi	Universidade Federal do Rio Grande do Sul, Brazil
Dragan Gasevic	Monash University, Australia
Isabela Gasparini	UDESC, Brazil
Elena Gaudioso	UNED, Spain
Michail Giannakos	Norwegian University of Science and Technology, Norway
Niki Gitinabard	North Carolina State University, USA
Ashok Goel	Georgia Institute of Technology, USA
Alex Sandro Gomes	Universidade Federal de Pernambuco, Brazil
Art Graesser	University of Memphis, USA
Monique Grandbastien	Universite de Lorraine, France
Nathalie Guin	Université de Lyon, France
Gahgene Gweon	Seoul National University, South Korea
Rawad Hammad	University of East London, UK
Jason Harley	McGill University, Canada
Yusuke Hayashi	Hiroshima University, Japan
Bastiaan Heeren	Open University of the Netherlands, the Netherlands
Martin Hlosta	The Open University, UK
Tomoya Horiguchi	Kobe University, Japan
Sharon Hsiao	Arizona State University, USA
Stephen Hutt	University of Pennsylvania, USA
Paul Salvador Inventado	California State University, Fullerton, USA
Seiji Isotani	University of São Paulo, Brazil
Patricia Jaques	UNISINOS, Brazil
Srecko Joksimovic	University of South Australia, Australia
Akihiro Kashihara	The University of Electro-Communications, Japan
Sandra Katz	University of Pittsburgh, USA
Carmel Kent	Educate Ventures, UK
Simon Knight	University of Technology Sydney, Australia
Ken Koedinger	Carnegie Mellon University, USA
Kazuaki Kojima	Teikyo University, Japan
Emmanuel Awuni Kolog	University of Ghana Business School, Ghana
Amruth Kumar	Ramapo College of New Jersey, USA
Tanja Käser	EPFL, Switzerland
Susanne Lajoie	McGill University, Canada
Sébastien Lallé	The University of British Columbia, Canada
Andrew Lan	University of Massachusetts Amherst, USA
Jim Larimore	Riiid Labs, USA
Nguyen-Thinh Le	Humboldt-Universität zu Berlin, Germany
Blair Lehman	Educational Testing Service, USA
Sharona Levy	University of Haifa, Israel
Fuhua Lin	Athabasca University, Canada

Diane Litman	University of Pittsburgh, USA
Zitao Liu	TAL Education Group, China
Yu Lu	Beijing Normal University, China
Vanda Luengo	Sorbonne Université, France
Collin Lynch	North Carolina State University, USA
Laura Malkiewich	Columbia University, USA
Ye Mao	North Carolina State University, USA
Leonardo Brandão Marques	University of São Paulo, Brazil
Mirko Marras	EPFL, Switzerland
Roberto Martinez Maldonaldo	Monash University, Australia
Smit Marvaniya	IBM, India
Jeffrey Matayoshi	McGraw Hill ALEKS, USA
Manolis Mavrikis	University College London, UK
Katie McCarthy	Georgia State University, USA
Danielle McNamara	Arizona State University, USA
Sein Minn	Inria, France
Kazuhisa Miwa	Nagoya University, Japan
Riichiro Mizoguchi	Japan Advanced Institute of Science and Technology, Japan
Camila Morais Canellas	Sorbonne Université, France
Bradford Mott	North Carolina State University, USA
Kasia Muldner	Carleton University, Canada
Anabil Munshi	Vanderbilt University, USA
Susanne Narciss	TU Dresden, Germany
Benjamin Nye	University of Southern California, USA
Ruth Okoilu	North Carolina State University, USA
Korinn Ostrow	Worcester Polytechnic Institute, USA
Ranilson Paiva	Universidade Federal de Alagoas, Brazil
Radek Pelánek	Masaryk University, Czech Republic
Elvira Popescu	University of Craiova, Romania
Thomas Price	North Carolina State University, USA
Ramkumar Rajendran	IIT Bombay, India
Genaro Rebolledo-Mendez	Tecnologico de Monterrey, Mexico
Steven Ritter	Carnegie Learning, USA
Ido Roll	Technion - Israel Institute of Technology, Israel
Rod Roscoe	Arizona State University, USA
Rinat Rosenberg-Kima	Technion - Israel Institute of Technology, Israel
José A. Ruipérez Valiente	University of Murcia, Spain
Vasile Rus	University of Memphis, USA
Demetrios Sampson	Curtin University, Australia
Mohammed Saqr	University of Eastern Finland, Finland
Zahava Scherz	Weizmann Institute of Science, Israel
Flippo Sciarrone	Roma Tre University, Italy
Shitian Shen	North Carolina State University, USA
Yu Shengquan	Beijing Normal University, China

Lei Shi	Durham University, UK
Sean Siqueira	Federal University of the State of Rio de Janeiro, Brazil
Caitlin Snyder	Vanderbilt University, USA
Trausan-Matu Stefan	Politehnica University of Bucharest, Romania
Angela Stewart	Carnegie Mellon University, USA
Thepchai Supnithi	NECTEC, Thailand
Pierre Tchounikine	University of Grenoble, France
K. P. Thai	Age of Learning, USA
Craig Thompson	The University of British Columbia, Canada
Armando Toda	University of São Paulo, Brazil
Richard Tong	Yixue Education Inc, China
Maomi Ueno	The University of Electro-Communications, Japan
Hedderik van Rijn	University of Groningen, the Netherlands
Kurt Vanlehn	Arizona State University, USA
Felisa Verdejo	Universidad Nacional de Educación a Distancia, Spain
Rosa Vicari	Universidade Federal do Rio Grande do Sul, Brazil
Elle Yuan Wang	Arizona State University, USA
Chris Wong	University of Technology Sydney, Australia
Simon Woodhead	Eedi, UK
Sho Yamamoto	Kindai University, Japan
Xi Yang	NCSU, USA
Bernard Yett	Vanderbilt University, USA
Ningyu Zhang	Vanderbilt University, USA
Qian Zhang	University of Technology Sydney, Australia
Guojing Zhou	University of Colorado Boulder, USA
Jianlong Zhou	University of Technology Sydney, Australia
Gustavo Zurita	Universidad de Chile, Chile

Additional Reviewers

Abdelshiheed, Mark
Afzal, Shazia
Anaya, Antonio R.
Andres-Bray, Juan Miguel
Arslan, Burcu
Barthakur, Abhinava
Bayer, Vaclav
Chung, Cheng-Yu
Cucuiat, Veronica
Demmans Epp, Carrie
Diaz, Claudio
DiCerbo, Kristen
Erickson, John
Finocchiaro, Jessica
Fossati, Davide

Frost, Stephanie
Gao, Ge
Garg, Anchal
Gauthier, Andrea
Gaweda, Adam
Green, Nick
Gupta, Itika
Gurung, Ashish
Gutiérrez Y. Restrepo, Emmanuelle
Haim, Aaron
Hao, Yang
Hastings, Peter
Heldman, Ori
Jensen, Emily
Jiang, Weijie

John, David
Johnson, Jillian
Jose, Jario
Karademir, Onur
Landes, Paul
Lefevre, Marie
Li, Zhaoxing
Liu, Tianqiao
Lytle, Nick
Marwan, Samiha
Mat Sanusi, Khaleel Asyraaf
Matsubayashi, Shota
McBroom, Jessica
Mohammadhassan, Negar
Monaikul, Natawut
Munshi, Anabil
Paredes, Yancy Vance
Pathan, Rumana
Prihar, Ethan
Rodriguez, Fernando

Segal, Avi
Serrano Mamolar, Ana
Shahriar, Tasmia
Shi, Yang
Shimmei, Machi
Singh, Daevesh
Stahl, Christopher
Swamy, Vinitra
Tenison, Caitlin
Tobarra, Llanos
Xhakaj, Franceska
Xu, Yiqiao
Yamakawa, Mayu
Yang, Xi
Yarbro, Jeffrey
Zamecnick, Andrew
Zhai, Xiao
Zhou, Guojing
Zhou, Yunzhan

International Artificial Intelligence in Education Society

Management Board

President

Rose Luckin   University College London, UK

President-Elect

Vania Dimitrova   University of Leeds, UK

Secretary/Treasurer

Bruce M. McLaren   Carnegie Mellon University, USA

Journal Editors

Vincent Aleven   Carnegie Mellon University, USA
Judy Kay   The University of Sydney, Australia

Finance Chair

Ben du Boulay   University of Sussex, UK

Membership Chair

Benjamin D. Nye   University of Southern California, USA

Publicity Chair

Manolis Mavrikis   University College London, UK

Executive Committee

Ryan S. J. d. Baker   University of Pennsylvania, USA
Min Chi   North Carolina State University, USA
Cristina Conati   The University of British Columbia, Canada
Jeanine A. Defalco   CCDC-STTC, USA
Vania Dimitrova   University of Leeds, UK
Rawad Hammad   University of East London, UK
Neil Heffernan   Worcester Polytechnic Institute, USA
Christothea Herodotou   The Open University, UK
Akihiro Kashihara   University of Electro-Communications, Japan
Amruth Kumar   Ramapo College of New Jersey, USA
Diane Litman   University of Pittsburgh, USA
Zitao Liu   TAL Education Group, China
Rose Luckin   University College London, UK
Judith Masthoff   Utrecht University, the Netherlands
Noboru Matsuda   Texas A&M University, USA
Tanja Mitrovic   University of Canterbury, New Zealand
Amy Ogan   Carnegie Mellon University, USA
Kaska Porayska-Pomsta   University College London, UK
Ma. Mercedes T. Rodrigo   Ateneo De Manila University, Philippines
Olga Santos   UNED, Spain
Ning Wang   University of Southern California, USA

–

Keynotes

Scrutability, Control and Learner Models: Foundations for Learner-Centred Design in AIED

Judy Kay ⓘ

The University of Sydney, Australia
judy.kay@sydney.edu.au

Abstract. There is a huge, and growing, amount of personal data that has the potential to help people learn. There is also a growing and broad concern about the ways that personal data is harvested and used. This makes it timely to draw on the decades of AIED research towards creating systems and interfaces that enable learners to truly harness and control their learning data. This invited keynote will present a whirlwind tour of my learner modelling research and a selection of other work that has influenced my own towards the goal of putting people in control of their own learning data and its use. I will explain the rationale for my focus on scrutability, as a foundation for users to harness and control their learning data, especially for learning contexts.

I will share key lessons from my work for creating AIED systems that are deeply learner centred. Building on this, I will present a vision for AIED, one that takes a learner-centred perspective to designing AIED systems and recognises the inherent limitations of learning data. This is a broad view of AIED that returns its founding goals to create advanced learning technologies.

Keywords. AIED · Learner models · Personalised learning systems · Scrutability · User control · User-centred design · Holistic design · Software engineering · Human-computer interaction

Augmenting Learning with Smart Design, Smart Systems, and Intelligence

Daniel M. Russell

Google Inc., Mountain View, CA, USA

Abstract. We all want better educational systems, no matter what the implementation might be. We tend to think of building ever more capable AI systems as the way to do this, but what is AI? It's rapidly becoming fancy software engineering: the definition continues to shift over time. What CAN we do in education to help students? My answer: Provide great, well-designed content; put it in a framework where others can use it; wrap it within a social system that lets students learn effectively, no matter the place or time; teach students how to learn. From my perspective, we have already built enormously effective information providing systems, but teaching students how to teach themselves remains key.

Daniel Russell is Google's Senior Research Scientist for Search Quality and User Happiness in Mountain View. He earned his PhD in computer science, specializing in Artificial Intelligence. These days he realizes that amplifying human intelligence is his real passion. His day job is understanding how people search for information, and the ways they come to learn about the world through Google. Dan's current research is to understand how human intelligence and artificial intelligence can work together to better than either as a solo intelligence. His 20% job is teaching the world to search more effectively. His MOOC, PowerSearchingWithGoogle.com, is currently hosting over 3,000 learners / week in the course. In the past 3 years, 4.5 million students have attended his online search classes, augmenting their intelligence with AI. His instructional YouTube videos have a cumulative runtime of over 350 years (24 hours/day; 7 days/week; 365 weeks/year). His new book, The Joy of Search, tells intriguing stories of how to be an effective searcher by going from a curious question to a reliable answer, showing how to do online research with skill and accuracy. Please note that the first paragraph of a section or subsection is not indented. The first paragraphs that follows a table, figure, equation etc. does not have an indent, either.

Invited Panels

Mind the Gap: The Bidirectional Relationship Between Diversity, Equity, and Inclusion (DEI) and Artificial Intelligence (AI)

Shima Salehi[1] and Rod D. Roscoe[2]

[1] Stanford University, Stanford, CA 94305, USA
salehi@stanford.edu
[2] Arizona State University, Mesa, AZ 85212, USA
rod.roscoe@asu.edu

Abstract. This panel discussion session explores the potential bidirectional relationship between (a) artificial intelligence (AI) methods and (b) diversity, equity, and inclusion (DEI) approaches in education.

Keywords. Artificial Intelligence · Inclusion · Equity

1 A Bidirectional Relationship

This panel discussion session explores the potential bidirectional relationship between (a) artificial intelligence (AI) methods and (b) diversity, equity, and inclusion (DEI) approaches in education. Participants will consider how AI methods can promote DEI in learning environments (AI for DEI) and how DEI approaches can improve AI analysis and interpretation to better meet the needs of diverse learners (DEI for AI).

1.1 AI for DEI

AI methods are particularly powerful for investigating complex relationships among variables, and have the potential to characterize, analyze, and make predictions regarding diverse learners in various contexts. These affordances can empower educators and researchers to more accurately monitor and identify learners' needs and progress. In turn, these insights might inform more equitable learning. For example, AI techniques enable the rapid analysis of rich data (e.g., interactions with simulations) that can inform formative assessments and feedback that are personalized to individual learners.

1.2 DEI for AI

As a potential paradigm shift, artificial intelligence in education (AIED) experts are increasingly attending to aspects of diversity, equity, and inclusion in theie conceptualizations, methods, and applications. For instance, there is a growing awareness of algorithmic bias, such that algorithms and automated systems can create or exacerbate discriminatory or prejudicial outcomes. Similarly, there is increasing awareness that conclusions based on statistical means can be misleading or exclusionary for learners who do not conform to "average" or majority demographics.

To address such concerns, AIED scholars must consider alternative approaches to studying educational phenomena, analyzing data, and drawing meaningful conclusions. For example, models may need to be disaggregated to include more nuanced variables and effects related to demographic factors and social identities. Simultaneously, intersectional approaches are needed to represent learners' multiple identities (and associated power, privilege, and history), and to interpret these effects within our findings and models. Consequently, this paradigm shift in AIED is not only poised to contribute to personalized learning, but to do so for a much broader diversity of learners.

2 Panel Organization

The panel comprises four presenters and two organizers who represent diverse yet complementary backgrounds related to DEI and AIED. Presenters (alphabetical order) include **Nia Dowell** (Assistant Professor, School of Education, University of California-Irvine [1]; **Rose Luckin** (Professor of Learner Centered Design, UCL Knowledge Lab, London) [2]; **Chris Piech** (Assistant Professor, Computer Science and Education, Stanford University) [3]; and **Marcelo Worsley** (Assistant Professor, Education and Social Policy, Northwestern University) [4]. The organizers include **Shima Salehi** (Assistant Professor, Graduate School of Education, Stanford University) [5]; and **Rod D. Roscoe** (Associate Professor, Fulton Schools of Engineering, Arizona State University) [6].

Presenters will first share their experiences regarding the bidirectional nature of DEI and AI in various contexts. Next, presenters and organizers will discuss questions submitted by the audience and questions emerging from the panelists. This interactive format will allow for a more inclusive session by incorporating opinions and experience of the wide-ranging audience. This diversity is crucial as the topic is emerging, nascent, but of significance to the future of the AIED community.

References

1. Dowell, N., Lin, Y., Godfrey, A., Brooks, C.: Promoting inclusivity through time-dynamic discourse analysis in digitally-mediated collaborative learning. In: Isotani, S. et al. (eds.)

AIED 2019. LCNS, vol. 11625, pp. 207–219. Springer, Cham (2019). https://doi.org/10.1007/978-3-030-23204-7_18

2. Holmes, W., Bektik, D., Woolf, B., Luckin, R.: Ethics in AIED: who cares? In: Isotani, S. et al. (eds.) AIED 2019. LCNS, vol. 11625, pp. 25–29. Springer, Cham (2019)

3. Piech, C. et al.: Co-teaching computer science across borders: human-centric learning at scale. In: 7th ACM Conference on Learning @ Scale 2020, pp. 103–113. ACM (2020)

4. Worsley, M, Bar-El, D.: Inclusive making: designing tools and experiences to promote accessibility and redefine making. Comput. Sci. Educ. (2020).

5. Salehi, S., Cotner, S., Ballen, C.: Variation in incoming academic preparation: consequences for minority and first-generation students. Front. Educ. 5, 552364 (2020)

6. Roscoe, R.D., Chiou, E.K., Wooldridge, A.R.: Advancing Diversity, Inclusion, and Social Justice Through Human Systems Engineering. CRC Press, Boca Raton (2020)

Research-Based Digital-First Assessments and the Future of Education

Alina A. von Davier[1], Valerie Shute[2], Jill Burstein[3],
Michelle Barrett[4], and Saad Khan[5]

[1] Duolingo
[2] Florida State University
[3] Educational Testing Service
[4] Edmentum
[5] FineTune Learning

Abstract. AI, learning engineering, computational Psychometrics, and big data coupled with numerous technology breakthroughs propose a new paradigm for education. From adaptive learning systems to digital-first -testing with automated content generation and automatic scoring - the possibilities for efficiency, scalability, and access are promising. The unprecedented disruption of COVID-19 leaves little doubt that advances in learning sciences and technology can augment the in-classroom educational experience. Digital-first assessments, sometimes called intelligent assessments are a new generation of tests where the technological advances and AI affordances are used to (re)create comprehensive assessments that are adaptive, efficient, rigorous, valid, and, most distinctively, attuned to perfect the user's experience. Digital-first assessments may be integrated into other systems (school systems, LMS, etc) being part of the new Internet of Education (IoE), where through integrative frameworks and standards one can optimize the support for each student while protecting their privacy. Stealth assessments through the use of process data from interactive tasks and multimodal data sources are moving from research labs into practice.

The panelists will share their research, provide evidence of how these new methodologies work, and engage the audience in a thought-provoking discussion on the impact of the new tests on education in general.

Keywords. Computational psychometrics · Stealth assessment · Automated writing evaluation · Digital-first assessment · Generating assessment

1 Computational Psychometrics as an Integrative Framework for Digital-First Assessments

In 2015, von Davier coined the term "computational psychometrics" (CP) to describe the fusion of psychometric theories and data-driven algorithms for improving the inferences made from technology-supported learning and assessment systems (LAS). Meanwhile, "computational" [insert discipline] has become a common occurrence. In CP the process data collected from virtual environments should be intentional: we should design & provide ample opportunities for people to display the skills we want to measure. CP uses the expert-developed theory as a map for the measurement efforts using process data. CP is also interested in the knowledge discovery from the (little, big) process data. Psychometric theories and data-driven algorithms are fused to make accurate and valid inferences in complex, virtual learning and assessment environments.

2 Stealth Assessment—What, Why, and How?

Proposed summary of the presentation: Games can be powerful vehicles to support learning, but this hinges on getting the assessment part right. In the past several years, we have designed, developed, and evaluated a number of stealth assessments in games to see: (a) if they provide valid and reliable estimates of students' developing competencies (e.g., in the areas of qualitative physics understanding, creativity, and persistence); (b) if students can actually learn anything as a function of gameplay; (c) the added value of inserting engaging learning supports (cognitive and affective) into the mix; and (d) if the games are still fun with the embedded assessments and supports. My presentation will cover the topic of stealth assessment in games to measure and support important 21st-century competencies. I'll describe why it's important, what it is, and how to develop/accomplish it. Time permitting, I'll also provide examples and videos in the context of a game we developed called Physics Playground.

3 Extending Automated Writing Evaluation for Integrative Frameworks

I will speak to systems and systems of systems that provide a digital-first assessment of the evidence of learning (either with or without testing) suitable for informing multiple adaptive decision-making loops in the educational ecosystem, including those at the learner, educator, school, district, and/or state levels. I will share a few exemplar theories of action and a conceptual model for such systems. I will provide an overview of industry standards that have been designed to facilitate the implementation of such systems to date and describe gaps and challenges that remain. Finally, I will reflect on research findings to date on hybrid systems that integrate digital adaptive assessment and adaptive instruction and describe a few elements I believe to be important for the research agenda moving forward.

4 Platforms and Standards in Support of Digital-First (Adaptive) Assessments

I will speak to systems and systems of systems that provide a digital-first assessment of the evidence of learning (either with or without testing) suitable for informing multiple adaptive decision-making loops in the educational ecosystem, including those at the learner, educator, school, district, and/or state levels. I will share a few exemplar theories of action and a conceptual model for such systems. I will provide an overview of industry standards that have been designed to facilitate the implementation of such systems to date and describe gaps and challenges that remain. Finally, I will reflect on research findings to date on hybrid systems that integrate digital adaptive assessment and adaptive instruction and describe a few elements I believe to be important for the research agenda moving forward.

5 Generating Assessment Items and Content with Artificial Intelligence

Educational assessment, learning, and publishing companies dedicate significant resources for the creation of original content for use in formative and summative tests, as well as in-classroom learning or open educational resources. Manual content creation can be laborious, highly dependent on domain expertise, and difficult to scale up. This bottleneck has come into sharper focus during the current pandemic, which has accelerated the shift to remote learning and heightened concerns of assessment items exposure.

I will share my experiences in artificial intelligence-based automated item and content generation. I will speak to the advances in natural language processing (models such as BERT [1], GPT3 [2]) that have enabled progress in this exciting field as well as current limitations to this technology and share thoughts on future directions. I will also discuss how AI-based automated item and content generation can result in scalable quality standardization, and open new possibilities for formative assessments and personalized learning experiences.

References

1. Devlin, J., Chang, M.W., Lee, K., Toutanova, K.: Bert: pre-training of deep bidirectional transformers for language understanding, arXiv preprint arXiv:1810.04805 (2018). Author, F.: Article title. Journal 2(5), 99–110 (2016)
2. Brown, T.B., Mann, B., Ryder, N., Subbiah, M., Kaplan, J., Dhariwal, P., Agarwal, S.: Language models are few-shot learners, arXiv preprint arXiv:2005.14165 (2020)

Workshops

Supporting Lifelong Learning

Oluwabunmi (Adewoyin) Olakanmi[1], Oluwabukola Mayowa Ishola[2],
Julita Vassileva[2], Ifeoma Adaji[2], and Zapata-Rivera Diego[3]

[1] Department of Computing Science, Concordia University of Edmonton
[2] Department of Computer Science, University of Saskatchewan
[3] Educational Testing Service Princeton, NJ
oluwabunmi.olakanmi@concordia.ab.ca, {bukola.ishola,
jiv, ifeoma.adaji}@usask.ca, dzapata@ets.org

Workshop Description

To achieve the theme of AIED 2021 "Mind the Gap: AIED for Equity and Inclusion", advanced learning technology research needs to support lifelong learners with the knowledge and skills needed to succeed in a rapidly changing world. The proliferation of social media and the recent need for everyone to transit to online learning due to the pandemic have made millions of lifelong learners turn to online learning communities (OLCs). With the availability of big data about learners from the OLCs and the availability of the enabling technologies, opportunities arise to provide personalized support to learners. During the first international workshop on supporting lifelong learning (SLL) co-located with the 20th international conference on Artificial intelligence in education (AIED 2019) some emerging themes were discussed in the areas of learner models, learner feedback, privacy and sustainability of lifelong learning systems.

The goal of the second workshop on supporting lifelong learning is to build on the first workshop by fostering further discussions around optimizing the learner models of lifelong learners to achieve their learning goals. SLL 2021 workshop aims at providing a forum for researchers to critically discuss ways to advance research in supporting lifelong learning beyond the walls of traditional educational systems. The second workshop will cover areas that address the application of advanced technologies like social recommendation, adaptive technologies, collaborative tools, persuasive strategies, learning analytics and educational data mining to support lifelong learners. This workshop aims at enhancing lifelong learning through collaboration, educational games, personalized recommendation, self-motivated learning and educational diagnosis of lifelong learners; and also, to review studies addressing lifelong learning.

Based on the category of papers, time will be allotted for presentation and questions. At the end of the workshop, there will be a discussion on workshop presentations, challenges and the ways forward, and we will develop a co-authored document to summarize the workshop papers. In summary, SLL 2021 will serve to expand the frontiers of knowledge within the advanced learning technology community, by providing opportunities for researchers to establish long term collaborations that can help to expand on studies that support lifelong learning. In addition, we look forward to the possibility of publishing a Special Issue in a relevant journal with extended versions of the accepted papers in the workshop from SLL 2019 and SLL 2021.

The First International Workshop on Multimodal Artificial Intelligence in Education

Daniele Di Mitri[1], Roberto Martínez-Maldonado[2], Olga C. Santos[3], Jan Schneider[1], Khaleel Asyraaf Mat Sanusi[4], Mutlu Cukurova[5], Daniel Spikol[6], Inge Molenaar[7], Michail Giannakos[8], Roland Klemke[4,9], and Roger Azevedo[10]

[1] DIPF - Leibniz Institute for Research and Information in Education, Frankfurt, Germany
[2] Monash University, Melbourne, Australia
[3] Spanish National University for Distance Education, Madrid, Spain
[4] Cologne Game Lab, TH Köln, Cologne, Germany
[5] University College London, UK
[6] University of Copenhagen, Denmark
[7] Radboud University, Nijmegen, The Netherlands
[8] Norwegian University of Science and Technology, Trondheim, Norway
[9] Open University of the Netherlands, Heerlen, The Netherlands
[10] University of Central Florida, USA

Abstract. This workshop aims at gathering new insights around the use of Artificial Intelligence (AI) systems and autonomous agents for education and learning leveraging multimodal data sources. The workshop is entitled Multimodal Artificial Intelligence in Education (MAIEd). It builds upon the Cross-MMLA workshop series at the Learning Analytics & Knowledge conference. The workshop calls for new empirical studies, even if in their early stages of developments. It also welcomes novel experimental designs, theoretical contributions and practical demonstrations which can prove the use of multimodal and multi-sensor devices ``beyond mouse and keyboard" in learning contexts with the purpose of automatic feedback generation, adaptation and personalisation in learning. Through a call for proposals, we seek to engage the scientific community in opening up the scope of AI in Education towards novel and diverse data sources.

1 Introduction

At the MAIEd workshop, we want to discuss which scientific, state-of-the-art ideas and approaches are being pursued and which impacts we expect on educational technologies and education. We are especially interested in contributions targeting the intersection of these two fields of AI and multimodal interaction. We are looking for original contributions that advance the state of the art in theories, technologies, methods, and knowledge towards the development of multimodal intelligent tutors, multimodal intelligence augmentation in teaching and learning and multimodal applications for self-regulated learning. The full text of the Call for Proposal and more information about the MAIEd 2021 workshop can be found on the workshop website http://maied.edutec.science/http://maied.edutec.science/.

Challenges and Advances in Team Tutoring Workshop

Anne M. Sinatra, Benjamin Goldberg, and Jeanine A. DeFalco

U.S. Army Combat Capabilities Development Command (DEVCOM) Soldier
Center
{anne.m.sinatra.civ,benjamin.s.goldberg.civ,
jeanine.a.defalco.civ}@mail.mil

Workshop Description

The "Challenges and Advances in Team Tutoring" workshop is a follow on to two previous AIED conference workshops held in person in 2018 and 2019 [1, 2]. It was clear from the workshops that team tutoring is a diverse and on-going field of study that is in constant development. Therefore, the current workshop specifically focuses on the Challenges and Advances in Team Tutoring. In line with one of those familiar challenges experienced this last year, the current workshop is virtual instead of in-person. With education and work settings shifting to distributed environments, understanding these impacts on collaborative learning and team development through tutoring are critical. The current virtual workshop covers all topic areas related to team tutoring, and provides an opportunity to discuss advances in the field that have been made by both new and returning presenters.

The workshop has three topic areas/themes: 1) Towards Intelligent Tutoring Systems for Teams in Distributed Environments, 2) Challenges and Lessons Learned in Creating Intelligent Tutoring Systems for Teams, and 3) Intelligent Tutoring System based Collaborative Problem Solving and Learning. Each topic area will include presentations of work and periods of open discussion to identify commonalities in approaches. Further gaps will be identified and addressed for future attention.

The workshop is expected to be of interest to those in academia, industry, and government in the field of team tutoring, along with those who would like to learn more about it. The expected outcomes of the workshop include an identification of current gaps and challenges in team tutoring, addressing those challenges across varying contexts and use cases, and defining next steps for the AIED community as they work towards maturing team tutoring solutions.

Acknowledgement. The statements and opinions expressed do not necessarily reflect the position or the policy of the United States Government, and no official endorsement should be inferred.

References

1. Sinatra, A.M., DeFalco, J.A. (eds): Proceedings of the Assessment and Intervention during Team Tutoring Workshop, London, England, 30 June 2018. http://ceur-ws.org/Vol-2153
2. Sinatra, A.M., DeFalco, J.A. (eds): Proceedings of the Approaches and Challenges in Team Tutoring Workshop, Chicago, IL, 29 June 2019. http://CEUR-WS.org/Vol-2501

Third Workshop on Intelligent Textbooks

Sergey Sosnovsky[1], Peter Brusilovsky[2], Richard G. Baraniuk[3],
and Andrew S. Lan[4]

[1] Utrecht University, Princetonplein 5, Utrecht 3584 CC, the Netherlands
s.a.sosnovsky@uu.nl
[2] University of Pittsburgh, 135 North Bellefield Ave., Pittsburgh, PA. 15260,
USA
peterb@pitt.edu
[3] Rice University, 6100 Main Street, Houston, TX 77005, USA
richb@rice.edu
[4] University of Massachusetts Amherst, 140 Governors Dr., Amherst, MA
01003, USA
andrewlan@cs.umass.edu

Abstract. Textbooks have evolved over the last several decades in many aspects. Most textbooks can be accessed online, many of them freely. They often come with libraries of supplementary educational resources or online educational services built on top of them. As a result of these enrichments, new research challenges and opportunities emerge that call for the application of AIEd methods to enhance digital textbooks and learners' interaction with them. Intelligent textbooks have the potential to benefit a large number of learners in online learning settings, especially after the COVID-19 pandemic. However, a number of research challenges have to be addressed before this vision become a reality. How to facilitate the access to textbooks and improve the reading process? What can be extracted from textbook content and data-mined from the logs of students interacting with it? The Third Workshop on Intelligent Textbooks focuses on these and other research questions related to intelligent textbooks. It seeks to bring together researchers working on different aspects of learning technologies to establish intelligent textbooks as a new, interdisciplinary research field.

Keywords. Digital and online textbooks · Open educational resources (OER) · Modelling and representation of textbook content · Assessment generation · Adaptive presentation and navigation · Content curation end enrichment

The transition of textbooks from printed copies to digital formats has facilitated numerous attempts to enrich them with various kinds of interactive functionalities including search and annotation, interactive content modules, and automated assessments. New research challenges and opportunities emerge that call for the application of AI methods to enhance digital textbooks and learners' interaction with them. Intelligent digital textbooks have the potential to significantly enhance the online learning experience, the importance of which is highlighted by the COVID-19 pandemic. Our workshop seeks to unify research efforts across several different fields, including AI, human-computer interaction, information retrieval, intelligent tutoring systems, and user modeling. This workshop brings together researchers working on different aspects of intelligent textbook technologies in these fields and beyond to establish intelligent textbooks as a new, interdisciplinary research field.

Advancing AI-Powered Education through Industry-Academia Cooperation

Richard Tong[1,2], Avron Barr[1], Xiangen Hu[1,3], Robby Robson[1,4],
and Brandt Redd[1,5]

[1] IEEE Learning Technology Standards Committee
[2] Squirrel AI Learning
[3] University of Memphis
[4] Eduworks
[5] MatchMaker Education Labs

The goal of "Advancing AI-Powered Education through Industry-Academia Cooperation" workshop co-sponsored by IEEE Learning Technology Standard Committee and Artificial Intelligence Standards Committee is to explore opportunities to empower educational systems with the most advanced AI technologies through industry and academia collaboration and to explore how to standardize on these systems, technologies, and practices, including adaptive learning systems, virtual classrooms, and systems that use machine learning to model student interactions and preferences to improve learning outcomes.

Programs:

- S01 Workshop Opening Remarks and Introduction
- S02* How technical standards and infrastructure support equity and inclusion. ("Mind the Gap: AIED for Equity and Inclusion")
- S03* How Learning Technology Standards Committee and Artificial Intelligence Committee can work together to bring AI to the forefront of education innovation - IEEE LTSC and AISC
- S4** How Industry and Research Community can benefit from advanced Virtual Classroom Technology and IEEE Standards
- S05 Explainable AI
- S06 Digital Textbook and Mobile Learning
- S07 Adaptive Instructional System @LTSC
- S08 Enterprise Learning Record
- S09 Interoperable Learning Record
- S10 LTSC standards Alpha Soup (xAPI, Virtual Classroom, Competencies, ..)
- S11* Cutting-Edge real-world projects. Where the industry is going?
- S12 AIS Consortium Overview and Practices
- S13* Academia and Industry Joint Research - Trend and Applications
- S14* Joint research with Industry and Academia
- S15 AI Architecture in Action

Contents – Part II

xlvi Contents – Part II

Industry and Innovation

Doctoral Consortium

Contents – Part I

Keynotes

Scrutability, Control and Learner Models: Foundations for Learner-Centred Design in AIED

Judy Kay(✉)

The University of Sydney, Sydney, Australia
judy.kay@sydney.edu.au

Abstract. There is a huge, and growing, amount of personal data that has the potential to help people learn. There is also a growing and broad concern about the ways that personal data is harvested and used. This makes it timely to draw on the decades of AIED research towards creating systems and interfaces that enable learners to truly harness and control their learning data. This invited keynote will present a whirlwind tour of my learner modelling research and a selection of other work that has influenced my own towards the goal of putting people in control of their own learning data and its use. I will explain the rationale for my focus on scrutability, as a foundation for users to harness and control their learning data, especially for learning contexts.

I will share key lessons from my work for creating AIED systems that are deeply learner centred. Building on this, I will present a vision for AIED, one that takes a learner-centred perspective to designing AIED systems and recognises the inherent limitations of learning data. This is a broad view of AIED that returns its founding goals to create advanced learning technologies.

Keywords: AIED · Learner models · Personalised learning systems · Scrutability · User control · User-centred design · Holistic design · Software engineering · Human-computer interaction

What is Scrutability in AIED?

In 2021, the global pandemic accelerated the already established growth in the use of technology in our lives, and particularly in learning. This has resulted in a large and growing amount of personal data that has the potential to be valuable for learning. We have also seen a growth in public concern at the ways that personal data is used, and misused. This is reflected in legislation over many jurisdictions, notable the EU for broad uses of personal data[1] and, FERPA in the case of education[2]. This makes it timely to identify some key lessons from

---

[1] European Union's General Data Protection Regulation (GDPR) https://gdpr-info.eu/.

[2] US Family Education Rights and Privacy Act (FERPA) https://www2.ed.gov/policy/gen/guid/fpco/ferpa/index.html.

© Springer Nature Switzerland AG 2021
I. Roll et al. (Eds.): AIED 2021, LNAI 12749, pp. 3–8, 2021.
https://doi.org/10.1007/978-3-030-78270-2_1

the decades of AIED research on learner modelling, now viewed as a principled way to harness personal data to support learning. In this keynote, I will focus on the notion of scrutable user modelling broadly, usually called learner modelling in the AIED research.

When I set about designing and building learner models in the early 1990s, I wanted to support the learner's agency and responsibility. It was also clear that AIED systems need to be design to take account of the typical deficiencies of learning data. I searched for a suitable term that was in use and could not find one. For example, explainable AI (XAI) has a long history; but that term is techno-centric, with a focus on the system explaining itself. XAI, and other terms, such as transparency, failed to capture the reality of the effort that the learner would need to invest if they want to actually understand a learner model at the level needed for learner agency in the learning processes that make use of technology. There are many terms including understandable and, from personal informatics research, intelligible, that are more user-centred [7]. These also fail to reflect how inherently challenging it actually is to understand a learner model which is based on the noisy, incomplete and changing data that is typical in learning contexts. So I chose the term *scrutable user model* as a more modest, but still challenging, goal for designers of advanced technology for learning.

Normal English uses the word inscrutable when we describe a person as inscrutable because we cannot understand them. Normal English also uses scrutinise as in this dictionary definition:

Scrutinse: Close, careful examination or observation.[3]

A starting point of my work was that human teachers may well be inscrutable to their students, but machines should not be. I set out to design AIED systems to be scrutable, as described in this dictionary definition:

Capable of being understood through study and observation. open to or able to be understood by scrutiny.[4]

The Central Role of Learner Models for Scrutability

In the earliest AIED work, with a key mission to create personalised tutors for every learner, the learner model was identified as one of the four core elements of AIED. The others were: domain expertise, teaching expertise and the interface. Such learner models are the drivers for personalisation that is a defining aspect of AIED. So opening them to the learner was a starting point for scrutability in that it makes at least some aspects of the learner model available to the learner. Even in 1997, I argued for repurposing classic learner models for valuable learning benefits [9] since an open learner model can:

1. Build shared understanding between the learner and the system builder about the goals of the teaching system;
2. Help the learner become aware of their current knowledge and progress;

---

[3] https://www.thefreedictionary.com/scrutiny.
[4] https://www.thefreedictionary.com/scrutable.

3. Identify suitable learning goals and facilitate the learner's planning to achieve them.

This can be operationalised in terms of the questions that an open learner model could enable a learner to answer — the questions map to the above purposes:

1. What does this system teach?
2. What have I demonstrated that I know? How well do I know a particular aspect, X? Am I making steady progress? Or not?
3. What should I do next?

Early learner models were deeply embedded within a particular teaching system, such as a cognitive tutor. It has become increasingly clear that the learner model, as a systematic way to interpret learning data, can also be seen as a "first class citizen", what Susan Bull described as an independent learner model. This means that it does not need to be tied to any single system. This is increasingly appealing today, for example, in a typical university course that uses diverse educational technology tools, each able to produce data that might be useful for a learner model. This could also be part of "Personal User Models for Life-long, Life-wide Learners" (PUMLs) [12] where learning data is stored in the learner's own storage space for long term modelling.

A Learner-Centred Definition of Scrutablity: Competency Questions

The normal English use of scrutable is a useful starting point for the designers of AI systems. But if we are to build such systems and evaluate whether they are, indeed, scrutable, we need systematic ways to evaluate scrutability of a system. My talk will explain a deeply user-centred approach to tackling the design, building and evaluation of scrutable AIED, starting with scrutable, independent learner models. This starts with the questions that a learner should be able to answer by scrutinising the system ([10]).

- Why does the system think I know Y (or do not know Z, or like A ...)?
- What else does the system think I know? Or don't know?
- *How can I tell the system I don't know Y (or do know Z) or dislike A ...) ?*
- Why did the system do X?
- What would the system do if I knew Z?
- What does the system do for other people?

To evaluate the scrutability of a system, we need to do studies to determine whether learners actually can find the answers to these questions. (See [12] for a more comprehensive list.) The first three are about the learner model and the other three about a system that uses the learner model. The third, italicised question, is one simple and elegant, learner centred and pedagogically grounded way for a learner to control the system. If a learner says they know Y, this can be seen as valuable information about the learner's reported self-assessment.[5]

---

[5] Of course, it is the nature of AIED systems that the learner may be more or less self-aware and they may be more or less honest about sharing their self-assessment with the system.

In my work, this is one valuable form of evidence for a learner model. Importantly, it can be combined with other evidence, such as the learner's performance on learning tasks that may indicate the learner's competence on Y. The accretion/resolution (AR) approach [11] to building learner and user models was driven by the goal of scrutability. AR is based on the view that, over time, learning evidence "accretes"[6] about each model component and when we need to "resolve" a value for the component, a piece of code inteprets the available evidence. Resolvers should be designed with awareness of the profound limitations of learning data for learner modelling and data mining [1]. Importantly, there can be multiple interpretations of the same evidence. With AR, learner control means that there should be interfaces that enable the learner to scrutinise details of: the raw data evidence allowed into their model; the inferred data evidence; and the resolver used to determine the value of a model component [10–12].

Why Do We Need Scrutability and Control in AIED Systems?

Scrutablity is important for AIED because:

- It affirms the role of the machine as the servant or aid of the user - reflecting asymmetry in the human-machine relationships;
- It supports the learner's right to see and appreciate the meaning of personal information systems hold about them;
- It enhances programmer accountability for personal data their systems collect and the way it is intepreted and used;
- It should enable the learner to determine the correctness and acceptability of the model and so, determine how much they trust it;
- If there are multiple resolvers, the scrutable learner model can enable the learner to decide which they want a system to use to interpret their learning data;
- It may motivate learners to share user model data because they feel confident about its meaning and use;
- The learner model can support valuable meta-cognitive processes such as self-monitoring, reflection and planning [4].

Scrutablity is a foundation for learners to control their learner model and AI system. This certainly relates to the growing call for users to be able to understand and control the increasing pervasive technology and personal data. But for learning contexts, it is has the many additional benefits outlined above. Achieving these benefits will require interfaces, such as those my group has explored [6,11] to discover whether learners do scutinise (they did), when they scrutinise a learner model and its use for personalisation (mostly after a quiz), whether they scrutinise when the system makes an error (mostly they did not,

---

[6] Where accretes means that it builds up over time − in my work [10,11], this means that timestamped raw data is added to the learner model, potentially triggering inferences that add more timestamped inferred data which also accretes.

some indicating they are used to accepting personalisation errors). At that time, we concluded that users were not really ready or hungry for scrutability and control of our learning system. Recent events and the ubiquity of personalisation and concern over the way our personal data is used sets the stage for this to change. A promising research agenda is create and evaluate interfaces scrutable AIED systems [7].

The Peculiar Place of OLMs and Visions for AIED

This abstract outlines the ideas that underpin the examples I will present in my keynote, with selected pointers to published work that provides the technical details. I will share some of my favourite Open Learner Models (OLMs), each with valuable lessons for building scrutable AIED: the huge body of work by Susan Bull, notably her OLMlets, as independent, practical and widely deployed OLMs [3]; the elegant and rigorously evaluated learner models by Mitrovic's group [14] the seminal work of Brusilovsky's teams on OLMs that support scrutiny [2,8] the use of self-assessment combined with an OLM from Aleven's group [13]; and the recent work by Conati's team to support scrutiny of an AIED system [5]. The aspects above are tightly focused on scrutability, control and learner models.

But one other theme of my keynote relates to the nature of AIED. In line with the earliest AIED researchers, I see AIED as multi-disciplinary research that strives to create advanced learning technology. As one of the Editors-in-Chief of the International Journal of Artificial Intelligence in Education, I am deeply aware of, and troubled by, the very narrow view that some researchers and reviewers have for AIED – seeing it as limited to creating new AI tools, albeit within an educational context and driven by educational needs. That narrow view fails to embrace the exciting technology and profound HCI challenges that will be central to creating future innovative educational technology. Nor does it embrace the current situation where many techniques, methods and tools that people think of as AI are now readily available as powerful off-the-shelf tools ready to be integrated into a system. These make it increasingly easy to build learning systems that make use of machine learning, speech understanding, robots, avatars and much else. There is certainly a need to innovate in the creation of new AIED techniques and tools and our community is the natural home for that work. But we need to go beyond that.

It is a curious artefact of history that Open Learner Modelling research is an accepted part of AIED, even when the core research goals relate to the HCI challenges and there is quite simple algorithmic reasoning underpinning the learner model. This is good news for scrutability since the simpler a learner model is, the more likely we are to succeed in building interfaces that enable learners to scrutinise them effectively. My talk will share the lessons for embracing simplicity as a goal. I see it as important that AIED takes a broad scope that includes multi-disciplinary perspectives and making progress in systems, software engineering and especially Human-Computer Interaction aspects. These will be critical to

ensuring our relevance in creating advanced technology for teachers and learners
for formal learning and for lifelong, life-wide learning.

References

1. Baker, R.S.: Stupid tutoring systems, intelligent humans. Int. J. Artif. Intell. Educ.
26(2), 600–614 (2016)
2. Brusilovsky, P., Schwarz, E., Weber, G.: ELM-ART: an intelligent tutoring system
on world wide web. In: Frasson, C., Gauthier, G., Lesgold, A. (eds.) ITS 1996.
LNCS, vol. 1086, pp. 261–269. Springer, Heidelberg (1996). https://doi.org/10.
1007/3-540-61327-7_123
3. Bull, S.: There are open learner models about!. IEEE Trans. Learn. Technol. **13**(2),
425–448 (2020)
4. Bull, S., Kay, J.: Open learner models as drivers for metacognitive processes.
In: Azevedo, R., Aleven, V. (eds.) International Handbook of Metacognition and
Learning Technologies. SIHE, vol. 28, pp. 349–365. Springer, New York (2013).
https://doi.org/10.1007/978-1-4419-5546-3_23
5. Conati, C., Barral, O., Putnam, V., Rieger, L.: Toward personalized XAI: a case
study in intelligent tutoring systems. Artif. Intell. **298**, 103503 (2021)
6. Czarkowski, M., Kay, J.: Giving learners a real sense of control over adaptivity,
even if they are not quite ready for it yet. In: Advances in Web-Based Education:
Personalized Learning Environments, pp. 93–126. IGI Global (2006)
7. Eiband, M., Buschek, D., Hussmann, H.: How to support users in understanding
intelligent systems? Structuring the discussion. In: 26th International Conference
on Intelligent User Interfaces, pp. 120–132 (2021)
8. Guerra-Hollstein, J., Barria-Pineda, J., Schunn, C.D., Bull, S., Brusilovsky, P.:
Fine-grained open learner models: complexity versus support. In: Proceedings of
the 25th Conference on User Modeling, Adaptation and Personalization, pp. 41–49
(2017)
9. Kay, J.: Learner know thyself: student models to give learner control and responsibility. In: Proceedings of International Conference on Computers in Education,
pp. 18–26 (1997)
10. Kay, J.: A scrutable user modelling shell for user-adapted interaction. Ph.D. thesis,
Basser Department of Computer Science, Faculty of Science, University of Sydney
(1998)
11. Kay, J., Kummerfeld, B.: Creating personalized systems that people can scrutinize
and control: Drivers, principles and experience. ACM Trans. Interact. Intell. Syst.
(TiiS) **2**(4), 1–42 (2013)
12. Kay, J., Kummerfeld, B.: From data to personal user models for life-long, life-wide
learners. Br. J. Edu. Technol. **50**(6), 2871–2884 (2019)
13. Long, Y., Aleven, V.: Enhancing learning outcomes through self-regulated learning
support with an open learner model. User Model. User-Adap. Inter. **27**(1), 55–88
(2017)
14. Mitrovic, A., Martin, B.: Evaluating the effect of open student models on self-
assessment. Int. J. Artif. Intell. Educ. **17**(2), 121–144 (2007)

Short Papers

Short Paper

Open Learner Models for Multi-activity Educational Systems

Solmaz Abdi$^{(\boxtimes)}$, Hassan Khosravi, Shazia Sadiq, and Ali Darvishi

The University of Queensland, Brisbane, Australia
solmaz.abdi@uq.edu.au

Abstract. In recent years, there has been an increasing trend in the use of student-centred approaches within educational systems that engage students in various higher-order learning activities such as creating resources, creating solutions, rating the quality of resources, and giving feedback. In response to this trend, this paper proposes an interpretable and open learner model called MA-Elo that capture an abstract representation of a student's knowledge state based on their engagement with multiple types of learning activities. We apply MA-Elo to three data sets obtained from an educational system supporting multiple student activities. Results indicate that the proposed approach can provide a higher predictive performance compared with baseline and some state-of-the-art learner models.

Keywords: Learnersourcing · Open learner model · Higher-order learning activity

1 Introduction

Learner models capture an abstract representation of a student's knowledge state. There are two main use cases for learner models: they are (1) employed as a key component of adaptive educational systems to provide personalised feedback or adaptivity functionalities and (2) externalised as open learner models (OLMs) [7,8] to students with the aim of incentivising, and regulating learning. Commonly, learner models estimate a student's knowledge state only based on their performance on attempting (answering) assessment items. As a point of reference, many well-known approaches for learner modelling including Bayesian Knowledge Tracing (BKT) [11], Item Response Theory (IRT) [22], Adaptive Factor Models (AFM) [9], Performance Factor Analysis (PFA) [23], deep knowledge tracing (DKT) [25], and DAS3H [10], as well as various rating based learner models [2,5,21,24] only employ students' performance on assessment items in their modelling. The reliance on only the performance of students on attempting assessment items can probably be explained by the fact that in many educational systems, students are prominently involved in just answering assessment items.

© Springer Nature Switzerland AG 2021
I. Roll et al. (Eds.): AIED 2021, LNAI 12749, pp. 11–17, 2021.
https://doi.org/10.1007/978-3-030-78270-2_2

In recent years, contemporary models of learning have placed a great emphasis on the use of learner-centred approaches that involve students in higher-order learning activities. A well-recognised approach for doing so is to employ learner-sourcing, which refers to a pedagogically supported form of crowdsourcing that partners with students to contribute novel content to teaching and learning while engaging in a meaningful learning experience themselves [17,20]. Prior studies on learnersourcing, as well as evidence from the learning sciences, indicate that students have the ability to meaningfully contribute to teaching and learning activities such as creating and evaluating learning resources [3,12,13,16,29,30] and that engaging with these activities enhances student learning [6,14,18,28].

So, how can educational systems that engage students in a range of activities openly and accurately model student learning? Some of the recently proposed learner models employ data from student engagement with multiple activities towards more accurately modelling learners [1,31]; however, they employ complex machine learning algorithms such as knowledge tracing machines [1] or tensor factorisation [31] which are not interpretable. We aim to address this limitation by proposing a multi-activity open and interpretable approach for modelling learners based on engagement with multiple types of learning activities.

2 Multi-activity Knowledge Modelling

Problem Formulation. We denote students by $s_n \in \{s_1 \ldots s_N\}$, learning resources (items) by $q_m \in \{q_1 \ldots q_M\}$, and knowledge components (concepts) by $\delta_c \in \{\delta_1 \ldots \delta_C\}$. Each item can be tagged with one or more concepts. We denote the relationship between items and concepts by $\omega_{mc} \in \Omega_{M \times C}$, where ω_{mc} is $1/f$ if item q_m is tagged with f concepts including δ_c, and 0 otherwise. Let $A = \{a_1 \ldots a_k\}$ denote the different types of activities that students are allowed to perform (e.g., creating, evaluating, linking or attempting items). Finally, let's assume that the system records the interaction log for s_n on each type of activity a_k as $i_t^k = (s_n, q_m, a_k, t, r_{nmt}^k)$, where t index the timestamp of the interaction and r_{nmt}^k indicates the outcome of the interaction. If it is a graded activity and the outcome of the interaction is success then $r_{nmt}^k = 1$ and zero otherwise. For a non-graded activity, the outcome is always considered as success. Our aim is to employ interpretable methods to (1) infer a learner model for estimating s_n's knowledge state on each concept δ_c and (2) infer the difficulty of each item q_m.

Proposed Approach. Employing the popular method of using rating systems for modelling learners [2,4,5,21,24,27], we present the Multi-Activity Elo-based learner model (MA-Elo), which is an extension over the multivariate Elo-based system [5], enabling interactions with multiple types of activities. To keep track of students' mastery, MA-Elo uses a two-dimensional array $\Lambda_{N \times C}$, where λ_{nc} represents student s_n's knowledge state on concept δ_c. For each item q_m, MA-Elo uses a global difficulty d_m approximating the difficulty of the item. For learning activities, MA-Elo considers two high-level categories. The first category incorporates activities in which the difficulty of learning items impacts the chance of a student's success. Examples of activities that fall into this category include attempting a learning item and creating a sample solution for an existing item. For each activity a_k in the first category, MA-Elo uses d_m of the item q_m associated in the activity to estimate the overall hardness of that activity for students. The second category consists of activities in which the chance of a student's success is independent of the difficulty level of the learning item (e.g., liking a resource). For each activity a_k in the second category, MA-Elo uses a global parameter h_k estimating the overall hardness of that activity. In practice, there are two options to calibrate the value of h_k: (1) a data-driven approach that treats h_k as a hyper-parameter and set it via cross-validation, or (2) the domain expert determines the relative difficulty of each of the learning activities. Whenever a student s_n performs a learning activity related to item q_m, MA-Elo first investigates if the activity comes from the first category or not and then uses the following equation to compute the chance of s_n's success:

$$P(r^k_{nmt} = 1) = \begin{cases} \sigma(\sum_{l=1}^{L} \lambda_{nc} \times \omega_{mc} - d_m), & \text{if the activity is from the first category} \\ \sigma(\sum_{l=1}^{L} \lambda_{nc} \times \omega_{mc} - h_k), & \text{otherwise} \end{cases}$$

where $\sigma(.)$ is the sigmoid function and $\sum_{l=1}^{L} \lambda_{nc} \times \omega_{mc}$ estimates s_n's weighted average competency on the concepts that are associated with q_m. MA-Elo then updates the student's mastery on each concept δ_l the question is tagged with based on the type of activity that is performed using $\lambda_{nl} := \lambda_{nl} + \zeta_k \cdot (r^k_{nmt} - P(r^k_{nmt} = 1))$, where r^k_{nmt} is the outcome of the interaction and ζ_k is a constant determining the sensitivity of the estimations based on the student's last interaction of the activity of type a_k. In addition, if the interaction was from the first category of activities, concurrent with updating the estimations of the student's knowledge state, the estimations of the model about the difficulty of the item q_m is also updated using $d_m := d_m + U(n) \cdot (P(r^k_{nmt} = 1) - r^k_{nmt})$, where $U(n)$ is an uncertainty function used for stabilising the estimates of item difficulty and is computed as $U(n) = \frac{\gamma}{1+\beta*n}$, where γ and β are constant hyper-parameters determining the starting value and slope of changes, respectively, and n indicates the number of prior updates on the item difficulty [24].

3   Evaluations

To evaluate MA-Elo, we use three historical data sets obtained from an educational system called RiPPLE and compare the predictive performance of MA-Elo with five existing learner models. At its core, RiPPLE is learnersourcing adaptive educational system that recommends learning items to students based on their estimated mastery level from a pool of items learnersourced by their peers [19]. RiPPLE enables students to engage with three main types of activities within the system, namely (1) practising learning items, (2) creating new items to be added to the repository of the system, and (3) moderating learning items in which students are involved in reviewing and evaluating learning items. Please refer to [19]

Table 1. RiPPLE Data sets

Data set	Students	Items	Concepts	Practice	Create	Moderate	Interactions
InfoSys	422	2008	7	47,122	940	4,586	52,648
NEUR	519	2,836	7	26,933	2,852	628	30,413
AI	322	1,312	12	19,031	1,305	6,475	26,811

Table 2. AUC and RMSE for the RiPPLE data sets.

Model	InfoSys		NEUR		AI	
	AUC	MSE	AUC	MSE	AUC	MSE
IRT	0.688	0.203	0.740	0.189	0.726	0.197
AFM	0.571	0.222	0.533	0.225	0.550	0.229
PFA	0.619	0.216	0.610	0.218	0.592	0.224
DAS3H	0.719	0.197	0.747	0.183	0.724	0.203
Multivariate-Elo	0.722	0.199	0.741	0.187	0.726	0.205
MA-Elo	**0.730**	**0.193**	**0.758**	**0.183**	**0.737**	**0.200**

for the detailed information about RiPPLE, the interface used for learning item creation and learning item moderation, and the formulation of the consensus approaches used by RiPPLE for each of these tasks. The three data sets used in the experiment as outlined in Table 1 are named (1) Introduction to Information Systems (InfoSys), (2) The Brain and Behavioural Sciences (NEUR) and, (3) Artificial Intelligence (AI). For our analysis to be consistent with the prior works (e.g., [10,26,31]), we evaluated the predictive performance of the models using 5-fold cross-validation where each data set split was done at the student-level. We compare the predictive performance of MA-Elo to IRT, PFA, AFM, and DAS3H. For this comparison, we use the implementation of these models provided by [15]. We also compare the predictive performance of MA-Elo to Multivariate-Elo [5], which is the most similar single-activity Elo-based learner model to our proposed model. Given the three main learning activities that students are engaged within RiPPLE, without loss of generalisability, we implemented MA-Elo based on these three activities namely attempt (a_1), create (a_2), moderate (a_3). In addition, we only used interactions related to learning items of type MCQ. We conducted a grid search to determine the hyper-parameters of MA-Elo. Across all experiments, for MA-Elo, the value of ζ_1 (determining the sensitivity of the estimations when attempting learning items), is set to 0.4, the value of ζ_2 is set to 0.25, and the value of ζ_3 is set to 0.15. For each model, we report the area under the curve (AUC) and mean squared error (MSE).

As it is presented in Table 2, on all of the data sets, MA-Elo outperforms other learner models in terms of predictive performance. This outcome is aligned with findings from the existing literature on learnersourcing (e.g., [14]) that suggest engaging students in higher-order activities impacts their learning. MA-Elo is followed by both Multivariate-Elo and the state-of-the-art DAS3H model, which are ranked as the second best-performing models on the RiPPLE data sets. This finding shows that, in spite of simplicity, ease of implementation, and without necessitating pre-calibration on big samples of data, the models developed based on Elo rating system could perform as well as or even better than the best-performing learner models known in the literature and can be considered as practical models for the implementation of real-world educational systems.

4 Conclusion

The overarching goal of this paper is to address the problem of learner modelling in educational systems where in addition to answering assessment items, students are also engaged with multiple types of learning activities. To do so, we proposed a learner model called MA-Elo that leverages data from students engagement with different types of learning activities other than answering assessment items when modelling their learning. The results of our conducted experiment on three data sets obtained from an adaptive learnersourcing educational system suggest that MA-Elo provides higher predictive performance compared with conventional learner models. Future work aims to investigate the impact of opening MA-Elo to students and its potential impact on self regulation and student learning.

References

1. Abdi, S., Khosravi, H., Sadiq, S.: Modelling learners in crowdsourcing educational systems. In: Bittencourt, I.I., Cukurova, M., Muldner, K., Luckin, R., Millán, E. (eds.) AIED 2020. LNCS (LNAI), vol. 12164, pp. 3–9. Springer, Cham (2020). https://doi.org/10.1007/978-3-030-52240-7_1
2. Abdi, S., Khosravi, H., Sadiq, S.: Modelling learners in adaptive educational systems: a multivariate glicko-based approach. In: 11th International Learning Analytics and Knowledge Conference, LAK21, pp. 497–503. Association for Computing Machinery (2021)
3. Abdi, S., Khosravi, H., Sadiq, S., Demartini, G.: Evaluating the quality of learning resources: a learner sourcing approach. IEEE Trans. Learn. Technol. 14(1), 81–92 (2021)
4. Abdi, S., Khosravi, H., Sadiq, S., Gasevic, D.: Complementing educational recommender systems with open learner models. In: Proceedings of the Tenth International Conference on Learning Analytics & Knowledge, pp. 360–365 (2020)
5. Abdi, S., Khosravi, H., Sadiq, S., Gasevic, D.: A multivariate Elo-based learner model for adaptive educational systems. In: Proceedings of the Educational Data Mining Conference, pp. 462–467 (2019)
6. Boud, D., Soler, R.: Sustainable assessment revisited. Assess. Eval. High. Educ. 41(3), 400–413 (2016)
7. Bull, S., Ginon, B., Boscolo, C., Johnson, M.: Introduction of learning visualisations and metacognitive support in a persuadable open learner model. In: Proceedings of the Sixth International Conference on Learning Analytics & Knowledge, pp. 30–39. ACM (2016)
8. Bull, S., Kay, J.: Open learner models. In: Nkambou, R., Bourdeau, J., Mizoguchi, R. (eds.) Advances in Intelligent Tutoring Systems. SCI, vol. 308, pp. 301–322. Springer, Heidelberg (2010). https://doi.org/10.1007/978-3-642-14363-2_15

9. Cen, H., Koedinger, K., Junker, B.: Learning factors analysis – a general method for cognitive model evaluation and improvement. In: Ikeda, M., Ashley, K.D., Chan, T.-W. (eds.) ITS 2006. LNCS, vol. 4053, pp. 164–175. Springer, Heidelberg (2006). https://doi.org/10.1007/11774303_17

10. Choffin, B., Popineau, F., Bourda, Y., Vie, J.J.: DAS3H: modeling student learning and forgetting for optimally scheduling distributed practice of skills. arXiv preprint arXiv:1905.06873 (2019)

11. Corbett, A.T., Anderson, J.R.: Knowledge tracing: modeling the acquisition of procedural knowledge. User Model. User-Adap. Inter. **4**(4), 253–278 (1994)

12. Darvishi, A., Khosravi, H., Sadiq, S.: Utilising learnersourcing to inform design loop adaptivity. In: Alario-Hoyos, C., Rodríguez-Triana, M.J., Scheffel, M., Arnedillo-Sánchez, I., Dennerlein, S.M. (eds.) EC-TEL 2020. LNCS, vol. 12315, pp. 332–346. Springer, Cham (2020). https://doi.org/10.1007/978-3-030-57717-9_24

13. Denny, P., Hamer, J., Luxton-Reilly, A., Purchase, H.: Peerwise: students sharing their multiple choice questions. In: Proceedings of the Fourth International Workshop on Computing Education Research, pp. 51–58 (2008)

14. Doroudi, S., et al.: Crowdsourcing and Education: Towards a Theory and Praxis of Learnersourcing. International Society of the Learning Sciences, Inc. [ISLS] (2018)

15. Gervet, T., Koedinger, K., Schneider, J., Mitchell, T., et al.: When is deep learning the best approach to knowledge tracing? JEDM—J. Educ. Data Min. **12**(3), 31–54 (2020)

16. Guo, P.J., Markel, J.M., Zhang, X.: Learnersourcing at scale to overcome expert blind spots for introductory programming: a three-year deployment study on the python tutor website. In: Proceedings of the Seventh ACM Conference on Learning@ Scale, pp. 301–304 (2020)

17. Khosravi, H., Demartini, G., Sadiq, S., Gasevic, D.: Charting the design and analytics agenda of learnersourcing systems. In: 11th International Learning Analytics and Knowledge Conference, LAK21, pp. 32–42. Association for Computing Machinery, New York (2021)

18. Khosravi, H., Gyamfi, G., Hanna, B.E., Lodge, J.: Fostering and supporting empirical research on evaluative judgement via a crowdsourced adaptive learning system. In: Proceedings of the Tenth International Conference on Learning Analytics & Knowledge, pp. 83–88 (2020)

19. Khosravi, H., Kitto, K., Williams, J.J.: Ripple: a crowdsourced adaptive platform for recommendation of learning activities. J. Learn. Anal. **6**(3), 91–105 (2019)

20. Kim, J., et al.: Learnersourcing: improving learning with collective learner activity. Ph.D. thesis, Massachusetts Institute of Technology (2015)

21. Klinkenberg, S., Straatemeier, M., van der Maas, H.L.: Computer adaptive practice of maths ability using a new item response model for on the fly ability and difficulty estimation. Comput. Educ. **57**(2), 1813–1824 (2011)

22. Lord, F.M.: Applications of Item Response Theory to Practical Testing Problems. Routledge, London (2012)

23. Pavlik Jr., P.I., Cen, H., Koedinger, K.R.: Performance factors analysis-a new alternative to knowledge tracing. Online Submission (2009)

24. Pelánek, R., Papoušek, J., Řihák, J., Stanislav, V., Nižnan, J.: Elo-based learner modeling for the adaptive practice of facts. User Model. User-Adap. Inter. **27**(1), 89–118 (2017)

25. Piech, C., et al.: Deep knowledge tracing. In: Advances in Neural Information Processing Systems, pp. 505–513 (2015)

26. Piech, C., et al.: Deep knowledge tracing. In: Advances in Neural Information Processing Systems, vol. 28, vol. 505–513 (2015)

27. Reddick, R.: Using a glicko-based algorithm to measure in-course learning. In: Proceedings of the Educational Data Mining Conference, pp. 754–759. ERIC (2019)
28. Tai, J., Ajjawi, R., Boud, D., Dawson, P., Panadero, E.: Developing evaluative judgement: enabling students to make decisions about the quality of work. High. Educ. **76**(3), 467–481 (2018)
29. Wang, X., Talluri, S.T., Rose, C., Koedinger, K.: Upgrade: sourcing student open-ended solutions to create scalable learning opportunities. In: Proceedings of the 6th ACM Conference on Learning@ Scale, pp. 1–10, June 2019
30. Zahirović Suhonjić, A., Despotović-Zrakić, M., Labus, A., Bogdanović, Z., Barać, D.: Fostering students' participation in creating educational content through crowdsourcing. Interact. Learn. Environ. **27**(1), 72–85 (2019)
31. Zhao, S., Wang, C., Sahebi, S.: Modeling knowledge acquisition from multiple learning resource types. arXiv preprint arXiv:2006.13390 (2020)

Personal Vocabulary Recommendation to Support Real Life Needs

Victoria Abou-Khalil[✉], Brendan Flanagan, and Hiroaki Ogata

Academic Center for Computing and Media Studies, Kyoto University, Kyoto, Japan

Abstract. The vocabulary taught in language classes or through digital language learning tools is disconnected from the real-life needs of many language learners. Immigrants, refugees, students abroad learn a language to navigate through their daily lives and often need words that are missing from their curricula they study. Today's language learners rely heavily on digital translators and dictionaries, creating a database of words they need in their everyday life. The availability of this data could allow personal vocabulary suggestions that meet real-life needs. To show the unsuitability of commonly provided vocabulary lists, we compare them to the vocabulary needed by 37 Syrian refugees living in Lebanon and Germany. We show that the vocabulary provided by the Cambridge English List and Duolingo has low usefulness and low efficiency and discuss future directions for personal vocabulary recommendations.

Keywords: Vocabulary learning · Language learning · Immigration · Recommendation system · Personalization

1 Introduction

Since the 70s, language learning instruction has shifted from situational teaching of sentence patterns to a communicative teaching approach [12]. While the teaching methods followed the needs of the modern world, the vocabulary curriculum did not. Vocabulary learning materials are standardized to a certain extent and usually provide words based on their frequency and usefulness [11]. Frequency is measured based on the most frequent words used by native speakers [5]. Typically, a beginner's language course includes words frequently used by native speakers, whereas an advanced language course includes words rarely used by native speakers. On the other hand, the usefulness of the vocabulary is judged by teachers and curriculum writers [11] instead of the learners themselves. Thus, the curriculum ends up reflecting the perception that curriculum writers have about the a learner vocabulary needs rather than their actual needs [22]. This choice of vocabulary curriculum still places native speakers at the center of language education by considering that their needs and lifestyle are the references to aim for. For example, a Syrian refugee in Germany would end up learning how to say *hike* (from the Goethe A1 exam), six months before learning how to

© Springer Nature Switzerland AG 2021
I. Roll et al. (Eds.): AIED 2021, LNAI 12749, pp. 18–23, 2021.
https://doi.org/10.1007/978-3-030-78270-2_3

say *migrant* (from the Goethe B1 exam), and eight months before learning how to say *refugee*(Aspekt Neu B2 vocabulary list).

The Digital Trace of Language Learners. Vocabulary curriculum today are still standardized to fit the needs of the "average" learner within formal education settings. However, learners learn mostly informally, and have complete control over the content and the learning activities [9,16]. This is particularly true today considering that dictionaries and online translators are the preferred language learning tools [10] and learners choose to translate and learn words inspired by their surroundings, interests, and goals [21]. Through these interactions, learners leave behind them a digital trace of all the words and sentences that they translated or searched for in a digital dictionary.

The digital trace of a learner or a group of learners can be valuable for the recommendation and learning of the vocabulary they need. This data has been mostly overlooked and little research has explored it for the benefit of learning, possibly due to the difficulty of accessing it. Jung and Graf (2008) proposed a system that recommends target words based on the learner's lexical knowledge [14]. However, the lexical knowledge is formed by the words taught by the system, and not extracted from the real-life data of a learner. Abou-Khalil et al. used the past vocabulary logs to provide learners with personalized vocabulary recommendation and personalized translation [1,2]. Except sparse examples, language learner data has been mainly used for cognitive personalization [13] like the levels of difficulty of the vocabulary [17], learning memory cycles [8], or medium of teaching based on learning styles [7]. The semantic aspect of the vocabulary and the indications it gives about a learner's needs, activities, and interests is yet to be explored.

Personal Vocabulary Learning. The corpus formed by words registered in a digital dictionary can be used to create a personal vocabulary learning experience. The term personal language learning has been introduced by Kukulska-Hulme in 2016 to draw a distinction with personalized learning [15]. The term personalization presumes the adaptation of teaching methods to reach predetermined education goals, whereas personal learning highlights the learner's control over their own learning and the setting of their own goals. In a world where mobility and migration are a part of many people's lives, people learn a new language to achieve different goals and needs. Personal vocabulary learning would allow language learners to access the appropriate vocabulary based on their real-life needs. In this work we aim to 1) demonstrate that the vocabulary needs of language learners are not met by available language learning curriculum and tools and 2) Discuss directions for personal vocabulary recommendation.

2   Case-Study: Real-Life Vocabulary Needs of Syrian Refugees

In this section, we analyze the real-life vocabulary needed by Syrian refugees living in Lebanon and in Germany and compare it to the vocabulary they encounter through language courses and mobile language learning applications.

Data Collection. 25 Syrian refugees residing in Lebanon, and 12 Syrian refugees residing in Germany were recruited to participate in this study. We collected the vocabulary needed by the refugees using the language learning environment SCROLL for a period of ten days. Through SCROLL, users can translate and save words that they wish to learn [19]. We asked the Syrian refugees in Germany to input in SCROLL unknown words that they encounter and need in their daily lives. On the other hand, the participants in Lebanon were asked to input the words they want to learn. Syrian refugees in Lebanon speak Arabic, Lebanon's official language but many of them are studying English [6,20] to facilitate their immigration to another country with better education opportunities, more safety, and higher respect for human rights [23]. The participants in Lebanon logged 1525 words whereas the participants in Germany logged 674 words.

Analysis of the Vocabulary. To determine the similarity between the vocabulary needed by the refugees in their real life and the vocabulary available to them, we calculate the usefulness and efficacy of the word lists of Cambridge English List (CEL) and Duolingo's whole course, compared to the vocabulary needed by the Syrian refugees. We chose the CEL by the Cambridge Assessment English as it develops one of the most common standardized tests like IELTS and its qualifications are aligned with the levels of the Common European Framework of Reference for Languages (CEFR). On the other hand, with more than 300 million registered users around the word, Duolingo is one of the most used language learning applications, and is specifically widespread among refugees [3].

The usefulness U of a vocabulary list represents how much the list meets the needs of the learners, i.e. the percentage of the words that they searched in SCROLL that are present in the list. A usefulness of 100% means that every word searched by the learner is included in the vocabulary list. L represents the set of words needed by the learner i.e. the list of words recorded by the refugees in SCROLL; and V the set of words in the vocabulary list, in this case Duolingo (all modules) or the CEL (whole list). The efficiency E of a vocabulary list represents the portion of the list that is searched by the learners. An efficiency of 100% means that every word in the vocabulary list was searched by the learner. In the following, we will use $L_{Lebanon}$ and $L_{Germany}$ to refer to the vocabulary searched by the refugees in Lebanon and Germany, respectively.

$$U = \frac{|L \cap V|}{|L|} \qquad \text{(1a)} \qquad\qquad E = \frac{|L \cap V|}{|V|} \qquad \text{(1b)}$$

Results and Discussion. Table 1 shows the usefulness and efficiency of CEL and Duolingo word lists by comparing them to the words searched by the participants in Lebanon and Germany. Duolingo's whole course covers a bigger part of the vocabulary needed by the Syrian refugees than the CEL. On the other hand, Syrian refugees in Lebanon can find a bigger portion of their needed vocabulary in available word lists compared to refugees in Germany. This may be due to the fact that refugees in Lebanon imagine the words that they will need in the future, whereas refugees in Germany have a clearer idea of the vocabulary they actually need as they already lived for some time in their target country, and used the target language. For example, the most frequent vocabulary searched by the Syrian refugees in Lebanon includes words relating to healthcare (doctor, pharmacy), to family (sister, brother, children), to food (zucchini, potato, tomatoes) and words like embassy and swimming pool whereas most frequent vocabulary searched by the Syrian refugees in Germany includes vocabulary representing everyday life activities (labour office, children section, department of obstetrics, foreigner office), ingredients of levantine cuisine (parsley), bureaucracy (work contract, telephone contract, internet contract, financial penalty) and words like baby carriage and home country. Finally, both Duolingo and CEL have a low efficiency and usefulness, which means that refugees have to learn a big number of words that only covers a portion of their needs.

Table 1. Usefulness and efficiency of vocabulary lists available to refugees

	CEL		Duolingo	
	U	E	U	E
L_{Lebanon}	30%	28%	40%	20%
L_{Germany}	15%	9%	25%	9%

3   Conclusions and Future Directions

In this paper, we showed the unsuitability of commonly available vocabulary lists and the need for personal vocabulary recommendation that answers a learner real life needs. This can be facilitated by the increasing availability of the learner digital trace that creates opportunities for a 1) purpose-based, 2) demographic-based, and 3) content-based personal vocabulary recommendation. The vocabulary suggestion can be enabled by a recommendation system that is connected to the learners' preferred dictionary to collect words that learners need in their daily life. Both purpose and demographic factors were shown to be important in previous user modeling and language learning research [4,18]. Combining these factors with the learners' data can provide personal vocabulary recommendation by suggesting vocabulary that other people with the same demographics or purpose searched for in their digital dictionaries. Learners' data can also enable content-based recommendations by suggesting vocabulary that is thematically similar to the one translated by the learner in the past. Additionally, the digital trace can be used to detect a change in purpose or recommend words based on the situated context of the learner. Future work could focus on implementing a personal vocabulary recommendation framework, validating it and the identifying the drawbacks and challenges in adopting it.

References

1. Abou-Khalil, V., et al.: Vocabulary recommendation approach for forced migrants using informal language learning tools (2021, in press)
2. Abou-Khalil, V., Helou, S., Flanagan, B., Chen, M.-R.A., Ogata, H.: Learning isolated polysemous words: identifying the intended meaning of language learners in informal ubiquitous language learning environments. Smart Learn. Environ. 6(1), 1–18 (2019). https://doi.org/10.1186/s40561-019-0095-0
3. Abou-Khalil, V., Helou, S., Flanagan, B., Pinkwart, N., Ogata, H.: Language learning tool for refugees: identifying the language learning needs of Syrian refugees through participatory design. Languages 4(3), 71 (2019). https://doi.org/10.3390/languages4030071
4. Abou-Khalil, V., Helou, S., Khalifé, E., Majumdar, R., Ogata, H.: Emergency remote teaching in low-resource contexts: how did teachers adapt? In: Proceedings of the 28th International Conference on Computers in Education, vol. I, pp. 686–688, November 2020
5. Adolphs, S., Schmitt, N.: Lexical coverage of spoken discourse. Appl. Linguist. 24(4), 425–438 (2003). https://doi.org/10.1093/applin/24.4.425
6. Casalone, M., Puig, N.: ENFANCES EN MIGRATION Une étude sur les enfants syriens réfugiés au Liban, vol. 7, pp. 23–24. Université Paris Diderot-Paris, Paris (2015)
7. Chen, C.P., Wang, C.H.: The effects of learning style on mobile augmented-reality-facilitated English vocabulary learning. In: 2015 2nd International Conference on Information Science and Security (ICISS), pp. 1–4. IEEE (2015)
8. Chen, C.M., Chung, C.J.: Personalized mobile English vocabulary learning system based on item response theory and learning memory cycle. Comput. Educ. 51(2), 624–645 (2008)
9. Comas-Quinn, A., Mardomingo, R., Valentine, C.: Mobile blogs in language learning: making the most of informal and situated learning opportunities. ReCALL 21(1), 96–112 (2009). https://doi.org/10.1017/S0958344009000032
10. Demouy, V., Jones, A., Kan, Q., Kukulska-Hulme, A., Eardley, A.: Why and how do distance learners use mobile devices for language learning? EuroCALL Rev. 24(1), 10–24 (2016)
11. He, X., Godfroid, A.: Choosing words to teach: a novel method for vocabulary selection and its practical application. TESOL Q. 53(2), 348–371 (2019). https://doi.org/10.1002/tesq.483

12. Howatt, A.P., Smith, R.: The history of teaching English as a foreign language, from a British and European perspective. Lang. History **57**(1), 75–95 (2014). https://doi.org/10.1179/1759753614Z.00000000028
13. Ismail, H.M., Harous, S., Belkhouche, B.: Review of personalized language learning systems. In: 2016 12th International Conference on Innovations in Information Technology (IIT), pp. 1–6. IEEE (2016)
14. Jung, J., Graf, S.: An approach for personalized web-based vocabulary learning through word association games. In: 2008 International Symposium on Applications and the Internet, pp. 325–328. IEEE (2008)
15. Kukulska-Hulme, A.: Personalization of language learning through mobile technologies (2016)
16. Livingstone, D.W.: Exploring the icebergs of adult learning: findings of the first Canadian survey of informal learning practices (1999)
17. Nascimento, H., Marques, L.B., de Souza, D.G., Salgado, F.M., Bessa, R.Q.: A AIED Game to help children with learning disabilities in literacy in the Portuguese language. In: SBC-Proceedings of SBGames (2012)
18. Niu, X., McCalla, G., Vassileva, J.: Purpose-based user modelling in a multi-agent portfolio management system. In: Brusilovsky, P., Corbett, A., de Rosis, F. (eds.) UM 2003. LNCS (LNAI), vol. 2702, pp. 398–402. Springer, Heidelberg (2003). https://doi.org/10.1007/3-540-44963-9_56
19. Ogata, H., Li, M., Hou, B., Uosaki, N., El-Bishouty, M.M., Yano, Y.: SCROLL: supporting to share and reuse ubiquitous learning log in the context of language learning. Res. Pract. Technol. Enhanced Learn. **6**(2) (2011)
20. Riller, F., Deployee, I.R.: On the resettlement expectations of Iraqi refugees in Lebanon, Jordan, and Syria. UNHCR and ICMC, Geneva (2009)
21. Sharples, M., Taylor, J., Vavoula, G.: Towards a theory of mobile learning. In: Proceedings of mLearn, vol. 1, no. 1, pp. 1–9 (2005)
22. Tollefson, J.W.: Functional competencies in the US refugee program: theoretical and practical problems. TESOL Q. **20**(4), 649–664 (1986). https://doi.org/10.2307/3586516
23. UNHCR: Document - Vulnerability Assessment of Syrian Refugees in Lebanon - VASyR 2017 (2017). https://data2.unhcr.org/en/documents/details/61312

Artificial Intelligence Ethics Guidelines for K-12 Education: A Review of the Global Landscape

Cathy Adams$^{(\boxtimes)}$, Patti Pente, Gillian Lemermeyer, and Geoffrey Rockwell

University of Alberta, Edmonton, Canada

{cathy.adams,pente,gillianl,geoffrey.rockwell}@ualberta.ca

Abstract. To scope the global landscape of ethical issues involving the use of AI in K-12 education, we identified relevant ethics guidance documents, and then compared and contrasted concerns raised and principles applied. We found that while AIEdK-12 ethics guidelines employed many principles common to non-AIEd policy statements (e.g., transparency), new ethical principles were being engaged including pedagogical appropriateness and children's rights.

Keywords: Artificial intelligence in education · AI and ethics · AI ethics guidelines · AI literacy · Children's rights · K-12 education · Teacher well-being

1 Introduction

Advances in Artificial Intelligence (AI) are providing K-12 teachers with a wealth of new tools and smart services to facilitate student learning and to augment their professional practice. Meanwhile, growing public concern over potentially harmful societal effects of AI has prompted a publication flurry of more than 160 AI ethics guidelines and policy documents authored by national and international government agencies, academic consortia and industrial stakeholders [1]. AI ethics policy guidance specific to K-12 education has lagged behind, even though the ethical issues involving AI in the classroom are "equally, if not more acute" than those troubling AI in larger society [2]. Recent initiatives suggest that this AI ethics policy gap in K-12 education is swiftly being addressed.

In this paper, we take stock of the current landscape of AI ethics policy development for children and K-12 education. Employing AI ethical principles identified by Jobin *et al.* via a recent review of cross-sectoral AI ethics guidelines [3], we organized and compared five AI Ethics K-12 Education policy documents as a means to discern patterns as well as any new ethical and pedagogical principles informing their development. Our approach served to (1) highlight similarities with and differences among principles common to AIEd and non-AIEd ethics statements, (2) highlight similarities with and differences among the five AIEd ethics documents, and (3) provide a clearer overview of current coverage. Our intent is to help educational policymakers, AIED researchers, K-12 teachers and other stakeholders make sense of the complex issues before us and to suggest possible ways forward.

© Springer Nature Switzerland AG 2021

I. Roll et al. (Eds.): AIED 2021, LNAI 12749, pp. 24–28, 2021.

https://doi.org/10.1007/978-3-030-78270-2_4

2   Methodology

We performed a targeted literature search to locate AI ethics guidelines that were international or national in scope, focused on children (even if not directly school-related), and recently published (2015 and later). We searched the academic database ERIC (via OVID) using multiple combinations of the following search terms and key words: ("AI" OR "artificial intelligence") AND ("ethics") AND ("guidelines" OR "Code of ethics") with and without ("children"). Less than ten results were returned for all searches, with no documents meeting eligibility requirements. Similar search terms and combinations were then tried with Google Scholar, which also did not produce any eligible documents; then Google, where some eligible documents were finally located. The first 40 resulting hits per search query were reviewed by title and brief description. All the AI K-12 ethics guidance documents identified were published in 2019 or later, indicating the nascence of this policy field.

Ultimately, we identified only five AI ethics guidelines directly relevant to children and K-12 Education [4–8]. One document was a report commissioned by the Government of Australia [4]; all others intended global scope [5–8]. Two documents reflected the work of a single international workshop [5] and a conference [6]. The other two documents represented preliminary or first versions [7, 9], with final versions scheduled for release in 2021. The final report of one of these has since been released [8] with an annex [10], which we swapped in for the last iteration of analysis.

In order to compare and contrast AIEd to non-AIEd ethics guidelines, we began by organizing the AIEd Ethics documents based on eleven ethical principles identified by Jobin *et al.* [3] in their content analysis of 84 AI ethics documents: Transparency, Justice and fairness, Non-maleficence, Responsibility, Privacy, Beneficence, Freedom and autonomy, Trust, Dignity, Sustainability, and Solidarity. After our first round of content analysis, we discarded the last four principles since they were not well represented in the AIEd ethics documents or had been included under other principles. Based on our analysis, we added four new principles: Pedagogical appropriateness, Children's rights, AI literacy and education, and Teacher well-being (see Table 1).

3   Findings and Discussion

All five sets of ethical guidelines promoted and upheld a vision of AI-rich educational environments as a source of teaching and learning innovation, of human capacity enhancement and empowerment, and as a "good" for children and youth. Ethical tensions and contradictions were also identified and in each case, guidelines were proposed to address these issues. Many of the concerns echoed those already raised regarding AI deployment in broader society (e.g., the automation of systemic racism via AI and big data). Other ethical and social justice issues, such as exacerbation of the digital divide, had previously been expressed regarding the use of digital technologies more generally. Nonetheless, some concerns identified were unique to AI in K-12 educational contexts; for example, worry over the erosion of valued skills such as "introspection, resilience and the ability to think for oneself" due to misuse of AI [9]. Below, we describe the four new ethical principles added during the content analysis.

Table 1. AI in K-12 education ethics guidelines summary

AI Ethics Guideline Document -> Ethical Principles \ Constituency	Southgate et al. [4] Australia	World Economic Forum [5] International	UNESCO Beijing [6] International	UNICEF [7] International	IEAIED [8] UK & beyond
Transparency Key words: Transparency, explainability, explicability, understandability, interpretability, communication, disclosure, showing, age-appropriate language	Transparency; Explainability	Ensuring algorithmic accountability	Ensuring ethical, transparent and auditable use of education data and algorithms	Provide transparency, explainability, and accountability for children	Transparency and Accountability
Justice & fairness Key words: Justice, fairness, consistency, inclusion, equality, equity, (non-) bias, (non-)discrimination, diversity, plurality, accessibility, reversibility, remedy, redress, challenge, access and distribution	Fairness	Accounting for marginalized groups; Ensuring fairness in machine learning	Promoting equitable and inclusive use of AI in education; Gender-equitable AI and AI for gender equality	Ensuring inclusion of and for children; Prioritize fairness & non-discrimination for children	Equity
Non-maleficence Key words: Non-maleficence, security, safety, harm, protection, precaution, prevention, integrity (bodily or mental), non-subversion	(addressed under other categories)	(addressed under other categories)	(addressed under other categories)	Ensure safety for children	Ethical Design
Responsibility Key words: Responsibility, accountability, liability, acting with integrity	Accountability	Consumer Protection	(addressed under other categories)	Provide transparency, explainability, and accountability for children	Transparency and Accountability
Privacy Key words: Privacy, personal or private information	(addressed under other categories)	Privacy	(addressed under other categories)	Protecting children's data and privacy	Privacy
Beneficence Key words: Benefits, beneficence, well-being, peace, social good, common good	Beneficence	Recognizing developmental science in policy	(addressed under other categories)	Support children's development and well-being	Achieving Educational Goals
Freedom & autonomy Key words: Freedom, autonomy, consent, assent, choice, self-determination, liberty, empowerment	Awareness	Agency	(addressed under other categories)	Ensure safety for children	Autonomy; Informed Participation
Pedagogical appropriateness Keywords: Appropriate use, educational research-based, evidence-based, alignment with learner needs, child-centred AI, developmentally appropriate	Learning with AI	Algorithms for Children; Assessment and Evaluation	AI for learning and learning assessment; Monitoring, evaluation and research	Create an enabling environment for child-centred AI	Achieving Educational Goals; Forms of Assessment
Children's rights Keywords: Children's or child rights	Human rights	Child Rights	In preamble, aligned with Universal Declaration of Human Rights	Empower governments and businesses with knowledge of AI & children's rights	(addressed under other categories)
AI literacy Keywords: Teacher well-being; teacher workload; teacher empowerment	Learning about AI	Public education	Development of values and skills for life and work in the AI era	Prepare children for present and future developments in AI	Informed Participation
Teachers' well-being Keywords: Teacher well-being; teacher workload; teacher empowerment	(addressed under other categories)	-	AI to empower teaching and teachers	-	Administration and Workload

Table Legend:	Major Category
	Subcategory
	Principle addressed under other sub/categories

Pedagogical appropriateness refers to a complex of educational values such as child-centeredness, differentiated and personalized learning, evidence-based school practices, as well as concerns about vulnerable educational outcomes [9]. As an ethical principle for AIEdK-12, pedagogical appropriateness also means ensuring that teachers retain their professional freedom and responsibility to choose and use AI with due regard for "what is good or right and what is life enhancing, just, and supportive" of children and youth in their local classroom and community contexts [11].

The inclusion of *children's rights*—as opposed to human rights—acknowledges the special rights that apply globally to all persons under the age of 18. Of the five documents examined, two advocated strongly for children's rights as a unique guiding principle [5, 7]. For example, UNICEF's draft guidelines stated that "AI systems can uphold or undermine children's rights" and that "a children's rights-based approach rejects a traditional welfare approach to children's needs and vulnerabilities and instead recognizes children as human beings with dignity, agency and a distinct set of rights and entitlements, rather than as passive objects of care and charity" [7]. The basic message is that in all things, including AI, "public and private stakeholders should always act in the best interests of the child." The World Economic Forum report similarly draws attention to children's rights with a highlighted section. Both embed these rights throughout [5].

AI literacy and education underlines the importance of children and youth learning about AI so that they may be critically informed, as well as the need to build teacher knowledge capacity and parental awareness. As an extension of digital literacy, AI literacy refers to "knowledge [of] basic AI concepts and data literacy, skills such as basic AI programming, and attitudes and values to understand the ethics of AI" [7]. All documents recognized that teachers also needed to develop AI literacy, underlining the requirement for professional development. AI literacy and education is not widely included in school curricula, but this was posed as a good.

Three of the AI ethics documents expressed concern for *teacher well-being* [4, 6, 8]. The other two did not because their primary focus was AI and children, not schooling [5, 7]. Teacher well-being includes sensitivity to increased workloads [8], changing work conditions [6], additional time needed for preparation [4], shifting relationships with students [8], and worries about technological unemployment due to AI. AI and other digital technologies inevitably change teacher roles and practices, including the quality of student-teacher relations, sometimes with unintended consequences. For example, one guideline insisted that, "human interaction and collaboration between teachers and learners must remain at the core of education" [6].

4 Conclusion

In his review of 22 AI ethics guidelines across sectors, Hagendorff concludes by calling for a move away from "action-restricting", principles-focused ethics that require adherence to universal rules [12]. We echo this call. The AIEd ethics documents of this present review point to the need for more context-sensitive, pedagogically responsive ethical approaches for K-12 education. Children are vulnerable but also adaptive to the disruptive, transformative effects that AI will certainly have on their cognitive, social, physical and emotional lives. Ethical approaches employed must be responsive not only to children's diverse abilities, cultural backgrounds and developmental needs, but also to "who-what" they are becoming as AI-human hybrids [13].

References

1. United Nations: Report of the Secretary-General Roadmap for Digital Cooperation, June 2020. https://www.un.org/en/content/digital-cooperation-roadmap/assets/pdf/Roadmap_for_Digital_Cooperation_EN.pdf
2. Luckin, R., Holmes, W., Griffiths, M., Forcier, L.B.: Intelligence Unleashed: An Argument for AI in Education. Pearson Education, London (2016). https://static.googleusercontent.com/media/edu.google.com/en//pdfs/Intelligence-Unleashed-Publication.pdf
3. Jobin, A., Ienca, M., Vayena, E.: The global landscape of AI ethics guidelines. Nat. Mach. Intell. 1(9), 389–393 (2019). https://doi.org/10.1038/s42256-019-0088-2
4. Southgate, E., Blackmore, K., Pieschl, S., Grimes, S., McGuire, J., Smithers, K.: Artificial intelligence and emerging technologies (virtual, augmented and mixed reality) in schools: a research report. Newcastle: University of Newcastle, Australia (2019). https://docs.education.gov.au/system/files/doc/other/aiet_final_report_august_2019.pdf
5. World Economic Forum: Generation AI: Establishing Global Standards for Children and AI. World Economic Forum (2019). https://www.weforum.org/reports/generation-ai-establishing-global-standards-for-children-and-ai
6. United Nations Educational, Scientific and Cultural Organization: Beijing Consensus on Artificial Intelligence and Education. UNESCO (2019). https://unesdoc.unesco.org/ark:/48223/pf0000368303
7. UNICEF: Policy Guidance on AI for Children. New York, NY (2020). https://www.unicef.org/globalinsight/media/1171/file/UNICEF-Global-Insight-policy-guidance-AI-children-draft-1.0-2020.pdf
8. The Institute for Ethical AI in Education: The Ethical Framework for AI in Education. University of Buckingham (2021). https://fb77c667c4d6e21c1e06.b-cdn.net/wp-content/uploads/2021/03/The-Institute-for-Ethical-AI-in-Education-The-Ethical-Framework-for-AI-in-Education.pdf
9. The Institute for Ethical AI in Education: Developing a shared vision of ethical AI in education: An invitation to participate. University of Buckingham (2020). https://fb77c667c4d6e21c1e06.b-cdn.net/wp-content/uploads/2020/09/Developing-a-Shared-Vision-of-Ethical-AI-in-Education-An-Invitation-to-Participate.pdf
10. The Institute for Ethical AI in Education: Annex: Developing the Ethical Framework for AI in Education. University of Buckingham (2021). https://www.buckingham.ac.uk/wp-content/uploads/2021/03/The-Institute-for-Ethical-AI-in-Education-Annex-Developing-the-Ethical-Framework-for-AI-in-Education-IEAIED-.pdf
11. Van Manen, M.: Pedagogical Tact: Knowing What to Do When You Don't Know What to Do. Left Coast Press (2015)
12. Hagendorff, T.: The ethics of AI ethics: an evaluation of guidelines. Mind. Mach. 30(1), 99–120 (2020). https://doi.org/10.1007/s11023-020-09517-8
13. Stiegler, B.: Technics and Time: The Fault of Epimetheus, vol. 1. Stanford University Press (1998)

Quantitative Analysis to Further Validate WC-GCMS, a Computational Metric of Collaboration in Online Textual Discourse

Adetunji Adeniran[1]([✉])([iD]) and Judith Masthoff[2]

[1] Carnegie Mellon University, Pittsburgh, PA, USA
[2] Utrecht University, Utrecht, The Netherlands
j.f.m.masthoff@uu.nl

Abstract. Online learning is increasingly prevalent; for its advantage of unhindered access to quality learning and its leverage for education during the pandemic. Improving social experience in online learning would potentially scaffold the cognitive benefits it provides. A potential strategy is to support online-groups in real-time, similar to how a teacher guides face-to-face (F2F) group-learning in traditional classroom. Previously, we introduced the Word-Count/Gini-Coefficient Measure of Symmetry (WC-GCMS) that can automatically reflect the collaboration level of online textual discourse. In this paper, we introduce Social Coherence (SC), another marker of collaboration, and our analysis shows that WC-GCMS is sensitive to the SC level of group discourse, further validating the potency of the metric.

Keywords: Collaboration · Collaboration-metric · WC-GCMS · Social-coherence · Online groups

1 Introduction

Collaborative learning aids students' cognition [5,26,30,38,43,52]. Group-learners however do not automatically collaborate well [13,29,35]; instructors often guide the group activities to maximise collaboration [26,30,53]. These kind of interventions are seen to be more efficient and effective with face-to-face (F2F) groups in traditional classrooms, where teachers are physically present to monitor and guide. Our research goal is to investigate mechanisms to orchestrate such real-time support to guide online group-learners and maximise socio-cognition.

One challenge is to provide a computational means to measure and reflect phenomena of collaboration within online groups (teachers in traditional classroom do this in their head in a fuzzy manner) and to apply this measure in regulating online collaboration in real-time. We previously developed and advanced WC-GCMS [2,3] to solve this challenge, specifically for text-based online interactions. We discussed the validation of WC-GCMS by comparing its measure

© Springer Nature Switzerland AG 2021
I. Roll et al. (Eds.): AIED 2021, LNAI 12749, pp. 29–36, 2021.
https://doi.org/10.1007/978-3-030-78270-2_5

with the discourse quality of 5-experimental groups, in [2]. In this paper, we re-explore the group discourse collected in the previous study [2], and we analyse to further demonstrate the potency of WC-GCMS in the context.

2   Background: Investigating Support for Group-Learners to Scaffold Collaboration

As we have mentioned, our core objective is to investigate reliable means to measure online collaboration, computationally, automatically, and to be able to reflect this measure in real-time. We previously explored data of *learner-learner* problem-solving interactions, for metrics of collaboration [1]. Leveraging findings from existing related investigations [4,9–12,14,15,18,21,22,24,25,32–34,36,40,41,46], we inferred indices of collaboration that are consensual in many studies, and this inspired our rationalised mathematical modelling of WC-GCM which we studied and advanced for assessing collaboration level in text-based discourse [2,3]. We validated the potency of WC-GCMS in a case-study of 5 experimental online groups by comparing it to the quality of the groups' discourse [2]. This paper presents a second analysis to validate WC-GCMS; we introduce and rationalize social-coherence (SC) as another marker of collaboration and we show that WC-GCMS is sensitive to the SC level of groups' discourse.

3   Social-Coherence: Characterisation, Computation and Comparison with WC-GCMS Measure

Coherent group discourse suggests socially shared meta-cognition, coordinated interaction, realisation of speech roles, and creation of social identity and in-group relationships [23,51]. It is easier to assess and regulate coherency with F2F group-learners (or online interactions via multimedia), where verbal and visual cues are conveyed [8,19,28,42,44,47–49]. Contrarily, when online interaction is text-based, which is our scope group environment, the timing and rhythm of contributions are less visible, which can affect and distort the sequence of contribution [28] in a manner such as the illustration of the *in-coherent* text chat in Simpson [45]. This limitation, peculiar to textual interactions, is referred to as "quasi-synchronous" communication in [16,17] and complicates evaluation of actual coherency of textual discourse [27].

Most existing strategies for assessing coherency in textual discourse emphasise *syntax* and *semantics* which relate to the cognitive aspect of the discourse [31,37,50,54]. We separate the *cognitive* from the *social* aspect of discourse coherency to focus on the latter which we term *social-coherence* (SC). We advance a simple computation of SC in discourse and argue that this measure indicates mutual understanding and interest of collaborators. While we understand that assessing collaborative-learning encompasses the cognitive and social aspects of interactions, we adopt the idea of layered evaluation [6,7,39], focusing on the social aspect, to elicit in the context "what is/not working" and "what is influencing what". Next, we introduce SC as a marker of collaboration, characterise its computation, and re-explore the discourse data from our previous study [1,3] to compare the SC-level with the WC-GCMS measure, across studied groups.

We define the SC level of a group discourse based on the *unit-segments* contained in the discourse. A unit-segment in a text-based discourse is a sequence of N consecutive text contributions, where N is equal to the group size. We adapt the *handshake equation* [20] which computes the number of possible handshakes H_{sh} among n people given by Eq. 1:

$$HS_p = \frac{n \times (n-1)}{2} \tag{1}$$

We assume that HS_p is the ideal in a *unit-segment* of discourse (see illustration in Fig. 1 (a), where every member of the group contributes). However, in actual interaction, all *unit-segments* may not be perfect, such that the number of contributors in the segment c is smaller than the group-size n (illustrated in Fig. 1 (b), (c) and (d)), resulting in an actual HS_a given by Eq. 2:

$$HS_a = \frac{c \times (c-1)}{2} \tag{2}$$

Based on these assumptions we characterise the SC-level[1] in the sequence of *unit-segments* of textual discourse as given by Eq. 3:

$$SC_l = \frac{HS_a}{HS_p} \tag{3}$$

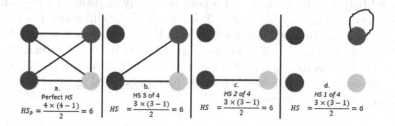

Fig. 1. Adaptation of the handshake equation to illustrate turn-taking and mutual contribution in textual discourse

Assuming our reasoning about and characteristic computation of the SC level is correct, we re-explore the data (discourse of 5 experimental online group[2])

---

[1] SC-level is between 0 to 1, and possible levels of SC_l between 0 and 1 dependent on the group size.

[2] See chat of each group here: http://colab-learn.herokuapp.com/Study3/groupX.php - $X \in \{1, 2, 3, 4, 5\}$.

32   A. Adeniran and J. Masthoff

from our previous study [2,3] and compute the SC level in the sequence of *unit-segments* of each group's discourse (see Fig. 2). We investigate the hypothesis: *that the dimension of SC-level difference correlates with the gradient of the WC-GCMS measure of collaboration, between-groups.*

SC levels

US	Grp1	Grp2	Grp3	Grp4	Grp5
1	0.5	0	0.167	0.167	0.5
2	0.167	0.167	0.5	0.167	0.167
3	0.167	0.5	0.5	0	0.167
4	0.167	0.5	0.167	0	0.167
5	0.167	0.5	0.167	0	0.167
6	0.167	0.5	0.167	0.167	0.167
7	0.167	0.167	0.5	0	0.167
8	0.5	0.5	0.167	0.167	0.167
9	0.167	0.167	0.5	0.167	0.5
10	0.167	0.167	0.5	0.167	0.5
11	0.167	0.167	0.5	0.5	0.5
12	0.167	0.5	0.5	0.5	0.5
13	0.5	0.5	0.5	0.167	0.5
14	0.167	0.167	0.5	0.167	0.5
15	0.167	0	0.5	0.5	0.5
16	0.5	0.167	0.167	0.167	0.5
17	0.167	0.167	0.5	0.167	0.5
18	0.167	0.5	0.167	0.167	0.167
19	0.167	0.167	0.5	0.167	0.5
20	0.167	0	0.5	0.5	0.5
21	0.167	0.167	0.5	0.167	0.167
22	0.5			0.167	0.167
23	0.5			0.5	0.167
24	0.5			0.167	0.167
25,26				0.167	0.5
27,29				0.167	0.5
28				0.5	0.167
30,31					0.5
32,34					0.167
33,35					0.5

Post Hoc

G1	G2	Mean Diff	n(G1)	n(G2)	SE	q	q(5,120)=3.91
Grp1	Grp2	0.006	24	21	0.035	0.169	Reject
Grp1	Grp3	0.125	24	21	0.035	3.542	Reject
Grp1	Grp4	0.051	24	29	0.033	1.572	Reject
Grp1	Grp5	0.093	24	35	0.031	2.979	Reject
Grp2	Grp3	0.119	21	21	0.036	3.266	Reject
Grp2	Grp4	0.057	21	29	0.034	1.69	Reject
Grp2	Grp5	0.087	21	35	0.033	2.678	Reject
Grp3	Grp4	0.176	21	29	0.034	5.208	Accept
Grp3	Grp5	0.032	21	35	0.033	0.974	Reject
Grp4	Grp5	0.144	29	35	0.03	4.873	Accept

Correlation data

	WC-GCMS	Mean SC-Levels
Grp1	842.24	0.2639
Grp2	1029.94	0.2698
Grp3	1537.96	0.3889
Grp4	796.24	0.2126
Grp5	1411.95	0.3571

Fig. 2. SC levels per unit segment (US) (*left*), ANOVA Post hoc analysis (*right*), data mean SC level versus WC-GCMS (*lower-right*)

An ANOVA ($F(4,129) = 4.94$, $p < 0.001$) indicates a substantial difference in SC-level between the groups, and a post-hoc analysis (using Turkey-Kramer Post Hoc test: see *upper right* of Fig. 2) shows that the difference lies between Groups 3 and 4, and between Groups 4 and 5. This corresponds to the gradient of the WC-GCMS measure, between-groups as reported in [2,3] (see visualisation in Fig. 3) [2,3]). A correlation analysis of the mean SC-level versus WC-GCMS measure shows: $r(3) = 0.97$, $p < .005$ (see *lower right* of Fig. 2). These results provide support for our hypothesis.

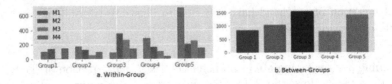

Fig. 3. Visualising online groups' collaboration-level with WC-GCMS

4   Conclusion

This paper demonstrates that WC-GCMS is sensitive to collaborative phenomena such as SC level of text-based discourse. Although currently limited in capturing the cognitive aspect of collaboration, it provides a simpler mechanism to assess social aspects of interactivity within groups. There are ample research opportunities to upgrade contextual sensitivity and enhance WC-GCMS for generic application.

References

1. Adeniran, A.: Investigating feedback support to enhance collaboration within groups in computer supported collaborative learning. In: Penstein Rosé, C., et al. (eds.) AIED 2018. LNCS (LNAI), vol. 10948, pp. 487–492. Springer, Cham (2018). https://doi.org/10.1007/978-3-319-93846-2_91
2. Adeniran, A., Masthoff, J., Beacham, N.: An appraisal of a collaboration-metric modelbased on text discourse. In: AIED 2019 TeamTutoring Workshop. CEUR WS (2019)
3. Adeniran, A., Masthoff, J., Beacham, N.: Model-based characterization of text discourse content to evaluate online group collaboration. In: Isotani, S., Millán, E., Ogan, A., Hastings, P., McLaren, B., Luckin, R. (eds.) AIED 2019. LNCS (LNAI), vol. 11626, pp. 3–8. Springer, Cham (2019). https://doi.org/10.1007/978-3-030-23207-8_1
4. Alvarez, C., Brown, C., Nussbaum, M.: Comparative study of netbooks and tablet pcs for fostering face-to-face collaborative learning. Comput. Hum. Behav. **27**(2), 834–844 (2011)
5. Bruffee, K.A.: Collaborative learning and the "conversation of mankind". Coll. Engl. **46**(7), 635–652 (1984)
6. Brusilovsky, P., Karagiannidis, C., Sampson, D.: The benefits of layered evaluation of adaptive applications and services. In: Empirical Evaluation of Adaptive Systems. Proceedings of workshop at the Eighth International Conference on User Modeling, UM2001, pp. 1–8 (2001)
7. Brusilovsky, P., Karagiannidis, C., Sampson, D.: Layered evaluation of adaptive learning systems. Int. J. Continuing Eng. Educ. Life Long Learn. **14**(4–5), 402–421 (2004)
8. Couper-Kuhlen, E.: English Speech Rhythm: Form and Function in Everyday Verbal Interaction, vol. 25. John Benjamins Publishing (1993)
9. Cukurova, M., Luckin, R., Millán, E., Mavrikis, M.: The NISPI framework: analysing collaborative problem-solving from students' physical interactions. Comput. Educ. **116**, 93–109 (2018)
10. Daniel, B.K., McCalla, G.I., Schwier, R.A.: Social network analysis techniques: implications for information and knowledge sharing in virtual learning communities. Int. J. Adv. Media Commun. **2**(1), 20 (2008)
11. Daradoumis, T., Martínez-Monés, A., Xhafa, F.: A layered framework for evaluating on-line collaborative learning interactions. Int. J. Hum. Comput. Stud. **64**(7), 622–635 (2006)
12. De Laat, M., Lally, V.: It's not so easy: researching the complexity of emergent participant roles and awareness in asynchronous networked learning discussions. J. Comput. Assist. Learn. **20**(3), 165–171 (2004)

13. Dillenbourg, P.: What do you mean by collaborative learning? (1999)
14. Dringus, L.P., Ellis, T.: Using data mining as a strategy for assessing asynchronous discussion forums. Comput. Educ. **45**(1), 141–160 (2005)
15. Duensing, A., Stickler, U., Batstone, C., Heins, B.: Face-to-face and online interactions-is a task a task? J. Learn. Des. **1**(2), 35–45 (2006)
16. Garcia, A.C., Baker Jacobs, J.: The eyes of the beholder: understanding the turn-taking system in quasi-synchronous computer-mediated communication. Res. Lang. Soc. Interact. **32**(4), 337–367 (1999)
17. Garrison, D.R., Anderson, T., Archer, W.: Critical inquiry in a text-based environment: computer conferencing in higher education. Internet High. Educ. **2**(2–3), 87–105 (1999)
18. Gunawardena, C.N., Lowe, C.A., Anderson, T.: Analysis of a global online debate and the development of an interaction analysis model for examining social construction of knowledge in computer conferencing. J. Educ. Comput. Res. **17**(4), 397–431 (1997)
19. Hall, E.T., Hall, T.: The Silent Language, vol. 948. Anchor Books (1959)
20. Hedegaard, R.: Handshake problem. From mathworld-a wolfram web resource, created by Eric W. Weisstein. http://mathworld.wolfram.com/HandshakeProblem.htm
21. Heo, H., Lim, K.Y., Kim, Y.: Exploratory study on the patterns of online interaction and knowledge co-construction in project-based learning. Comput. Educ. **55**(3), 1383–1392 (2010)
22. Hewitt, J.G.: Progress toward a knowledge-building community (1996)
23. Iiskala, T., Vauras, M., Lehtinen, E., Salonen, P.: Socially shared metacognition of dyads of pupils in collaborative mathematical problem-solving processes. Learn. Instr. **21**(3), 379–393 (2011)
24. Järvelä, S., Häkkinen, P.: Levels of web-based discussion: theory of perspective-taking as a tool for analyzing interaction, pp. 22–26 (2000)
25. Järvelä, S., Häkkinen, P.: Web-based cases in teaching and learning-the quality of discussions and a stage of perspective taking in asynchronous communication. Interact. Learn. Environ. **10**(1), 1–22 (2002)
26. Johnson, D.W., Johnson, R.T.: An educational psychology success story: social interdependence theory and cooperative learning. Educ. Res. **38**(5), 365–379 (2009)
27. Jones, R.: Inter-activity: how new media can help us understand old media. In: Language and New Media: Linguistic, Cultural, and Technological Evolutions, pp. 13–32. Hampton Press (2009)
28. Jones, R.H.: Rhythm and timing in chat room interaction (2013)
29. Kock, M., Paramythis, A.: Towards adaptive learning support on the basis of behavioural patterns in learning activity sequences. In: 2010 2nd International Conference on Intelligent Networking and Collaborative Systems (INCOS), pp. 100–107. IEEE (2010)
30. Laal, M., Ghodsi, S.M.: Benefits of collaborative learning. Procedia Soc. Behav. Sci. **31**, 486–490 (2012)
31. Lai, A., Tetreault, J.: Discourse coherence in the wild: a dataset, evaluation and methods. arXiv preprint arXiv:1805.04993 (2018)
32. Maldonado, R.M., Yacef, K., Kay, J., Kharrufa, A., Al-Qaraghuli, A.: Analysing frequent sequential patterns of collaborative learning activity around an interactive tabletop. In: Educational Data Mining 2011 (2010)
33. Mansur, A.B.F., Yusof, N., Othman, M.S.: Analysis of social learning network for wiki in moodle e-learning. In: The 4th International Conference on Interaction Sciences, pp. 1–4. IEEE (2011)

34. Martinez, R., Wallace, J.R., Kay, J., Yacef, K.: Modelling and identifying collaborative situations in a collocated multi-display groupware setting. In: Biswas, G., Bull, S., Kay, J., Mitrovic, A. (eds.) AIED 2011. LNCS (LNAI), vol. 6738, pp. 196–204. Springer, Heidelberg (2011). https://doi.org/10.1007/978-3-642-21869-9_27

35. Martinez Maldonado, R.: Analysing, visualising and supporting collaborative learning using interactive tabletops (2013)

36. Meier, A., Spada, H., Rummel, N.: A rating scheme for assessing the quality of computer-supported collaboration processes. Int. J. Comput.-Support. Collab. Learn. **2**(1), 63–86 (2007)

37. Mohiuddin, T., Joty, S., Nguyen, D.T.: Coherence modeling of asynchronous conversations: a neural entity grid approach. arXiv preprint arXiv:1805.02275 (2018)

38. Panitz, T.: Benefits of cooperative learning in relation to student motivation. Motivation from within: approaches for encouraging faculty and students to excel, New directions for teaching and learning. San Francisco, CA (1999)

39. Paramythis, A., Weibelzahl, S., Masthoff, J.: Layered evaluation of interactive adaptive systems: framework and formative methods. User Model. User-Adap. Inter. **20**(5), 383–453 (2010)

40. Perera, D., Kay, J., Yacef, K., Koprinska, I.: Mining learners' traces from an online collaboration tool. In: 13th International Conference on Artificial Intelligence in Education, Educational Data Mining Workshop, AIED07, pp. 60–69. Citeseer (2007)

41. Redondo, M.A., Bravo, C., Bravo, J., Ortega, M.: Applying fuzzy logic to analyze collaborative learning experiences in an e-learning environment. USDLA J. (US Distance Learn. Assoc.) **17**, 19–28 (2003)

42. Sajavaara, K., Lehtonen, J.: The silent Finn revisited. Silence: interdisciplinary perspectives, pp. 263–283 (1997)

43. Schmidt, H.G., Loyens, S.M., Van Gog, T., Paas, F.: Problem-based learning is compatible with human cognitive architecture: commentary on kirschner, sweller, and. Educ. Psychol. **42**(2), 91–97 (2007)

44. Sever, T.: Conversational style: analyzing talk among friends. ELT Res. J. **2**(3), 146–149 (2014)

45. Simpson, J.: Conversational floors in synchronous text-based CMC discourse. Discourse Stud. **7**(3), 337–361 (2005)

46. Talavera, L., Gaudioso, E.: Mining student data to characterize similar behavior groups in unstructured collaboration spaces. In: Workshop on Artificial Intelligence in CSCL. 16th European Conference on Artificial Intelligence, pp. 17–23 (2004)

47. Tannen, D.: Conversational Style: Analyzing Talk Among Friends. Ablex, Norwood (1984)

48. Tannen, D.: Talking Voices: Repetition, Dialogue, and Imagery in Conversational Discourse, vol. 26. Cambridge University Press, Cambridge (2007)

49. Tannen, D., et al.: Conversational Style: Analyzing Talk Among Friends. Oxford University Press, Oxford (2005)

50. Vakulenko, S., de Rijke, M., Cochez, M., Savenkov, V., Polleres, A.: Measuring semantic coherence of a conversation. In: Vrandečić, D., et al. (eds.) ISWC 2018. LNCS, vol. 11136, pp. 634–651. Springer, Cham (2018). https://doi.org/10.1007/978-3-030-00671-6_37

51. Van Leeuwen, T.: Introducing Social Semiotics. Psychology Press, London (2005)

52. Vygotsky, L.S.: Mind in Society: The Development of Higher Mental Processes (E. Rice, ed. & trans.) (1978)
53. Wood, D.F.: Problem based learning. BMJ **326**(7384), 328–330 (2003)
54. Yannakoudakis, H., Briscoe, T.: Modeling coherence in ESOL learner texts. In: Proceedings of the Seventh Workshop on Building Educational Applications Using NLP, pp. 33–43 (2012)

Generation of Automatic Data-Driven Feedback to Students Using Explainable Machine Learning

Muhammad Afzaal[(✉)], Jalal Nouri, Aayesha Zia, Panagiotis Papapetrou,
Uno Fors, Yongchao Wu, Xiu Li, and Rebecka Weegar

Stockholm University, Stockholm, Sweden
{muhammad.afzaal,jalal,aayesha,panagiotis,uno,
yongchao.wu,xiu.li,rebeckaw}@dsv.su.se

Abstract. This paper proposes a novel approach that employs learning analytics techniques combined with explainable machine learning to provide automatic and intelligent actionable feedback that supports students self-regulation of learning in a data-driven manner. Prior studies within the field of learning analytics predict students' performance and use the prediction status as feedback without explaining the reasons behind the prediction. Our proposed method, which has been developed based on LMS data from a university course, extends this approach by explaining the root causes of the predictions and automatically provides data-driven recommendations for action. The underlying predictive model effectiveness of the proposed approach is evaluated, with the results demonstrating 90 per cent accuracy.

Keywords: Learning analytics · Explainable machine learning · Feedback provision · Recommendations generation · Dashboard

1 Introduction

Providing feedback is one of the many tasks that teachers perform to guide students towards increased learning and performance and it is viewed by many as one of the most powerful practices to enhance student learning [1,2]. A number of studies within the field of learning analytics have investigated how students' self-regulation and teachers feedback provision can be supported through for instance teacher and student dashboards that provide predictions of student's performance [3–5]. Although such feedback might be useful to some extent, it does not provide any meaningful insights or actionable information about the reasons behind the prediction – that is, students and teachers do not receive actionable feedback [6–9].

However, some studies have moved beyond the presentation of prediction results and extracted pivotal factors that could affect the students performance over time and utilizing those factors as feedback [10–15]. Nevertheless, the current approaches do not provide teachers and students with actual explanations

© Springer Nature Switzerland AG 2021
I. Roll et al. (Eds.): AIED 2021, LNAI 12749, pp. 37–42, 2021.
https://doi.org/10.1007/978-3-030-78270-2_6

of the predictions, and tools for students that provide automatic and intelligent feedback in the form of recommendations to students during ongoing courses. Such explanations would help both teachers and students to regulate their behaviour in a data-driven manner. Moreover, the existing approaches have overlooked the prediction and feedback of student performance at the assignment or quiz level on currently running courses. Such information can be helpful for the course staff who are planning interventions or other strategies to improve the student retention rate [9].

Against this background, this study addresses the following research questions:

1. How can we employ an explainable machine learning approach to compute data-driven feedback and generate actionable recommendations that are beneficial for students and teachers?
2. How can we predict students' academic performance accurately at the assignment or quiz level, in order to provide intelligent recommendations that support students' performance in quizzes and assignments?

Hence, in this paper, we propose an explainable machine learning-based approach that predicts student's performance and computes informative feedback along with actionable recommendations. The main contribution, though, is that we combine the prediction approach with an explainable machine learning approach that – in comparison with previous studies - allows for fine-grained insights that support the provision of detailed data-driven actionable feedback to students and teachers that explains the "why" of the predictions. That is, we present an approach that gives teachers and students more actionable information than what can be achieved through just informing these actors about a prediction.

2   Method

In this section, an automatic feedback provision approach is presented that consists of six main phases. First, students' data about different social and educational activities are collected from the learning management system (LMS). The data in this study concern a programming course, which was taught consecutively for two years (2019 and 2020). Overall, 157 students enrolled in the course and asked the students to give consent for research. Second, preprocessing is performed on the collected data to create a link between students and their activities and to remove irrelevant data. Third, features are generated and predictive models are built based on the developed features using established ML algorithms. Generated features could be grouped into four categories: (1) initial assessment, (2) quiz and assignment attributes, (3) resources access, and (4) practical exercise attempts. Fifth, informative feedback is computed and actionable recommendations are generated by employing an explainable ML-based approach that uses predictive models and students' data. Last, a dashboard prototype is presented to display the approach output (feedback and recommendations) in a usable and actionable way.

3   Results

In this section, we present the proposed explainable machine learning approach for the provision of intelligent and automatic feedback, along with an evaluation of the effectiveness of our proposed approach.

3.1   Evaluation of Predictive Models

In this section, we present the experimental results and evaluate our proposed approach's effectiveness in terms of its ability to predict student academic performance at the assignment level. Table 1 shows the performance of each ML algorithm in assessment activities of the course. The results show that the ANN dominated the ML algorithm in terms of all the evaluation measures for all the assessment activities. Although the RF and ANN presented quite similar results, the performance of the RF was lower than that of the ANN by 0.02. On the contrary, the lowest performance for all the evaluation measures in the assessment activities was provided by the LR.

Table 1. Results of the prediction for the assessment activities

Predictive model	Accuracy	Precision	Recall	F-measure
RF	0.88	0.88	0.88	0.88
ANN	**0.90**	**0.90**	**0.90**	**0.90**
KNN	0.87	0.87	0.87	0.87
SVM	0.87	0.87	0.87	0.87
BayesNet	0.85	0.86	0.86	0.86
LR	0.78	0.78	0.78	0.77

3.2   Feedback Provision and Recommendations Generation

Having built predictive models and evaluated their performance, we introduce an explainable ML-based approach that utilizes the best-performing model to compute informative feedback and generate actionable recommendations for each student individually. In this approach, we employed an example-based explainable ML technique called counterfactual explanation to compute feedback and generate recommendations. This technique selects particular instances of the data set to explain ML models' behaviour. The proposed approach consists of six steps. Firstly, to compute informative feedback, counterfactual explanations (CFs) are generated by passing the predictive model, student features, desired performance and threshold to the Diverse Counterfactual Explanations (DICE) library [16]. A CF describes the smallest change to the student feature values that changes the prediction to a predefined output (desired performance). Secondly, since few explanations, are difficult to achieve for a student, the correlation between the actual features' values and the CFs' values is calculated to select the closest CF. Now, students and teachers can know which set of features a student should work on to reach the desired performance level. However, it is not shown which feature has more importance than others. Therefore, to generate more actionable recommendations, each feature's impact is calculated and then the features are sorted accordingly.

The output of the proposed approach is the list of features and their recommended values; however, it is still complicated for a student to understand and interpret such information. Therefore, we propose a dashboard prototype that helps students to understand the feedback and make them easy to follow the recommendations. Figure 1 presents the dashboard for a student that consists of three main components (1) prediction that offers the predicted performance probability for each assessment activity, (2) progress that informs students in which areas they are lacking and how much progress is required, (3) recommendation that inform the student about the list of essential items to perform to reach high performance.

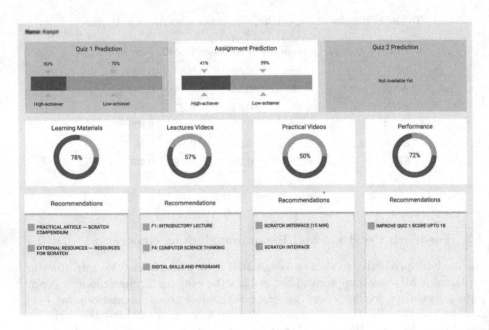

Fig. 1. Feedback dashboard

4   Conclusion

In this paper, we propose an automatic feedback provision approach using explainable machine learning to provide information feedback and actionable recommendations for both students and teachers. An explainable ML-based algorithm is developed that utilizes students' LMS data and that computes informative feedback at the student level. The proposed approach provides automatic data-driven feedback in the form of recommendations for action that can help students to self-regulate their learning. Such an approach opens doors to intelligent learning systems that automatically provide teachers and students with intelligent recommendations.

References

1. Hattie, J., Gan, M., Brooks, C.: Instruction based on feedback. In: Handbook of Research on Learning and Instruction, pp. 249–271 (2011)
2. Hattie, J., Timperley, H.: The power of feedback. Rev. Educ. Res. **77**(1), 88–118 (2007)
3. Choi, S.P., Lam, S.S., Li, K.C., Wong, B.T.: Learning analytics at low cost: at-risk student prediction with clicker data and systematic proactive interventions. J. Educ. Technol. Soc. **21**(2), 273–290 (2018)
4. Howard, E., Meehan, M., Parnell, A.: Contrasting prediction methods for early warning systems at undergraduate level. Internet High. Educ. **37**, 66–75 (2018)
5. Marbouti, F., Diefes-Dux, H.A., Madhavan, K.: Models for early prediction of at-risk students in a course using standards-based grading. Comput. Educ. **103**, 1–15 (2016)
6. Baneres, D., Rodríguez-Gonzalez, M.E., Serra, M.: An early feedback prediction system for learners at-risk within a first-year higher education course. IEEE Trans. Learn. Technol. **12**(2), 249–263 (2019)
7. Bennion, L.D., et al.: Early identification of struggling learners: using prematriculation and early academic performance data. Perspect. Med. Educ. **8**(5), 298–304 (2019). https://doi.org/10.1007/s40037-019-00539-2
8. Rosenthal, S., et al.: Identifying students at risk of failing the USMLE step 2 clinical skills examination. Fam. Med. **51**(6), 483–499 (2019)
9. Kuzilek, J., Hlosta, M., Herrmannova, D., Zdrahal, Z., Vaclavek, J., Wolff, A.: OU analyse: analysing at-risk students at The Open University. Learn. Anal. Rev. 1–16 (2015)
10. Akhtar, S., Warburton, S., Xu, W.: The use of an online learning and teaching system for monitoring computer aided design student participation and predicting student success. Int. J. Technol. Des. Educ. **27**(2), 251–270 (2015). https://doi.org/10.1007/s10798-015-9346-8
11. Xie, K., Di Tosto, G., Lu, L., Cho, Y.S.: Detecting leadership in peer-moderated online collaborative learning through text mining and social network analysis. Internet High. Educ. **38**, 9–17 (2018)
12. Lu, O.H., Huang, A.Y., Huang, J.C., Lin, A.J., Ogata, H., Yang, S.J.: Applying learning analytics for the early prediction of Students' academic performance in blended learning. J. Educ. Technol. Soc. **21**(2), 220–232 (2018)
13. Bibi, M., Abbas, Z., Shahzadi, E., Kiran, J.: Identification of factors behind academic performance: a case study of University of Gujrat students. J. ISOSS **5**(2), 103–114 (2019)
14. Kamal, P., Ahuja, S.: An ensemble-based model for prediction of academic performance of students in undergrad professional course. J. Eng. Des. Technol. (2019)

15. Ahuja, S., Kaur, P., Panda, S.: Identification of influencing factors for enhancing online learning usage model: evidence from an Indian university. Int. J. Educ. Manag. Eng. **9**(2), 15 (2019)
16. Mothilal, R.K., Sharma, A., Tan, C.: Explaining machine learning classifiers through diverse counterfactual explanations. In: Proceedings of the 2020 Conference on Fairness, Accountability, and Transparency, pp. 607–617 (2020)

Interactive Personas: Towards the Dynamic Assessment of Student Motivation within ITS

Ishrat Ahmed[1]([✉]), Adam Clark[2], Stefania Metzger[2], Ruth Wylie[2], Yoav Bergner[3], and Erin Walker[1]

[1] University of Pittsburgh, Pittsburgh, USA
{isa14,eawalker}@pitt.edu
[2] Arizona State University, Tempe, USA
{Adam.T.Clark,skmetzg1,Ruth.Wylie}@asu.edu
[3] New York University, New York, USA
ybb2@nyu.edu

Abstract. An intelligent system can provide sufficient collaborative opportunities and support yet fail to be pedagogically effective if the students are unwilling to participate. One of the common ways to assess motivation is using self-report questionnaires, which often do not take the context and the dynamic aspect of motivation into account. To address this, we propose *personas*, a user-centered design approach. We describe two design iterations where we: identify motivational factors related to students' collaborative behaviors; and develop a set of representative personas. These personas could be embedded in an interface and be used as an alternative method to assess motivation within ITS.

Keywords: Assessment of motivation · Collaborative learning · Intelligent Tutoring Systems (ITS)

1 Introduction

Adaptive collaborative learning support (ACLS) aims to design efficacious support that models students' interactions [12,15]. Student motivation is a key factor to consider as it contributes to learning from collaboration [8,11]. In an ITS context, one common way of assessing motivation is self-report. This is often done prior to an interaction, which has two drawbacks: 1) motivation is influenced by the environment [10], so it should be examined in the context of events; and 2) motivation is dynamic and fluctuates over time [6], so it should be assessed as such. In addition, responding to long questionnaires or to multiple surveys can lower student response rates as well as the quality of the responses [13].

We propose a novel method that captures student motivation dynamically during collaborative interactions. To achieve this, we describe an application of the *Persona* method [4], a user-centered design approach for understanding important end-users characteristics like preferences and goals. A persona is a

© Springer Nature Switzerland AG 2021
I. Roll et al. (Eds.): AIED 2021, LNAI 12749, pp. 43–47, 2021.
https://doi.org/10.1007/978-3-030-78270-2_7

fictitious representative target user created from a large number of heterogeneous users [4] consisting of a name, a picture or illustration, and a short narrative. The main purpose of this method is to provide a better characterization of the target audience for product design. Personas have also been used in educational research as part of the design of both technological and non-technological pedagogical interventions [2,3,14,16,18]. We believe that personas can be adapted to make them contextually sensitive, dynamic, and easy-to-use motivational assessments. In this paper, we focus on a primary research question: *How can we design a persona that represents student motivation?* We use co-design to develop and iterate on a set of representative personas using multiple motivational factors from interviews. Ultimately, these personas could be used to deliver adaptive support based on motivation within collaborative learning.

This work is part of a broader project to design an ACLS system focusing on middle school students help-giving [9,17] across three different collaborative learning platforms. We investigate why students gave help across these platforms with the goal of supporting individual students' needs in each platform [1].

2   Persona Design Process

We developed personas that represented clusters of student motivational factors and evaluated them in two co-design sessions. We wanted to determine how students responded to the personas as indicators of their motivation and get students' input on the personas' language. The two co-design sessions were conducted with 13 middle school students from the Southwestern United States (F = 4, M = 7, 2 did not report) in an after-school two-hour workshop. Participants were in 7th and 8th grade and reported their race and ethnicity as follows: Hispanic (6), Mexican (4), White (2), did not report (1).

We chose four factors for the personas: math self-concept, help-giving self-concept, familiarity, and contextual factors (e.g., off-topic comments). We selected these factors from an initial thematic analysis on interviews with 16 middle school students about their help-giving behaviors and motivations. These factors are also related to learning in literature [5,7]. Each persona included a name, an age, a goal, a quote, and a narrative describing the persona's help-giving interactions in mathematics using these factors. Six personas (Gracie, Maurice, Sarah, Tobi, Lisa, Harry) were designed to approximate a specific type of student participation and fit the characteristics of students in our study.

In the first co-design session, each student was given the six persona documents and asked to determine how much they were or were not like the persona answering with a likert scale ranging from 1 ("exactly like me") to 7 ("not like me at all"). 3 students rated themselves most like Gracie, 5 most like Harry, 3 most like Sarah, and the other 2 students were spread across the other three personas. This suggests that while five of our six personas resonated with at least one student, three appeared to particularly match the students in the session. Next, the students selected the persona they resembled the most and edited that persona characteristics to be more like them, e.g., (1) adding intermediary options when talking about math performance, e.g., 'one of the top performers' to 'good performer' (7 students); (2) major editing of statements, e.g., 'during collaboration, he fears giving the *wrong* answer' to 'during collaboration, he *normally gives the* answer' (10 students); (3) minor editing of statements (5 students, e.g., modifying gender).

The second co-design session happened two weeks later with eleven students (2 from the first session were absent). Because there were many personas that students did not match to and because students made multiple edits to their persons, we decided to have students build their own personas. We gave a template of a persona to the students with two parts: a persona narrative and a persona figure. The persona narrative included free inputs (e.g., for persona hobbies) and fixed-choice inputs with a set of options to select from related to our four factors. For example, related to math self-concept, students had three options to choose: "Really good at doing math problems", "Just ok at doing math problems", "Not great at doing math problems". The intermediary statements were inspired by the edits observed during the first co-design session. After the session, we had eleven personas created by the students and analyzed them to look for common themes, an approach often used in persona design [4]. We first used math self-concept to group the students as a determining factor in our particular learning environment, resulting in three clusters: low (4 students), medium (4 students), high (3 students). However, from co-design session 1, we observed students move from high to medium math self-concept, e.g., 'good at math' to 'almost good at math', so we combined medium and high into a single group. Then, we chose 2 personas from the low group and 2 personas from the high group such that we had at least one persona from each group with a preference towards familiarity. We chose familiarity due to its importance in designing our learning environment, which had a public and a private collaboration space. Thus, we had a total of four representative personas, two with similar characteristics to the personas developed by the researchers in co-design session 1, and two more influenced by the students in this session.

We then created finalized personas from these four representative personas. As described above, the four representative personas had a range of values of math self-concept (MSC), help-giving self-concept (HSC), and familiarity (Fam)based on student responses. We decided to eliminate the contextual factors dimension from the personas because we wanted to focus on individual motivation factors. However, we replaced that dimension with a conscientious factor based on additional analysis of the interviews mentioned above. Since conscientiousness (Con) was added after the co-design sessions, we categorized each of the interviewed students under one of the four personas and then chose the level of conscientiousness that best described all the students in that persona category. The final characteristics for each of the four personas are: *Seel* (MSC:low, HSC:high, Fam:low, Con:high), *Abra* (MSC:low, HSC:high, Fam:high, Con:high), *Bellsprout* (MSC:high, HSC:high, Fam:low, Con:high), *Caterpie* (MSC:high, HSC: low, Fam:high, Con:low).

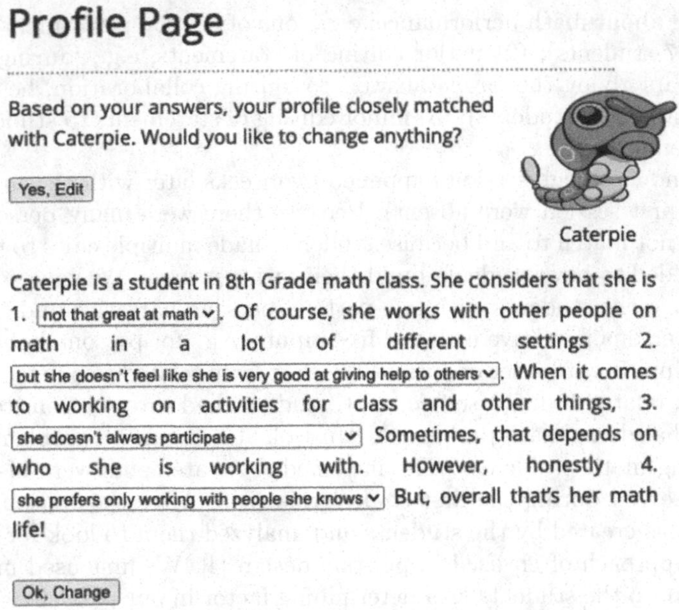

Profile Page

Based on your answers, your profile closely matched
with Caterpie. Would you like to change anything?

[Yes, Edit]

Caterpie

Caterpie is a student in 8th Grade math class. She considers that she is
1. [not that great at math ∨]. Of course, she works with other people on
math   in   a   lot   of   different   settings   2.
[but she doesn't feel like she is very good at giving help to others ∨]. When it comes
to working on activities for class and other things, 3.
[she doesn't always participate   ∨] Sometimes, that depends on
who   she   is   working   with.   However,   honestly   4.
[she prefers only working with people she knows ∨] But, overall that's her math
life!

[Ok, Change]

Fig. 1. Interface demo with dropdowns for students to self-indicate motivation

We embedded the final four personas as an interactive tool in the digital
textbook interface with a name, a picture, and a short narrative following the
original design. The design will allow the students to modify each of the four
characteristic values using a dropdown menu (Fig. 1). The values are represented
with words to fit in the narrative, e.g., 'pretty good at math' is mapped with
high MSC, and 'not that great at math' is mapped to low MSC.

3   Discussion and Conclusion

In this paper, we used co-design to create personas for assessing motivation
dynamically and in context. The students validated the factors used to develop
the personas and brought their own perspectives in the process [2]. We embedded these personas in the interface, allowing students to report their motivation
in context. This contextually embedded, easy to understand narrative may lead
the students to respond differently than to surveys. It represents a multidimensional perspective on motivation as it suggests motivation cannot be adequately
explained in terms of a single construct [10]. On a practical level, it may be
intractable for ACLS to respond differently to permutations of multiple interacting motivational factors, and thus leveraging personas can be a way for ACLS
to prioritize interventions based on logical clusters of individual characteristics.
Our vision is for this persona approach to be incorporated in ACLS as a contextually sensitive way of dynamically assessing and responding to motivation.

Acknowledgements. This work is supported by the National Science Foundation under Grant No 1736103.

References

1. Mawasi, A., et al.: Using design-based research to improve peer help-giving in a middle school math classroom. In: International Conference of the Learning Sciences (ICLS) (2020)
2. Albrechtsen, C., Pedersen, M., Pedersen, N.F., Jensen, T.W.: Proposing co-design of personas as a method to heighten validity and engage users: a case from higher education. Int. J. Sociotechnol. Knowl. Dev. (IJSKD) **8**(4), 55–67 (2016)
3. Ali Amer Jid Almahri, F., Bell, D., Arzoky, M.: Personas design for conversational systems in education. In: Informatics, vol. 6, no. 4, p. 46. Multidisciplinary Digital Publishing Institute, December 2019
4. Cooper, A.: The Inmates are Running the Asylum. Macmillan (1999)
5. Choi, N.: Self-efficacy and self-concept as predictors of college students' academic performance. Psychol. Sch. **42**(2), 197–205 (2005)
6. Dörnyei, Z.: Motivation in action: towards a process-oriented conceptualisation of student motivation. Br. J. Educ. Psychol. **70**(4), 519–538 (2000)
7. Hew, K.F., Cheung, W.S., Ng, C.S.L.: Student contribution in asynchronous online discussion: a review of the research and empirical exploration. Instr. Sci. **38**(6), 571–606 (2010)
8. Järvenoja, H., et al.: A collaborative learning design for promoting and analyzing adaptive motivation and emotion regulation in the science classroom. In: Frontiers in Education, vol. 5, p. 111. Frontiers, July 2020
9. Johnson, D.W., Johnson, R.T.: Cooperative Learning and Achievement (1990)
10. Marsh, H.W., Trautwein, U., Lüdtke, O., Köller, O., Baumert, J.: Integration of multidimensional self-concept and core personality constructs: construct validation and relations to well-being and achievement. J. Pers. **74**(2), 403–456 (2006)
11. Noponen, M.: Learners' motivation to collaborate in online learning environments: a situational and social network analysis (2016)
12. Olsen, J.K., Rummel, N., Aleven, V.: Investigating Effects of Embedding Collaboration in an Intelligent Tutoring System for Elementary School Students. Grantee Submission (2016)
13. Porter, S.R., Whitcomb, M.E., Weitzer, W.H.: Multiple surveys of students and survey fatigue. New Dir. Inst. Res. **2004**(121), 63–73 (2004)
14. Sankupellay, M., Mealy, E., Niesel, C., Medland, R.: Building personas of students accessing a peer-facilitated support for learning program. In: Proceedings of the Annual Meeting of the Australian Special Interest Group for Computer Human Interaction, pp. 412–416, December 2015
15. Walker, E., Rummel, N., Koedinger, K.R.: Adaptive intelligent support to improve peer tutoring in algebra. Int. J. Artif. Intell. Educ. **24**(1), 33–61 (2014)
16. Warin, B., Kolski, C., Toffolon, C.: Living persona technique applied to HCI education. In: 2018 IEEE Global Engineering Education Conference (EDUCON), pp. 51–59. IEEE, April 2018
17. Webb, N.M., Farivar, S.: Promoting helping behavior in cooperative small groups in middle school mathematics. Am. Educ. Res. J. **31**(2), 369–395 (1994)
18. Varela, S., Hall, C., Bang, H.J.: Creating middle school child-based personas for a digital math practice application. In: EdMedia+ Innovate Learning, pp. 532–537. Association for the Advancement of Computing in Education (AACE), June 2015

Agent-Based Classroom Environment Simulation: The Effect of Disruptive Schoolchildren's Behaviour Versus Teacher Control over Neighbours

Khulood Alharbi[1](✉), Alexandra I. Cristea[1], Lei Shi[1], Peter Tymms[2], and Chris Brown[2]

[1] Computer Science, Durham University, Durham, UK
{khulood.o.alharbi,alexandra.i.cristea,lei.shi}@durham.ac.uk
[2] School of Education, Durham University, Durham, UK
{p.b.tymms,chris.brown}@durham.ac.uk

Abstract. Schoolchildren's academic progress is known to be affected by the classroom environment. It is important for teachers and administrators to understand their pupils' status and how various factors in the classroom may affect them, as it can help them adjust pedagogical interventions and management styles. In this study, we expand a novel agent-based model of classroom interactions of our design, towards a more efficient model, enriched with further parameters of peers and teacher's characteristics, which we believe renders a more realistic setting. Specifically, we explore the effect of *disruptive neighbours* and *teacher control*. The dataset used for the design of our model consists of 65,385 records, which represent 3,315 classes in 2007, from 2,040 schools in the UK.

1 Introduction

The interactions that takes place in the classroom and how it affects school children achievement has received much attention by literature over the years [3, 4, 20]. Ingram and Brooks [8] simulated classroom environment to understand the effect of seating arrangement and friendships over attainment by considering factors like proximity to peers and teacher. Simulation of attainment was addressed in this work but not disruptive behaviour. Attainment, in sociological studies, refers to the long-term real educational gain [12] (computed in this study in Sect. 3).

In this paper we continue our work on understanding the effect of having disruptive pupils in a classroom through simulating *Inattentiveness* and *Hyperactivity* behaviour. Inattentiveness indicates moving between tasks, leaving one unfinished before losing interest, while Hyperactivity implies excessive movements in a situation where calmness is expected [25]. The two types are symptoms of the Attention-deficit hyperactivity disorder (ADHD) that has a clear negative impact on children's long term academic performance [17]. Our work considers a pupil's achievement and the influence of teachers' as well as peers' characteristics. In a previous work [1] we have considered peers

I. Roll et al. (Eds.): AIED 2021, LNAI 12749, pp. 48–53, 2021.
https://doi.org/10.1007/978-3-030-78270-2_8

Inattentiveness and teacher quality, in this work, we take into consideration the level of *teacher control* as an added influence on pupil state transitions. Specifically, we aim to answer the following research questions:

R1. *To what extent does the existence of disruptive pupils affect other pupils near them?*
R2. *How does Teacher Control along with peer characteristics contribute to the achievement of young pupils in a disruptive classroom?*

2   Related Work

Classroom interactions and environment have received attention by researchers due to their potential affect over attainment. Teacher-student interaction and student-student interaction have a significant impact over student achievement [9]. Interactions can be positive like social and pedagogical interactions [3] or negative like disruption [11] be it talking out of turn, aggression or leaving seat [7, 14]. The frequency of disruptive behaviours acts as the major problem for teachers rather than the intensity [7]. In class-rooms, we usually find a number of pupils, up to a quarter of a class, who display some form of disruptive behaviour [6]. These disruptive children can have a negative impact on their peers' achievement [11, 18]. Therefore, it is imperative to take the necessary measures to contain disruptive behaviour and one of such measures is modelling class-room interactions trough simulation. Agent *based modelling* (ABM) is a framework for modelling the simulation of interactions between agents in a defined environment with a set of behaviours that influences those interactions [15]. In the area of education, Agent Based Modelling has been utilised to serve different proposes, such as improving the educational process [22] or as a support of the learning activity [19]. [21] used it to improve engagement by simulating pupils and teacher's emotional state. Their findings suggest that pupil's negative emotions are influenced by the teacher's characteristics, such as poor communication skills and poor teaching. [10] proposed a proof-of-concept model of teacher's and pupils' interactions with educational content in a classroom. The model aimed to help educational researchers and stakeholders, to improve prediction of pupils learning outcomes and choice of interventions - but did not take into account pupils' social interactions or a pupil's disruptive behaviour effect.

3   Data

The main source of data was obtained from a monitoring system named PIPS the Per-formance Indicators in Primary Schools [23, 24][1]. PIPS measures the schoolchildren's development through a baseline assessment at the start and end of their first year in elementary school. The data we analysed was of the academic year 2007/2008. The cognitive assessment provides measure of early math development and the personal assessment measure elements of disruptive behaviour (i.e. Inattentiveness scale has 0 to 9, Hyperactivity scale has 0 to 6). The dataset contains 3,315 classes from 2,040 schools with an average of 26 pupils per class summing up to 65,385 records in total.

---

[1] RR344_-_Performance_Indicators_in_Primary_Schools.pdf (publishing.service.gov.uk).

4 Methodology

We have created a simulation of the learning process interactions using Agent Based Modelling (ABM). In this simulation, we present a classroom with 30 pupils where a pupil will change between three different states: *learning, passive* or *disruptive*. Functionality of this model and technical details follow the ones in our previous work [1]. The model offers first *switch variables*, **Disruptive behaviour** and **Teacher characteristics switches** that indicate a high or low level of pupils' disruptiveness and teacher characteristics [1]. Another switch was added for this work to explore the effect of disruptive pupils in close proximity [2], **Neighbours' Effect Threshold switch**: it reflects to which degree a pupil affects his neighbours, with a range of 1 (high) to 4 (low). The effect is high if one pupil is enough to change a neighbour's state and low if it takes 4 pupils to trigger an effect. Other variables are **Math attainment level** $A(s, c)$, which accounts for student learning differences, **Start Math** $Smath_{scaled}$, which can be taken from PIPS data or assigned randomly by the model. We use a logarithmic function to map the 'learning Minutes' into 'Score' [16] to compute the **End Math** variable, $Emath(s,c)$, computed in Eq. 1 as follows:

$$Emath(s, c) = \log(L(s, c, T_{end-time}) + Smath_{scaled}(s, c))^n + A(s, c) \qquad (1)$$

Where $L(s, c, T_{end-time})$ represents the total learning time until the last tick $T_{end-time}$ that student s from class c had during the simulated year. We present here 3 *runs* with different parameter inputs, to observe their different effect on the pupil End Math scores. In our previous work of [1], we presented the results of three parameters: all maximum values, low Inattentiveness and low Teacher Quality. In this paper, instead, we have examined the following parameters:

- **Neighbours' Effect Run:** In the first simulation run, we are exploring the effect of another pupil characteristic: *Neighbours' Effect*. We set this variable to one (out of its range 1 to 4) to understand the impact of very high neighbour's effect [13], when compared with other runs.
- **Hyperactivity Run:** Here, we switched off Hyperactivity and kept the rest of the parameters at maximum value, to understand the *no-Hyperactivity Effect*.
- **Teacher Control Run:** Here, all parameters had the maximum possible values of their ranges, except Teacher Control, which was given the lowest possible value of its range, i.e., 1 out of 5: to explore *no-Teacher Control Effect*.

5 Results and Discussion

As an initial step to answer **R1**, we explored the relationship between **disruptive behaviour** and **End Math scores** to understand the effect of *disruptive* pupils on other pupils and found a negative correlation between the percentage of disruptive students and average End Math score of the class [1]. This suggested an effect of the number of disruptive pupils in a class over the general attainment. We computed Cohen's d for the three runs and found the effect size to be is large or medium [5]. Table 1 shows the results of the average End Math score for the runs.

Table 1. Results of average End Math of three runs

Run	First tick (Start Math)	Last tick (End Math)
Neighbour's effect	27.43	28.71
Hyperactivity	27.43	64.32
Teacher control	27.43	30.60

Thus, for **R1**, when the Neighbours' Effect increased, the End Math results produced by the model were the lowest, with an average of *28.71*, indicating that pupils made the least progress in Maths of all runs which shows peers' disruptiveness over pupils' attainment. In contrast, the highest result was seen when the Hyperactivity switch was off, resulting in *64.32* for the average End Math score, and an average of *30.60* in the low Teacher Control run which provide an answer to **R2** by showing a positive effect of low disruptive pupils in a class and a negative effect of low Teacher Control over their attainment. To compare with the real-world PIPS data[2], we ran a Pearson correlation test for the three simulation runs (see Table 2).

Table 2. Correlation test between simulation runs results and model variables

	End Math (Neighbour's Effect Run)	End Math (Hyperactivity Run)	End Math (Teacher Control Run)	End Math (PIPS)
Start Math	0.98	0.40	0.69	0.70
Inattentiveness	−0.14	−0.17	−0.06	−0.34
Hyperactivity	−0.16	−0.19	−0.17	−0.18

The nearest correlation score to PIPS data can be seen in the third run, with *0.69*. A high correlation is seen in the first run with the highest degree of Neighbour Effect, due to low progress resulting in little difference of pupils' between End Math and Math score. These results can serve for further improving the use of the model by providing the simulation of several factors in the learning environment.

6   Conclusion

In this paper we improved the design of the ABM model to reflect the *effect of disruptive young pupils in a classroom environment over their neighbours*; supported via an experimentation with these parameters. The results present a positive link between attainment and reduced classroom disruptiveness and a negative link with high disruptiveness and low teacher control. A limitation of this study would be bypassing other pupils'

---

[2] Please note however that PIPS data is only available for Start Math and End Math, thus only the start and end of the simulation process.

characteristics that would influence disruptiveness in classrooms. Future work includes exploring and validating further additions to this model, such as teacher intervention to reduce disruptive behaviour and observing the impact over attainment.

References

1. Alharbi, K., et al.: Agent-based simulation of the classroom environment to gauge the effect of inattentive or disruptive students. In: 17th International Conference on Intelligent Tutoring Systems (2021)
2. Braun, S.S., et al.: Effects of a seating chart intervention for target and nontarget students. J. Exp. Child Psychol. **191**, 104742 (2020). https://doi.org/10.1016/j.jecp.2019.104742
3. Cardoso, A.P., et al.: Personal and pedagogical interaction factors as determinants of academic achievement. Procedia - Soc. Behav. Sci. **29**, 1596–1605 (2011). https://doi.org/10.1016/j.sbs pro.2011.11.402
4. Cobb, J.A.: Relationship of discrete classroom behaviors to fourth-grade academic achievement. J. Educ. Psychol. **63**(1), 74–80 (1972). https://doi.org/10.1037/h0032247
5. Cohen, J.: Statistical Power Analysis for the Behavioral Sciences. Academic Press (2013)
6. Esturgó-Deu, M.E., Sala-Roca, J.: Disruptive behaviour of students in primary education and emotional intelligence. Teach. Teach. Educ. **26**(4), 830–837 (2010). https://doi.org/10.1016/j.tate.2009.10.020
7. Houghton, S., et al.: Classroom behaviour problems which secondary school teachers say they find most troublesome. Br. Educ. Res. J. **14**(3), 297–312 (1988). https://doi.org/10.1080/0141192880140306
8. Ingram, F.J., Brooks, R.J.: Simulating classroom lessons: an agent-based attempt. In: Proceedings of the Operational Research Society Simulation Workshop 2018, SW 2018, pp. 230–240 (2018)
9. Jafari, S., Asgari, A.: Predicting students' academic achievement based on the classroom climate, mediating role of teacher-student interaction and academic motivation. Integr. Educ. **24**(1), 62–74 (2020). https://doi.org/10.15507/1991-9468.098.024.202001.062-074
10. Koster, A., Koch, F., Assumpção, N., Primo, T.: The role of agent-based simulation in education. In: Koch, F., Koster, A., Primo, T., Guttmann, C. (eds.) CARE/SOCIALEDU 2016. CCIS, vol. 677, pp. 156–167. Springer, Cham (2016). https://doi.org/10.1007/978-3-319-52039-1_10
11. Kristoffersen, J.H.G., et al.: Disruptive school peers and student outcomes. Econ. Educ. Rev. **45**, 1–13 (2015). https://doi.org/10.1016/j.econedurev.2015.01.004
12. Kuyper, H., et al.: Motivation, meta-cognition and self-regulation as predictors of long term educational attainment. Educ. Res. Eval. **6**(3), 181–205 (2000). https://doi.org/10.1076/1380-3611(200009)6:3;1-a;ft181
13. Lavasani, M.G., Khandan, F.: Maintaining the balance: teacher control and pupil disruption in the classroom. Cypriot J. Educ. **2**, 61–74 (2011)
14. Little, E.: Secondary school teachers' perceptions of students' problem behaviours. Educ. Psychol. **25**(4), 369–377 (2005). https://doi.org/10.1080/01443410500041516
15. Macal, C., North, M.: Introductory tutorial: agent-based modeling and simulation, pp. 6–20 (2014)
16. Mauricio, S., et al.: Analysing differential school effectiveness through multilevel and agent-based modelling multilevel modelling and school effectiveness research. **17**, 1–13 (2014)
17. Merrell, C., et al.: A longitudinal study of the association between inattention, hyperactivity and impulsivity and children's academic attainment at age 11. Learn. Individ. Differ. **53**, 156–161 (2017). https://doi.org/10.1016/j.lindif.2016.04.003

18. Müller, C.M., et al.: Peer influence on disruptive classroom behavior depends on teachers' instructional practice. J. Appl. Dev. Psychol. **56**, 99–108 (2018). https://doi.org/10.1016/j.app dev.2018.04.001
19. Ponticorvo, M., et al.: An agent-based modelling approach to build up educational digital games for kindergarten and primary schools. Expert Syst. **34**(4), 1–9 (2017). https://doi.org/ 10.1111/exsy.12196
20. Smith, D.P., et al.: Who goes where? The importance of peer groups on attainment and the student use of the lecture theatre teaching space. FEBS Open Bio **8**(9), 1368–1378 (2018). https://doi.org/10.1002/2211-5463.12494
21. Subramainan, L., et al.: A conceptual emotion-based model to improve students engagement in a classroom using agent-based social simulation. In: Proceeding of the - 2016 4th International Conference on User Science and Engineering, i-USEr 2016, pp. 149–154 (2017). https://doi. org/10.1109/IUSER.2016.7857951
22. Subramainan, L., Mahmoud, M., Ahmad, M., Yusoff, M.: A simulator's specifications for studying students' engagement in a classroom. In: Omatu, S., Rodríguez, S., Villarrubia, G., Faria, P., Sitek, P., Prieto, J. (eds.) DCAI 2017. AISC, pp. 206–214. Springer, Cham (2018). https://doi.org/10.1007/978-3-319-62410-5_25
23. Tymms, P.: Baseline Assessment and Monitoring in Primary Schools. David Fulton Publisher, London (1999)
24. Tymms, P., Albone, S.: Performance indicators in primary schools. In: School Improvement through Performance Feedback, pp. 191–218 (2002)
25. World Health Organization: The ICD-10 classification of mental and behavioural disorders. World Heal. Organ. **55**, 135–139 (1993). https://doi.org/10.4103/0019

Integration of Automated Essay Scoring Models Using Item Response Theory

Itsuki Aomi[✉], Emiko Tsutsumi, Masaki Uto, and Maomi Ueno

The University of Electro-Communications, Tokyo, Japan
{aomi,tsutsumi,uto,ueno}@ai.lab.uec.ac.jp

Abstract. Automated essay scoring (AES) is the task of automatically grading essays without human raters. Many AES models offering different benefits have been proposed over the past few decades. This study proposes a new framework for integrating AES models that uses item response theory (IRT). Specifically, the proposed framework uses IRT to average prediction scores from various AES models while considering the characteristics of each model for evaluation of examinee ability. This study demonstrates that the proposed framework provides higher accuracy than individual AES models and simple averaging methods.

Keywords: Automated essay scoring · Item response theory · Model averaging

1 Introduction

In recent years, various studies have examined automated essay scoring (AES) models to reduce the costs involved in scoring essays in mass testing. Most AES models can be roughly divided into two approaches: *feature-engineering approach* and *automatic feature extraction approach* [5, 7]. The features-engineering approach manually extracts features (e.g., essay length and number of spelling errors) from given essays and uses these features to predict scores. An important benefit of this approach is its explicability. The approach, however, generally requires careful feature creation and selection to achieve high accuracy. To obviate the need for feature engineering, the automatic feature extraction approach using neural networks has been recently proposed [1,2,4,13,14,21]. Such conventional AES models are known to provide different advantages. Therefore, averaging the scores of various AES models is expected to improve scoring accuracy. However, scores that are simply averaged might be inaccurate because each AES model has different accuracy for evaluating examinee ability.

To resolve this problem, we propose a framework that aggregates various AES models using item response theory (IRT) [10], which is a test theory based on mathematical models. In recent years, IRT models that are able to estimate scores while considering the characteristics of human raters, such as rater severity and consistency, have been proposed [3,8,11,15,17–19]. The present study

© Springer Nature Switzerland AG 2021
I. Roll et al. (Eds.): AIED 2021, LNAI 12749, pp. 54–59, 2021.
https://doi.org/10.1007/978-3-030-78270-2_9

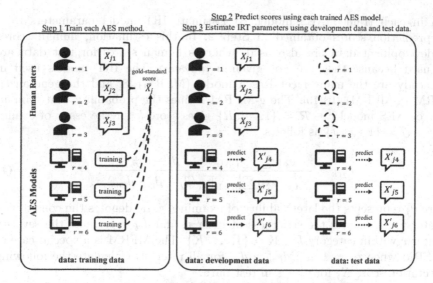

Fig. 1. Proposed framework for three human raters ($r = 1, 2, 3$) and three AES models ($r = 4, 5, 6$). X_{jr} indicates the score given by human-rater r for the essay of examinee j. $\bar{X}_j$ is the average of all the scores given by all the human-raters. X'_{jr} is the prediction score given by the r-th AES model for the essay of examinee j.

focuses on the use of such IRT models with AES models instead of human raters. The proposed framework is expected to provide scores that are more accurate than those obtained by simple averaging or a single AES model because the framework can integrate prediction scores from various AES models while considering the characteristics of each model for each examinee's ability level. Our experiments demonstrate that the proposed framework provides higher accuracy than individual AES models and than simply averaged scores.

Of note, Uto and Okano have recently proposed another AES framework that uses IRT [16]. They, however, use IRT to remove rater bias effects within training data to improve the robustness of the model training process. The research objective and the developed framework are completely different from those of the present study.

2   Proposed Framework

In this section, we propose a framework for averaging scores of various AES models in consideration of the characteristics of each model. Figure 1 shows the outline of the proposed framework. As shown in the figure, the proposed framework executes model training and score prediction through the following four steps: 1) Train each AES model individually using gold-standard scores in training data. 2) Predict scores for essays using development data and test data in each trained AES model. 3) Estimate IRT model parameters from the prediction scores obtained in Step 2. In this estimation, human scores for development data are also used, whereas human scores for test data are not used because they are not given in advance. The IRT models used in this study are the many facet Rasch model (MFRM) [8] and the generalized MFRM (g-MFRM) [18,19]. The g-MFRM defines the probability that human-rater or AES model $r \in \mathcal{R} = \{1, \ldots, R\}$ gives score k for the essay of examinee $j \in \mathcal{J} = \{1, \ldots, J\}$ as follows.

$$P_{jrk} = \frac{\exp \sum_{m=1}^{k} [\alpha_r(\theta_j - \beta_r - d_{rm})]}{\sum_{l=1}^{K} \exp \sum_{m=1}^{l} [\alpha_r(\theta_j - \beta_r - d_{rm})]}. \tag{1}$$

where θ_j represents the latent ability of examinee j, α_r denotes the consistency of rater r, β_r denotes the strictness of rater r, and d_{rk} represents the severity of rater r within category $k \in K = \{1, \ldots, K\}$. The MFRM is a special case of g-MFRM when $\alpha_r = 1$ and $d_{rm} = d_m$ for all rater. 4) Calculate the following expectation score $\hat{X}_j$ for essays in test data.

$$\hat{X}_j = \frac{1}{|\mathcal{R}_{\text{human}}|} \sum_{r \in \mathcal{R}_{\text{human}}} \sum_{k=1}^{K} k \cdot P_{jrk}, \tag{2}$$

where $\mathcal{R}_{\text{human}}$ is the set of human raters. This calculation is performed given IRT parameter estimates including the latent examinee ability $\hat{\theta}_j$, which are estimated from multiple AES model predictions in Step 3.

3   Experiments

We evaluate the effectiveness of the proposed framework using the Automated Student Assessment Prize (ASAP) dataset, which has been used in various AES studies [6,13,14,21] and Kaggle competitions[1]. We use five-fold cross validation to evaluate scoring accuracy in terms of quadratic weighted kappa (QWK) which is the common evaluation metric in the ASAP competition.

The following AES models are used in our experiment: Feature-engineering approach models, including **EASE (SVR)**, **EASE (BLRR)** [12], and **XGBoost** [6,9]. Automatic feature extraction approach models, including **LSTM**-based model [13] and **SkipFlow** model [14]. We also used a hybrid model **BERT+F** [20] that integrates the feature-engineering approach and automatic feature extraction approach. Model settings, including hyperparameter settings, are the same as those used in the original studies.

The present experiment compares the proposed framework incorporating the model described above with the individual AES models (hereinafter, BASE models), and with two simple model averaging methods; **MEAN** (arithmetic averaging of AES scores) and **VOTING** (hard voting of AES scores). Hereinafter, we call the simple averaging methods as AVG methods.

---

[1] https://www.kaggle.com/c/asap-aes.

In the proposed framework, we examine two IRT models: MFRM and g-MFRM. We refer to the proposed frameworks using these IRT models respectively as **Proposal (MFRM)** and **Proposal (g-MFRM)**. The IRT parameter estimation was conducted by Markov chain Monte Carlo following [19].

Table 1. QWK score of the BASE models and the AVG methods.

	AES models	Prompts								
		1	2	3	4	5	6	7	8	Avg.
BASE	EASE (SVR)	0.558	0.533	0.564	0.571	0.659	0.749	0.545	0.350	0.566
	EASE (BLRR)	0.804	0.603	0.656	0.717	0.784	0.761	0.730	0.675	0.716
	XGBoost	0.814	0.640	0.593	0.660	0.763	0.657	0.692	0.676	0.687
	LSTM	0.777	0.619	0.651	0.730	0.770	0.760	0.750	0.460	0.690
	SkipFlow	0.798	0.652	0.657	0.729	0.783	0.778	0.751	0.614	0.720
	BERT+F	0.827	0.637	0.672	0.620	0.780	0.673	0.720	0.681	0.701
AVG	MEAN	0.820	0.667	0.673	0.730	**0.805**	0.774	0.768	0.678	0.739*
	VOTING	0.833	0.660	**0.675**	0.731	0.794	0.770	0.745	0.666	0.734*
	Proposal (MFRM)	0.821	0.626	0.663	0.685	0.777	0.728	0.768	0.674	0.718*
	Proposal (g-MFRM)	**0.838**	**0.686**	0.668	**0.743**	0.796	**0.785**	**0.793**	**0.717**	**0.753**

Table 1 presents the experimentally obtained results. * indicates that the performance of Proposal (g-MFRM) is higher than that of the other AVG methods at the 5 % significance level by one-tailed paired t-test. The results show that Proposal (g-MFRM) provides a higher QWK score than that of all the BASE models except for only one case (BERT+F in prompt 3). Furthermore, Proposal (g-MFRM) achieves the highest QWK score on average.

In Table 1, simple averaging methods are shown to also outperform the BASE models for almost all prompts. Compared with the simple averaging methods, Proposal (g-MFRM) provides a higher QWK score for prompts 1, 2, 4, 6, 7, and 8, but it provides a slightly lower QWK score for prompts 3, and 5. The reason for this improvement is that the proposed framework can estimate scores while considering the characteristics of the respective BASE models. In prompts where Proposal (g-MFRM) provides higher QWK score, the difference in QWK score among the BASE models tends to be large. For example, EASE (SVR) in prompts 1 and 7, XGBoost and BERT+F in prompt 6, and EASE (SVR) and LSTM in prompt 8 show much lower QWK score. Thus, Proposal (g-MFRM) can maintain high scoring accuracy even when models with various characteristics exist, although simple averaging methods can not.

4   Conclusion

In this work, we proposed a new framework for integrating AES models that uses IRT. We described how simply averaged scores can lower evaluating accuracy because each AES model has a different assessment accuracies for scoring examinee ability. To resolve this issue, we presented the idea of estimating scores using

IRT models while considering the characteristics of the AES models. Based on experiment results, we demonstrated that the proposed framework with a latent IRT model provides higher accuracy than individual AES models and higher accuracy than simply averaged scores.

References

1. Alikaniotis, D., Yannakoudakis, H., Rei, M.: Automatic text scoring using neural networks. In: Proceedings of the 54th Annual Meeting of the Association for Computational Linguistics (Volume 1: Long Papers), pp. 715–725 (2016)
2. Dasgupta, T., Naskar, A., Dey, L., Saha, R.: Augmenting textual qualitative features in deep convolution recurrent neural network for automatic essay scoring. In: Proceedings of the Fifth Workshop on Natural Language Processing Techniques for Educational Applications, pp. 93–102 (2018)
3. Eckes, T.: Introduction to Many-Facet Rasch Measurement. Peter Lang, Bern (2015)
4. Farag, Y., Yannakoudakis, H., Briscoe, T.: Neural automated essay scoring and coherence modeling for adversarially crafted input. In: Proceedings of the 2018 Conference of the North American Chapter of the Association for Computational Linguistics: Human Language Technologies, Volume 1 (Long Papers), pp. 263–271 (2018)
5. Hussein, M.A., Hassan, H.A., Nassef, M.: Automated language essay scoring systems: a literature review. PeerJ Comput. Sci. **5** (2019)
6. Jin, C., He, B., Hui, K., Sun, L.: TDNN: a two-stage deep neural network for prompt-independent automated essay scoring. In: Proceedings of the 56th Annual Meeting of the Association for Computational Linguistics (Volume 1: Long Papers), pp. 1088–1097 (2018)
7. Ke, Z., Ng, V.: Automated essay scoring: a survey of the state of the art. In: Proceedings of the Twenty-Eighth International Joint Conference on Artificial Intelligence, IJCAI-19, pp. 6300–6308 (2019)
8. Linacre, J.M.: Many-Facet Rasch Measurement. MESA Press, Chicago (1989)
9. Liu, J., Xu, Y., Zhu, Y.: Automated Essay Scoring based on Two-Stage Learning. arXiv e-prints arXiv:1901.07744, January 2019
10. Lord, F.M.: Applications of Item Response Theory to Practical Testing Problems. Routledge, Abingdon-on-Thames (1980)
11. Myford, C.M., Wolfe, E.W.: Detecting and measuring rater effects using many-facet Rasch measurement: part I. J. Appl. Measur. **4**(4), 386–422 (2003)
12. Phandi, P., Chai, K.M.A., Ng, H.T.: Flexible domain adaptation for automated essay scoring using correlated linear regression. In: Proceedings of the 2015 Conference on Empirical Methods in Natural Language Processing, pp. 431–439 (2015)
13. Taghipour, K., Ng, H.T.: A neural approach to automated essay scoring. In: Proceedings of the 2016 Conference on Empirical Methods in Natural Language Processing, pp. 1882–1891 (2016)
14. Tay, Y., Phan, M., Luu, A.T., Hui, S.C.: SkipFlow: incorporating neural coherence features for end-to-end automatic text scoring. In: Thirty-Second AAAI Conference on Artificial Intelligence, pp. 5948–5955 (2018)
15. Ueno, M., Okamoto, T.: Item response theory for peer assessment. In: 2008 Eighth IEEE International Conference on Advanced Learning Technologies, pp. 554–558 (2008). https://doi.org/10.1109/ICALT.2008.118

16. Uto, M., Okano, M.: Robust neural automated essay scoring using item response theory. In: Artificial Intelligence in Education, pp. 549–561 (2020)
17. Uto, M., Ueno, M.: Item response theory for peer assessment. IEEE Trans. Learn. Technol. **9**(2), 157–170 (2016)
18. Uto, M., Ueno, M.: Item response theory without restriction of equal interval scale for rater's score. In: Artificial Intelligence in Education, pp. 363–368 (2018)
19. Uto, M., Ueno, M.: A generalized many-facet Rasch model and its Bayesian estimation using Hamiltonian Monte Carlo. Behaviormetrika **47**, 469–496 (2020)
20. Uto, M., Xie, Y., Ueno, M.: Neural automated essay scoring incorporating handcrafted features. In: Proceedings of the 28th International Conference on Computational Linguistics, pp. 6077–6088 (2020)
21. Wang, Y., Wei, Z., Zhou, Y., Huang, X.: Automatic essay scoring incorporating rating schema via reinforcement learning. In: Proceedings of the 2018 Conference on Empirical Methods in Natural Language Processing, pp. 791–797 (2018)

Towards Sharing Student Models Across Learning Systems

Ryan S. Baker[1]([⊠]), Bruce M. McLaren[2], Stephen Hutt[1], J. Elizabeth Richey[2],
Elizabeth Rowe[3], Ma. Victoria Almeda[3], Michael Mogessie[2],
and Juliana M. AL. Andres[1]

[1] Graduate School of Education, University of Pennsylvania, Philadelphia, USA
rybaker@upenn.edu
[2] Human-Computer Interaction Institute, Carnegie Mellon University, Pittsburgh, USA
[3] EdGE at TERC, Cambridge, USA

Abstract. Modern AIED systems develop sophisticated and multidimensional models of students. However, what is learned about students in one system—their skills, behaviors, and affect—is not carried over to other systems that could benefit students by using the information, potentially reducing both the effectiveness and efficiency of these systems. This challenge has been cited by a number of researchers as one of the most important for the field of AIED. In this paper, we discuss existing progress towards resolving this challenge, break down five sub-challenges, and propose how to address the sub-challenges.

Keywords: Student model sharing · AIED system integration · BLAP

1 Introduction

More and more students use learning technologies each year, a trend accelerated by COVID-19 [6, 14]. Schools often have students use several learning platforms, even within the same subject [4]. However, these learning technologies do not currently work together to support students. What one learning system determines about a student's skills and behaviors is generally not carried over to other learning systems, reducing both educational effectiveness and efficiency—if a student learns a topic several times, and multiple learning technologies need time to learn the same thing about a student.

This challenge, bringing together distinct learning technologies, has been repeatedly referred to as a key goal for learning technologies. Kay [11] argued for "lifelong user models…existing independently of any single application and controlled by the learner." It was also a key part of the fifth challenge, "Lifelong and Lifewide Learning," in the AI Grand Challenges proposed in [18]. Finally, it was one of six "Baker Learning Analytics Prizes" (BLAP) challenges [3]. This challenge, "Transferability: The Learning System Wall," was posed as not just transferring student information from learning system A to learning system B, but in improving a student model that is already successful in learning system B and using that improved model to change how system B supports students at runtime, improving learning outcomes. Intentionally conceived in a more specific fashion

I. Roll et al. (Eds.): AIED 2021, LNAI 12749, pp. 60–65, 2021.
https://doi.org/10.1007/978-3-030-78270-2_10

than previous challenges, the Transferability/Learning Wall challenge was designed to represent a stepping-stone to the visions proposed in [11] and [18]—while representing improvement for students in itself.

2 Prior Work

Although learning systems do not yet connect their student models, there has been some past work to integrate learning systems in other fashions. In this section, we review that literature and discuss why it remains a significant step to integrate two systems' student models in an actionable way.

One of the most well-known areas of relevant prior work is in standards for logging data and representing student models. The Caliper framework provides a large set of ways to represent data from a variety of types of learning activities seen in learning management systems but has less support for the types of activities seen in the more complex interactions in AIED systems [9]. xAPI attempts to offer support for representing and sharing the data from a broader range of learning activities [5].

Both these platforms can be used to integrate systems through connections such as the Learning Tools Interoperability (LTI) standards [10]. Still, the connections offered are very simple, such as specifying the correctness of an action. Neither framework provides functionality designed for sharing the type of complex student models used in modern AIED systems. One AIED project was able to develop a workaround for the LTI standard to support simple transfer of student model information between platforms [1], but the approach only worked in a single direction, for a single piece of information, and required a direct platform-to-platform connection. In another example, [7, 15] connected two learning environments into the same reporting system.

Other research has attempted to simulate a student model connection between different learning systems or activities, without actually connecting systems/activities to each other. [15, 16] developed a mapping between the skills in two different learning systems and then tested it by administering paper tests to students and analyzed the degree of agreement between the skills (but solely from the paper test data). [8] analyzed whether student knowledge model estimates from one lesson in a Cognitive Tutor would improve knowledge estimation on later lessons in a Cognitive Tutor. [17] asked twenty subjects to use both a research paper recommendation system and a scientific talk recommendation system (with order randomized) and then analyzed whether the second system's recommendations would have been more accurate if the first system's data had been used. These studies established the feasibility and potential usefulness of connecting student models across learning systems, paving the way for the next step: actually making the connection between learning systems.

3 The Problem, Broken Down into Its Constituent Parts

The problem of sharing student models between two learning systems in a meaningful way that improves student outcomes breaks down into five sub-challenges:

1) *Connection*: The two systems need to seamlessly and digitally connect to each other, whether via API, shared database, or another technical link, so that one system can use the other system's inferences to inform its behavior.
2) *Mapping Related Constructs*: The two systems need to have student models of similar or related constructs, each of sufficient accuracy to be practically useful, and a mapping between the constructs in each system is needed [16].
3) *Evidence Integration*: Each system needs to have a way to integrate evidence from the other system into its own estimates based on how strongly each system's evidence predicts behavior in the other system.
4) *A Good Reason*: There needs to be a practical reason for connecting the student models, e.g. the student model drives an automated intervention, or the student model helps with a teacher orchestration system.
5) *Demonstration of Benefit*: The intervention (whether automated or by teachers) driven by the shared student model needs to actually make a difference to student behavior and outcomes if properly delivered, but only for some students (i.e. a student model is actually needed; the intervention is not universally beneficial).

4   Potential Steps Towards an Architecture and Student Model Integration Algorithms

There are many possible approaches to connecting and sharing information between two or more learning systems (sub-challenge 1, Connection): these approaches can generally be grouped into two categories, system-to-system direct connections, and server-mediated connections. System-to-system direct connections are likely the quickest approach but are also hard to scale more broadly. It will be difficult to develop an ecosystem of learning systems working in concert through direct connections between individual learning systems. Instead, it will be more scalable to build a single server to facilitate connections between many learning systems. This could be achieved by an external web service, shown in Fig. 1, that different learning systems can post student model inferences to or request student model inferences from. This external service would also need to be able to securely maintain a mapping of student IDs in different learning systems, with some form of access control for school districts or learning system developers to authorize sharing between learning systems.

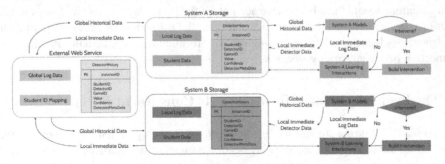

Fig. 1. A potential architecture for student model sharing

Assuming that the two platforms model similar or related constructs (sub-challenge 2, Mapping Related Constructs), and that these models drive practical interventions (sub-challenge 4, A Good Reason), the next step is to select an algorithm that each platform will use to integrate information from the other platform (sub-challenge 3, Evidence Integration), improving, replacing, or initializing the other system's estimates. Each system should take in the other system's evidence but make its own decision, rather than having a unified student model external to either system. This design choice keeps student model control local to each system—keeping system developers in control of their system's functioning. We propose investigating the following five approaches to information integration and selecting the most successful:

1) **System-weighted averaging**. Take each system's estimates, and average them together, weighting the other system's estimates lower than its own..

2) **System and evidence quantity weighted averaging.** Take each system's estimates and average them together. Each system weighs the other system's evidence in terms of the amount of evidence, penalized by a percentage due to the evidence not being from the local system.

3) **Performance Factors Analysis (PFA)** [13]. PFA is typically used in a single system. It computes a linear combination of weighted successes and failures for a skill so far (weights fit per skill) and then runs that combination through a logistic function to predict correctness on future items. PFA could be extended for a multi-system student model by fitting "successes" and "failures" for each system.

4) **Bayesian Network.** A Bayesian Network allows complex inter-relationships between skills [cf. 2, 12]. Both the current system's evidence and the other system's evidence can be integrated into a network, with the other system's evidence providing updates to the estimates of the current system's evidence.

5) **Deep Knowledge Tracing +** [19]. Deep Knowledge Tracing (DKT) can find complex relationships between multiple sources of evidence to predict future performance. DKT + is an extension based on regularization that fixes problems with the original formulation (such as correct performance leading to predictions of worse performance and wild swings in proficiency estimates). The other system's evidence and the current system's evidence can be integrated into DKT + to predict multiple student attributes or behaviors simultaneously.

Having integrated the two student models, the next step will be to test whether an intervention based on the integrated student model is beneficial for learners (sub-challenge 5, Demonstration of Benefit): beneficial only to students in need and better than an intervention from only a single system's data. One of the biggest areas of potential will be for "cold start" situations – where one system has evidence on student knowledge of a topic not yet encountered in the other learning system. There will also be potential around inferring constructs where considerable amounts of aggregate data are needed to draw a clear inference or where the behavior or state of interest only manifests occasionally.

5 Conclusion

In this article, we discuss the potential of sharing student models between learning systems. We frame this challenge in terms of five sub-challenges that need to be addressed in order to solve this challenge. We then offer an architecture to address a key sub-challenge and discuss algorithms that could potentially be used for student model integration. We encourage our AIED colleagues to join in solving this challenge.

References

1. Aleven, V., et al.: Towards deeper integration of intelligent tutoring systems: one-way student model sharing between GIFT and CTAT. In: Proceedings of the 7th Annual Generalized Intelligent Framework for Tutoring (GIFT) Users Symposium (2019)
2. Arroyo, I., Woolf, B.P.: Inferring learning and attitudes from a Bayesian Network of log file data. In: Proceedings of the International Conference on Artificial Intelligence and Education, pp. 33–40 (2005)
3. Baker, R.S.: Challenges for the future of educational data mining: the Baker learning analytics prizes. J. Educ. Data Mining 11(1), 1–17 (2019)
4. Baker, R.S., Gowda, S.M.: The 2018 technology & learning insights report: towards understanding app effectiveness and cost. BrightBytes, Inc., San Francisco, CA (2018)
5. Berking, P., Gallagher, S.: Choosing a learning management system. In: Advanced Distributed Learning (ADL) Co-Laboratories, pp. 40–62 (2013)
6. BrightBytes, Inc.: 2020 Remote Learning Survey Research Results (2020). https://www.brightbytes.net/rls-research . Accessed 12 Feb 2021
7. Brusilovsky, P., et al.: Database exploratorium: a semantically integrated adaptive educational system. In: 7th International Workshop on Ubiquitous User Modelling (UbiqUM 2009) (2009)
8. Eagle, M., et al.: Predicting individual differences for learner modeling in intelligent tutors from previous learner activities. In: Cena, F., Desmarais, M., Dicheva, D., Zhang, J. (eds.) Proceedings of the 24th Conference on User Modeling, Adaptation and Personalization (UMAP 2016), pp. 55–63. ACM, New York (2016)
9. IMS Global: Caliper Analytics. (n.d., a). http://www.imsglobal.org/activity/caliper. Accessed 4 Jan 2021
10. IMS Global: Learning Tools Interoperability. (n.d., b). https://www.imsglobal.org/activity/learning-tools-interoperability. Accessed 4 Jan 2021
11. Kay, J.: Lifelong learner modeling for lifelong personalized pervasive learning. IEEE Trans. Learn. Technol. 1(4), 215–228 (2008)
12. Kim, Y.J., Almond, R.G., Shute, V.J.: Applying evidence-centered design for the development of game-based assessments in physics playground. Int. J. Test. 16(2), 142–163 (2016)
13. Pavlik Jr., P.I., Cen, H., Koedinger, K.R.: Performance factors analysis--a new alternative to knowledge tracing. In: Proceedings of the International Conference on Artificial Intelligence and Education (2009)
14. Reich, J., et al.: Remote Learning Guidance from State Education Agencies during the COVID-19 Pandemic: A First Look (2020). Unpublished manuscript. http://osf.io/k6zxy/. Accessed 12 Feb 2021
15. Sosnovsky, S., Brusilovsky, P., Yudelson, M., Mitrovic, A., Mathews, M., Kumar, A.: Semantic integration of adaptive educational systems. In: Advances in Ubiquitous User Modelling, pp. 134–158 (2009)

16. Sosnovsky, S., Dolog, P., Henze, N., Brusilovsky, P., Nejdl, W.: Translation of overlay models of student knowledge for relative domains based on domain ontology mapping. In: Luckin, R., Koedinger, K.R., Greer, J. (eds.) Proceedings of the 13th International Conference on Artificial Intelligence in Education, IOS, pp. 289–296 (2007)
17. Wongchokprasitti, C., Peltonen, J., Ruotsalo, T., Bandyopadhyay, P., Jacucci, G., Brusilovsky, P.: User model in a box: cross-system user model transfer for resolving cold start problems. In: Proceedings of the International Conference on User Modeling, Adaptation, and Personalization, pp. 289–301 (2015)
18. Woolf, B.P., Lane, H.C., Chaudhri, V.K., Kolodner, J.L.: AI grand challenges for education. AI Mag. **34**(4), 66–85 (2013)
19. Yeung, C.K., Yeung, D.Y.: Addressing two problems in deep knowledge tracing via prediction-consistent regularization. In: Proceedings of the Fifth Annual ACM Conference on Learning at Scale, pp. 1–10 (2018)

Protecting Student Privacy with Synthetic Data from Generative Adversarial Networks

Peter Bautista$^{(\boxtimes)}$ ⓘ and Paul Salvador Inventado$^{(\boxtimes)}$ ⓘ

California State University, Fullerton, CA 92831, USA
pinventado@fullerton.edu

Abstract. Educational data requires layers of protection that prohibit easy access to sensitive student data. However, the additional layers of security hinder research that relies on educational data to progress. In this paper, a Least Squares GAN (LSGAN) is proposed to create synthetic student performance datasets based on a master dataset without recreating samples. Synthetic data is less likely to be traced back to a student thereby reducing privacy issues. Two feature subsets were considered in the study: sequential, and all features. GANs trained on the sequential data produced new datasets that were representative of student performance from the training dataset, while the GAN trained on all features was not able to capture characteristics from the dataset. Based on the results, the synthetic dataset can provide an alternative unrestricted source of data without compromising student privacy.

1 Introduction

Student data is now easier to collect with the advent of learning platforms that make it easy to track learner behavior and performance [6]. Such data allows instructors and researchers to apply learning analytics and educational data mining techniques to analyze student learning that inform teaching [1]. However, as more data about students are collected, consumers of this data need to be more mindful about how it is used and shared to maintain student privacy [11].

Several policies exist to protect student data. In the United States, the Family Educational Rights and Privacy Act (FERPA) of 1974, 20 U.S.C. § 1232g (1974) requires federally funded institutions to get parental or student consent before disclosing personal information. The Children's Online Privacy Protection Act of 1998, 15 U.S.C. § 6501- 6506 (1998) requires web hosts and content providers to seek parental consent to store data about children under 13. The Student Digital Privacy and Parental Rights Act of 2015, H.R.2092, 114th Cong. (2015–2016) prohibits operators from selling personal information to third parties or collecting student information for purposes unrelated to educational activities. In academic settings, researchers are required to get approval from an institutional review board (IRB) in addition to the restrictions set by FERPA before collecting student information [15,16].

© Springer Nature Switzerland AG 2021
I. Roll et al. (Eds.): AIED 2021, LNAI 12749, pp. 66–70, 2021.
https://doi.org/10.1007/978-3-030-78270-2_11

This work aims to leverage generative adversarial networks (GANs) to produce synthetic data based on real student data. Generated data cannot be traced back to an individual thereby reducing privacy issues and satisfying privacy policy requirements [2].

2 Related Works

There are two approaches commonly considered in data privacy: privacy-preserving data publishing, and privacy-preserving data mining [8]. Privacy-preserving data publishing involves sharing sensitive information about individuals without violating their privacy. Privacy-preserving data mining involves the application of data mining without using sensitive information. This work focuses on privacy-preserving data publishing.

The simplest approach to protect students' privacy is to remove unique, identifying information such as names and student IDs. However, non-identifying student information can be checked against other data sources, like social media, and possibly uncover students' identities through implied relationships [10,14].

Another approach to preserving privacy is k-anonymity and l-diversity. According to k-anonymity we can protect privacy by ensuring that each distinct pattern of key variables is possessed by at least a minimum of k records [13]. k-anonymity can be implemented by generalizing attributes with few observations into categories (e.g., age ranges instead of actual age), or injecting missing values (suppression) to data that is easy to distinguish. l-diversity measures the frequency of values in sensitive attributes (e.g., 2 instances of age 13). Records can be removed or further generalized and suppressed to maintain an l-diversity threshold [8]. A major drawback in this anonymization approach is the loss of data fidelity. Model performance may suffer as more data is lost or modified. There is a tradeoff between privacy and the utility of learning analytics and data mining [6].

Generative adversarial networks create completely synthetic data based on real data [5] and can be used to protect privacy. Baowaly et al., investigated different types of GANs to generate synthetic patient electronic health records (EHR), which are highly restricted data [2,4]. They developed a medical boundary-seeking GAN (medBGAN) to produce patient data containing International Classification of Diseases (ICD) codes that are used for medical diagnoses. Their evaluation showed that the performance of logistic regression models trained on both real and synthetic data were comparable. GANs are also able to create similar characteristics to the real data such as distribution and mean [3,9].

3 Data and Methodology

We developed a GAN and used it with student performance data collected from introductory computer science classes at a public Hispanic-serving institution to investigate how well it protected student privacy. The data set contained information from 104 students who were enrolled in two sections of the same class taught by the same professor. The data consisted of 77 attributes describing grades for multiple assessments across the semester including quizzes, in-class activities, homework, projects, midterm exams, final exams, and their final grade.

Our GAN's discriminator is a recurrent neural network and its generator is a multi-layer densely connected neural network. Both networks utilized the least squares loss (LS Loss) for the loss function. We added regularizers to prevent the generator from producing the same outputs, also known as mode collapse. The output layer is the same shape as a row within the student data with ReLU activation function bounded from 0 to 100.

4   Results and Discussion

Two subsets of the original dataset were used for generation. The first included quizzes, midterms, and finals which were considered as sequential data where an earlier element of a row preceded those later in the same row. For example, students' performance in quiz 1 precedes the same student's performance for quiz 2 and quiz 3 in the same data point. There were 14 input features in total. The second subset contained all 77 initial graded attributes. The original dataset was split into 70 students for training and 34 for validation. The dataset generated by the GAN is the same size as the original with 104 students.

Figure 1 shows the average values of the 14 attributes from the synthetic data produced by our GAN and real student data. We measured the residuals to determine how well the generated data matched the real data. The average residual was 1.2% indicating that the synthetic data closely resembled real student data, but did not fit it exactly. We want to avoid residuals that are 0 as this might indicate overfitting. If there is overfitting, there is a possibility the GAN is recreating samples from the original dataset.

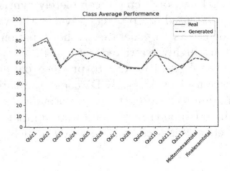

Fig. 1. Class average for features within sequential data

Looking closer at the individual rows, we find that in addition to creating a good fit for the data, the generator created unique rows. Figure 2 shows a heatmap with attribute values from 14 randomly chosen real and synthetic students. The two heatmaps are essentially indistinguishable. Compared to real student data, the synthetic student data attribute values were close to their means, within the same range, followed similar patterns, but did not duplicate real student data.

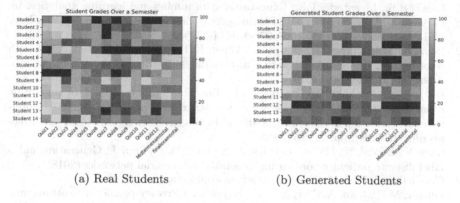

(a) Real Students   (b) Generated Students

Fig. 2. Heatmap of student grades for quizzes, midterm, and final

Unfortunately, increasing the number of features caused the GAN to destabilize and resulted in the generator producing near zero outputs for a majority of features. Neither the generator nor discriminator were able to reach an acceptable level of performance even after increasing the number of training data.

5   Conclusion and Future Work

Overall, the GAN successfully generated synthetic student data that did not replicate real student data. It was able to recreate similar characteristics such as the averages and distributions of real data with only minor deviations. Therefore it can protect students' privacy.

Further work is necessary to generate synthetic data with more attributes without destabilization. Possible next steps include applying GANs on multiple attribute subsets or using autoencoders to reduce destabilization [2]. The data used in this work focused on continuous data, but we plan to explore its performance on categorical features such as ethnicity, gender, and major.

Other than protecting student privacy, GANs' generative nature increases data set sizes. As we saw in our experiment, the generator produced realistic data even with limited training data. Data generation can be useful for small scale-studies such as those conducted by instructors in their classes which often have limited data. Since model performance will likely depend on recent student performance data, data sources will contain information on students who are still in school. Therefore, these studies will benefit from GANs' privacy-preserving nature. Small-scale studies are important because they inform researchers on how they might scale their research and it is also a common activity conducted by teachers to inform their teaching [7,12].

References

1. Baker, R.S., Inventado, P.S.: Educational data mining and learning analytics. In: Larusson, J.A., White, B. (eds.) Learning Analytics, pp. 61–75. Springer, New York (2014). https://doi.org/10.1007/978-1-4614-3305-7_4
2. Baowaly, M.K., Lin, C.C., Liu, C.L., Chen, K.T.: Synthesizing electronic health records using improved generative adversarial networks. J. Am. Med. Inform. Assoc. 26(3), 228–241 (2019)
3. Behjati, R., Arisholm, E., Bedregal, M., Tan, C.: Synthetic test data generation using recurrent neural networks: a position paper. In: 2019 IEEE/ACM 7th International Workshop on Realizing Artificial Intelligence Synergies in Software Engineering (RAISE), pp. 22–27 (2019)
4. Choi, E., Biswal, S., Malin, B., Duke, J., Stewart, W.F., Sun, J.: Generating multi-label discrete patient records using generative adversarial networks (2018)
5. Goodfellow, I.J., et al.: Generative adversarial networks (2014)
6. Gursoy, M.E., Inan, A., Nergiz, M.E., Saygin, Y.: Privacy-preserving learning analytics: challenges and techniques. IEEE Trans. Learn. Technol. 10(1), 68–81 (2016)
7. Kitchin, R., Lauriault, T.P.: Small data in the era of big data. GeoJournal 80(4), 463–475 (2014). https://doi.org/10.1007/s10708-014-9601-7
8. Kyritsi, K.H., Zorkadis, V., Stavropoulos, E.C., Verykios, V.S.: The pursuit of patterns in educational data mining as a threat to student privacy. J. Interact. Media Educ. 2019(1) (2019)
9. Lan, J., Guo, Q., Sun, H.: Demand side data generating based on conditional generative adversarial networks. Energy Procedia 152, 1188–1193 (2018). https://doi.org/10.1016/j.egypro.2018.09.157. http://www.sciencedirect.com/science/article/pii/S187661021830701X. Cleaner Energy for Cleaner Cities
10. Narayanan, A., Shmatikov, V.: Robust de-anonymization of large sparse datasets. In: Proceedings of the 2008 IEEE Symposium on Security and Privacy. SP 2008, pp. 111–125. IEEE Computer Society, USA (2008). https://doi.org/10.1109/SP.2008.33
11. Pardo, A., Siemens, G.: Ethical and privacy principles for learning analytics. Br. J. Edu. Technol. 45(3), 438–450 (2014)
12. Rodríguez-Triana, M.J., Martínez-Monés, A., Villagrá-Sobrino, S.: Learning analytics in small-scale teacher-led innovations: ethical and data privacy issues. J. Learn. Anal. 3(1), 43–65 (2016)
13. Samarati, P., Sweeney, L.: Protecting privacy when disclosing information: k-anonymity and its enforcement through generalization and suppression (1998)
14. Swenson, J.: Establishing an ethical literacy for learning analytics. In: Proceedings of the Fourth International Conference on Learning Analytics and Knowledge, pp. 246–250 (2014)
15. Willis, J.E., Slade, S., Prinsloo, P.: Ethical oversight of student data in learning analytics: a typology derived from a cross-continental, cross-institutional perspective. Educ. Tech. Res. Dev. 64(5), 881–901 (2016). https://doi.org/10.1007/s11423-016-9463-4
16. Zeide, E.: Student privacy principles for the age of big data: moving beyond FERPA and FIPPS. Drexel L. Rev. 8, 339 (2015)

Learning Analytics and Fairness: Do Existing Algorithms Serve Everyone Equally?

Vaclav Bayer(✉)[iD], Martin Hlosta[iD], and Miriam Fernandez[iD]

The Open University, Milton Keynes, UK
{vaclav.bayer,martin.hlosta,miriam.fernandez}@open.ac.uk

Abstract. Systemic inequalities still exist within Higher Education (HE). Reports from Universities UK show a 13% degree-awarding gap for Black, Asian and Minority Ethnic (BAME) students, with similar effects found when comparing students across other protected attributes, such as gender or disability. In this paper, we study whether existing prediction models to identify students at risk of failing (and hence providing early and adequate support to students) do work equally effectively for the majority vs minority groups. We also investigate whether disaggregating of data by protected attributes and building individual prediction models for each subgroup (e.g., a specific prediction model for females vs the one for males) could enhance model fairness. Our results, conducted over 35 067 students and evaluated over 32,538 students, show that existing prediction models do indeed seem to favour the majority group. As opposed to hypothesise, creating individual models does not help improving accuracy or fairness.

Keywords: Learning analytics · Degree-awarding gap · Fairness

1 Introduction

The latest statistics from UniversitiesUK and AdvanceHE [3,4] show a 13% degree-awarding gap for BAME students in UK universities. Similar issues are found for female students in Science Technology Engineering and Maths (STEM) subjects. In terms of disability, 14.5% of undergraduate students in the UK declared that they had a disability in 2017. However, disabled students are less likely to obtain a degree-level qualification (21.8%) compared to non-disabled students [2]. Degree-awarding gaps in HE translate into socio-economic gaps and further inequality. Educated people are less dependent on public aid and are more resistant to economic downturns [1].

Learning Analytics (LA) have been widely applied in HE to improve the ways in which the learning processes are supported [11,13]. We aim to study whether existing LA prediction models to identify students at risk of failing are fair in their predictions and serve majority and minority groups with the same effectiveness. We focused on assessing the prediction models currently used by

© Springer Nature Switzerland AG 2021
I. Roll et al. (Eds.): AIED 2021, LNAI 12749, pp. 71–75, 2021.
https://doi.org/10.1007/978-3-030-78270-2_12

The Open University [7]. These models are currently deployed in 530 courses and are used by more than 1,500 teachers who receive weekly alerts of students at risk of failing, so that interventions to support them can be planned. We evaluate existing LA prediction models based on protected attribute (ethnicity, gender, disability) and study several new variations of the models to assess whether the proposed variations could enhance their fairness. We address this work by two Research Questions (RQs): **RQ1:** Do existing LA prediction models work equally effectively for all types of students? **RQ2:** Do LA population-specific prediction models, trained with data from particular protected groups, perform better than general LA prediction models trained on all students?

2   Methods

Learning Analytics Prediction Models: The LA prediction models generate predictions per each course and study week whether students will successfully submit upcoming assignment, i.e. a student will have more predictions in each course they are enrolled in. The models are built based on the Gradient Boosting Machine Learning Model [6] which has been selected as the best performing model [7]. Training is based on a combination of dynamic and static features. Static features are focused on socio-demographic data, such as gender, age, ethnicity, education, occupation, disability, index of multiple deprivation and country. Dynamic features capture the students' progress, as well as their weekly activity in a Virtual Learning Environment.

Data Selection: We selected data from the largest fourteen courses taught across all faculties to ensure a large and balanced sample of students across different disciplines. To train our models we used data from the selected courses for the 2018/19 academic year (35,067 unique students), and we tested with data from the 2019/20 academic year (32,538 unique students).[1] We used test data in the first 15 weeks with the latest prediction in Jan'20, therefore the data are not affected by the Covid-19 pandemic.

Experiment Design: We depart from a set of LA prediction models (one per course per week, trained in the same manner) called *Baseline*. To address our research questions we consider three protected attributes (ethnicity, gender and disability) and split student data into different subgroups based on these attributes: (i) *Black, Asian, White* and *Rest* for ethnicity, (ii) *Male* and *Female* for gender and, (iii) *Disabled* and *Non-Disabled* for disability. Note that *Rest* refers to an aggregated list of ethnic groups that are neither White, Black, or Asian and occurs only in RQ1 as a distribution of the training data is not suitable for computing a separate model for RQ2.

To address RQ1 we first compute the predictions for all students in the test set and assess the performance of those predictions for each of the above mentioned subgroups. Then we compare the performance of those predictions between the

---

[1] More details about data samples at https://doi.org/10.6084/m9.figshare.14444567.v1.

majority and the minority subgroups from the same protected attribute, e.g. White vs Asian. To address RQ2 we compute population-specific prediction models that use only training data of each of the subgroups. The hypothesis is that specific models can learn the specific patterns of minority subgroups and provide more accurate predictions [12]. The results of these population-specific models are then compared against *Baseline*.

Metrics: Following the work of [5], we have used three metrics to compute the models' performance: (i) Area Under the ROC Curve (AUC), (ii) False Positive rate $FPR = FP/(FP + TN)$ and (iii) False Negative Rate $FNR = FN/(TP + FN)$. AUC indicates the overall accuracy of the model. FPR, in our context, indicates those instances where the model predicts that the student will not submit (i.e., the student is at risk) but the prediction is false. In this case, the teacher may follow up on the student and provide her support while the support is not needed. FNR indicates those instances where the model predicts that the student will submit but the prediction is wrong. This is a much more problematic error since the teacher will not be alerted, and therefore, won't be able to provide support to the student when needed. For each RQ, we compute the significance test across chosen metrics using paired Wilcoxon signed-rank test. The selection was influenced by the Kruskal-Walis test [10] that indicated that the underlying data do not follow a normal distribution.

3   Results

RQ1: Fairness of the Existing Models. In terms of ethnicity, the *Baseline* model shows (see Table 1) the highest accuracy for White ethnicity across ethnic groups, without high disparity in AUC. FPR is significantly lower for White students than for all other groups. That means the model erroneously predicts with a higher frequency than students from Asian, Black and other Ethnic backgrounds will not submit their assignments. This is a less problematic error since students will still receive support. In terms of FNR, the model makes fewer errors for Black and Rest students than for White students. When looking at gender, the model is more accurate (AUC) for Male students than for Female students, with Female students having higher FNR, i.e. the more problematic error. In terms of disability, the model predicts 3% more accurately (AUC) for Non-Disabled than for Disabled students. The model also presents higher FPR for Disabled students. In summary, the *Baseline* model seems to consistently perform slightly better in terms of accuracy for White, Male, and Non-Disabled students. The model predicts most erroneously Black, Male, Disabled students that they will Not Submit an assignment (FPR) and that Asian, Female, Non-disabled students will Submit an assignment (FNR).

RQ2: Fairness Through Population-Specific Models. The comparison of corresponding protected subgroups between Table 1 and Table 2 reveals that the only individual model showing better performance in terms of AUC is for White students. All other models have lower AUC, and show a higher ratio of errors, particularly for FNR - the more problematic type of errors.

We also investigated **fairness by removing the protected attribute** from the *Baseline* model. The accuracy for ethnic minorities did not change much, but Asian students have significantly lower FNR and higher FPR; for Black, the trend is the opposite. For Females, FPR and FNR stayed nearly the same. For Males, removing the attribute worsened the FPR significantly but lowered the FNR. For disabled students, the FNR significantly increased, while the FPR significantly decreased with the overall accuracy decreased. As such, we recommend removing the ethnicity attribute and keep gender and disability.

Table 1. Results of the Baseline model across subgroups. *p < 0.1 **p < 0.05 ***p < 0.01

Protected attr.	Protected subgroup	AUC	FPR	FNR
Ethnicity	Asian	0.8588	0.0479***	0.5078***
	Black	0.8743	0.0721***	0.3912***
	Rest	0.8730**	0.0407***	0.4847
	White	0.8771	0.0287	0.5003
Gender	Female	0.8714	0.0303	0.5186
	Male	0.8880***	0.0340	0.4419***
Disability	NO	0.8816	0.0278	0.4967
	YES	0.8588***	0.0437***	0.4913***

Table 2. Results of individual models across subgroups.*p < 0.1 **p < 0.05 ***p < 0.01

Protected attr.	Protected subgroup	AUC	FPR	FNR
Ethnicity	Asian	0.8287***	0.0423***	0.5725***
	Black	0.8413***	0.0916***	0.4290***
	White	0.8776	0.0303***	0.4948***
Gender	Female	0.8702***	0.0318***	0.5171
	Male	0.8814***	0.0335	0.4560***
Disability	NO	0.8802***	0.0284***	0.4991*
	YES	0.8472***	0.0449**	0.5063***

4   Discussion and Conclusions

This paper investigates whether existing LA prediction models serve everyone equally. This is important and timely research considering existing educational degree-awarding gaps and the impact that LA could have on either perpetuating or reducing these gaps. The results of our study show that existing prediction models seem to slightly favour the majority groups. Among the tested configurations, creating population-specific models harmed the accuracy and fairness of the predictions, which is in line with the results of [5] and [8]. The presented work can find its practical utility as a part of the evaluation process when existing models are being modified, e.g. by integrating new features. More research in terms of different adaptations and definitions of fairness [9] is needed to ensure that the technology we generate does not perpetuate existing educational gaps. It is also important to note that our research has been conducted over 14 largest courses, and on LA prediction models that aim to identify students at risk. More extensive research, e.g. increasing the number of courses, should be conducted to achieve a more general understanding of the problem. Qualitative research is also needed to complement these studies and assess how different fairness definitions affect the problem. While there are still many challenges to solve, this work constitutes an important step towards the understanding of LA algorithmic decision-making, its fairness and potential impact on minority groups.

References

1. Equity and quality in education: supporting disadvantaged students and schools. OECD (2012)
2. Disability and education, UK: 2019 (2019). https://www.ons.gov.uk/peoplepopu lationandcommunity/healthandsocialcare/disability/bulletins/disabilityandeducat ionuk/2019. Accessed 07 Feb 2021
3. AdvanceHE: Degree attainment gaps (2017). https://www.advance-he.ac.uk/ guidance/equality-diversity-and-inclusion/student-recruitment-retention-and-attainment/degree-attainment-gaps. Accessed 08 Feb 2021
4. Amos, V.: Black, Asian and minority ethnic student attainment at UK universities: closing the gap (2019). https://www.universitiesuk.ac.uk/policy-and-analysis/ reports/Documents/2019/bame-student-attainment-uk-universities-closing-the-gap.pdf
5. Anderson, H., Boodhwani, A., Baker, R.S.: Assessing the fairness of graduation predictions. In: EDM 2019, pp. 488–491 (2019)
6. Greenwell, B., Boehmke, B., Cunningham, J., Developers, G., Greenwell, M.B.: Package 'gbm' (2020)
7. Hlosta, M., Zdrahal, Z., Bayer, V., Herodotou, C.: Why predictions of at-risk students are not 100% accurate? Showing patterns in false positive and false negative predictions (2020)
8. Hutt, S., Gardner, M., Duckworth, A.L., D'Mello, S.K.: Evaluating fairness and generalizability in models predicting on-time graduation from college applications. International Educational Data Mining Society (2019)
9. Kizilcec, R.F., Lee, H.: Algorithmic fairness in education (2021)
10. Kruskal, W.H., Wallis, W.A.: Use of ranks in one-criterion variance analysis. J. Am. Stat. Assoc. **47**(260), 583–621 (1952)
11. Leitner, P., Khalil, M., Ebner, M.: Learning analytics in higher education—a literature review. In: Peña-Ayala, A. (ed.) Learning Analytics: Fundaments, Applications, and Trends. SSDC, vol. 94, pp. 1–23. Springer, Cham (2017). https://doi. org/10.1007/978-3-319-52977-6_1
12. Perez, C.C.: Invisible Women: Exposing Data Bias in a World Designed for Men. Random House, New York (2019)
13. Viberg, O., Hatakka, M., Bälter, O., Mavroudi, A.: The current landscape of learning analytics in higher education. Comput. Hum. Behav. **89**, 98–110 (2018)

Exploiting Structured Error to Improve Automated Scoring of Oral Reading Fluency

Beata Beigman Klebanov$^{(\boxtimes)}$ and Anastassia Loukina

Educational Testing Service, Princeton, UK
{bbeigmanklebanov,aloukina}@ets.org

Abstract. In order to track the development of young readers' oral reading fluency (ORF) at scale, it is necessary to move away from hand-scoring responses to automating the assessment of ORF, while retaining the quality of the scores. We present a method for improving automated ORF scoring that utilizes an observed systematicity in machine error, namely, that cases with low estimated reading accuracy are harder to score correctly for fluency. We show that the method yields an improved performance, including on out-of-domain data.

Keywords: Automated scoring · Oral reading fluency · Speech processing

1 Introduction

Automating scoring is a promising technology for delivering educational assessment results swiftly and at a lower cost per student at a large scale. Furthermore, with the development of educational applications that log continuous stream of user data, automated assessment would allow for very frequent, if not continuous, formative feedback to the learner and/or to the teacher.

The case in point for this study is assessment of oral reading fluency (**ORF**), typically measured as words read correctly per minute of reading out loud (**wcpm**). For example, a median 4th grader in the Fall semester in an U.S. school is expected to read at a rate of about 94 wcpm [8]. Currently, ORF evaluations are typically carried out by teachers in a 1:1 fashion, three times a year [4,7]; automated scoring of such tests is an emergent technology [1,3]. Moreover, new educational technology might make specialized assessment of ORF unnecessary, if the system can "listen" while children read out loud as part of regular reading activities, and provide a stream of valuable formative assessment data to the teacher. We explore data collected from just such an application.

For evaluating automated ORF scoring systems, Pearson's correlation is a commonly used metric. While providing critical formative information such as which students in a class are stronger and weaker readers, it might be insufficient to determine whether a given student reads at the 4th grade level. That is, if an

© Springer Nature Switzerland AG 2021
I. Roll et al. (Eds.): AIED 2021, LNAI 12749, pp. 76–81, 2021.
https://doi.org/10.1007/978-3-030-78270-2_13

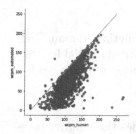

automated system systematically produces lower scores than a human rater does, the correlation might be high but the alignment to grade-level norms might be compromised. The possibility of a systematic under-scoring is not a hypothetical one – the figure on the left shows the plot for the state-of-the-art scoring system used in this study. A similar trend is reported in [6].

Systematic, uniform under-scoring is easy to correct, by, for example, shifting the automated scores by the difference between the means of human and automated scores. However, it becomes more complicated if the system's error is not uniform. In this paper, we investigate the error structure of an automated ORF assessment system, with the goal of improving the system's ability to provide accurate ORF scores. In particular, we examine properties of the text being read and the reader's accuracy. To our knowledge this is the first study to effectively exploit a pattern of systematic error in automated assessment of ORF.

2   Data

The data come from RELAY READER™ (https://relayreader.org), an app that elicits and records children's oral reading in a shared reading environment where the reader is taking turns reading out loud from a book with an adult virtual partner (audiobook) [11]. For this study, we use the following two datasets (see Table 1). For Dataset 1, the children read *Harry Potter and the Sorcerer's Stone* at school during dedicated independent reading time in the 2018–2019 school year and in summer school in 2018; for Dataset 2, children read two other books in summer schools in 2019 (in person) and 2020 (virtually). The data collections occurred at various sites in the greater New York and Washington DC areas. The children in both datasets were of comparable ages (predominantly 3–5 graders) and demographics. For both datasets, recordings of passages shorter than 50 words, of insufficiently long duration, and those with a quiet audio were removed from the study. For each of the 3,476 recordings, we used a human ORF score (namely, ORF score calculated using a manual transcription of the audio sample produced by a professional agency), automated ORF score, as well as three additional variables described in Sect. 3.2.

Table 1. Datasets 1 and 2.

Dataset	Book	Students	Samples
Dataset 1	Harry Potter and the Sorcerer's Stone	115	3,026
Dataset 2	The Adventures of Pinocchio	78	410
	The Wonderful Wizard of Oz	7	40

3   Method

We describe the variables and the models we used for predicting human ORF scores, using automatically predicted ORF scores and additional variables that characterize aspects of the reader's performance or of the passage being read.

3.1   Automated Estimation of ORF

The speech processing pipeline used for ORF scoring includes three major components: automated speech recognition (ASR), off-task speech identification, and computation of ORF. The ASR which converts speech to text is a state-of-the-art system described in [9]. The ASR transcription is then processed through the off-task speech identification module [10], since many of the recordings contain off-task speech, especially before and after the reading. Finally, the ASR transcript and the associated timestamps for beginning and end of on-task speech are used to compute wcpm: the number of correctly read words divided by the time it took the child to read the passage.

3.2   Additional Variables for Exploration of Prediction Error

Accuracy of a reader on a given passage is the proportion of the words in the passage the student read correctly. The number of correctly read words is calculated as part of the ORF estimation; for accuracy, it is normalized by the total number of words in the passage.

Text complexity is a well-established predictor of ORF [2]. We used the literary mode of TextEvaluator [12,15], a state-of-the-art tool that combines a range of linguistic features such as concreteness, narrativity, lexical cohesion, syntactic complexity, word unfamiliarity, into a complexity score.

Oral production To account for text properties that predict how fast it would be read, such as the distribution of stressed syllables and prosodic boundaries, we follow the literature [14,16–18] and use a state-of-the-art text-to-speech synthesizer (male Alex voice in Apple Inc's OS X 10.11.6 built-in TTS engine) [5] and measure its reading rate per minute.

3.3   Models

To allow for flexible relationships between the predictors and the outcome, we used generalized boosted regression models [13] as implemented in the *gbm* package for R. We considered three models for the outcome (human-produced ORF score), based on the following automatically estimated indices:

Model 1 (baseline): ORF
Model 2 (+accuracy): ORF and accuracy
Model 3 (+accuracy+text): ORF, accuracy, text complexity, oral production

For each model, for Evaluation 1, we used 5-fold cross-validation on Dataset 1, and selected the optimal number of trees by minimizing cross-validated prediction error. For Evaluation 2, we used Dataset 2 to evaluate the generalization of the most promising models from Evaluation 1 to reading data from new books.

4 Results

Evaluation 1: Model 1 made less accurate predictions than Models 2 and 3, by a large margin ($R^2 = 0.56$ vs 0.64). While the automated estimate of ORF dominates Model 2 as expected (relative influence of 0.63), accuracy also has a large influence (0.37). This is because in cases where the automatically estimated accuracy is low, the automated ORF estimate is a much poorer proxy for human ORF scores than cases where the accuracy is high. In contrast, accounting for properties of text known to impact ORF provided no tangible gain over Model 2 in terms of R^2, and the cumulative relative weight of the two text variables in Model 3 was only 5%. We will thus use Model 2 for the next evaluation.

Evaluation 2: We evaluate generalization of the model to a new dataset, collected at different sites (therefore in different acoustic environments), from different children, and, most importantly, reading from different books than the original dataset. Table 2 shows strong generalization for Model 2. The improvement in MSE over Model 1 specifically suggests not only improved correlation but also better precision of the numerical estimates themselves. Additionally, we evaluated Model 2 on the subset of Dataset 2 with medium-to-high estimated accuracy – above 0.7, namely, cases where the system found at least 7 out of every 10 words in the passage, on average, in the student's oral response. Model 2 showed strong performance, at $r = 0.936$, MSE = 193 (N = 225).

Table 2. Performance on Dataset 2.

Model	r	Mean Squared Error (MSE)
Model 1	0.655	951
Model 2	0.863	538

5 Conclusion

We investigated performance of an automated system built to support unobtrusive formative assessment of oral reading fluency (ORF) while children are reading great works of fiction in a specially designed shared oral reading app. We found that accounting for the structure of the error of the automated ORF estimate, namely, for the tendency of the error to be larger when the estimated accuracy of the reader is lower, allows for an improved overall prediction, as well as a strong generalization to completely new data – different children reading different books at different sites from the original training data. This generalization is particularly important, as we are looking to extend the library of books offered to the readers as well as to support a variety of implementation sites, including elementary school classroom, summer camp, and home environments. The new model is especially strong on data with medium-to-high estimated reading accuracy, allowing for filtering of data to ensure more precise fluency estimations.

Acknowledgement. We thank J. R. Lockwood for expert advice and help with the analyses; T.O'Reilly, A. Misra, K. Zechner for their helpful comments.

References

1. Balogh, J., Bernstein, J., Cheng, J., Van Moere, A., Townshend, B., Suzuki, M.: Validation of automated scoring of oral reading. Educ. Psychol. Measur. **72**(3), 435–452 (2012)
2. Barth, A.E., Tolar, T.D., Fletcher, J.M., Francis, D.: The effects of student and text characteristics on the oral reading fluency of middle-grade students. J. Educ. Psychol. **106**(1), 162–180 (2014)
3. Bernstein, J., Cheng, J., Balogh, J., Rosenfeld, E.: Studies of a self-administered oral reading assessment. In: Proceedings of the 7th ISCA Workshop on Speech and Language Technology in Education, Stockholm, Sweden, pp. 180–184 (2017)
4. Biancarosa, G., Kennedy, P.C., Park, S., Otterstedt, J., Gearin, B., Yoon, H.: 8th edition of dynamic indicators of basic early literacy skills (DIBELS®): administration and scoring guide. University of Oregon, Technical report (2019)
5. Capes, T., et al.: Siri on-device deep learning-guided unit selection text-to-speech system. In: Proceedings of the Annual Conference of the International Speech Communication Association, INTERSPEECH 2017, pp. 4011–4015 (2017)
6. Godde, E., Bailly, G., Bosse, M.L.: Reading prosody development: automatic assessment for a longitudinal study. In: Speech & Language Technology for Education (SLaTE) (2019)
7. Good, R., Kaminski, R.: Dynamic indicators of basic early literacy skills. Institute for the Development of Educational Achievement, Eugene, OR (2002)
8. Hasbrouck, J., Tindal, G.: An update to compiled ORF norms. Behavioral Research and Teaching, University of Oregon, Technical report (2017)
9. Loukina, A., et al.: Automated estimation of oral reading fluency during summer camp e-Book reading with MyTurnToRead. In: Proceedings of the Annual Conference of the International Speech Communication Association, INTERSPEECH 2019, pp. 21–25 (2019). https://doi.org/10.21437/Interspeech.2019-2889
10. Loukina, A., Klebanov, B.B., Lange, P., Gyawali, B., Qian, Y.: Developing speech processing technologies for shared book reading with a computer. In: Proceedings of WOCCI 2017: 6th International Workshop on Child Computer Interaction, pp. 46–51 (2017). https://doi.org/10.21437/WOCCI.2017-8
11. Madnani, N., et al.: MyTurnToRead: an interleaved e-book reading tool for developing and struggling readers. In: Proceedings of the 57th Conference of the Association for Computational Linguistics: System Demonstrations, pp. 141–146. Association for Computational Linguistics, Florence (2019). https://www.aclweb.org/anthology/P19-3024
12. Napolitano, D., Sheehan, K., Mundkowsky, R.: Online readability and text complexity analysis with TextEvaluator. In: Proceedings of the North American Chapter of the Association for Computational Linguistics, pp. 96–100 (2015)
13. Ridgeway, G.: The state of boosting. Comput. Sci. Stat. **31**, 172–181 (1999)
14. van Santen, J.P.: Assignment of segmental duration in text-to-speech synthesis. Comput. Speech Lang. **8**(2), 95–128 (1994)
15. Sheehan, K.M., Kostin, I., Napolitano, D., Flor, M.: The TextEvaluator tool: helping teachers and test developers select texts for use in instruction and assessment. Elementary School J. **115**(2), 184–209 (2014)

16. Tokuda, K., Hashimoto, K., Oura, K., Nankaku, Y.: Temporal modeling in neural network based statistical parametric speech synthesis. In: 9th ISCA Speech Synthesis Workshop, no. 2, pp. 106–111 (2016)
17. Yoshimura, T., Tokuda, K., Kobayashi, T., Masuko, T., Kitamura, T.: Simultaneous modeling of spectrum, pitch and duration in HMM-based speech synthesis. EUROSPEECH **1999**, 2347–2350 (1999)
18. Zen, H., Tokuda, K., Black, A.W.: Statistical parametric speech synthesis. Speech Commun. **51**(11), 1039–1064 (2009)

Data Augmentation for Enlarging Student Feature Space and Improving Random Forest Success Prediction

Timothy H. Bell[1,2]([envelope]) [ID], Christel Dartigues-Pallez[1] [ID], Florent Jaillet[1] [ID], and Christophe Genolini[2] [ID]

[1] Université Côte d'Azur, CNRS, I3S, Sophia Antipolis, France
{Timothy.BELL,Christel.DARTIGUES-PALLEZ,
Florent.JAILLET}@univ-cotedazur.fr
[2] R++, 5 place Jean Deschamps, 31100 Toulouse, France
cg@rplusplus.com

Abstract. One of the main problems encountered when predicting student success, as a tool to aid students, is the lack of data used to model each student. This lack of data is due in part to the small number of students in each university course and also, the limited number of features that describe the educational background for each student. In this article, we introduce new features by augmenting the student feature space to obtain an improved model. These features are divided into several groups, namely, external added data, metric and counter data, and evolutive data. We will then assess the quality of the augmented data to classify at-risk students in their first year of university. For this article, the classifiers are built using Random Forests. As this learning method measures variable importance, we can enquire on the relevance of the augmented data, as well as the data groups that allow a more significant collection of features.

Keywords: Student success · Random forest · Data augmentation · Educational data mining · Student metrics

1 Introduction

In France among the students in their first year of university one in two will either repeat the year, change major, or drop out mid-year [14]. In 2017 only 29% got their first cycle degree without repeating or changing major. Many approaches to predict student success have been investigated through means of grade prediction or dropout prevention [2,3,7,8]. Generic data such as secondary education grades but also sociodemographic indicators [6,8,10] are used to predict student outcome. This ends in having a small amount of features usable by learning algorithms to output predictions. For this article, we will augment our initial set of data by performing operations on the existing features to obtain ratios or time-series coefficients. We also have metrics on the various high-schools. To

I. Roll et al. (Eds.): AIED 2021, LNAI 12749, pp. 82–87, 2021.
https://doi.org/10.1007/978-3-030-78270-2_14

classify at-risk students we are using Random Forests [4]. We first introduce the
data to train the model, then our method for augmenting the given data, and
lastly, before concluding, we discuss the obtained results.

2 Data

2.1 Initial Data

The data used comes from students studying in a University Institute of Technology. Students enter UITs after completing secondary studies. This particular
set of data is pooled from first-year students of 18 different majors. All the data
is thoroughly anonymized beforehand to respect student privacy within the General Data Protection Regulation [1]. Among the different majors, classes vary in
size and display a very heterogeneous distribution of students. The particular
set we are working on is of the year 2019 with a high of 169 students in one
promotion and as low as 10 for another. All the data used for training is taken
from the students' curriculum during their secondary education at high-school.

The French high-school system is divided into 3 years, and each school year is
divided into 3 trimesters. For each student we have data from the first trimester
of the second year up to the second trimester of the third year. In total five
trimesters. We also have the results for the Baccalauréat (the end of high-school
exam). Each year, a variety of subjects are taught, some common core courses
(e.g., Physical Education) and some speciality courses (e.g., Economy). For every
subject we have (see Table 1) the student's grade (Stu), the class's mean (Avg),
the class's highest grade (Max), and the class's lowest grade (Min). Most of the
augmented data derive from these features.

Table 1. Stored information for each subject.

Stu	Avg	Min	Max
13	8.5	7	17
...	...	...	...
16	10	6	17

Additional data consist of: professor comments for every subject and each
trimester, a cover letter, the student's high-school name, and comments from the
high-school on the student's potential for succeeding in further studies. For this
article only numeric data is used, disregarding all non-ordinal or non-categorical
textual data. Therefore the professor comments, high-school comments, and
cover letters are omitted in this work.

Lastly since optional courses can be taken at school, we get rid of features
with a high number of missing data (>70%) during a pre-processing step.

To train the model we are doing supervised learning, and the label for each
student is whether the student passed or failed. This is done by discretizing

their weighted mean grade in the first university semester. This weighted mean attributes more weight to more important courses depending on the chosen major.

2.2 Augmented Data

We separated the augmented data into 3 groups to attribute changes in the models' outcomes to the different data. The augmented data is divided as such: pre-processed initial data(PPD), external data(G1), metrics and counters(G2), evolving data(G3). Although most of the features are numeric, i.e., grades, some are nominal such as chosen language courses and some ordinal (Good, Very Good, etc.). These features are encoded respectively by one-hot encoding and ordinal encoding.

The first group of external data (G1) consists of various metrics for French high-schools: the percentage of students that repeat years, the percentage of graduated-with-honours students, the percentage of students that pass the final examination, and lastly, the added-value which indicates how well the high-school performed given its sociodemographic context.

The metrics and counters group (G2) holds features obtained from simple calculations: Stu-Avg, the student's highest grade - lowest grade for any given trimester, Stu-Min, Stu-Max. It also has the number of: repeated years, top marks, lowest marks, times Stu<Avg.

For the last group of data (G3), we apply linear regression, by k-combinations of all trimesters, to extract the regression coefficients β. These coefficients depict the evolution of G1 and G2 data. For instance, the evolution of: the student's grades, number of top marks, the difference Stu-Avg.

3 Methodology

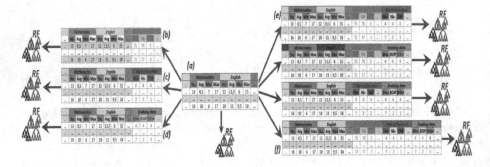

Fig. 1. Data group evaluation process. Blue: PPD. Yellow: G1. Green: G2. Grey: G3. (Color figure online)

For this article we chose Random Forest (RF) due to its high classification accuracy rates seen in [9,12,13,15]. To assess the efficiency of the applied methodology we use 'Zero-Rule' [5] as a baseline for this classification task. This predicts the class as the majority class, in this case the majority will always be students that have passed the first semester. This objective avoids using false model accuracy due to class imbalance mentioned in [3]. When running our algorithms, majors are trained separately for this paper as certain features (e.g., French and mathematics grades) vary in relevancy depending on the chosen field of studies. Additionally, Random Forests' built-in Gini importance will be used to score each feature and its importance. The Gini importance will allow us to assess if our augmentation creates any relevant features.

For each configuration and its corresponding RF model we run the model 10 times with 10-fold cross validation to test the performance. The metrics used are the accuracy in classification and the F1-score.

4 Results

For the majors with less than 50 students, the results were inconclusive. The prediction didn't, or barely, perform better than the baseline. This was expected as the sample population is too low.

The higher scores on average were obtained with a combination of all groups, (f) in Fig. 1.

Table 2. Resulting classification scores for each model on the Computer Science course.

Scores	PPD	PPD+G1	PPD+G2	PPD+G3	PPD+G1+G2	PPD+G1+G3	PPD+G2+G3	PPD+G1+G2+G3
Accuracy	0.76	0.77	0.85	0.78	0.88	0.77	0.85	**0.89**
F1-score	0.73	0.69	0.79	0.75	0.81	0.70	0.79	**0.84**

For the particular major in Table 2, we notice that G1 and G3 only marginally improve the classification. Whereas G2 improves it by quite a lot. But some features in both G1 and G3 can have substantial importance regarding the classification, therefore it might be interesting to perform feature selection [11] on all the augmented features. There were only 350 features before augmentation, and 1500 features total after augmentation. Some examples of augmented features that figure in the top 10 most relevant features are: Stu-Min for 3rd trimester French (G2), Stu-Min for 3rd trimester French (G2), regression on the 2nd, 3rd, and 5th trimesters in mathematics (G3), regression of Stu-Avg on trimesters 2 and 3 in English (G3). Interestingly for the statistics major all 10 top features are augmented features from groups G2 and G3 with mostly regressions on the student's relative grades to the class's highest grades in mathematics.

5 Conclusion

This work sought to extend the feature space to improve student failure prediction, allowing a better understanding of what features may best represent students. Data augmentation improved prediction on classes of more than 50 students. It can also be used as a tool for Random Forest Feature Selection prior to inputting this into any learning model.

In future works, extra textual data could be exploited. Our dataset also provides for each subject professor comments. These comments usually hold information such as regular absenteeism and class disruption. The next step will be to incorporate these comments in the model as well, and further increase the prediction accuracy of our model.

References

1. Regulation (EU) 2016/679 of the European Parliament and of the Council of 27 April 2016 on the protection of natural persons with regard to the processing of personal data and on the free movement of such data, and repealing Directive 95/46/EC (General Data Protection Regulation) (Text with EEA relevance). http://data.europa.eu/eli/reg/2016/679/2016-05-04
2. Balakrishnan, G.: Predicting student retention in massive open online courses using hidden Markov models. Ph.D. thesis, EECS Department, University of California, Berkeley (2013)
3. Barros, T.M., Souza Neto, P.A., Silva, I., Guedes, L.A.: Predictive models for imbalanced data: a school dropout perspective. Educ. Sci. **9**(4), 275 (2019). https://doi.org/10.3390/educsci9040275. Number: 4 Publisher: Multidisciplinary Digital Publishing Institutenumber: 4 Publisher: Multidisci-plinary Digital Publishing Institute
4. Breiman, L.: Random forests. Mach. Learn. **45**(1), 5–32 (2001). https://doi.org/10.1023/A:1010933404324
5. Choudhary, R., Gianey, H.K.: Comprehensive review on supervised machine learning algorithms. In: 2017 International Conference on Machine Learning and Data Science (MLDS) (2017). https://doi.org/10.1109/MLDS.2017.11
6. Cortez, P., Silva, A.: Using data mining to predict secondary school student performance. EUROSIS (2008)
7. Del Bonifro, F., Gabbrielli, M., Lisanti, G., Zingaro, S.P.: Student dropout prediction. In: Bittencourt, I.I., Cukurova, M., Muldner, K., Luckin, R., Millán, E. (eds.) AIED 2020. LNCS (LNAI), vol. 12163, pp. 129–140. Springer, Cham (2020). https://doi.org/10.1007/978-3-030-52237-7_11
8. Gonçalves, O., Beltrame, W.: Socioeconomic data mining and student dropout: analyzing a higher education course in Brazil. Int. J. Innov. Educ. Res. **8**, 505–518 (2020). https://doi.org/10.31686/ijier.vol8.iss8.2554
9. Hussain, S., Dahan, N.A., Ba-Alwi, F.M., Ribata, N.: Educational data mining and analysis of students' academic performance using WEKA. Indonesian J. Electr. Eng. Comput. Sci. **9**(2), 447–459 (2018). https://doi.org/10.11591/ijeecs.v9.i2.pp447-459
10. Kovacic, Z.: Early prediction of student success: mining students enrolment data, pp. 647–665 (2010). https://doi.org/10.28945/1281
11. Ma, C., Yao, B., Ge, F., Pan, Y., Guo, Y.: Improving prediction of student performance based on multiple feature selection approaches. In: Proceedings of the 2017 International Conference on E-Education, E-Business and E-Technology. ICEBT 2017, pp. 36–41. Association for Computing Machinery, New York (2017). https://doi.org/10.1145/3141151.3141160

12. Mahboob, T., Irfan, S., Karamat, A.: A machine learning approach for student assessment in E-learning using Quinlan's C4.5, Naive bayes and random forest algorithms. In: 2016 19th International Multi-Topic Conference (INMIC), pp. 1–8 (2016). https://doi.org/10.1109/INMIC.2016.7840094

13. Miguéis, V.L., Freitas, A., Garcia, P.J.V., Silva, A.: Early segmentation of students according to their academic performance: a predictive modelling approach. Decis. Supp. Syst. **115**, 36–51 (2018). https://doi.org/10.1016/j.dss.2018.09.001

14. Razafindratsima, N.: État de l'Enseignement supérieur, de la Recherche et de l'Innovation en France. Les parcours et la réussite en Licence, Licence professionnelle et Master à - l'université - État de l'Enseignement supérieur, de la Recherche et de l'Innovation en France **n°13**, 50–51 (2020). https://publication. enseignementsup-recherche.gouv.fr/eesr/FR/T149/les_parcours_et_la_reussite_en_ licence_licence_professionnelle_et_master_a_l_universite/

15. Sorour, S.E., Mine, T.: Building an interpretable model of predicting student performance using comment data mining. In: 2016 5th IIAI International Congress on Advanced Applied Informatics (IIAI-AAI), pp. 285–291 (2016). https://doi.org/ 10.1109/IIAI-AAI.2016.114

The School Path Guide: A Practical Introduction to Representation and Reasoning in AI for High School Students

Sara Guerreiro-Santalla, Francisco Bellas$^{(\boxtimes)}$, and Oscar Fontenla-Romero

CITIC Research Center, Universidade da Coruña, A Coruña, Spain
{sara.guerreiro,francisco.bellas,oscar.fontenla}@udc.es
http://www.gii.udc.es

Abstract. This paper presents a structured activity to introduce high school students in the topics of representation and reasoning in Artificial Intelligence, which are completely new for them at this educational level. The activity has been designed in the scope of the Erasmus+ project called AI+, which aims to develop a curriculum of Artificial Intelligence (AI) for high school students in Europe. As established in the AI+ principles, all the teaching activities are based on the use of the student's smartphone as the core element to introduce a practical approach to AI in classes. In this case, a smartphone app is developed by students using the MIT App Inventor software. The topics of representation and reasoning are introduced to students by means of topological maps and graph-like representations, which are used later to perform a simple probabilistic reasoning over them.

Keywords: AI curriculum · AI for K12 · AI resource for classroom · Representation and reasoning · Smartphone app

1 Introduction

The activity that is presented in this work has been created in the scope of the AI+ project [1], which aims to develop a curriculum of AI for high school students in Europe. Starting from the remarkable work carried out in the AI4K12 initiative [2] and including the own experience of the University of Coruña (UDC) experts, eight AI topics that must be covered at this educational level have been already established, namely: perception, actuation, representation, reasoning, learning, artificial collective intelligence, motivation and SEL (sustainability, ethics and legal aspects of AI) [3]. These topics are organized in teaching units, designed for the teacher, that make up a two-year subject, targeted to students with a technical background.

The AI+ curriculum follows the STEM methodology [4], and each teaching unit presents a challenge or project that must be faced through cooperative Project Based Learning (CPBL) [5]. To support this practical approach, it has been established to rely on the student's Smartphone as the core educational tool. Current smartphones have the technological level required for AI teaching in terms of sensors, actuators, computing

I. Roll et al. (Eds.): AIED 2021, LNAI 12749, pp. 88–92, 2021.
https://doi.org/10.1007/978-3-030-78270-2_15

power and communications. In addition, a large majority of high school students have their own Smartphone, so they can use it. This reduces the cost of introducing this discipline and equalizes regions with different economic capacity [6].

The activity described in this paper corresponds to the challenge that students must face in the third teaching unit of the AI+ project, devoted with the introduction to representation and reasoning in AI. This teaching unit is carried out after one focused in perception and actuation, so students are already familiar with the main sensors in this scope, like cameras, microphones and tactile screens [1]. Specifically, in this activity they will develop an intelligent Smartphone app, "The School Path Guide", using App Inventor. The global duration of the activity is 5 h, approximately, and all the teaching material related with the activity is available to download [7].

Fig. 1. Left: Initial screen. Middle: User capturing QR. Right: Guidance screen

2 School Path Guide App

The app will guide users in the school building from their current location to a destination and should work as follows: there are different *location points* in different places of the building identified by a QR code, which has been encoded with the location name. When the user arrives at one of these points, he/she scans the QR code through the app. Once scanned, the app shows a list of possible destinations and the user selects the desired one. From this moment on, the app shows the optimal path to reach the destination through photos and instructions displayed in the screen (Fig. 1).

In this introductory teaching unit, it was decided to teach both concepts, representation and reasoning, in the same practical case. This way, students will understand how a proper representation facilitates reasoning. Specifically, the following topics will be addressed: 2D representation, topological vs metric maps, first person representation, basic graph definition, probabilistic reasoning and basic route searching over graphs.

2.1 Representation

As a first approach, it is proposed to use a representation of the school based on images, that is, photographs of locations the user can easily identify (*location points*). The first step for students is to define these location points, which will be marked with a QR code placed in a wall at a visible position. To simplify programming, it has been decided to create a division of the school into *path sections*. They are defined by *crossing points*, which are relevant points where more than one path coincides, or by *location points*. The possible routes are the result of the union of path sections. To clarify this representation, Fig. 2 shows the floorplan of an example building.

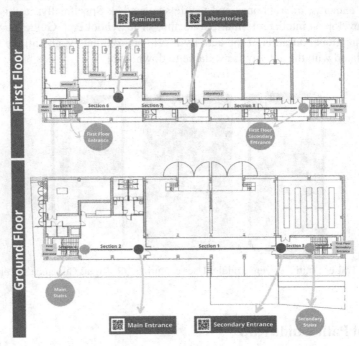

Fig. 2. Map with location points (blue), crossing points (orange) and path sections (green) (Color figure online)

The set of location points, crossing points and path sections make up the topological map representation that will be applied [8]. To simplify it, it is proposed to use a graph, which reduces the 2D (flat) map to a graphic based on nodes and links like the one shown in Fig. 3. Therefore, the task the students have to carry out is to create a graph for their particular school. To do it, it is recommended to start from a floorplan of the school, if possible, and to perform the process "by hand" (in a printed paper or a tablet), so the concept of representation change becomes clearer.

To finish this first part, students must take photographs of the beginning and end of each path section with their smartphone, which will be displayed in their app version.

2.2 Reasoning

Students will be introduced in the basics of probabilistic reasoning [9]. The first step to do is to calculate the time required to walk through each path section in their school (T_{ij}, where i is the origin and j the destination). We recommend them to obtain these data empirically. These time values must be included in the links of the graph, as displayed in the left diagram of Fig. 3. The programming to be carried out implies to create a function that calculates the optimal route from an origin to a destination by adding the times of each of the path sections that constitute the route. The result of this function should be the direction of the route (clockwise or counter clockwise).

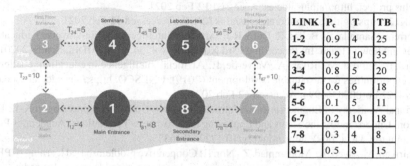

LINK	P_c	T	TB
1-2	0.9	4	25
2-3	0.9	10	35
3-4	0.8	5	20
4-5	0.6	6	18
5-6	0.1	5	11
6-7	0.2	10	18
7-8	0.3	4	8
8-1	0.5	8	15

Fig. 3. Left: Topological map graph representation. Right: link parameters for reasoning

Second, we will consider a realistic situation derived from the break time of the school. In this period, some paths con be overcrowded, affecting the walking time, mainly those that imply using stairs. To deal with it, the current clock time must be introduced in the app and two new variables must be included on each link of the graph. The first one is the probability of congestion (P_c), a value between 0 and 1 that represents the probability of finding people in a given path section. It can be empirically estimated, for instance, by observing the congestion in the break period during different days and establishing an average probability for each path section. The second one is the average time required to travel each section during the break time (TB_{ij}). To calculate it empirically, students should walk the different sections in many congestion periods and compute an average time. Figure 3 (right) shows a possible set of values for the 3 parameters to consider in the case of the UDC building.

Considering these parameters in the graph, depending on the clock time the user executes the guidance, apparently slower paths can be provided by the app, but the resulting travelling time is lower because congestions are avoided.

3 Conclusions

This paper presents a structured activity to introduce high school students in the topics of representation and reasoning in a practical way. Topological maps and graphs have been used as simple cases of internal representation in AI. Then, basic probabilistic reasoning has been performed to find the fastest path in the graph. With this activity, students obtain a first idea of how an appropriate representation leads to a simpler reasoning process, and how including probabilities makes the solution smarter.

Acknowledgments. The authors wish to acknowledge the CITIC research center, funded by Xunta de Galicia and European Regional Development Fund (grant ED431G 2019/01), and the Erasmus+ Programme of the European Union (grant 2019–1-ES01-KA201–065742).

References

1. AIplus project. http://aiplus.udc.es. Accessed 12 Feb 2021
2. AI4K12. https://github.com/touretzkyds/ai4k12/wiki. Accessed 12 Feb 2021
3. Guerreiro-Santalla, S., Bellas, F., Duro, R.J.: Artificial intelligence in pre-university education: what and how to teach. In: Proceedings 3rd XoveTIC, vol. 54, p. 48. MDPI, Spain (2020)
4. Pedró, F., Subosa, M., Rivas, A., Valverde, P.: Artificial intelligence in education: challenges and opportunities for sustainable development (2019). UNESCO. https://unesdoc.unesco.org/ark:/48223/pf0000366994. Accessed 12 Feb 2021
5. National Academy of Engineering and National Research Council: Engineering in K-12 Education: Understanding the Status and Improving the Prospects. The National Academies Press, Washington, DC (2009). https://doi.org/10.17226/12635
6. Khairiyah, M., Syed, A., Mohammad, Z., Nor, F.: Cooperative problem-based learning (CPBL): framework for integrating cooperative learning and problem-based learning. Procedia – Soc. Behav. Sci. **56**, 223–232 (2012)
7. Resources of the teaching unit 3 (AI+ project) – Representation and Reasoning, https://cutt.ly/ckKKE1N. Accessed 12 Feb 2021
8. Gelfond, M., Kahl, Y.: Knowledge Representation, Reasoning, and the Design of Intelligent Agents: The Answer-Set Programming Approach. Cambridge University Press, Cambridge (2014)
9. Pearl, J.: Probabilistic Reasoning in Intelligent Systems. Morgan Kaufmann, Universidad de California (2014)

Kwame: A Bilingual AI Teaching Assistant for Online SuaCode Courses

George Boateng[✉]

ETH Zurich, Zürich, Switzerland
gboateng@ethz.ch

Abstract. Introductory hands-on courses such as our smartphone-based coding course, SuaCode require a lot of support for students to accomplish learning goals. Online environments make it even more difficult to get assistance especially more recently because of COVID-19. Given the multilingual context of SuaCode students—learners across 42 African countries that are mostly Anglophone or Francophone—in this work, we developed a bilingual Artificial Intelligence (AI) Teaching Assistant (TA)—Kwame—that provides answers to students' coding questions from SuaCode courses in English and French. Kwame is a Sentence-BERT (SBERT)-based question-answering (QA) system that we trained and evaluated offline using question-answer pairs created from the course's quizzes, lesson notes and students' questions in past cohorts. Kwame finds the paragraph most semantically similar to the question via cosine similarity. We compared the system with TF-IDF and Universal Sentence Encoder. Our results showed that fine-tuning on the course data and returning the top 3 and 5 answers improved the accuracy results. Kwame will make it easy for students to get quick and accurate answers to questions in SuaCode courses.

Keywords: Virtual teaching assistant · Question answering · NLP · BERT · SBERT · Machine learning · Deep learning

1 Introduction

Introductory hands-on courses such as our smartphone-based coding course, SuaCode [4,5,10] require a lot of support for students to accomplish learning goals. Offering assistance becomes even more challenging in an online course environment which has become important recently because of COVID-19 with students struggling to get answers to questions. Hence, offering quick and accurate answers could improve the learning experience of students. However, it is difficult to scale this support with humans when the class size is huge—hundreds of thousands. There has been some work to develop virtual teaching assistants (TA) such as Jill Watson [6,7], Rexy [2], and a physics course TA [11] (see [3] for a detailed description of related work). All of these TAs have focused on logistical questions, and none have been developed and evaluated using coding courses

© Springer Nature Switzerland AG 2021
I. Roll et al. (Eds.): AIED 2021, LNAI 12749, pp. 93–97, 2021.
https://doi.org/10.1007/978-3-030-78270-2_16

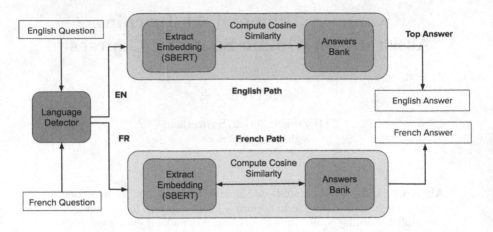

Fig. 1. System architecture of Kwame

in particular. Also, they have used one language (e.g. English). Given the multilingual context of our students—learners across 42 African countries that are mostly Anglophone or Francophone—in this work, we developed a bilingual Artificial Intelligence (AI) TA—Kwame—that provides answers to students' coding questions from SuaCode courses in English and French. Kwame is named after Dr. Kwame Nkrumah the first President of Ghana and a Pan Africanist. Kwame is a Sentence-BERT-based question-answering (QA) system that is trained using the SuaCode course material [1] and evaluated offline using accuracy and time to provide answers. Kwame finds the paragraph most semantically similar to the question via cosine similarity. We compared Kwame with other approaches and performed a real-time implementation. Our work offers a unique solution for online learning, suited to the African context.

2   Kwame's System Architecture

Kwame's system model is Sentence-BERT (SBERT), a modification of the BERT architecture with siamese and triplet network structures that was shown to outperform state-of-the-art sentence embedding methods such as BERT and Universal Sentence Encoder for semantic similarity tasks[8]. We used the multilingual version [9]. We also created a basic real-time implementation of Kwame using the SBERT model fine-tuned with the course data. A user types a question, Kwame detects the language automatically (using a language detection library) and then computes cosine similarity scores with a bank of answers (described next) corresponding to that language, retrieves, and displays the top answer along with a confidence score which represents the similarity score (Fig. 1).

3   Dataset and Preprocessing

We used the course materials from our "Introduction to Programming" SuaCode course written in English and French [1]. Each document contains text organized by paragraphs that explain concepts along with code examples, tables, and figures which were removed during preprocessing. Also, the course has multiple choice quizzes for each lesson and the answer to each question has a corresponding paragraph in the course material. In this work, we used lesson 1, "Basic Concepts in the Processing Language" to create 2 types of question-answer pairs (1) quiz-based *(n = 20)* using the course' quiz questions and (2) student-based *(n = 12)* using real-world questions from students in past cohorts along with the corresponding paragraph answers in the course materials. There were 39 paragraphs and hence a random baseline of 2.6% for answer accuracy.

4   Experiments and Evaluation

We evaluated the accuracy of our proposed 3 models and the duration to provide answers. The first model—SBERT (regular)—is the normal SBERT model with no course-specific customization. For the second model—SBERT (trained) —, we trained the SBERT model using weakly-labeled triplet sentences from the course materials like in [8] to learn the semantics of the course text. For each paragraph, we used each sentence as an anchor, the next sentence after it in that paragraph as a positive example, and a random sentence in a random paragraph in the document as a negative example. We created a train-test split (75%:25%) and trained the model using the triplet objective in [8]. For the third model, we fine-tuned the SBERT model separately using the quiz QA pairs and student QA pairs using the same triplet objective. We compared these models with TF-IDF with bi-grams and Universal Sentence Encoder. The models were evaluated separately with the quiz and student QA pairs. To evaluate, we extracted each question's embedding and then computed the cosine similarity between the question's embedding and all the possible answers' embeddings, and returned the answer with the biggest similarity score. We then computed the accuracy of the predictions and the average duration per question. We precomputed and saved the embeddings. Evaluations were performed on a MacBook Pro with 2.9 GHz Dual-Core Intel Core i5 processor. In addition to this top 1 accuracy evaluation, we computed top 3, and 5 accuracy results for the SBERT models similar to Zylich et al. [11] in which Kwame returns the top 3 or 5 answers and we check if the correct answer is any of those 3 or 5 answers.

Table 1. Accuracy and duration results

Model	Accuracy (%)				Duration (secs per question)	
	English		French		English	French
	Quiz	Student	Quiz	Student		
TF-IDF (Baseline)	30%	16.7%	45%	8.3%	0.03	0.02
Universal Sentence Encoder (USE)	40%	25%	35%	16.7%	3.7	3.2
SBERT (regular)	50%	25%	65%	8.3%	3.0	2.7
SBERT (trained)	50%	25%	60%	8.3%	6.0	5.5
SBERT (fine-tuned with Quiz)	**65%**	16.7%	**70%**	8.3%	6.0	6.0
SBERT (fine-tuned with Student)	60%	**58.3%**	65%	**58.3%**	5.8	5.6

Table 2. Top 1, 3, and 5 accuracy results for SBERT

Model	Accuracy (%)											
	English						French					
	Quiz			Student			Quiz			Student		
	Top 1	Top 3	Top 5	Top 1	Top 3	Top 5	Top 1	Top 3	Top 5	Top 1	Top 3	Top 5
SBERT (regular)	50	75	75	25	50	75	65	75	75	8.3	50	75
SBERT (trained)	50	75	75	25	33	91.7	60	75	75	8.3	58	75
SBERT (fine-tuned with Quiz)	65	80	75	16.7	50	83.3	**70**	75	75	8.3	33	91.7
SBERT (fine-tuned with Student)	60	80	85	58.3	83.3	100	65	80	80	58.3	91.7	91.7

5   Results and Discussion

The duration results show that TF-IDF is the fastest method, followed by SBERT regular, USE, SBERT trained, and SBERT fine-tuned taking the most time of 6 s (Table 1) which is not long compared to the 6 min average response time in our recent SuaCode course. For the quiz data, TF-IDF has the worst performance of 30% and 45% for English and French respectively (but better than the random baseline of 2.6%) with USE and SBERT (Regular) having better performance. Our SBERT model that we trained using the weakly-labeled data from the course materials did not perform better than SBERT regular. The SBERT models that were fine-tuned on the QA task had the highest accuracies as expected. Overall, the models performed better for the quiz data than the student data. The real-world questions were noisy with various phrases like "Any idea please?", "Good day class" etc.; see [3] for a more thorough discussion. The accuracy results improved for top 3 and top 5 even getting up to 100% with similar results for English and French questions (Table 2). In a course setting, returning 3 answers should not be overwhelming. Zylich et al. [11] attempted to correctly answer 18 factual physics questions which are similar to our coding content questions (20 quiz type and 12 student type). Their document retrieval approach which is analogous to our QA system had 44.4%, 88.9% and 88.9% top

1, 3 and 5 accuracies respectively. Our best system had 58.3% (58.3%), 83.3% (80%) and 100% (91.7%) top 1, 3 and 5 accuracies for the student QA type for English (French). Hence, our results are overall slightly better than theirs.

6   Conclusion

In this work, we developed a bilingual AI TA—Kwame—to answer students' questions from our online introductory coding course, SuaCode in English and French. Our results showed that fine-tuning on the course data and returning the top 3 and 5 answers improved the results. Future work will improve the accuracy for real-world questions, deploy and evaluate Kwame in a real-world course, and explore offering answers in various African languages such as Twi (Ghana).

Acknowledgement. We thank Professor Elloitt Ash and Victor Kumbol for helpful discussions.

References

1. SuaCode - smartphone-based coding course. https://github.com/Suacode-app/ Suacode/blob/master/README.md
2. Benedetto, L., Cremonesi, P.: *Rexy*, a configurable application for building virtual teaching assistants. In: Lamas, D., Loizides, F., Nacke, L., Petrie, H., Winckler, M., Zaphiris, P. (eds.) INTERACT 2019. LNCS, vol. 11747, pp. 233–241. Springer, Cham (2019). https://doi.org/10.1007/978-3-030-29384-0_15
3. Boateng, G.: Kwame: a bilingual AI teaching assistant for online SuaCode courses. arXiv preprint arXiv:2010.11387 (2020)
4. Boateng, G., Kumbol, V.: Project iSWEST: promoting a culture of innovation in Africa through stem. In: 2018 IEEE Integrated STEM Education Conference (ISEC), pp. 104–111. IEEE (2018)
5. Boateng, G., Kumbol, V.W.A., Annor, P.S.: Keep calm and code on your phone: a pilot of suacode, an online smartphone-based coding course. In: Proceedings of the 8th Computer Science Education Research Conference, pp. 9–14 (2019)
6. Goel, A.: Ai-powered learning: making education accessible, affordable, and achievable. arXiv preprint arXiv:2006.01908 (2020)
7. Goel, A.K., Polepeddi, L.: Jill Watson: a virtual teaching assistant for online education. Technical report, Georgia Institute of Technology (2016)
8. Reimers, N., Gurevych, I.: Sentence-BERT: sentence embeddings using Siamese BERT-networks. arXiv preprint arXiv:1908.10084 (2019)
9. Reimers, N., Gurevych, I.: Making monolingual sentence embeddings multilingual using knowledge distillation. In: Proceedings of the 2020 Conference on Empirical Methods in Natural Language Processing. Association for Computational Linguistics (2020). https://arxiv.org/abs/2004.09813
10. Nsesa runs SuaCode Africa – Africa's first smartphone-based online coding course (2020). https://nsesafoundation.org/nsesa-runs-suacode-africa/
11. Zylich, B., Viola, A., Toggerson, B., Al-Hariri, L., Lan, A.: Exploring automated question answering methods for teaching assistance. In: Bittencourt, I.I., Cukurova, M., Muldner, K., Luckin, R., Millán, E. (eds.) AIED 2020. LNCS (LNAI), vol. 12163, pp. 610–622. Springer, Cham (2020). https://doi.org/10.1007/978-3-030-52237-7_49

Early Prediction of Children's Disengagement in a Tablet Tutor Using Visual Features

Bikram Boote[1]([✉]), Mansi Agarwal[2], and Jack Mostow[3]

[1] Jadavpur University, Kolkata, India
[2] Delhi Technological University, Rohini, India
[3] Carnegie Mellon University, Pittsburgh, USA

Abstract. Intelligent tutoring systems could benefit from human teachers' ability to monitor students' affective states by watching them and thereby detecting early warning signs of disengagement in time to prevent it. Toward that goal, this paper describes a method that uses input from a tablet tutor's user-facing camera to predict whether the student will complete the current activity or disengage from it. Training a disengagement predictor is useful not only in itself but also in identifying visual indicators of negative affective states even when they don't lead to non-completion of the task. Unlike prior work that relied on tutor-specific features, the method relies solely on visual features and so could potentially apply to other tutors. We present a deep learning method to make such predictions based on a Long Short Term Memory (LSTM) model that uses a target replication loss function. We train and test the model on screen capture videos of children in Tanzania using a tablet tutor to learn basic Swahili literacy and numeracy. We achieve balanced-class-size prediction accuracy of 73.3% when 40% of the activity is still left. We also analysed how prediction accuracy varies among tutor activities, revealing two distinct causes of disengagement.

Keywords: Intelligent tutoring systems · Student disengagement · Computer vision · Deep learning

1 Introduction

Analyzing the dynamics of students' affective states over the course of a learning process is important in order to create a more engaging environment. Human teachers can monitor students' facial expressions, behavior, and performance to detect disengagement and address it. Ideally, intelligent tutors should likewise detect and respond to early warning signs of disengagement [7].

B. Boote and M. Agarwal—Both student authors contributed equally.

I. Roll et al. (Eds.): AIED 2021, LNAI 12749, pp. 98–103, 2021.
https://doi.org/10.1007/978-3-030-78270-2_17

This paper describes a method to monitor affective state, and evaluates its ability to predict disengagement. We investigate this problem in the context of RoboTutor [11], a tablet tutor that teaches basic Swahili literacy and numeracy with thousands of educational activities, too many to detail here. We train and test a method that analyzes screen capture videos that include tablet-camera input and predicts whether the child will complete the current activity. Space precludes discussing the relevance of prior work on student engagement [4,5] and inferring it from video [6,13]), student behavior [2,8], and physiological sensors [9,12].

2 Dataset Description

The data for this study consists of screen-capture recordings of 200 RoboTutor sessions of children aged 6–12. Each video has temporal resolution of 48 frames per second and spatial resolution of 1024 × 720 pixels. The 192 × 136 pixel window at the top right of the screen displays the input from the tablet's user-facing camera, including the user's face.

We segmented each video into one clip per activity and labeled its outcome as *Complete* if the child completed the activity or *Bailout* if the child tapped on the Back button to stop the activity. Labeling affective states manually is subjective, unreliable, and costly [1]. Hence we used objective labels to represent engagement, excluding the final clip of each session, where children typically bailed out because their time was up. The 200 videos yielded 1195 clips, 803 labeled *Complete* and 392 labeled *Bailout*.

3 Methodology

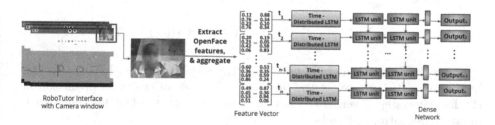

Fig. 1. Model architecture

Automatic Segmentation and Labeling: To segment a video into activities, we needed to know where each activity started and ended. Unfortunately, the videos were from a version of RoboTutor that did not yet log this information. Instead, we inferred activity boundaries from the videos themselves by detecting the selector screen displayed before an activity and the rating screen displayed afterwards, as follows. To detect if a video frame shows one of these screens, compare it to a reference image of the screen type and decide if fewer than 5% of their pixels differ. This threshold is low enough to detect a screen type accurately but flexible enough to tolerate normal variations in its appearance, such as which item in a menu is highlighted.

The screen videos show taps as small white dots. To detect taps on the Back button, we used OpenCV's HoughCircles [14] method to look for a white circle in the 64×36 pixel area at the top left of the screen where the Back button is located. Our automated labeller segmented the videos into clips, each starting at an activity selector screen and ending either at the activity rating screen (labeled *Complete*) or tap on the Back button (labeled *Bailout*).

It is important to emphasize that we used this information about RoboTutor only to segment and label the videos, which could be done instead from times-tamped log entries if available. The subsequent process of training and testing a disengagement prediction model did not use any tutor-specific information.

Feature Engineering: We used visual features computed by OpenFace [3], a facial behavior analysis tool. We used the same set of static features as [1], namely head proximity, pitch, yaw, roll, eye gaze, blink, pupil dilation, and Facial Action Units. To smoothe noisy measurements, we averaged static features over 4 frames, and normalized each feature to the interval $[0, 1]$.

Model Architecture and Training: To predict whether a child will complete an activity, we train a Long Short Term Memory (LSTM) based deep learning model using the target replication technique [10] to reward early prediction. The model consists of a Time Distributed LSTM layer with 4 units followed by 2 regular LSTM layers each of 64 units and finally a dense (fully connected) network. The dense network has 3 layers, with 200 neurons in the first layer, 50 neurons in the second layer, and a single output neuron in the final layer to represent a probabilistic binary output. Figure 1 shows the model architecture.

The target replication objective function is:

$$\alpha . \frac{1}{T} \sum_{t=1}^{T} loss(y'(t), y(t))) + (1 - \alpha) . loss(y'(T), y(T)) \qquad (1)$$

Here T represents the total number of timesteps and α is a hyperparameter in the range $[0, 1]$ to weight the relative importance of errors on prediction at the last time step T versus at the individual time steps t. Our loss function is binary cross-entropy given by:

$$- (y(i) . log(y'(i)) + (1 - y(i)) . log(1 - y'(i))), \qquad (2)$$

where $y(i)$ is the true label (*Complete* or *Bailout*) and $y'(i)$ is the predicted probability of completing the activity.

We use Stochastic Gradient Descent optimization with a learning rate of 0.001 to train the model for 500 epochs. The clips vary in length, so each epoch updates the gradient separately for one clip at a time, rather than zero-padding shorter clips to a common clip length.

4   Results

We tested our model on a balanced held-out test set of 47 *Complete* and 43 *Bailout* instances, with no overlap between the test and training data

With 40% of each clip left, the model had accuracy 73.3% at $\alpha = 0.6$. Table 1 reports the variation in accuracy with different values of alpha (α).

Table 1. $\alpha = 0.6$ got the top accuracy of the values tried.

alpha (α)	0.1	0.2	0.3	0.4	0. 5	**0.6**	0.7	0.8	0.9
Accuracy (%)	52.2	61.1	66.7	65.6	71.1	**73.3**	66.7	67.8	68.9

Disengagement Modeling for Early and Later Bailouts: The test set consisted of 22 Bubble Pop activities, 9 Story activities, 7 Writing activities, and 5 Arithmetic activities. We analysed the test set and realised most bailouts in Bubble Pop and Writing activities were early. We hypothesize that early bailouts are due to children recognizing activity types they dislike and tapping on the *Back* button to escape from the activity.

To understand the reasons behind later bailouts, we analysed the 10 longest *Bailout* clips from the test set. We found that 4 of them were Arithmetic, reflecting the fact that bailouts in Arithmetic activities occurred later rather than immediately when the activity started as in Bubble Pop and Writing activities.

Knowledge of which tutor interactions lead to student disengagement and frustration could help tutor developers address them. By inspecting *Bailout* clips to investigate frustrating events during tutor interactions, we found the following:

- In Arithmetic activities, the tutor's inability to correctly recognise a digit written by the student sometimes led to bailing out. For example, in one of the activities, the child wrote the answer 8 correctly. Still, the tutor misrecognized it as 5 and then as 6, leading to the child's confusion and eventual bailout. Our data set came from an early version of RoboTutor; by the time of this study, RoboTutor's developers had already improved RoboTutor's writing recognizer to curb this cause of disengagement.
- In one of the Writing activities, the child missed the correct location to write the letter and could not undo his mistake. Subsequently, he filled in the letters at incorrect locations and finally bailed out. This suggests that the tutor should be more flexible, allowing the child to erase mistakes easily.
- One of the Bubble Pop activities exposed a bug where no bubbles were displayed after the activity started. The child waited for a while but eventually bailed out. Such bugs should be removed for a smoother learning experience.

In summary, early bailouts are apparently caused by recognizing a type of activity the child dislikes, and could be addressed by making them more engaging or choosing different activities. In contrast, later bailouts occur after tutorial interactions with a negative impact on the student's affective state, leading to disengagement, and could be addressed by repairing such interactions.

5   Conclusion

This paper presents a deep learning model to predict task completion as an indicator of disengagement in children using a tablet tutor. Our model only uses visual cues extracted from the tablet's user-facing camera input, so it could potentially be generalized to other tutors more easily than methods that rely on tutor-specific features. We analyzed the test set to shed light on different causes of disengagement, namely recognizing a disliked type of activity on sight versus experiencing displeasure during the activity. This work contributes to the automated identification of early visual harbingers of disengagement. It should help improve tutors at design time, for example, by pinpointing specific situations and tutor actions that tend to elicit the sort of visible behavior that presages bailout. Furthermore, it should eventually help guide pedagogical decisions at runtime by detecting disengagement early and triggering actions.

References

1. Agarwal, M., Mostow, J.: Semi-supervised learning to perceive children's affective states in a tablet tutor. In: Thirty-Fourth AAAI Conference on Artificial Intelligence, pp. 13350–13357 (2020)
2. Arroyo, I., Woolf, B.P.: Inferring learning and attitudes from a Bayesian network of log file data. In: International Conference on Artificial Intelligence in Education, pp. 33–40 (2005)
3. Baltrušaitis, T., Robinson, P., Morency, L.P.: OpenFace: an open source facial behavior analysis toolkit. In: 2016 IEEE Winter Conference on Applications of Computer Vision (WACV), pp. 1–10. IEEE (2016)
4. Bosch, N., D'Mello, S.: Automatic detection of mind wandering from video in the lab and in the classroom. IEEE Trans. Affect. Comput. (2019)
5. Bosch, N., D'mello, S.K., Ocumpaugh, J., Baker, R.S., Shute, V.: Using video to automatically detect learner affect in computer-enabled classrooms. ACM Trans. Interact. Intell. Syst. (TiiS) 6(2), 1–26 (2016)
6. Chang, C., Zhang, C., Chen, L., Liu, Y.: An ensemble model using face and body tracking for engagement detection. In: Proceedings of the 20th ACM International Conference on Multimodal Interaction, pp. 616–622 (2018)
7. Dewan, M.A.A., Murshed, M., Lin, F.: Engagement detection in online learning: a review. Smart Learn. Environ. 6(1), 1 (2019)
8. González-Brenes, J.P., Mostow, J.: Predicting task completion from rich but scarce data (2010)
9. Liang, W.C., Yuan, J., Sun, D.C., Lin, M.H.: Changes in physiological parameters induced by indoor simulated driving: effect of lower body exercise at mid-term break. Sensors 9(9), 6913–6933 (2009)
10. Lipton, Z.C., Kale, D.C., Elkan, C., Wetzel, R.: Learning to diagnose with LSTM recurrent neural networks. arXiv preprint arXiv:1511.03677 (2015)
11. McReynolds, A.A., Naderzad, S.P., Goswami, M., Mostow, J.: Toward learning at scale in developing countries: lessons from the global learning XPRIZE field study. In: Proceedings of the Seventh ACM Conference on Learning@ Scale, pp. 175–183 (2020)

12. Mostow, J., Chang, K., Nelson, J.: Toward exploiting EEG input in a reading tutor. In: Biswas, G., Bull, S., Kay, J., Mitrovic, A. (eds.) AIED 2011. LNCS (LNAI), vol. 6738, pp. 230–237. Springer, Heidelberg (2011). https://doi.org/10.1007/978-3-642-21869-9_31

13. Thomas, C., Jayagopi, D.B.: Predicting student engagement in classrooms using facial behavioral cues. In: Proceedings of the 1st ACM SIGCHI International Workshop on Multimodal Interaction for Education, pp. 33–40 (2017)

14. Yuen, H., Princen, J., Illingworth, J., Kittler, J.: Comparative study of Hough transform methods for circle finding. Image Vis. Comput. 8(1), 71–77 (1990)

An Educational System for Personalized Teacher Recommendation in K-12 Online Classrooms

Jiahao Chen, Hang Li, Wenbiao Ding, and Zitao Liu[✉]

TAL Education Group, Beijing, China
{chenjiahao,lihang4,dingwenbiao,liuzitao}@tal.com

Abstract. In this paper, we propose a simple yet effective solution to build practical teacher recommender systems for online one-on-one classes. Our system consists of (1) a pseudo matching score module that provides reliable training labels; (2) a ranking model that scores every candidate teacher; (3) a novelty boosting module that gives additional opportunities to new teachers; and (4) a diversity metric that guardrails the recommended results to reduce the chance of collision. Offline experimental results show that our approach outperforms a wide range of baselines. Furthermore, we show that our approach is able to reduce the number of student-teacher matching attempts from 7.22 to 3.09 in a five-month observation on a third-party online education platform.

Keywords: Teacher recommendation · Recommender systems · K-12 education · Online education

1 Introduction

Because of the better accessibility and immersive learning experience, one-on-one class stands out among all the different forms of online courses [1–3,5,11]. In one-on-one courses, teacher recommender systems play an important role in helping students find their most appropriate teachers [10,13]. However, teacher recommendation presents numerous challenges that come from the following special characteristics of real-world educational scenarios:

- *Limited sizes of demand and supply*: The number of teachers in supply side is incredibly smaller compared to Internet-scaled inventories. Moreover, different from item based recommendation where popular items can be suggested to millions of users simultaneously, a teacher can only take a very limited amount of students and students may only take one or two classes at each semester.
- *Lack of gold standard*: There is no ground truth showing how good a match is between a teacher and a student. The rating based mechanism doesn't work since ratings from K-12 students are very noisy and unreliable.

© Springer Nature Switzerland AG 2021
I. Roll et al. (Eds.): AIED 2021, LNAI 12749, pp. 104–108, 2021.
https://doi.org/10.1007/978-3-030-78270-2_18

- *Cold-start teachers*: The online educational marketplace is dynamic and there are always new teachers joining. It is important to give such new teachers opportunities to let them take students instead of keeping recommending existing best performing teachers.
- *High-demand diversity*: It is undesirable to recommend the same set of teachers to students and the teacher recommender systems are supposed to reduce chances that two students want to book the same teacher at the same time.

The objective of this work is to study and develop approaches that can be used for personalized teacher recommendation for online classes. More specifically, we design techniques to (1) generate robust pseudo training labels as ground truth for learning patterns of good matches between students and teachers; (2) boost newly arrived teachers by giving incentives to their ranking scores when generating the recommended candidates; and (3) fairly evaluate and guard the diversity of recommendation results by the proposed measure of teacher diversity. We compare our approach with a wide range of baselines and evaluate its benefits on a real-world online one-on-one class dataset. We also deploy our algorithm into the real production environment and demonstrate its effectiveness in terms of number of matching attempts.

2   The Framework

In this section we will discuss the details about our teacher recommendation framework for the online one-on-one courses. Our framework is made up of four key components: (1) the pseudo matching scores module; (2) the ranking model; (3) the novelty boosting module; and (4) the diversity metric.

Pseudo Matching Scores. One of the most challenging problems in building teacher recommender systems is the missing of ground truth. To remedy above problem, we choose to generate the pseudo matching scores from students' dropouts. Our mechanism relies on the assumption that matching scores reflect student preferences, which are approximated by the number of one-on-one courses between each teacher and student. In addition, we capture of the recency effect of dropout cases by using an exponential function. We design the pseudo matching scores as follows:

Definition 1. POSITIVE PSEUDO MATCHING SCORE. *For student s_i who has completed the class, let $\mathbf{T}_i = \{t_1, t_2, \cdots, t_{p_i}\}$ be the collection of teachers who have ever taught student s_i and t_j denotes the jth teacher, p_i denotes the total number of teachers who have taught student s_i. Let $M_i(t_j)$ be the number of courses taught by teacher t_j. The positive pseudo matching scores of (s_i, t_j) is defined as $\mathcal{P}(s_i, t_j) = M_i(t_j) / \sum_{j=1}^{p_i} M_i(t_j)$, where $\mathcal{P}(\cdot, \cdot)$ is the positive matching score function.*

Definition 2. NEGATIVE PSEUDO MATCHING SCORE. *For student s_k who has dropped the class, with similar notations in Definition 1, the negative pseudo matching scores of (s_k, t_j) is defined as $\mathcal{N}(s_k, t_j) = -\exp(1 - M_k(t_j))$, where $\mathcal{N}(\cdot, \cdot)$ is the negative matching score function.*

According to Definitions 1 and 2, the pseudo matching scores range from -1 to 1. It reaches the maximum value of 1 when a student never requests a change of teacher and completes the entire class and it goes to the minimum value of -1 when a student immediately quits after the first course.

The Ranking Model. The ranking model learns from a collection of teacher-student pairs with pseudo matching scores. We design the following three categories of features: (1) *demographic features:* the demographic information of both students and teachers, such as gender, schools, etc. (2) *in-class features:* the class behavioral features from both students and teachers, such as lengths of talking time, the number of spoken sentences, etc. (3) *historical features:* the historical features aggregate each teacher's past teaching performance, which includes total numbers of courses and historical dropout rates, etc. In this work, we choose to use gradient boosting decision tree (GBDT) [6] as our ranking model due to its robustness and generalization capability.

Novelty Boosting for New Teachers. We design a novelty boosting component that gives extra ranking incentives to new teachers and enhances the chances of successful matches for new teachers. The novelty boosting score for teacher t_j is defined as follows:

$$b_j = \begin{cases} \frac{\alpha}{\sqrt{Z_j+\beta}} & Z_j < \delta \\ 0 & \text{otherwise} \end{cases}, \quad Z_j = \sum_{i \in \mathbf{I}_j} M_i(t_j) \tag{1}$$

where $\mathbf{I}_j$ represents the index set of all taught students and Z_j represents the total number of courses taught by teacher t_j. α, β and δ are positive hyper parameters. Moreover, we measure the overall effect of novelty boosting by computing the overall new teacher ratios $r = \frac{1}{N} \sum_{i=1}^{N} \sum_{t_j \in \hat{\mathbf{T}}_i} (\mathbb{1}_{b_j>0}/|\hat{\mathbf{T}}_i|)$ in the top-recommended candidates, where $\hat{\mathbf{T}}_i$ represents the set of recommended teachers for student s_i and $\mathbb{1}_{b_j>0}$ is the indicator function that indicates whether the teacher t_j is a new teacher.

Diversity Measurement. Diversity is important when conducting teacher recommendations in K-12 online scenarios. In this work, we propose a diversity guardrail measurement $d = 1 - \frac{2}{|\mathbf{S}|(|\mathbf{S}|-1)} \sum_{i=1}^{|\mathbf{S}|-1} \sum_{j=i+1}^{|\mathbf{S}|} \frac{|\hat{\mathbf{T}}_i \cap \hat{\mathbf{T}}_j|}{|\min(|\hat{\mathbf{T}}_i|,|\hat{\mathbf{T}}_j|)|}$, where $\mathbf{S}$ represents the set of students needed online one-on-one instructors. The diversity scores d range from 0 to 1. It reaches the maximum value of 1 when each student's recommendation results don't overlap.

3   Experiments

Offline Evaluation. The offline evaluation of recommendation is different from standard binary classification tasks where we can only partially observe the ground truth. When designing an effective offline evaluation environment, we will only focus on the "good" matches between students and teachers and ensure that a positively matched teacher exists in our recommended candidate list. Therefore, the performance is mainly evaluated by recall. Besides, we measure the effects of new teacher ratio and diversity. The hyper parameters in novelty boosting score function are $\alpha = 0.04$, $\beta = 1$, and for each student we select top 200 teachers as our recommended candidates. We collect a real-world dataset with 3,672 students, 2,139 teachers, and 8,072 student-teacher matches. Here, to simulate "good" matches, we first compute the pseudo matching scores for all 8,072 student-teacher matches and randomly select 821 pairs whose positive scores are over 0.5 as our test data. We compare our approach with ItemCF [12], SVD [7], NMF [8], DeepFM [9],W&D [4]. The results are shown in Table 1. The proposed approach has competitive performance against the widely used recommendation models. Please note that the matrix factorization based baselines, such as ItemCF, SVD and NMF, cannot seamlessly integrate new teachers into their corresponding rating matrices and hence fail to recommend new teachers.

Table 1. Results on our offline educational dataset.

Model	Precision	Recall	Diversity	New Teacher Ratio
Our	**0.0017**	**0.2545**	**0.7454**	**0.0333**
Wide& Deep	0.0016	0.2335	0.7446	0.0070
DeepFM	0.0016	0.2351	0.7438	0.0013
ItemCF	0.0014	0.2011	0.7232	N/A
SVD	0.0013	0.1909	0.7233	N/A
NMF	0.0013	0.1924	0.7233	N/A

Online Experiments. We deploy our algorithm to a real production environment. We continuously observe the change of the mean value of the number of times a student requires to change their teacher. Over the five-month observation period (2020/01 - 2020/05), we found that the number of matching attempts decreased from 7.22 to 3.09, reflecting that our algorithm can accurately make more good recommendations to teachers.

4   Conclusion

In this paper, we present an end-to-end teacher recommendation framework for online one-on-one classes in the real-world scenario. The results on the real-world educational teacher recommendation dataset show that our proposed system can not only accurately recommend teachers in terms of recall but give more opportunities to new teachers in terms of new teacher ratios. Meanwhile, we guardrail the overall recommendation quality in terms of diversity experimentally. In online experiments, the proposed model is deployed in the real production system and the results show that the proposed approach is able to greatly reduce the number of matching attempts.

Acknowledgment. This work was supported in part by National Key R&D Program of China, under Grant No. 2020AAA0104500 and in part by Beijing Nova Program (Z201100006820068) from Beijing Municipal Science & Technology Commission.

References

1. Blatchford, P., Bassett, P., Brown, P.: Examining the effect of class size on class-room engagement and teacher-pupil interaction: differences in relation to pupil prior attainment and primary vs. secondary schools. Learn. Instr. **21**(6), 715–730 (2011)
2. Blatchford, P., Bassett, P., Goldstein, H., Martin, C.: Are class size differences related to pupils' educational progress and classroom processes? Findings from the institute of education class size study of children aged 5–7 years. Br. Edu. Res. J. **29**(5), 709–730 (2003)
3. Chen, J., Li, H., Wang, W., Ding, W., Huang, G.Y., Liu, Z.: A multimodal alerting system for online class quality assurance. In: Isotani, S., Millán, E., Ogan, A., Hastings, P., McLaren, B., Luckin, R. (eds.) AIED 2019. LNCS (LNAI), vol. 11626, pp. 381–385. Springer, Cham (2019). https://doi.org/10.1007/978-3-030-23207-8_70
4. Cheng, H.T., et al.: Wide & deep learning for recommender systems. In: Proceedings of the 1st Workshop on Deep Learning for Recommender Systems, pp. 7–10 (2016)
5. Finn, J.D., Achilles, C.M.: Tennessee's class size study: findings, implications, misconceptions. Educ. Eval. Policy Anal. **21**(2), 97–109 (1999)
6. Friedman, J.H.: Greedy function approximation: a gradient boosting machine. Ann. Stat. **29**, 1189–1232 (2001)
7. Golub, G.H., Reinsch, C.: Singular value decomposition and least squares solutions. Numer. Math. **14**(5), 403–420 (1970). https://doi.org/10.1007/BF02163027
8. Gu, Q., Zhou, J., Ding, C.H.Q.: Collaborative filtering: Weighted nonnegative matrix factorization incorporating user and item graphs. In: Proceedings of the SIAM International Conference on Data Mining, SDM 2010, Columbus, Ohio, USA, 29 April –1 May 2010, pp. 199–210. SIAM (2010). https://doi.org/10.1137/1.9781611972801.18
9. Guo, H., Tang, R., Ye, Y., Li, Z., He, X.: DeepFM: a factorization-machine based neural network for CTR prediction. In: Sierra, C. (ed.) Proceedings of the Twenty-Sixth International Joint Conference on Artificial Intelligence, IJCAI 2017, Melbourne, Australia, 19–25 August 2017, pp. 1725–1731. ijcai.org (2017). https://doi.org/10.24963/ijcai.2017/239
10. Li, H., Ding, W., Yang, S., Liu, Z.: Identifying at-risk K-12 students in multimodal online environments: a machine learning approach. In: International Conference on Educational Data Mining (2020)
11. Liang, J.K.: A few design perspectives on one-on-one digital classroom environment. J. Comput. Assist. Learn. **21**(3), 181–189 (2005)
12. Sarwar, B.M., Karypis, G., Konstan, J.A., Riedl, J.: Item-based collaborative filtering recommendation algorithms. In: Shen, V.Y., Saito, N., Lyu, M.R., Zurko, M.E. (eds.) Proceedings of the Tenth International World Wide Web Conference, WWW 10, Hong Kong, China, 1–5 May 2001, pp. 285–295. ACM (2001). https://doi.org/10.1145/371920.372071
13. Xu, S., Ding, W., Liu, Z.: Automatic dialogic instruction detection for K-12 online one-on-one classes. In: Bittencourt, I.I., Cukurova, M., Muldner, K., Luckin, R., Millán, E. (eds.) AIED 2020. LNCS (LNAI), vol. 12164, pp. 340–345. Springer, Cham (2020). https://doi.org/10.1007/978-3-030-52240-7_62

Designing Intelligent Systems to Support Medical Diagnostic Reasoning Using Process Data

Elizabeth B. Cloude[1](✉) [ID], Nikki Anne M. Ballelos[2], Roger Azevedo[1],
Analia Castiglioni[3], Jeffrey LaRochelle[4], Anya Andrews[3],
and Caridad Hernandez[3]

[1] Department of Learning Sciences and Educational Research,
University of Central Florida, Orlando, FL, USA
elizabeth.cloude@knights.ucf.edu, roger.azevedo@ucf.edu
[2] Burnett School of Biomedical Sciences, University of Central Florida,
Orlando, FL, USA
nballelos28@knights.ucf.edu
[3] Department of Internal Medicine, University of Central Florida,
Orlando, FL, USA
{analia.castiglioni,anya.andrews,caridad.hernandez}@ucf.edu
[4] Medical Education Department, University of Central Florida,
Orlando, FL, USA
jeffrey.larochelle@ucf.edu

Abstract. We captured 36 medical professionals' process data across five medical cases using CResME, a multimedia system designed to activate illness scripts. Findings showed medical expertise was unrelated to diagnostic performance when illness scripts were disrupted, and that process data was predictive of diagnostic performance for some medical cases. Implications of our study illustrate ways to design AIEd systems capable of scaffolding diagnostic reasoning to reduce medical errors.

Keywords: Diagnostic reasoning · Process data · AIEd systems

1 Introduction

To study diagnostic reasoning and its relation to diagnostic performance, researchers are building theoretically-guided AIEd systems to enhance training using illness scripts [1,2]. Studies find using illness scripts was related to higher diagnostic performance [3–5]. Yet, a study found when illness scripts were disrupted, i.e., random order, it led to lower diagnostic performance regardless of medical expertise [6]. We investigated disrupted illness scripts by studying 36 medical professionals' process data on diagnostic reasoning and performance using CResME [8], a hypermedia environment. Our research questions include (1)

Thanks to University of Central Florida College of Medicine for funding this research.

I. Roll et al. (Eds.): AIED 2021, LNAI 12749, pp. 109–113, 2021.
https://doi.org/10.1007/978-3-030-78270-2_19

are there differences in the distribution of diagnostic performance across the five medical cases presented within CResME? and (2) are there associations between the number and time spent using tools, correctly matching critical findings to cases[1], and diagnostic performance across the five cases with CResME?,

2   Methods

Thirty-six medical professionals (56% female; $M_{Age} = 29.36, SD = 10.89$) were recruited from a College of Medicine at a North American university and completed a 2-hour remote study with CResME. The sample consisted of 23 s- (64%) and 6 third-year medical students (17%), 5 board-certified physicians (14%) and 2 residents (6%). Participants received a $40 electronic gift card for completing the study. CResME presented patient information for five cases related to the common cough via nodes (see [8] for details). Participants needed to connect nodes to each history and write a final diagnosis for each case. Each node contained information that related to enabling conditions, fault, and consequences [8,10]. Tools were built into CResME to scaffold diagnostic reasoning: (1) lab values illustrating normal ranges for lab results; (2) a legend explaining medical abbreviations; and (3) chest X-rays or Spirometry tests.

Participants were recruited via email and completed screening items and a series of questionnaires at least 24 h before the study. Next, participants were trained on how to conduct a think-aloud. Afterwards, they began solving medical cases in CResME and the researcher recorded the session. After providing a diagnosis for each case, the recording was stopped and participants completed post-test items. Last, participants were debriefed and compensated. All participants received diagnostic-performance scores for each medical case rated by a domain expert: 0 = incorrect, 1 = plausible, 2 = correct/lacking precision (P-), and 3 = correct/precise (P+). Process data were extracted using video files to calculate number and seconds spent using the legend and lab value tools as well as viewing the X-ray and Spirometry images. We also calculated pathway scores to represent whether participants matched nodes with the correct history and diagnosis: 0 = incorrect nodes, 1 = one correct node, and 2 = all correct nodes.

3   Results

Prior to conducting our analysis, we examined the effect of medical expertise level on diagnostic-performance scores. We calculated an ANOVA for each of the five cases using a Bonferroni correction ($p< 0.05/5 = 0.01$) across three groups: (1) practicing physicians, (2) residents, and (3) medical students. We did not find differences in diagnostic-performance scores across the expertise levels ($ps> 0.01$), suggesting expertise did not have an effect on diagnostic performance. **To what extent are there differences in the distribution of diagnostic performance scores across the five medical cases presented**

---

[1] We refer to the correct matching as pathway scores in the rest of the paper.

within CResME? A chi-square was calculated to examine differences in the distribution of diagnostic-performance scores across the five medical cases within CResME. The analysis revealed there were differences in diagnostic-performance scores between the five cases, $\chi^2(12, \text{N} = 36) = 87.35$, $p < 0.05$ where case 4 had the highest diagnostic-performance scores. This finding suggested differences exist in medical professionals' ability to diagnose cases across the five medical cases. **To what extent are there associations between the frequency and time using tools, pathway scores, and diagnostic-performance scores across the five cases within CResME?** Separate ordinal logistic regressions (ORL) were calculated for each medical case using a Bonferroni correction ($p < 0.05/5 = 0.01$. Results revealed relationships between pathway scores and diagnostic-performance scores for cases 1 ($\beta = -1.138$, $p < .01$) and 2 ($\beta = -0.951$, $p < .01$). An OLR model also showed relationships between tool use ($\beta = -0.415$, $p < 0.01$), pathway scores ($\beta = -0.947$, $p < 0.01$), and diagnostic-performance scores ($\beta = -0.865$, $p < 0.01$) for case 4.

4 Discussion

AIEd systems have been built to enhance diagnostic performance using illness scripts [7,9]. Yet, few systems use illness scripts representing a disrupted structure, a common theme in real-world medicine [1,6]. In this study, we capturing 36 medical professionals' diagnostic reasoning and performance for 5 medical cases using process data with CResME, a multimedia system representing disrupted illness scripts [8]. Results found no differences in diagnostic performance between medical expertise levels. This finding was consistent with previous literature [6,7]. We also found differences in the distribution of diagnostic-performance scores across the five medical cases, where case 4 (median = 3) demonstrated the highest diagnostic performance, followed by case 1 (median = 2) and case 5 (median = 2). We suspect these findings were due to differences in critical findings illustrated in each node, requiring medical professionals to use reasoning strategies to reach a diagnosis without relying on the common illness script structure [9]. The second research question examined whether tool use (frequency and duration) and pathway scores were predictive of diagnostic performance. The models showed pathway scores were a consistent (and positive) predictor of diagnostic performance for cases 1, 2, and 4, suggesting that medical professionals' ability to identify illness scripts associated with each diagnosis (i.e., correctly match enabling conditions, fault, and consequences) was related to the correct diagnosis. However, this finding was not present for cases 3 and 5. The models also found that tool use (e.g., duration of viewing chest X-rays or Spirometry tests) was a positive predictor for diagnostic performance for case 4. We did not find these results for cases 1–3 and 5. We expected tool use and pathway scores to predict diagnostic performance [9,10], and a possible explanation for this partial inconsistency could be that there were inefficiencies in medical professionals' illness scripts or diagnostic-reasoning strategies related to the diseases for cases 1–3 and 5.

These findings provide implications for building AIEd systems by monitoring process data generating during diagnostic tasks, such as what multiple sources of information are being matched to particular histories and tool use, including the time spent engaging with images (e.g., xrays) to assess whether the correct illness script is activated to provide just-in-time intervention. For instance, if a medical professional was spending less time viewing an chest xray that illustrates a critical finding related to a particular disease (e.g., tumor) and matching incorrect physical exams and data with a history, then it may triggers the system to intervene and redirect the individual to raise awareness to their reasoning process (e.g., *did you view symptomology associated with node X instead Y?*). In future work, we aim to examine multimodal data (i.e., concurrent verbalizations) to pinpoint *when, where,* and *how* medical professionals may have been reasoning through each medical case and its relation to particular illness scripts and diagnostic performance across the medical cases. We recommend future work utilize multichannel data (e.g., eye tracking and facial expressions of emotions) to assess the extent to which these data streams suggest discrepancies in diagnostic reasoning and illness script activation. Implications of this research may inform ways to build AIEd systems that capture medical professionals' multimodal data to provide just-in-time scaffolding and feedback to augment diagnostic reasoning and performance.

References

1. Patel, V.-L., Kaufman, D.-R., Kannampallil, T.-G.: Diagnostic reasoning and decision making in the context of health information technology. Rev. Human Factors Ergon. **8**(1), 149–190 (2013)
2. Chan, K.-S., Zary, N.: Applications and challenges of implementing artificial intelligence in medical education: integrative review. JMIR Med. Educ. **5**(1), e13930 (2019)
3. Lubarsky, S., Dory, V., Audétat, M.-C., Custers, E., Charlin, B.: Using script theory to cultivate illness script formation and clinical reasoning in health professions education. Can. Med. Educ. J. **6**(2), e61–e70 (2015)
4. Keemink, Y., Custers, E.-J., van Dijk, S., Ten Cate, O.: Illness script development in pre-clinical education through case-based clinical reasoning training. Int. J. Med. Educ. **9**, 35–41 (2018)
5. Hayward, J., et al.: Script-theory virtual case: a novel tool for education and research. Med. Teach. **38**(11), 1130–1138 (2016)
6. Norman, G.-R., Grierson, L.-E.-M., Sherbino, J., Hamstra, S.-J., H.-G., Mamede, S.: Expertise in medicine and surgery. In: Ericsson, K.-A., Hoffman, R.-R., Kozbelt, A., Williams, A.-M. (eds.) Cambridge Handbooks in Psychology. The Cambridge Handbook of Expertise and Expert Performance, pp. 331–355. Cambridge University Press, New York (2018)
7. Crowley, R., Medvedeva, O., Jukic, D.: SlideTutor: a model-tracing intelligent tutoring system for teaching microscopic diagnosis. In: Proceedings of the 11th International Conference on Artificial Intelligence in Education, pp. 157–164 (2003)

8. Torre, D.-M., et al.: The clinical reasoning mapping exercise (CResME): a new tool for exploring clinical reasoning. Perspect. Med. Educ. **8**, 47–51 (2019). https://doi.org/10.1007/s40037-018-0493-y

9. Braun, L.-T., et al.: Representation scaffolds improve diagnostic efficiency in medical students. Med. Educ. **51**(11), 1118–1126 (2017)

10. Custers, E.-J.: Thirty years of illness scripts: theoretical origins and practical applications. Med. Teach. **37**(5), 457–462 (2015)

Incorporating Item Response Theory into Knowledge Tracing

Geoffrey Converse[1]([✉]) [iD], Shi Pu[2] [iD], and Suely Oliveira[1] [iD]

[1] University of Iowa, Iowa City, IA 52246, USA
suely-oliveira@uiowa.edu
[2] ETS Canada Inc., 240 Richmond Street W, Toronto, ON M5V1V6, Canada
spu@etscanada.ca

Abstract. The popularity of artificial neural networks has brought high predictive power to many difficult machine learning problems. Knowledge tracing (KT), the task of tracking students' understanding of various concepts over time, is included in this category. But the deep learning methods which have performed best in knowledge tracing are hard to explain in a statistical sense.

In this work, we leverage the psychological theory from Item Response Theory (IRT) to build interpretable neural networks for knowledge tracing which are competitive with other deep learning methods. This presents a trade-off between a small loss in predictive power and an increase in interpretability. The advantage of IRT-inspired knowledge tracing is that it transforms the high-dimensional student ability representation from deep learning models into an explainable IRT representation at each timestep. Further, the item parameters from IRT models can be directly recovered from the trained neural network weights.

Keywords: Knowledge tracing · Item response theory · Neural networks

1 Introduction

Knowledge tracing has become an important feature in electronic learning environments. Intelligent tutoring systems are designed to tailor an education experience to the specific needs of individual students. Corbett and Anderson [2] first introduced Bayesian Knowledge Tracing (BKT), which placed a student in either a learned or unlearned state for each knowledge concept.

Recently, deep neural networks have been proposed for the knowledge tracing problem, including Deep Knowledge Tracing (DKT) [7], Dynamic Key-Valued Memory Networks (DKVMN) [12], and Self-Attentive Knowledge Tracing (SAKT) [6], each of which have yielded significantly higher accuracy than

---

This work was completed prior to joining the company.

© Springer Nature Switzerland AG 2021
I. Roll et al. (Eds.): AIED 2021, LNAI 12749, pp. 114–118, 2021.
https://doi.org/10.1007/978-3-030-78270-2_20

BKT yet lack interpretability. Black-box neural network models do not explicitly track the state of concept mastery for students, and instead the only output is the probability of students answering questions correctly at each timestep.

Item response theory (IRT) [5] provides statistical models for the probability of students answering questions correctly. IRT asserts that K student abilities can be quantified by a continuous vector $\Theta \in \mathbb{R}^K$, and that different items exercise student abilities in different ways. The multidimensional logistic 2-parameter (ML2P) model [9] gives the probability of a student j answering item i correctly as

$$P(u_{ij} = 1 | \Theta_j; a_i, b_i) = \frac{1}{1 + \exp\left(\sum_{k=1}^{K} -a_{ik}\theta_{jk} + b_i\right)} \tag{1}$$

where θ_{jk} is student j's value of the skill k, the discrimination parameter a_{ik} quantifies how much of skill k is required to answer item i correctly, and the difficulty of item i is quantified by b_i.

The connection between knowledge tracing and IRT has been explored before. The Deep-IRT [11] method modifies DKVMN to include two neural networks for student ability and *concept* difficulty (rather than item difficulty), respectively.

In this work, we present a trade-off between predictive power and interpretability, but the proposed method remains competitive with other deep learning methods. While sacrificing a small amount of AUC, IRT-inspired knowledge tracing provides an explicit representation of student knowledge Θ at each timestep. Additionally, parameters of the neural network can be interpreted as approximations to the item parameters a_{ik} and b_i in Eq. 1. In this sense, our proposed models function as both a knowledge tracing and a parameter estimation method.

2   Model Description

Given a tutoring system with n available items, K skills under assessment, and the skill association of each item given as a binary matrix $Q \in \{0,1\}^{n \times K}$ [10], each of the possible $2n$ student interactions (q_t, c_t) is represented as a learned d-dimensional vector $x_t \in \mathbb{R}^d$ and fed through a time-series neural network such as LSTM (similar to DKT [7]) or an attention-based model (similar to SAKT [6]).

$$v_t = \text{LSTM}(x_t, x_{t-1}, \ldots, x_0), \quad v_t \in \mathbb{R}^h \tag{2}$$

Next, each v_t is sent through a linear layer feed-forward-network with output size K (the number of latent concepts), yielding a vector s_t.

$$s_t = W_s v_t + y, \quad W_s \in \mathbb{R}^{K \times h}, y \in \mathbb{R}^K \tag{3}$$

The matrix W_s and vector y are trainable parameters. Each node in this "skill layer" represents a knowledge concept.

Finally, the output layer of the model has n nodes and a sigmoid activation function $\sigma(\cdot)$, with each node representing the probability of the student answering that item correctly.

$$p_t = \sigma(W_p s_t + z) = \frac{1}{1 + \exp(-W_p s_t - z)}, \quad W_p \in \mathbb{R}^{n \times K}, z \in \mathbb{R}^n \quad (4)$$

W_p and z are trainable and importantly, W_p is modified so that the nonzero values of W_p are determined by the Q-matrix [1,3,4]. If item i does not require skill k, then the weight between those nodes is fixed to be zero. In this way, we write

$$W_p \leftarrow W_p \odot Q, \quad (5)$$

where $\odot$ is element-wise multiplication of matrices. Then the probability that the student will answer question i correctly at timestep t is given by

$$p_{ti} = \frac{1}{1 + \exp\left(-\sum_{k=1}^{K} w_{ik} q_{ik} s_{tk} - z_i\right)} \quad (6)$$

where w_{ik}, q_{ik}, s_{tk}, and z_i are entries in W_p, Q, s_t, and z, respectively.

This constraint allows for interpretation of the final neural network layers as an approximate ML2P model: note the similarity between Eq. 6 and Eq. 1. The weights between the skill and output layer ($w_{ik} q_{ik}$) serve as estimates to the discrimination parameters a_{ik}, and the bias parameters in the output layer z_i are estimates to the difficulty parameters b_i. The student's k-th latent ability θ_k is estimated at timestep t via s_{tk}.

3   Experiments and Discussion

We use three datasets[1, 2, 3] which are standard in the knowledge tracing litera-ture. We also include a new data set, Sim200[4], which differs from Synth-5 in a few important ways. First, there are more items and more latent skills (200 and 20, respectively). Second, the Q-matrix is more dense – items require multiple skills in order for students to answer correctly (Synth-5 only includes simple items associated with a single skill). Lastly, Sim200 generates responses according to the ML2P model in Eq. 1, as opposed to the Rasch model [8].

As seen in Table 1, the two IRT-inspired methods (DKT-IRT and SAKT-IRT) are able to produce AUC values competitive with other deep learning methods. As expected, the sacrifice in accuracy is smaller in simulated datasets. In Synth-5 and Sim200, the responses were generated with known IRT models which match the architecture of IRT-inspired methods. When looking at the two real-world datasets, the trade-off in AUC is more significant, as it is not known if the student responses follow the ML2P model. Additionally, there could be inaccuracies in the given item-skill association Q-matrix.

---

[1] https://github.com/chrispiech/DeepKnowledgeTracing/tree/master/data/synthetic.
[2] https://pslcdatashop.web.cmu.edu/DatasetInfo?datasetId=507.
[3] https://sites.google.com/view/assistmentsdatamining.
[4] https://github.com/converseg/irt_data_repo/tree/master/sim200.

Table 1. Test AUC values for various models on each data set.

Method	Synth-5	Sim200	Statics2011	Assist2017
DKT	0.803	0.838	0.793	0.731
SAKT	0.801	0.834	0.791	0.754
DKVMN	0.827	0.829	0.805	0.796
DKT-IRT	0.799	0.824	0.777	0.724
SAKT-IRT	0.798	0.833	0.775	0.728

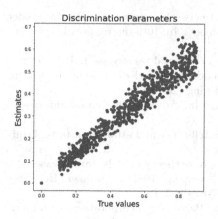 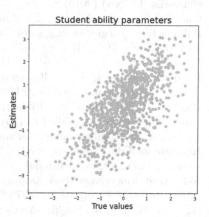

Fig. 1. Correlation between true and estimated Sim-200 discrimination parameters (left), and student ability parameters at $t = 200$ (right).

In Fig. 1, we can see the true values of item parameters a_{ik} and student ability parameters θ_{jk} plotted against estimates given by SAKT-IRT at the final timestep. Recall that the discrimination parameter estimates are the trained weights connecting the skill layer to the output layer and the explicit representation of knowledge Θ can be conveniently computed. For a student j's response sequence of length $L-1$, IRT-inspired knowledge tracing methods return a $K \times L$ matrix, where the entry (k, t) gives the latent trait estimate to the k-th skill at time t, θ_{jkt}.

The connection between IRT and knowledge tracing presented in this work presents a trade-off between accuracy and interpretability. Further work to increase AUC to the level of DKVMN while maintaining explainability is worth exploring. Our proposed method's ability to function as both a knowledge tracing model while also estimating item parameters gives it a unique interpretation rooted in Item Response Theory. This link with IRT is helpful in practice, because it provides an explicit and easy-to-obtain quantity for a student's latent abilities. This approximation of a student ability can be interpreted in the frame of IRT, as opposed to only a prediction of correctness for each item. This is a clear advantage that IRT-inspired knowledge tracing has over conventional non-interpretable deep learning methods.

References

1. Converse, G., Curi, M., Oliveira, S.: Autoencoders for educational assessment. In: International Conference on Artificial Intelligence in Education (AIED) (2019)
2. Corbett, A., Anderson, J.: Knowledge tracing: modeling the acquisition of procedural knowledge. User Model. User-Adap. Inter. **4**, 253–278 (1995). https://doi.org/10.1007/BF01099821
3. Curi, M., Converse, G., Hajewski, J., Oliveira, S.: Interpretable variational autoencoders for cognitive models. In: 2019 International Joint Conference on Neural Networks (IJCNN) (2019)
4. Guo, Q., Cutumisu, M., Cui, Y.: A neural network approach to estimate student skill mastery in cognitive diagnostic assessments. In: 10th International Conference on Educational Data Mining (2017)
5. Lord, F., Novick, M.R.: Statistical theories of mental test scores. IAP (1968)
6. Pandey, S., Karypis, G.: A self-attentive model for knowledge tracing. In: International Conference on Educational Data Mining (2019)
7. Piech, C., et al.: Deep knowledge tracing. In: Advances in Neural Information Processing Systems, pp. 505–513 (2015)
8. Rasch, G.: Probabilistic models for some intelligence and attainment tests. Danish Institute for Educational Research (1960)
9. Reckase, M.D.: Multidimensional item response theory models. In: Reckase, M.D. (ed.) Multidimensional Item Response Theory, pp. 79–112. Springer, Heidelberg (2009). https://doi.org/10.1007/978-0-387-89976-3_4
10. da Silva, M., Liu, R., Huggins-Manley, A., Bazan, J.: Incorporating the q-matrix into multidimensional item response theory models. Educ. Psychol. Measur. **79**, 665–687 (2018)
11. Yeung, C.: Deep-IRT: make deep learning based knowledge tracing explainable using item response theory. ArXiv abs/1904.11738 (2019)
12. Zhang, J., Shi, X., King, I., Yeung, D.Y.: Dynamic key-value memory networks for knowledge tracing. In: 26th International World Wide Web Conference, WWW 2017, pp. 765–774 (2017). https://doi.org/10.1145/3038912.3052580

Automated Model of Comprehension V2.0

Dragos-Georgian Corlatescu[1], Mihai Dascalu[1,2(✉)], and Danielle S. McNamara[3]

[1] University Politehnica of Bucharest, 313 Splaiul Independentei, 060042 Bucharest, Romania
{dragos.corlatescu,mihai.dascalu}@upb.ro
[2] Academy of Romanian Scientists, Street Ilfov, Nr. 3, 050044 Bucharest, Romania
[3] Department of Psychology, Arizona State University, PO Box 871104,
Tempe, AZ 85287, USA
dsmcnama@asu.edu

Abstract. Reading comprehension is key to knowledge acquisition and to reinforcing memory for previous information. While reading, a mental representation is constructed in the reader's mind. The mental model comprises the words in the text, the relations between the words, and inferences linking to concepts in prior knowledge. The automated model of comprehension (AMoC) simulates the construction of readers' mental representations of text by building syntactic and semantic relations between words, coupled with inferences of related concepts that rely on various automated semantic models. This paper introduces the second version of AMoC that builds upon the initial model with a revised processing pipeline in Python leveraging state-of-the-art NLP models, additional heuristics for improved representations, as well as a new radiant graph visualization of the comprehension model.

Keywords: Comprehension model · Natural language processing · Semantic links · Lexical dependencies

1 Introduction

Comprehension is fundamental to learning. While there is much more to learning (e.g., discussion, project building, problem solving), understanding text and discourse represents a key starting point when attempting to learn or relearn information. How well a reader understands text or discourse depends on many factors, including individual differences such as reading skill, prior knowledge of the domain or world, motivation, and goals. Comprehension also depends on the nature of the text – the difficulties imposed by the words in the text, the complexity of the syntax, and the flow of the ideas, or cohesion.

Cohesion between ideas can emerge from overlap between explicit words (e.g., nouns, verbs), implied words (anaphor), semantically related words, semantically related ideas, and the underlying parts of speech (i.e., parts of speech, syntactic overlap). When there is greater overlap, text is easier to understand. Cohesion gaps, by contrast, require inferences to make connections between the ideas. If the reader has little knowledge of the domain or the world, low cohesion text impedes comprehension [1]. For example,

© Springer Nature Switzerland AG 2021
I. Roll et al. (Eds.): AIED 2021, LNAI 12749, pp. 119–123, 2021.
https://doi.org/10.1007/978-3-030-78270-2_21

if the text is too complex, readers may struggle to understand it or event abandon the process; and on the other side, if too simple, readers may quickly lose focus or interest. Thus, designing reading materials suited for learners is an important aspect for educators as well as writers when targeting a specific audience.

The automated model of comprehension (AMoC) simulates the mental representation constructed by hypothetical readers, by building syntactic and semantic relations between words, coupled with inferences of related concepts that rely on various semantic models. AMoC offers the user the ability to model various aspects of the reader by modifying various parameters related to readers' knowledge, reading skill, and motivation (i.e., activation threshold, maximum active number of concepts per sentence, maximum number of semantically related concepts, and the type of knowledge model). This paper introduces an updated version of the automated model of comprehension (AMoC version 2.0), which is freely available online at http://readerbench.com/demo/amoc.

AMoC builds on the Construction Integration (CI) model [2], which introduced a semi-automated cyclical process to simulate reading, as well as the Landscape Model [3], which inherited the ideas from the CI model and provided a visual representation of the activation scores belonging to the concepts in the text. We describe a revised version of AMoC that provides several enhancements: a) an improved processing pipeline rewritten in Python, b) additional heuristics introduced to better model human constructs, and c) a radiant graph visualization to highlight the model's capabilities.

2   Method

The codebase for AMoC version 2 is developed in Python, rather than Java. This decision was influenced by the progress and the interest of the artificial intelligence community into libraries written in this programming language such as Tensorflow [4] and Pytorch [5], that are frequently used in general neural network projects, and SpaCy [6], which is an open source tool for NLP tasks. Additionally, the *ReaderBench* framework [7], previously implemented in Java and used in first version of AMoC, migrated to Python, offering improved functionalities based on state-of-the-art NLP models.

AMoC uses three customizable parameters: *minimum activation score* (the activation score required by a word to be active in the mental model), *maximum active concepts* (the maximum number of words that can be retrieved in the mental model) and *maximum dictionary expansion* (the number of words that can be inferred each sentence). Those three parameters and the target text are processed by the model. The processing begins by automatically splitting the text into sentences using ReaderBench. For the current study, *ReaderBench* Python uses SpaCy to split and store the relations between words. Next, the syntactic graph for each sentence is computed and stored in the model's memory. The difference between the AMoC v1 and v2 is that the coreferences are obtained and replaced using NeuralCoref [8] in the later version, while in the older version a Stanford Core NLP [9] module was applied; Wolf [10] argues that NeuralCoref obtains better overall performance. Additionally, the syntactic parser from SpaCy performs slightly better than Stanford CoreNLP [11, 12]: SpaCy UAS 92.07, LAS 90.27 versus Stanford CoreNLP UAS 92.0, LAS 89.7. The process includes:

1. Each sentence is processed iteratively and each content word (noun or verb) in the sentence is added to the graph if it was not present before, or its activation score is incremented by 1 if the reader has previously encountered the concept in the text.
2. When processing a sentence, the top 5 similar concepts are inferred using WordNet (to extract synonyms and hypernyms) and word2vec [13]. The word2vec models trained on TASA [14], COCA [15], or Google News [13] are considered to reflect different levels of reading proficiency in terms of exposure to language
3. The inferred words from a sentence are filtered based on two criteria: they must have a semantic similarity with the sentence of at least .30 (a value argued by Ratner [16]) and they must have a Kuperman Age of Aquisition [17] score < 9 (i.e., the word is accessible to an average reader).
4. Finally, all of the semantic links in the graph are removed, and the semantic nodes are sorted based on the similarity with the current sentence. Then, only the top *maximum dictionary expansion* concepts are added to the graph.

The key differences between the two versions of AMoC are in the second and third steps. The first version of AMoC used only the synonyms extracted from WordNet; the current version also uses the hypernyms and words extracted from a word2vec language model. Also, the filtering process in the older AMoC version did not include the Age of Acquisition score to represent the potential difficulty of the words.

After the semantic nodes are added to the graph, a modified PageRank algorithm [18] is run in order to spread activation between concepts and then a normalization step is applied. Lastly, after all these operations, nodes become or remain active if they have a score above the *minimum activation score*; otherwise, they are deactivated.

3 Results

A demo of AMoC v2 is available on the ReaderBench website with varying parameters. Since the release of the first model the UI was updated with a highly customizable radiant graph. Figure 1 illustrates the last sentence from the "Knight" story used to showcase the Landscape Model – http://www.brainandeducationlab.nl/downloads; the caption uses TASA as semantic model, a minimum activation threshold of .30, 20 maximum active concepts per sentence, and 2 maximum semantically related concepts introduced for each word in the original text. The inner circle depicts in blue text-based information that is still active, while the outer circle contains semantically inferred concepts in red and grayed out inactive concepts. When hovering over a node, the corresponding edges are colored, and the related concepts are marked in bold. While considering text-based information, "princess" is related to "dragon", "armor", "marry", and "knight", whereas from a semantic perspective, "princess" is linked to "damsel", "prince", and "sword"; all concepts and underlying links are adequate concepts for the selected story.

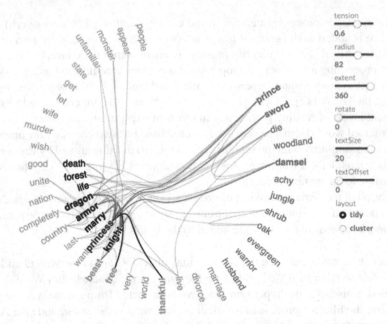

Fig. 1. AMoCv2 Radiant graph visualization of the last sentence from the Knight story.

4   Conclusions and Future Work

AMoC provides a fully automated means to model comprehension by leveraging current techniques in the Natural Language Processing field. The second version of AMoC described in this research provides an improved method and optimizations at the code base level in comparison to its predecessor, combined with a more rapid execution time. Additionally, a new and highly customizable method for concept graph visualization was added to the ReaderBench website.

In future research, we will further test the predictiveness of AMoC by applying it to previous studies that examined text comprehension. From a more technical perspective, we intend to evaluate potential advantages of using BERT contextualized embeddings [19], rather than word2vec. Our overarching objective is to comprehensively account for word senses and their contexts within sentences, paragraphs, texts, and language.

Acknowledgments. This research was supported by a grant of the Romanian National Authority for Scientific Research and Innovation, CNCS – UEFISCDI, project number TE 70 *PN-III-P1-1.1-TE-2019-2209, ATES* – "Automated Text Evaluation and Simplification", the Institute of Education Sciences (R305A180144 and R305A180261), and the Office of Naval Research (N00014-17-1-2300; N00014-20-1-2623). The opinions expressed are those of the authors and do not represent views of the IES or ONR.

References

1. McNamara, D.S., Kintsch, E., Songer, N.B., Kintsch, W.: Are good texts always better? Interactions of text coherence, background knowledge, and levels of understanding in learning from text. Cogn. Instr. **14**(1), 1–43 (1996)
2. Kintsch, W., Welsch, D.M.: The construction-integration model: a framework for studying memory for text. In: Relating Theory and Data: Essays on Human Memory in Honor of Bennet B. Murdock., pp. 367–385. Lawrence Erlbaum Associates, Inc., Hillsdale (1991)
3. Van den Broek, P., Young, M., Tzeng, Y., Linderholm, T.: The Landscape Model of Reading: Inferences and the Online Construction of a Memory Representation. The Construction of Mental Representations during Reading, pp. 71–98 (1999)
4. Abadi, M., et al.: Tensorflow: a system for large-scale machine learning. In: 12th {USENIX} Symposium on Operating Systems Design and Implementation ({OSDI} 16), pp. 265–283. {USENIX} Association, Savannah, GA, USA (2016)
5. Paszke, A., et al.: Pytorch: an imperative style, high-performance deep learning library. In: Advances in Neural Information Processing Systems, pp. 8026–8037, Vancouver, BC, Canada (2019)
6. Honnibal, M., Montani, I.: Spacy 2: natural language understanding with bloom embeddings. In: Convolutional Neural Networks and Incremental Parsing, vol. 7, no. 1 (2017)
7. Dascalu, M., Dessus, P., Trausan-Matu, Ş, Bianco, M., Nardy, A.: ReaderBench, an environment for analyzing text complexity and reading strategies. In: Lane, H.C., Yacef, K., Mostow, J., Pavlik, P. (eds.) AIED 2013. LNCS (LNAI), vol. 7926, pp. 379–388. Springer, Heidelberg (2013). https://doi.org/10.1007/978-3-642-39112-5_39
8. huggingface: Neuralcoref, Accessed 30 Dec 2020. https://github.com/huggingface/neural coref (2020)
9. Manning, C.D., Surdeanu, M., Bauer, J., Finkel, J.R., Bethard, S., McClosky, D.: The Stanford CoreNLP natural language processing toolkit. In: Proceedings of 52nd Annual Meeting of the Association for computational Linguistics: System Demonstrations, pp. 55–60. The Association for Computer Linguistics, Baltimore, MD, USA (2014)
10. Wolf, T.: State-of-the-art neural coreference resolution for chatbots, Accessed 30 Dec 2020. https://medium.com/huggingface/state-of-the-art-neural-coreference-resolution-for-chatbots-3302365dcf30 (2017)
11. Explosion: SpaCy, Accessed 9 Feb 2021. https://spacy.io/usage/facts-figures#benchmarks (2016–2021).
12. The Stanford Natural Language Processing Group: Neural Network Dependency Parser, Accessed 9 Feb 2021. https://nlp.stanford.edu/software/nndep.html
13. Google: word2vec, Accessed 30 Nov 2020. https://code.google.com/archive/p/word2vec/ (2013)
14. Deerwester, S., Dumais, S.T., Furnas, G.W., Landauer, T.K., Harshman, R.: Indexing by latent semantic analysis. J. Am. Soc. Inf. Sci. **41**(6), 391–407 (1990)
15. Davies, M.: The corpus of contemporary American English as the first reliable monitor corpus of English. Literary Linguist. Comput. **25**(4), 447–464 (2010)
16. Ratner, B.: The correlation coefficient: its values range between+1/−1, or do they? J. Target. Meas. Anal. Mark. **17**(2), 139–142 (2009)
17. Kuperman, V., Stadthagen-Gonzalez, H., Brysbaert, M.: Age-of-acquisition ratings for 30,000 English words. Behav. Res. Methods **44**(4), 978–990 (2012)
18. Brin, S., Page, L.: The anatomy of a large-scale hypertextual web search engine. Comput. Netw. **30**(1–7), 107–117 (1998)
19. Devlin, J., Chang, M.-W., Lee, K., Toutanova, K.: Bert: pre-training of deep bidirectional transformers for language understanding. arXiv preprint arXiv:1810.04805 (2018)

Pre-course Prediction of At-Risk Calculus Students

James Cunningham[1]([✉]), Raktim Mukhopadhyay[1], Rishabh Ranjit Kumar Jain[1],
Jeffrey Matayoshi[2], Eric Cosyn[2], and Hasan Uzun[2]

[1] Arizona State University, Tempe, AZ, USA
jim.cunningham@asu.edu
[2] McGraw Hill ALEKS, Irvine, CA, USA

Abstract. Identifying students who are at-risk of failing a mathematics course at the earliest possible moment allows for support and scaffolding to be applied when it can have greatest impact. However, because risk of non-success can arise from a complex interaction of factors, early detection of struggling students is difficult. Machine learning is particularly suited to modeling this challenging interplay of variables. In this study, we measure how well machine learning models can identify at-risk students before an entry-level university calculus course begins. Five classification algorithms were applied to data combined from the student information system, an adaptive placement test, and a student survey. We were able to produce predictions before class start that were competitive with other studies using course activity data after coursework began. In addition, important features of the model provided insights into possible causes of student non-success.

Keywords: Predictive modeling · Machine learning · Placement tests · Calculus

1 Introduction

Using machine learning (ML) for early alert systems of students at-risk in higher education has been an important application of learning analytics [1–5]. While there are many levels of early alert systems, our focus has been on course-level predictions especially in mathematics. The importance of completing calculus in the student's first attempt is critical in university majors related to science, technology, engineering, and math (STEM) [6–9]. However, calculus represents a substantial barrier to completing these majors especially for female students and students from underrepresented populations [10, 11]. Early detection of students who may be struggling in these calculus courses is critical for intervention, scaffolding and support.

Often by the time gradebook data, used by many instructors for assessing risk, has made it clear that a student is failing calculus, over half the course is complete, making changing the outcome of that student difficult. While machine learning models are a viable alternative to instructor gradebooks for assessing risk of failure, most studies featuring predictive models depend heavily on course activity data from a learning management system (LMS) or a mathematics learning platform to classify students, making

© Springer Nature Switzerland AG 2021
I. Roll et al. (Eds.): AIED 2021, LNAI 12749, pp. 124–128, 2021.
https://doi.org/10.1007/978-3-030-78270-2_22

early, accurate predictions difficult when interventions are critical but course activity data is sparse [12, 13]. For this reason, there is a lack of studies using machine learning predictive models before higher education courses begin [2–4].

This research contributes to this field of study by using placement assessment data as an alternative to course activity data for very early predictions in a higher education mathematics course. While a placement exam is a predictive model in itself, we hypothesized that combining data from the student information system (SIS) with placement data could yield classifications related to risk of failure with similar predictive power to classifications made in other studies with course activity data. Because these predictions would be available before the course begins, early intervention and allocation of limited resources for support and scaffolding could be strategically targeted when they could have the greatest impact.

One other source of pre-course data that was available to us for this study was a survey regarding math background that students filled out before the placement exam. Although this data was self-report, we were interested to see if this could also make a significant contribution to our model.

To this end, we conducted this study with these three research questions in mind:

- **RQ1:** How would a machine learning predictive model, limited to data only available before the course starts, compare to predictive models using course activity data in accurately identifying at-risk calculus students?
- **RQ2:** Which features of the model would be most important in predicting student risk?
- **RQ3:** How much lift would be contributed to the predictions of the model from data derived from the SIS, placement test, and the self-report student survey?

2 Method

The data used in this study came from historical data of 6,380 undergraduate students enrolled in the course, Calculus for Engineers I. The label used for the supervised learning classification was "At-Risk" for students who achieved a final letter grade of 'C' or below, and "Not At-Risk" for students who achieved a final letter grade 'C+' or above with the reasoning that students who barely passed the course with a 'C' grade might be underprepared and have more in common with at-risk students than not at-risk students. Of our total population, 2,739 (43%) were labeled "At-Risk," and 3,641 (57%) were labeled "Not At-Risk." So, a baseline model for this data based on the majority class should be considered 57%.

The genders of the students in this study were 23% female and 77% male, and the age breakdown was 83% at or below 22 years old, 11% 23–30 years old, and 6% over the age of 30. Two proxies were used for socio-economic status. Students who were first generation college students made up 28% of our sample, and Pell eligible students made up 33%. The ethnicity breakdown was Asian 15%, Black 4%, Hawaii/Pac <1%, Hispanic 21%, Native Am. 1%, No Report 3%, Two or more 5% and White 50%.

This data was merged with other academic and grade information from the SIS that would have been available before students began the calculus course, data from the placement test and the student self-report survey, and some engineered features from the SIS data that we thought might be predictive of risk.

The placement data used was from the ALEKS Placement, Preparation, and Learning (PPL), a specialized adaptive placement test developed to offer recommendations for placing students in post-secondary mathematics courses [14]. All this data was merged with three features from the self-report survey that was attached to the placement test. These three features were "last math level," "last math class," and "last math grade."

Five ML methods were used for classification comparison: Logistic Regression, Support Vector Machines (SVM), K-Nearest Neighbors, Random Forest, and CatBoost. All of these methods except CatBoost were accessed through the Scikit-learn Python machine learning library [15]. CatBoost, a newer method, seeks to mitigate prediction shifts that are present in most implementations of gradient boosting by means of ordered target statistics associated with categorical variables [16, 17]. The dataset was split into subsets, 80% for training and 20% for testing. Ten-fold cross validation was used to limit overfitting in our training set and increase generalization. The GridSearch CV library from Scikit-learn, which exhaustively considers all parameter combinations, was used for hyperparameter tuning.

A post-hoc algorithm was employed to extract feature importance from the black-box models and measure the lift of the different datasets. There are several new methods for model explainability; however, we chose Permutation Feature Importance (PMI) because it does not suffer from bias toward categorical variables as do some other methods [18–20].

3   Results

A comparison of the performance of the differing ML methods is presented in Fig. 1.

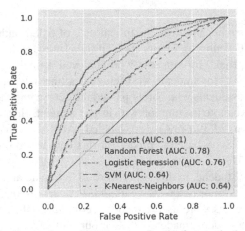

Fig. 1. Comparison of ROC curves for the five methods tested.

The receiver operating characteristic (ROC) is a performance measure of models at various threshold settings and is used to summarize the performance of models over a wide range of conditions. Of the five methods tested, CatBoost outperformed the other four machine learning methods with an area under the curve (AUC) of 0.81. The overall accuracy of our best model was 0.74, with a recall of 0.73, and an F_1 score of 0.71.

The Permutation Feature Importance algorithm scored each of the 46 features in terms of the contribution to the predictive power of our best model. The top five features were previous term GPA, last math class, part-time, placement test, and faculty difficulty with PFI scores of 0.049, 0.038, 0.035, 0.029 and 0.024 respectively.

By grouping the features from each dataset and using the PFI algorithm, we were able to score the impact of each dataset on the predictive power of the model. The scores for the three datasets were: SIS (0.148), ALEKS PPL (0.051), and survey (0.047).

4 Discussion and Conclusions

RQ1: Our first research question was aimed at measuring how well the models could predict who was at-risk of failing calculus before the course started without any course activity data. The ROC curves in Fig. 1 demonstrate that our predictions are comparable to other models using activity data in the first few weeks of a course [12, 13]. **RQ2:** Using a post-hoc, model explainability algorithm, we were able to determine five features that were most important in early prediction of Calculus for Engineers I: previous term GPA, the last math class taken, official part-time status of the student, placement test data, and how hard an instructor typically grades their students. All three datasets used in this study had features represented in the top five. **RQ3:** Of the three datasets, features derived from the student information system were most predictive, with the placement test being second, and survey data coming in third.

Because accurate early detection is possible, scarce resources, scaffolding and support can be targeted to students who need it the most when the impact of those interventions can help students at-risk get off to a strong start. Moreover, because this kind of data is typically available for all entry level math courses at the university, it is possible to construct similar models for other critical math courses as well. Future work will focus on developing these models for other courses and combining these models with course activity data after classes start for even more accurate student modeling and weekly predictions that can guide interventions throughout the course for increased student success.

References

1. Arnold, K., Pistilli, M.: Course signals at purdue: using learning analytics to increase student success. In: Proceedings International Learning Analytics & Knowledge Conference (LAK'12). ACM (2012)
2. Hellas, A., et al.: Predicting academic performance: a systematic literature review. In: Proceedings Companion of the 23rd Annual ACM Conference on Innovation and Technology in Computer Science Education (ITiCSE'18 Companion), pp. 175–199 (2018)
3. Namoun, A., Alshanqiti, A.: Predicting student performance using data mining and learning analytics techniques: a systematic literature review. Appl. Sci. **11**, 1–28 (2021)
4. Rastrollo-Guerrero, J., Gómez-Pulido, J., Durán-Domínguez, A.: Analyzing and predicting students' performance by means of machine learning: a review. Appl. Sci. **10**, 1–16 (2020)

5. Romero, C., Ventura, S.: Educational data mining and learning analytics: an updated survey. WIREs Data Min. Knowl. Disc. **10**, 1–21 (2019)
6. Ayerdi, J.: Relative rates of success of students in Calculus I. Honors Theses. 2904. https://scholarworks.wmich.edu/honors_theses/2904
7. Bigotte de Almeida, M., Queiruga-Dios, A., Cáceres, M.: Differential and integral calculus in first-year engineering students: a diagnosis to understanding the failure. Mathematics **9**(61), 1–18 (2021)
8. Dibbs, R.: Forged in failure: engagement patterns for successful students repeating calculus
9. Garaschuk, K.: Predicting failure in first-term calculus courses. UBC Faculty Research and Publications. https://open.library.ubc.ca/cIRcle/collections/facultyresearchandpublications/52383/items/1.0357414. Accessed 06 Feb 2021
10. Nortvedt, G., Siqveland, A.: Int. J. Math. Educ. Sci. Technol. **50**(3), 325–343 (2019)
11. Sanabria, T., Penner, A.: Weeded out? Gendered responses to failing calculus. Soc. Sci. **6**(47), 1–14 (2017)
12. MacFadyen, L., Dawson, S.: Mining LMS data to develop an "early warning system" for educators: a proof of concept. Comput. Educ. **54**, 588–599 (2010)
13. Hlosta, M., Zdrahal, Z., Zendulka, J., Ouroboros: early identification of at-risk students without models based on legacy data. In: Proceedings International Learning Analytics & Knowledge Conference (LAK'17). ACM (2017)
14. McGraw Hill ALEKS PPL: Pave the path to graduation with placement, preparation, and learning. Accessed 11 Feb 2021
15. Pedregosa, et al.: Scikit-learn: machine learning in Python. JMLR **12**, 2825–2830 (2011)
16. CatBoost is a high-performance open-source library for gradient boosting on decision trees. catboost.ai. Accessed 11 Feb 2021
17. Prokhorenkova et al.: CatBoost: unbiased boosting with categorical features. arXiv preprint arXiv:1706.09516 (2017)
18. Permutation feature importance. https://scikit-learn.org/stable/modules/permutation_importance.html. Accessed 11 Feb 2021
19. Strobl, C., Boulesteix, A., Kneib, T., Augustin, T., Zeileis, A.: Conditional variable importance for random forests. BMC Bioinform. **9**(1), 1–11 (2008)
20. Strobl, C., Boulesteix, A., Zeileis, A., Hothorn, T.: Bias in random forest variable importance measures: illustrations, sources, and a solution. BMC Bioinform. **8**(1), 1–21 (2007)

Examining Learners' Reflections over Time During Game-Based Learning

Daryn A. Dever[✉], Elizabeth B. Cloude, and Roger Azevedo

University of Central Florida, Orlando, USA
{daryn.dever,elizabeth.cloude,roger.azevedo}@ucf.edu

Abstract. Reflections are critical components of game-based learning environments (GBLEs) as learners must accurately use and monitor self-regulatory pro- while learning with instructional materials. Within this study, we examined how middle-school students (N = 35) learned with Crystal Island, a microbiology-based GBLE where learners are required to diagnose a disease infecting researchers on an island. This study aimed to identify how learners' time reflecting changed during gameplay and is related to learners' scientific reasoning actions (e.g., information gathering, note-taking, hypothesis formation and testing) and whether this was related to learning gains. Results from a multilevel growth model indicated that time spent reflecting increased over time, but the specific timing of reflection prompts (e.g., after submitting a diagnosis) was related to the time learners reflected over time. Further, time engaging in scientific actions and learning gains moderated the relationship between time spent reflecting between different reflection prompts but does not have a main effect on time spent reflecting. This paper discusses implications for when and how reflection prompts should be triggered during game-based learning and designing GBLEs capable of intelligently and dynamically modeling, scaffolding, and fostering reflective thinking.

Keywords: Reflection · Game-based learning · Log-file data

1 Introduction

Game-based learning environments (GBLEs) promote engagement and learning by simulating real-world scenarios with clear learning goals and instructional materials to increase domain knowledge [1–5]. To effectively learn within GBLEs, learners must engage in reflection as learners need to be aware of their knowledge, previous experiences, and future actions needed to achieve their goals (e.g., learning about microbiology and successfully solving a problem) [6,7].

Our study used McAlpine et al.'s [6] model of reflection describing the interaction between reflection and size other components (i.e.,g goals, knowledge, action, monitoring, decision making, corridor of tolerance) to temporally examine learners' time spent reflecting while problem-solving within Crystal Island, a GBLE used to promote microbiology knowledge by tasking learners with diagnosing an illness infecting learners on a virtual island. This study also examined

© Springer Nature Switzerland AG 2021
I. Roll et al. (Eds.): AIED 2021, LNAI 12749, pp. 129–133, 2021.
https://doi.org/10.1007/978-3-030-78270-2_23

whether the temporal dynamics of reflections was related to learners' time spent engaging in scientific reasoning actions (e.g., information gathering, note-taking, hypothesis testing) and if its stability or change was related to learning gains.

2   Methods

2.1   Participants and Exprimental Procedure

Data from 35 middle-school students ages ranging from 12 to 15 ($M_{age} = 13.5$, $SD_{age} = 0.70$; 57% female) were included for this study. Participants were tasked with reflecting on their progress and goals while interacting with Crystal Island, a microbiology-centered GBLE set on a remote island, over a period of three days. Participants completed a microbiology pre- and post-test and log files were collected (e.g., duration of activities) as they learned with Crystal Island. Participants identified an unknown disease infecting residents on the virtual island by engaging in actions (e.g., reading books, talking to non-player characters). Throughout gameplay, the system prompted participants to reflect based on a set of production rules that were associated with the actions a participant could take. Prompts were (1) time-based (e.g., 30 and 60 min); (2) movement-based (e.g., activating plot points throughout the environment); and (3) action-based (e.g., talking to the camp nurse, scanning multiple food items, testing a contaminated food item). Additionally, reflections were prompted when participants did or did not solve the mystery, or self-prompted. Within this study, we refer to these reflections as context-timed reflection prompts.

2.2   Coding and Scoring

Durations were defined as the time participants spent engaging in scientific actions and reflecting. Log files captured the amount of time participants read microbiology content (i.e., gather information), took notes via the worksheet, and scanned food items for pathogens (i.e., develop and test hypotheses). *Normalized change scores* calculated learning gains using pre- and post-test microbiology quiz scores while controlling for prior knowledge. To control for the total duration of participants' gameplay, *relative game time* was calculated by dividing the time at which a reflection began by the participants' total duration in game. This scales the reflection instances from 0 to 1 so all times can be uniformly compared.

2.3   Model Building and Estimation

Relative game time values were manipulated to represent the intercept as the first time a learner interacted with their first reflection. The dataset consisted of 218 observations nested within 35 individuals. Observation-level variables included a latent time variable (i.e., relative game time), the reflection prompt, and the total goal status responses that learners initiated during the reflection. Individual-level variables included learners' normalized change scores as well as their time completing scientific actions (i.e., gathering information by reading microbiology texts, constructing and testing hypotheses via scanning food items for diseases, and taking notes using the diagnosis worksheet).

A two-level multilevel linear growth model analyzed our data where observations (i.e., level one, N = 218) were nested within individual learners (i.e., level two, N = 35). Each learner had an average of 6.23 observations (Range: 4–9). Five models were calculated using maximum likelihood estimation within R: (1) an unconditional means model; (2) an unconditional growth model; (3) all observation-level variables and their interactions; (4) predictors from (3) and individual-level variables; and (5) variables from (4) and cross-level interactions.

3   Results

3.1   Research Question 1: To What Extent Does Game Time, Context-Timed Reflection Prompts, and Goal Status Responses Relate to Reflection Durations?

The average duration at participants' initial reflection was 34.78 s (SE = 6.37) and increased by 109.32 s (SE = 14.98) for every unit increase in game time. There was a main effect for reflection where, with the exception of prompts following multiple instances of scanning food items and testing a contaminated object, all prompted reflection durations were significantly smaller than self-initiated reflections ($ps < .01$). Durations on reflections prompted following multiple instances of scanning food items did not significantly differ from instances where learners initiated the reflection ($p > .05$). Durations on reflections after testing a contaminated object was significantly greater than self-initiated reflections by 2570.52 s (SE = 430.75; $t(181.57) = 5.97$, $p < .01$). When holding all other variables constant, goal status responses were positively related to reflection durations ($t(194.73) = 1.97$, $p = .05$). Reflection durations increased by 47.24 s (SE = 24.02) for every additional goal status response across all reflections.

Reflection durations prompted after thirty minutes in game ($t(203) = 3.95$, $p < .01$), activating plot points ($t(215.91) = 2.48$, $p < .05$), and talking to the camp nurse ($t(212.15) = 2.99$, $p < .01$), significantly increased over time. Reflection durations after testing a contaminated object significantly decreased over time compared to self-initiated reflections by 2604.58 s (SE = 451.5; $t(183.93) = -5.77$, $p < .01$). There were no significant relationships between relative game time and the number of goal status responses as well as the timing of the reflection and the goal status responses ($ps > .05$).

3.2   Research Question 2: Does Time Spent Reflecting Relate to Learning Gains and Scientific Actions?

There was no significant relationship between participants' normalized change scores, time reading in game, total time scanning food items, and time working on the diagnosis worksheet and reflection durations ($ps > .05$).

3.3 Research Question 3: To What Extent Does Learning Gains and Scientific Actions Moderate Relationships Between Reflection Duration and Game Time, Context-Timed Prompts, and Goal Status Responses?

While there was no moderating effect on the total time learners took notes ($ps > .05$), there are moderating effects of learners' total time gathering information as well as constructing and testing hypotheses. The total time learners read about microbiology weakened the relationship between durations on reflections prompted after 60 min ($t(138.02) = -3.53$, $p < .01$) and the activation of plot points ($t(179.53) = -2.67$, $p < .05$). However, learners' total time gathering information strengthens the relationship between durations on reflections prompted after solving the mystery ($t(62.22) = 2.04$, $p = .05$). The total time learners construct and test hypotheses strengthened the relationship between durations on reflections prompted after not solving a mystery ($t(52.24) = 3.38$, $p < .01$).

There was no significant relationship between normalized change scores and reflection durations ($ps > .05$). Yet, there was a significant moderating effect of normalized change scores on prompt timings and durations over time. Greater durations on reflections prompted at the 60-min mark of learners' gameplay are associated with greater normalized change scores where an increase of durations on these types of reflections resulted in greater normalized change scores ($t(180.89) = 2.39$, $p < .05$). There was a significant moderating effect of normalized change scores on durations where the reflection was prompted by activating plot points. Specifically, as game time increased, durations on reflections prompted by activating plot points increased and were associated with an increase on normalized change scores ($t(207.22) = 2.48$, $p = .01$).

4   Discussion

Results showed many promising avenues for further exploration of modifying McAlpine et al.'s [6] model to accommodate the complex reflective thinking necessary to learn within GBLEs. This study emphasized the need to consider the timings of context-timed reflection prompts in relation to learners' update on progress during the game as well as their interactions with scientific actions and overall learning gains. Specifically, time-, movement-, and success-based reflections are critical for learners to engage in scientific actions as well as increase their knowledge on a complex topic. Such considerations are important to integrate into GBLEs that incorporate prompted reflective thinking as well as future iterations of Crystal Island. This study also reveals the need for adaptive reflection prompts over time within GBLEs as learning gains moderated the relationship between durations and the types of reflection prompts over time. As such, future GBLEs should intelligently and dynamically model, scaffold, and foster reflections during complex learning over time within GBLEs.

References

1. Chen, Z.-H., Chen, H., Dai, W.-J.: Using narrative-based contextual games to enhance language learning: a case study. J. Educ. Technol. Soc. **86**, 18–29 (2018)
2. Plass, Y.L., Homer, B.D., Mayer, R.E., Kinzer, C.K.: Theoretical foundations of game-based and playful learning. In: The Handbook of Game-based Learning, pp. 3–24. The MIT Press, Cambridge (2020)
3. Taub, M., Sawyer, R., Smith, A., Rowe, J., Azevedo, R., Lester, J.: The agency effect: the impact of student agency on learning, emotions, and problem-solving behaviors in a game-based learning environment. Comput. Educ. **147**, 103781 (2020)
4. Azevedo, R., Mudrick, N., Taub, M., Bradbury, A.: Self-regulation in computer assisted learning systems. In: The Cambridge Handbook of Cognition and Education, pp. 587–618. Cambridge University Press (2019)
5. Winne, P.H., Azevedo, R.: Metacognition. In: Sawyer, K. (ed.) Cambridge Handbook of the Learning Sciences, 3rd edn., Cambridge University Press, Cambridge (in press)
6. McAlpine, L., Weston, C., Beauchamp, C., Wiseman, C., Beauchamp, J.: Building a metacognitive model of reflection. High. Educ. **37**, 105–131 (1999)
7. Taub, M., Azevedo, R., Bradbury, A., Mudrick, N.: Self-regulation and reflection during game-based learning. In: The Handbook of Game-Based Learning, pp. 239–262. The MIT Press (2020)

Examining the Use of a Teacher Alerting Dashboard During Remote Learning

Rachel Dickler[1]([✉]), Amy Adair[1], Janice Gobert[1,2], Huma Hussain-Abidi[1],
Joe Olsen[1], Mariel O'Brien[1], and Michael Sao Pedro[2]

[1] Rutgers University, New Brunswick, NJ 08901, USA
rachel.dickler@gse.rutgers.edu
[2] Apprendis, Berlin, MA 01503, USA

Abstract. Remote learning in response to the COVID-19 pandemic has introduced many challenges for educators. It is important to consider how AI technologies can be leveraged to support educators and, in turn, help students learn in remote settings. In this paper, we present the results of a mixed-methods study that examined how teachers used a dashboard with real-time alerts during remote learning. Specifically, three high school teachers held remote synchronous classes and received alerts in the dashboard about students' difficulties on scientific inquiry practices while students conducted virtual lab investigations in an intelligent tutoring system. Quantitative analyses revealed that students significantly improved across a majority of inquiry practices during remote use of the technologies. Additionally, through qualitative analyses of the transcribed audio data, we identified five trends related to dashboard use in a remote setting, including three reflecting effective implementations of dashboard features and two reflecting the limitations of dashboard use. Implications regarding the design of dashboards for use across varying contexts are discussed.

Keywords: Dashboard · Intelligent tutoring system · Remote learning

1 Introduction

The COVID-19 pandemic has impacted educators and students around the world, resulting in a shift in instructional contexts and methods [19]. As such, teachers require technologies that can help them to overcome common challenges with teaching remotely (e.g., assessing and monitoring student learning [2, 6, 8, 16]), particularly in STEM contexts. Fortunately, several innovative technologies exist for teacher monitoring in STEM [3], such as learning analytics dashboards [20] that provide educators with data on student progress based on an open learning model (OLM; [4, 5]). Several dashboards align with STEM learning environments (e.g., Lumilo [11, 12], Snappet [13], HOWARD [14], MTFeedback [17]). Researchers, however, have not explored the use of these technologies in remote synchronous contexts and few dashboards provide real-time alerts to teachers on students' difficulties on complex STEM practices.

In our recent work, we developed Inq-Blotter, a teacher dashboard that provides real-time alerts about students' difficulties on inquiry practices exhibited in a virtual science

© Springer Nature Switzerland AG 2021
I. Roll et al. (Eds.): AIED 2021, LNAI 12749, pp. 134–138, 2021.
https://doi.org/10.1007/978-3-030-78270-2_24

lab in the Inquiry Intelligent Tutoring System, Inq-ITS [9, 10]. Recent studies [1, 15] have shown the technologies to be effective in supporting student learning of inquiry practices, but researchers have yet to investigate their use in a remote synchronous setting. In the present paper, we conducted a mixed-methods study to answer the following research questions (RQs): RQ1) Do students improve on inquiry practices when Inq-Blotter is used with Inq-ITS in a remote synchronous setting? and RQ2) What common trends appear in terms of *how* Inq-Blotter was used in a remote synchronous setting?

2   Methods

The participants in the present study included three high school STEM teachers and their students (N = 121 students) from three high schools in the northeastern United States. All teachers used Inq-Blotter synchronously while their students completed an Inq-ITS lab remotely during a class period between December 2020 and January 2021.

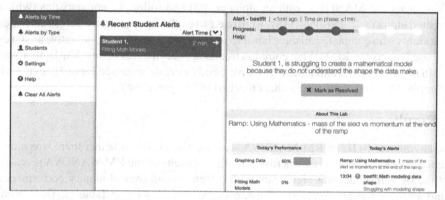

Fig. 1. Screenshot of Inq-Blotter with an alert for the Building Models stage.

In terms of materials, the *Inq-ITS* investigation that students completed in the present study was the Ramp: Using Mathematics virtual lab set (i.e., Ramp Lab). In the lab, students complete three investigations to identify the mathematical relationships between variables related to a sled going down a ramp. Each lab investigation includes six stages that align to inquiry practices including: 1) Hypothesizing (making a hypothesis), 2) Collecting Data (running experimental trials using a simulation), 3) Graphing Data (creating a graph), and 4) Building Models (selecting the type of mathematical relationship in the graph and creating a best-fit line). Students then summarize their findings. *Inq-Blotter* provides real-time alerts to teachers on students' difficulties and progress within Inq-ITS virtual labs (see Fig. 1). The alerts are triggered based on educational data-mined and knowledge-engineered scoring algorithms in Inq-ITS in stages 1–4 (see Measures section for further details). The individual student alerts that appear contain details on the specific difficulty a student is having with a practice, as well as other contextual information (see Fig. 1). There are also "Whole Class" alerts that appear when more than 50% of the class is struggling with a practice and "Slow Progress" alerts when a student has been on a stage for more than 5 min.

For the measures, *log data from Inq-ITS* were used to capture student performance. In particular, students' competencies with the science inquiry practices in stages 1–4 were automatically scored (from 0 to 1) by educational data-mined and knowledge-engineered algorithms as described in prior work [10]. *Log data from Inq-Blotter* were used by researchers to identify the types of alerts viewed by the teacher, the students who were helped by a teacher in response to the dashboard, and the inquiry practices on which they were helped. *Audio-recordings* from each of the remote dashboard implementations were transcribed and timestamped. The transcribed audio data was segmented by speaker turn and only segmented transcripts related to dashboard use were included in the analyses (N = 49 transcript segments). The data from Inq-ITS, Inq-Blotter, and transcript segments were triangulated based on timestamps for analyses.

In terms of the analyses, mixed-methods were used to examine student performance as well as to understand *how* the dashboard was used in the remote synchronous context. To answer RQ1 (Do students improve on inquiry practices when Inq-Blotter is used with Inq-ITS in a remote synchronous setting?), a Repeated Measures Multivariate Analysis of Variance (RM MANOVA; with an alpha = .05) and follow-up comparisons (with a corrected alpha = .0125 (.05/4; [18]) were used to explore performance across activities for students who completed all three lab activities ($N = 86$ students). Qualitative analyses were used to answer RQ2 (What common trends appear in terms of *how* Inq-Blotter was used in a remote synchronous setting?). Five trends were defined (see Table 2), reviewed, and applied to transcripts (researchers reached 90% agreement).

3 Results

First, to answer RQ1, an RM MANOVA was used to explore whether there was a difference in student performance across activities. Results of the RM MANOVA revealed that the overall model was significant with differences in overall inquiry performance found across activities, $F(8, 78) = 7.68, p < .001, n^2 = .44$ (see Table 1). There were also significant within-subjects main effects found for each of the inquiry practices with students improving from the first to third activity for all practices except Applying Equations (which is a particularly difficult practice [7]; see Table 1).

Table 1. Average inquiry practice scores across activities and results of RM MANOVA.

Practice stage	Lab 1 M (SD)	Lab 2 M (SD)	Lab 3 M (SD)	Within-subjects effects
Hypothesizing	.83 (.31)	.95 (.16)	.97 (.14)	$F(2, 170) = 10.99, p < .001$
Collecting data	.90 (.23)	.95 (.16)	.97 (.15)	$F(2, 170) = 9.34, p < .001$
Graphing data	.72 (.24)	.80 (.24)	.83 (.22)	$F(2, 170) = 9.68, p < .001$
Building models	.62 (.37)	.88 (.26)	.69 (.26)	$F(2, 170) = 14.56, p < .001$
Overall	**.77 (.19)**	**.90 (.14)**	**.86 (.16)**	$F(8, 78) = 7.68, p < .001$

To answer RQ2, we explored trends that reflected effective use of the design features of the dashboard in the remote synchronous context. The most commonly occurring trend across the transcribed audio segments was that teachers used the dashboard to Identify Student Difficulties followed by using the dashboard to Identify Trends in Class Data and Identify Inactive Students (see Table 2). We also identified two trends related to limitations of remote dashboard use including Communication Limitations and General Technical Challenges (see Table 2), which could be addressed in future design iterations to better support synchronous remote instruction.

Table 2. Trends in dashboard use during remote learning, definitions, and examples.

Category	Trend	Definition	Example (Segment ID)
Effective Use of Dashboard Features	Identifying Student Difficulties ($N = 18$)	Individual support to a student on an inquiry practice	T: I am seeing that you are probably having some trouble graphing? And you only have three data points...You must at the very minimum have 5 so you can actually see how the data... line up...(52)
	Identifying Trends in Class Data ($N = 12$)	Class support based on pattern across multiple students' inquiry performance	T: I see a whole bunch of them having trouble with the modeling because they don't have enough data points to see the fit ... (28)
	Identifying Inactive Students ($N = 7$)	Addressing students working on the wrong lab or not actively completing the lab	T: Flower growth?....Well I think one of my student groups is working on the flower lab instead of this [Ramp] one (18)
Limitations	Communication Limitations ($N = 15$)	Limitation related to modes of communication during remote dashboard use	T: This would be so much easier if I could take a glance over their shoulder. It takes so much extra time to get them to share everything to take a look... (17)
	General Technical Challenges ($N = 11$)	Internet, computer, or meeting programs interfering with dashboard use	T: I don't understand, sometimes [the meeting] breakout room allows me to move them to main session and sometimes they don't... so I cannot help her...(67)

4 Discussion

Remote learning involves a number of challenges for instructors [2, 6, 8, 16]. This study provides initial evidence that these challenges can be addressed by carefully-designed alerting dashboards that enable teachers to monitor and support students during synchronous instruction. Quantitative results showed that students improved across activities for the majority of science practices in Inq-ITS when teachers used Inq-Blotter remotely. Additionally, qualitative analyses further demonstrated *how* Inq-Blotter alerts and features enabled teacher monitoring within a remote synchronous context. Future designs might consider integrating a functionality to directly view student work or communicate through the dashboard to address some of the challenges identified. Additional studies are needed with a greater number of participants to better understand how these findings generalize across contexts . Overall, these initial implementation studies are essential for informing the iterative design of technologies to meet the needs of teachers and students across contexts.

References

1. Adair, A., Dickler, R., Gobert, J.: Intelligent tutoring system supports students maintaining their science inquiry competencies during remote learning due to COVID-19. Am. Educ. Res. Assoc. (AERA): Learn. Instr. (2021)
2. Archambault, L.: Identifying and addressing teaching challenges in K-12 online environments. Distance Learn. 7(2), 13 (2010)
3. Arnett, T.: Breaking the mold: how a global pandemic unlocks innovation in k-12 instruction. Christensen Institute Report (2021)
4. Bull, S.: There are open learner models about! IEEE Trans. Learn. Technol. 13(2), 425–448 (2020)
5. Bull, S., Kay, J.: SMILI: a framework for interfaces to learning data in open learner models, learning analytics and related fields. Int. J. Artif. Intell. Educ. 26(1), 293–331 (2016)
6. Cardullo, V., Wang, C.H., Burton, M., Dong, J.: K-12 teachers' remote teaching self-efficacy during the pandemic. J. Res. Innov. Teach. Learn. 14, 32–45 (2021)
7. De Bock, D., Neyens, D., Van Dooren, W.: Students' ability to connect function properties to different types of elementary functions: an empirical study on the role of external representations. Int. J. Sci. Math. Educ. 15(5), 939–955 (2017)
8. Garbe, A., Ogurlu, U., Logan, N., Cook, P.: Parents' experiences with remote education during COVID-19 school closures. Am. J. Qual. Res. 4(3), 45–65 (2020)
9. Gobert, J., Moussavi, R., Li, H., Sao Pedro, M., Dickler, R.: Scaffolding students' on-line data interpretation during inquiry with Inq-ITS. In: Cyber-Physical Laboratories in Engineering and Science Education. Springer (2018)
10. Gobert, J.D., Sao Pedro, M., Raziuddin, J., Baker, R.S.: From log files to assessment metrics: measuring students' science inquiry skills using educational data mining. J. Learn. Sci. 22, 521–563 (2013)
11. Holstein, K., McLaren, B. M., Aleven, V.: Student learning benefits of a mixed-reality teacher awareness tool in AI-enhanced classrooms. In: Proceedings of Artificial Intelligence in Education 2018, pp. 154–168. Springer, Cham (2018). https://doi.org/10.1007/978-3-319-93843-1_12
12. Holstein, K., McLaren, B.M., Aleven, V.: Co-designing a real-time classroom orchestration tool to support teacher-AI complementarity. J. Learn. Analytics 6(2), 27–52 (2019)
13. Knoop-van Campen, C., Molenaar, I.: How teachers integrate dashboards into their feedback practices. Frontline Learn. Res. 8(4), 37–51 (2020)
14. Lajoie, S.P., et al.: Toward quality online problem-based learning. In: Interactional Research into Problem-based Learning, pp. 367–390 (2020)
15. Li, H., Gobert, J., Dickler, R.: Testing the robustness of inquiry practices once scaffolding is removed. In: Proceedings of Intelligent Tutoring Systems, pp. 204–213 (2019)
16. Marshall, D.T., Shannon, D.M., Love, S.M.: How teachers experienced the COVID-19 transition to remote instruction. Phi Delta Kappan 102(3), 46–50 (2020)
17. Martinez-Maldonado, R., Kay, J., Yacef, K., Edbauer, M.T., Dimitriadis, Y.: MTClassroom and MTDashboard: supporting analysis of teacher attention in an orchestrated multi-tabletop classroom. In: Proceedings of the 10th International Conference on Computer Supported Collaborative Learning, pp. 320–327 (2013)
18. Oxford Reference: Bonferroni correction, https://www.oxfordrefer-ence.com/view/10.1093/oi/authority.20110803095517119. Accessed 10 Feb 2021
19. UNESCO COVID-19 educational disruption and response, https://en.unesco.org/covid19/educationresponse. Accessed 17 Apr 2020
20. Verbert, K., et al.: Learning dashboards: an overview and future research opportunities. Pers. Ubiquit. Comput. 18(6), 1499–1514 (2013). https://doi.org/10.1007/s00779-013-0751-2

Capturing Fairness and Uncertainty in Student Dropout Prediction – A Comparison Study

Efthyvoulos Drousiotis[1](✉), Panagiotis Pentaliotis[1](✉), Lei Shi[2](✉), and Alexandra I. Cristea[2](✉)

[1] Department of Electrical Engineering and Electronics, University of Liverpool, Liverpool, UK
{e.drousiotis,p.pentaliotis}@liverpool.ac.uk
[2] Department of Computer Science, Durham University, Durham, UK
{lei.shi,alexan-dra.i.cristea}@durham.ac.uk

Abstract. This study aims to explore and improve ways of handling a continuous variable dataset, in order to predict student dropout in MOOCs, by implementing various models, including the ones most successful across various domains, such as recurrent neural network (RNN), and tree-based algorithms. Unlike existing studies, we arguably fairly compare each algorithm with the dataset that it can perform best with, thus 'like for like'. I.e., we use a time-series dataset 'as is' with algorithms suited for time-series, as well as a conversion of the time-series into a discrete-variables dataset, through feature engineering, with algorithms handling well discrete variables. We show that these much lighter discrete models outperform the time-series models. Our work additionally shows the importance of handing the uncertainty in the data, via these 'compressed' models.

Keywords: Discrete variables · Capturing uncertainty · Time-series · LSTM · BART · Prediction · MOOCs · Learning analytics

1 Introduction

Over the years, an undeniable challenge in online learning became to find ways to reduce and predict students' dropout rates, which fall roughly at 77%–87% [3, 4]. The majority of the studies such as [3, 4], use the same dataset and variables to implement predictive models, without taking into consideration the type of variables each model uses for maximising its performance. For example, Tang *et al.* [3] trained a time-series Long Short-Term Memory (LSTM) model using the same dataset that was used to train other non-time series machine learning models, including Logistic Regression, Random Forest, and Gradient Boosting Decision Tree (GBDT) models. The results show that time-series models (LSTM) outperform other machine learning models (i.e., Linear Regression, Decision Tree), and achieve higher accuracy, precision and recall when they are compared to their natural environment (continuous/time-series variables). We argue, however, that previous methods do not take into account the target on which the algorithms are performing best. We thus aim to provide benchmarks for predicting the completers and non-completers and examine the following research question:

© Springer Nature Switzerland AG 2021
I. Roll et al. (Eds.): AIED 2021, LNAI 12749, pp. 139–144, 2021.
https://doi.org/10.1007/978-3-030-78270-2_25

Is it a good practice to use sequential time-series as-is, or first convert the dataset into a discrete-variables one, for obtaining enhanced metrics (precision, recall, accuracy) on predicting students' dropout with the appropriately tuned method?

2 Related Work

Many studies focused on classifying students into completers and non-completers. Some of them, such as [5, 6] use statistics, or traditional machine learning algorithms (e.g., Decision Trees, Logistic Regression, Random Forest, Support Vector Machines) [7–10], while others, such as [11, 12], used more advanced algorithms (e.g. Deep Learning), or even visualisation [13]. There are also a few studies [3, 4], that used both traditional machine learning algorithms and more advanced. However, they [3, 4] used the same dataset to train both Neural Networks and machine learning models (time-series), showing that NN outperformed the other machine learning techniques. In our case, we convert the time-series dataset through feature engineering into discrete variables and train each model on the type of dataset it can process best. For example, [14] indicates that if our aim were to train a Neural Network, it is better to use a time-series dataset, while [15] suggests that we should use discrete variables when we aim to train a tree-based algorithm (either categorical or continuous variables).

Interestingly, some papers [12, 13] show that Artificial Recurrent Neural Networks (RNN) with memory, such as Long-Short-Term-Memory (LSTM), are generally considered as superior models to solve time-series tasks, because of their nature – the way they operate and handle data. On the other hand, [18, 19] indicate that traditional machine learning algorithms, such as Logistic Regression, Random Forest and GBDT produce better results with discrete-variable data.

3 Method

The dataset used in this study is comprising 300,000 interactions and 2,000 unique registered students, extracted from XuetangX (launched in October 2013, one of the largest MOOC platforms in China). We converted the time-series dataset, which our LSTM model was trained on, into a discrete-variables dataset, which our tree-based models were trained on. For the construction of the discrete-variables dataset, we used the time-series dataset and we have counted for each student the number of unique actions. In total, there are 14 different types of unique actions and thus we engineered 14 features for 14 input variables for our predictive models. Considering the LSTM model's feature engineering in preparation of the dataset, the actions of each student were sequentially grouped together, according to the time they were performed. Thus, the essence of the time-series was preserved, while still considering the unique action performed. Afterwards, the actions were translated into a sequence of binary numbers, to retain the categorical nature of the actions. Here, we examined the effectiveness of converting a time-series dataset into a discrete dataset through feature engineering. We trained the predictive models with the initial raw datasets, aiming to produce a benchmark for future work. We implemented an LSTM model and several tree-based machine learning models, including Decision Tree, Random Forest, and BART.

For all the above models we used the basic parameters, including the basic split of the data into 70% train and 30% test sets. Moreover, to evaluate the machine learning models, we used the k-fold cross-validation technique, and we did not perform any hyperparameter optimisation. The purpose of this setting is to find a benchmark and compare the two datasets on their primitive forms, without any data pre-processing (sequential time-series and discrete). To evaluate our predictive model's performance, we utilised the following standard, comprehensive metrics:

- *Precision*: the proportion of positive identifications which was actually correct;
- *Recall*: the proportion of actual positives that were identified correctly;
- *F1 score*: the weighted average of Precision and Recall;
- *Accuracy*: the ratio of correctly predicted observations over the total observations.

4   Results and Discussions

Table 1 presents the result comprising three tree-based models (Decision Tree, Random Forest, BART) and an LSTM model. From the results, we can clearly determine the difference between the two types of datasets and draw some useful conclusions. BART outperforms the other models – achieving a very high accuracy of 90% for identifying students who might drop out from an online course. The Decision Tree and Random Forest models achieved relatively high accuracy of 83% and 89%, respectively. The LSTM model achieved the lowest accuracy of 77%. Table 1 especially showcases the performance of the BART model and its improved learning ability in comparison with the other models. From the four figures (Figs. 1, 2, 3 and 4), and the AUC scores, we observe that BART (Fig. 3) has an improved ability to discriminate the test values in comparison with the other models (Decision Tree, Random Forest, LSTM). Furthermore, we can identify the improved trained ability of the tree-based models, when the discrete dataset was used, by the recall metric, which shows a clear ability to select the most relevant items on the classification task with the highest percentage of 96% produced by BART. In comparison with the tree-based models, the LSTM model did not perform as well as the other models. That is partially because LSTMs are known to require a large amount of data, in order to be efficiently trained.

Table 1. Performance comparisons between the predictive models

Metric	DT	RF	BART	LSTM
Precision	0.83	0.88	**0.89**	0.77
Recall	0.83	0.88	**0.96**	0.76
F1	0.83	0.87	**0.92**	0.75
Accuracy	0.82	0.89	**0.90**	0.77

Our results suggest that, whenever possible, it could be beneficial to convert the time-series dataset into a discrete variable dataset, as it is highly probable to produce better performance, especially when the time-series datasets are not populated enough.

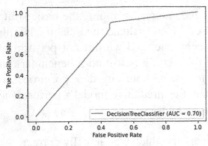

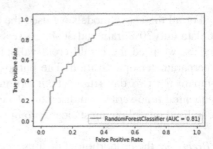

Fig. 1. Decision Tree ROC curve.   **Fig. 2.** Random Forest ROC curve.

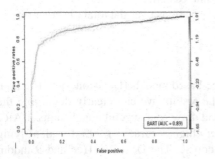

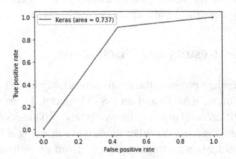

Fig. 3. BART ROC curve   **Fig. 4.** LSTM ROC curve.

5   Conclusions

In summary, this paper presents the results of a study aiming to discover whether it is efficient to convert a time-series dataset into discrete variables dataset, to train predictive models with better performance, in terms of predicting students' dropout. The research results have clearly indicated that we should convert a dataset into different forms when this is feasible. It has shown that this process assists different types of predictive models to obtain higher performance and enhance their learning ability. We have proven that it would be useful to manipulate the dataset for a variety of models first, thus enhancing the final results. We have also shown that BART, which includes a representation of uncertainty, outperforms all other tree-based methods.

Future work might include tuning the models' parameters and investigating the dataset further through data pre-processing and more sophisticated feature engineering techniques (i.e., Frequency count, Frequency Encoding) to achieve better performance. Also, it would be interesting to perform hyperparameter optimisation so that we can find out the optimal learning efficiency of the predictive models. In addition to improving the algorithms, more data could refine the results of this study.

References

1. Gütl, C., Rizzardini, R.H., Chang, V., Morales, M.: Attrition in MOOC: lessons learned from drop-out students. In: Uden, L., Sinclair, J., Tao, Y.-H., Liberona, D. (eds.) LTEC 2014. CCIS, vol. 446, pp. 37–48. Springer, Cham (2014). https://doi.org/10.1007/978-3-319-10671-7_4

2. Kloft, M., Stiehler, F., Zheng, Z., Pinkwart, N.: Predicting MOOC dropout over weeks using machine learning methods. In: Proceedings of the EMNLP 2014 Workshop on Analysis of Large Scale Social Interaction in MOOCs, Doha, Qatar, October 2014, pp. 60–65 (2014). https://doi.org/10.3115/v1/W14-4111
3. Tang, C., Ouyang, Y., Rong, W., Zhang, J., Xiong, Z.: Time series model for predicting dropout in massive open online courses. In: Penstein Rosé, C., et al. (eds.) AIED 2018. LNCS (LNAI), vol. 10948, pp. 353–357. Springer, Cham (2018). https://doi.org/10.1007/978-3-319-93846-2_66
4. Wang, L., Wang, H.: Learning behavior analysis and dropout rate prediction based on MOOCs data. In: 2019 10th International Conference on Information Technology in Medicine and Education (ITME), August 2019, pp. 419–423 (2019). https://doi.org/10.1109/ITME.2019.00100
5. Cristea, A., Alamri, A., Stewart, C., Alshehri, M., Shi, L.: Earliest predictor of dropout in MOOCs: a longitudinal study of futurelearn courses Mizue Kayama. Presented at the 27th International Conference on Information Systems Development (Isd2018 Lund, Sweden), August 2018
6. Zhu, M., Bergner, Y., Zhang, Y., Baker, R., Wang, Y., Paquette, L.: Longitudinal engagement, performance, and social connectivity: a MOOC case study using exponential random graph models. In: Proceedings of the Sixth International Conference on Learning Analytics & Knowledge - LAK 2016, Edinburgh, United Kingdom, 2016, pp. 223–230 (2016). https://doi.org/10.1145/2883851.2883934
7. Alamri, A., et al.: Predicting MOOCs dropout using only two easily obtainable features from the first week's activities. In: Coy, A., Hayashi, Y., Chang, M. (eds.) ITS 2019. LNCS, vol. 11528, pp. 163–173. Springer, Cham (2019). https://doi.org/10.1007/978-3-030-22244-4_20
8. Chen, J., Feng, J., Sun, X., Wu, N., Yang, Z., Chen, S.: MOOC dropout prediction using a hybrid algorithm based on decision tree and extreme learning machine. In: Mathematical Problems in Engineering, 18 March 2019. https://www.hindawi.com/journals/mpe/2019/8404653/. Accessed 02 Feb 2021
9. Jin, C.: MOOC student dropout prediction model based on learning behavior features and parameter optimization. In: Interactive Learning Environments, pp. 1–19, August 2020. https://doi.org/10.1080/10494820.2020.1802300
10. Pereira, F.D., et al.: Early dropout prediction for programming courses supported by online judges. In: Isotani, S., Millán, E., Ogan, A., Hastings, P., McLaren, B., Luckin, R. (eds.) AIED 2019. LNCS (LNAI), vol. 11626, pp. 67–72. Springer, Cham (2019). https://doi.org/10.1007/978-3-030-23207-8_13
11. Fei, M., Yeung, D.: Temporal models for predicting student dropout in massive open online courses. In: 2015 IEEE International Conference on Data Mining Workshop (ICDMW), November 2015, pp. 256–263 (2015). https://doi.org/10.1109/ICDMW.2015.174
12. Gardner, J., Yang, Y.: Modeling and experimental design for MOOC dropout prediction: a replication perspective. In: Proceedings of the 12th International Conference on Educational Data Mining (EDM 2019), p. 10 (2019)
13. Alamri, A., Sun, Z., Cristea, A.I., Senthilnathan, G., Shi, L., Stewart, C.: Is MOOC learning different for dropouts? A visually-driven, multi-granularity explanatory ML approach. In: Kumar, V., Troussas, C. (eds.) ITS 2020. LNCS, vol. 12149, pp. 353–363. Springer, Cham (2020). https://doi.org/10.1007/978-3-030-49663-0_42
14. Time series forecasting|TensorFlow Core, TensorFlow. https://www.tensorflow.org/tutorials/structured_data/time_series. Accessed 10 Feb 2021
15. Decision Tree - Overview, Decision Types, Applications, Corporate Finance Institute. https://corporatefinanceinstitute.com/resources/knowledge/other/decision-tree/. Accessed 10 Feb 2021

144   E. Drousiotis et al.

16. Gers, F.A., Eck, D., Schmidhuber, J.: Applying LSTM to time series predictable through time-window approaches. In: Neural Nets WIRN Vietri-01, London, 2002, pp. 193–200 (2002). https://doi.org/10.1007/978-1-4471-0219-9_20
17. Zhang, X., et al.: AT-LSTM: an attention-based LSTM model for financial time series prediction. IOP Conf. Ser.: Mater. Sci. Eng. **569**, 052037 (2019). https://doi.org/10.1088/1757-899X/569/5/052037
18. Sethi, I.K., Chatterjee, B.: Efficient decision tree design for discrete variable pattern recognition problems. Pattern Recogn. **9**(4), 197–206 (1977). https://doi.org/10.1016/0031-320 3(77)90004-8
19. Song, Y., Lu, Y.: Decision tree methods: applications for classification and prediction. Shanghai Arch. Psychiatr. **27**(2), 130–135 (2015). https://doi.org/10.11919/j.issn.1002-0829. 215044

Dr. Proctor: A Multi-modal AI-Based Platform for Remote Proctoring in Education

Ahmed E. Elshafey[2], Mohammed R. Anany[1], Amr S. Mohamed[2],
Nourhan Sakr[1(✉)], and Sherif G. Aly[1]

[1] The American University in Cairo, Cairo, Egypt
{moerefaat,sgamal}@aucegypt.edu, n.sakr@columbia.edu
[2] Data Science Hub, Cairo, Egypt
{ahmed_elshafey,amrsaeed}@aucegypt.edu

Abstract. Technological advancements have enabled remote exams as a viable alternative to in-person proctoring. In light of the COVID-19 pandemic, educational institutions relied heavily on remote operation. The sudden shift exposed the weaknesses in available proctoring solutions, as pertains to fairness, economic viability, data privacy, network issues and usability. Moreover, whether they are equal in function to physical proctoring is questionable. Based on extensive research, we establish the system requirements and design for Dr. Proctor, a non-commercial solution that addresses many of the exposed concerns about remote proctoring.

Keywords: AI · Remote proctoring · Remote assessment · Education

1 Introduction

The transition to remote education due to the COVID-19 pandemic exposed many weaknesses with existing educational solutions [23], especially for conducting remote exams. The absence of physical proctoring provided more cheating opportunities for students [29], necessitating the use of remote assessment platforms [2,3,5,9,27,28]. These platforms facilitate remote proctoring, yet have several limitations [8,10]. They impose resource requirements and often cause a stressful experience to students [20,21,26]. The fairness of the assessment process is also challenged [6]. Biases in the machine learning (ML) models used by these solutions may contribute to the fairness issue [17,33]. Table 1 illustrates a comparison of monitoring features offered by popular platforms.

Student privacy [8,10] is another concern due to unfettered access to students' data [18]. Some solutions try to combat cheating by requiring additional gadgets, e.g. [13]. These solutions come with their own concerns, such as causing further privacy infringement by using more invasive monitoring [25], as well as being expensive to implement. To the best of our knowledge, none of the available platforms clarify the intricacies of their models or demonstrate what biases they

I. Roll et al. (Eds.): AIED 2021, LNAI 12749, pp. 145–150, 2021.
https://doi.org/10.1007/978-3-030-78270-2_26

Table 1. AI-based features in current proctoring platforms.

Tool	Face recognition	Motion detection	Anomaly detection	Behavioral analysis	Eye tracking	Voice detection
Mettl	Yes	–	Yes	Yes	Yes	Yes
ProctorU	Yes	–	Yes	Yes	–	–
Respondus	Yes	Yes	Yes	Yes	Yes	–
Examity	Yes	Yes	Yes	Yes	–	Yes
ProctorTrack	Yes	Yes	Yes	Yes	–	–
HonorLock	Yes	Yes	Yes	–	–	Yes

exhibit. Furthermore, no available tool integrates all features listed in Table 1. Eye-tracking and voice detection also seem to be lacking in most. Dr. Proctor is a multi-modal, user-centric solution developed to tackle remote proctoring challenges, while integrating most of these monitoring features.

2 Identifying Requirements

Our extensive primary research[1] builds on 50 h of interviews and 100 survey responses from STEM faculty at the American University in Cairo. We augment this data with social media narratives from faculty and students worldwide, education and psychology literature, as well as market analysis. We establish that while faculty is mainly concerned with cheating, students are concerned about their privacy and face a lot of stress when using available software.

We distill this data to understand the challenges in remote proctoring and the best practices that address them. We also compile a list of cheating scenarios and technological solutions that combat them. We finally consolidate this into a framework of prioritized system requirements and useable features for an AI-based proctoring tool, infused with psychological means and economic considerations to present a multi-modal approach with novel offerings.

3 System Components

The main components of Dr. Proctor are the instructor hub, student hub, and cloud service. Figure 1 illustrates the system architecture. Cheating, data privacy, network vulnerability, tool usability and fairness are the main challenges we aim to tackle through a user-centric, multi-modal and economic design.

Cloud Service. Dr. Proctor cloud service offers the needed infrastructure for student and instructor hub. RabbitMQ is used for communication with clients, Kubernetes Service Mesh and AWS to serve APIs, Celery Task Queues [1] for task distribution, OpenVidu [7] and Kurento [4] for Web Real-Time Communication (WebRTC) streaming, and Timeboard [12] for proctoring reports.

---

[1] Details are out of scope of this paper and can be found here [31].

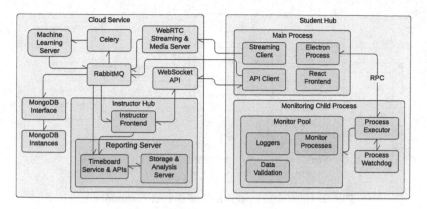

Fig. 1. System architecture overview.

Student Hub. The student hub aims to provide a stress-free, user-friendly environment while managing exam delivery and hedging against privacy and network issues. Dr. Proctor is designed by a diverse team of engineers, UI/UX designers, and psychologists to meet these objectives. During registration and prior to starting an exam, students read out the honor code, which reduces the probability of cheating [11]. The student hub is equipped with monitoring facilities such as screen and video monitoring, system monitoring, keystroke and activity watchers and other functionality to collect data and detect cheating attempts. It also allows offline operations, which instructors can limit. Exams feature MCQ, essay, oral, and coding questions with built-in code editor. Students can also view documents added by instructors to mimic an "open-book"-exam modality.

Instructor Hub. The instructor hub allows instructors to schedule exams, add student email lists for automatic invitations, approve sign-ups, create and customize exam features, edit and organize questions, and finally view proctoring reports. The instructor is given ample flexibility in setting the exam dyanmics and proctoring options. Reports are offered through Timeboard; a specialized fork of Timesketch [12]. It provides the necessary functionality to view and analyze proctoring data. Our research shows that instructors rely on oral exams heavily in suspected cheating cases. They find it effective but time-consuming. We provide the following automation: Instructors have the ability to schedule limited follow-up exams that primarily offer oral questions to select students.

4   AI-Powered Components

Dr. Proctor employs AI components that enable various proctoring and assessment features. Figure 2 illustrates these components.

Video Solutions. Face recognition identifies and authenticates an exam taker. We use FaceNet [32] and test on the CFPW dataset, producing an accuracy of

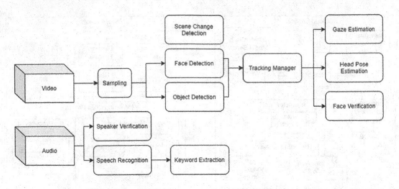

Fig. 2. Integration overview of the AI components.

96%. For tracking objects and faces, we adopt SORT [15]. Head pose estimation tracks the direction of the exam taker's attention. We use FSA-NET [34], which achieves the lowest mean average error on both AFLW2000 and BIWI datasets. Gaze estimation determines the direction of the user's gaze via RT-Gene [19], which reports the lowest angular error on both MPII and UT Multiview datasets. Scene change detection watches for changes in the student's environment using Gaussian Mixture Models [14], while object detection detects unauthorized objects in camera view. We fine-tune the YOLOv4 [16] model, pre-trained on the 80 COCO classes, to our set of desired classes to hone its focus.

Audio Solutions. Unlike most proctoring tools, Dr. Proctor provides audio solutions to enhance proctoring and offer new features. Speaker verification actively authenticates students. It uses Deep Speaker [24], which reports an accuracy of 86.1% on 1000 data points from VoxCeleb1 with an optimal threshold parameter of 0.4. It operates independently of spoken language. Speech recognition recognizes and transcribes captured audio. We use DeepSpeech [22], which achieves a word error rate (WER) of 5.97% on LibriSpeech clean test corpus with version 0.7.4, the lowest reported WER in relevant literature. Keyword extraction uses speech recognition module's output for keyword analysis. Rapid Automatic Keyword Extraction [30] is used and tested on 500 abstracts in the Inspec test set. It achieves the highest precision and F-measure, i.e. 33.7% and 37.2%, respectively.

5   Conclusion

Dr. Proctor is a multi-modal, user-centric solution developed to tackle existing remote proctoring challenges on both student and faculty sides. Dr. Proctor addresses network and privacy issues, affordability, student stress, and availability to underprivileged communities. It offers a comprehensive monitoring and analysis suite to prevent cheating. It integrates unique AI-powered audio solutions, high customizability for instructors, including different proctoring levels, embedded resources, IDE for coding exams, and visualized analysis reports.

Planned future work on Dr. Proctor includes improving the AI suite. Psychological analysis data, e.g. response time behavioral analysis, will be incorporated to examine their viability and efficiency as cheating detection solutions.

References

1. Celery: Distributed Task Queue. https://github.com/celery/celery
2. Examity Online Proctoring. https://www.examity.com/
3. Honorlock Proctoring. https://honorlock.com/
4. Kurento WebRTC media server. https://www.kurento.org/
5. Mettl. https://mettl.com/en-ae/
6. Online proctoring: Trust, transparency, and fairness. https://www.ecampusnews.com/2020/06/01/online-proctoring-trust-transparency-and-fairness/2/
7. OpenVidu: OpenVidu Platform main repository. https://github.com/OpenVidu/openvidu
8. ProctorU threatens UC Santa Barbara faculty over criticism during coronavirus crisis. https://www.thefire.org/proctoru-threatens-uc-santa-barbara-faculty-over-criticism-during-coronavirus-crisis/
9. Respondus Assessment Tools for Learning Systems. https://web.respondus.com/
10. Students express privacy and security concerns over proctoring software — The Charlatan, Carleton's independent newspaper. https://charlatan.ca/2020/10/students-express-privacy-and-security-concerns-over-proctoring-software/
11. The Best Ways to Prevent Cheating in College - The Atlantic. https://www.theatlantic.com/education/archive/2016/04/how-to-stop-cheating-in-college/479037/
12. Timesketch: Collaborative forensic timeline analysis. https://github.com/google/timesketch
13. Atoum, Y., Chen, L., Liu, A.X., Hsu, S.D., Liu, X.: Automated online exam proctoring. IEEE Trans. Multimed. **19**(7), 1609–1624 (2017). https://doi.org/10.1109/TMM.2017.2656064, https://www.cse.msu.edu/
14. Based, C.: Gabor Jets. In: Encyclopedia of Biometrics, pp. 627–627 (2009). https://doi.org/10.1007/978-0-387-73003-5_334
15. Bewley, A., Ge, Z., Ott, L., Ramos, F., Upcroft, B.: Simple online and realtime tracking. In: 2016 IEEE International Conference on Image Processing (ICIP), September 2016. https://doi.org/10.1109/icip.2016.7533003, http://dx.doi.org/10.1109/ICIP.2016.7533003
16. Bochkovskiy, A., Wang, C.Y., Liao, H.Y.M.: YOLOv4: Optimal Speed and Accuracy of Object Detection (2020). https://github.com/AlexeyAB/darknet
17. Buolamwini, J.: Gender Shades: Intersectional Accuracy Disparities in Commercial Gender Classification *. Technical report (2018)
18. Coghlan, S., Miller, T., Paterson, J.: Good proctor or "big brother"? AI ethics and online exam supervision technologies (2020)
19. Fischer, T., Chang, H.J., Demiris, Y.: RT-GENE: Real-time eye gaze estimation in natural environments. In: Lecture Notes in Computer Science (including subseries Lecture Notes in Artificial Intelligence and Lecture Notes in Bioinformatics), vol. 11214 LNCS, pp. 339–357 (2018). https://doi.org/10.1007/978-3-030-01249-6_21, www.imperial.ac.uk/PersonalRobotics

20. Gonzales, A.L., McCrory Calarco, J., Lynch, T.: Technology problems and student achievement gaps: a validation and extension of the technology maintenance construct. Commun. Res. **47**(5), 750–770 (2020). https://doi.org/10.1177/0093650218796366

21. van Halem, N., van Klaveren, C., Cornelisz, I.: The effects of implementation barriers in virtually proctored examination: a randomised field experiment in Dutch higher education. High. Educ. Q. p. hequ.12275 (2020). https://doi.org/10.1111/hequ.12275, https://onlinelibrary.wiley.com/doi/abs/10.1111/hequ.12275

22. Hannun, A., et al.: Deep speech: scaling up end-to-end speech recognition (2014). http://arxiv.org/abs/1412.5567

23. Hatzipanagos, S., Warburton, S.: Feedback as dialogue: exploring the links between formative assessment and social software in distance learning. Learn. Med. Technol. **34**(1), 45–59 (2009). https://doi.org/10.1080/17439880902759919

24. Li, C., et al.: Deep speaker: an end-to-end neural speaker embedding system (2017)

25. Migicovsky, A., Durumeric, Z., Ringenberg, J., Halderman, J.A.: Outsmarting proctors with smartwatches: a case study on wearable computing security. In: Lecture Notes in Computer Science (including subseries Lecture Notes in Artificial Intelligence and Lecture Notes in Bioinformatics), vol. 8437, pp. 89–96 (2014). https://doi.org/10.1007/978-3-662-45472-5_7, https://jhalderm.com/papers/

26. Moore, R., Vitale, D., Stawinoga, N.: The Digital Divide and Educational Equity: A Look at Students with Very Limited Access to Electronic Devices at Home. ACT Research & Center for Equity in Learning (August), 14 (2018). https://www.act.org/www.act.org/Divide/DigitalDivide6thMay/ED593163.pdf

27. Proctortrack: Trusted Exam Integrity—Remote Online Proctoring (2019). https://www.proctortrack.com/

28. ProctorU: ProctorU - The Leading Proctoring Solution for Online Exams (2020). https://www.proctoru.com/proctoru.com/academic-solutions

29. Rogers, C.: Faculty perceptions about e-cheating. J. Comput. Sci. Coll. **22**(2), 206–212 (2006). https://www.researchgate.net/publication/262311152

30. Rose, S., Engel, D., Cramer, N., Cowley, W.: Automatic keyword extraction from individual documents. Text Mining: Appl. Theory **2017**, 1–20 (2010). https://doi.org/10.1002/9780470689646.ch1

31. Sakr, N., Salama, A., Tameesh, N., Osman, G.: EduPal leaves no professor behind: supporting faculty via peer-powered recommender systems (2021). https://cutt.ly/EduPal-arxiv-link

32. Schroff, F., Kalenichenko, D., Philbin, J.: FaceNet: a unified embedding for face recognition and clustering. In: Proceedings of the IEEE Computer Society Conference on Computer Vision and Pattern Recognition, vol. 07–12 June, pp. 815–823 (2015). https://doi.org/10.1109/CVPR.2015.7298682

33. Selbst, A.D., Boyd, D., Friedler, S.A., Venkatasubramanian, S., Vertesi, J.: Fairness and abstraction in sociotechnical systems. In: FAT* 2019 - Proceedings of the 2019 Conference on Fairness, Accountability, and Transparency, pp. 59–68 (2019). https://doi.org/10.1145/3287560.3287598

34. Yang, T.Y., Chen, Y.T., Lin, Y.Y., Chuang, Y.Y.: FSA-net: Learning fine-grained structure aggregation for head pose estimation from a single image. In: Proceedings of the IEEE Computer Society Conference on Computer Vision and Pattern Recognition, vol. 2019-June, pp. 1087–1096 (2019). https://doi.org/10.1109/CVPR.2019.00118

Multimodal Trajectory Analysis of Visitor Engagement with Interactive Science Museum Exhibits

Andrew Emerson[✉], Nathan Henderson, Wookhee Min, Jonathan Rowe,
James Minogue, and James Lester

North Carolina State University, Raleigh, NC 27695, USA
{ajemerso,nlhender,wmin,jprowe,james_minogue,lester}@ncsu.edu

Abstract. Recent years have seen a growing interest in investigating visitor engagement in science museums with multimodal learning analytics. Visitor engagement is a multidimensional process that unfolds temporally over the course of a museum visit. In this paper, we introduce a multimodal trajectory analysis framework for modeling visitor engagement with an interactive science exhibit for environmental sustainability. We investigate trajectories of multimodal data captured during visitor interactions with the exhibit through slope-based time series analysis. Utilizing the slopes of the time series representations for each multimodal data channel, we conduct an ablation study to investigate how additional modalities lead to improved accuracy while modeling visitor engagement. We are able to enhance visitor engagement models by accounting for varying levels of visitors' science fascination, a construct integrating science interest, curiosity, and mastery goals. The results suggest that trajectory-based representations of the multimodal visitor data can serve as the foundation for visitor engagement modeling to enhance museum learning experiences.

Keywords: Museum learning · Visitor engagement · Multimodal trajectory analytics

1 Introduction

Visitor engagement plays a critical role in museum learning [1, 2]. By focusing on how to model and enhance core components of visitor engagement, museum exhibit designers can create meaningful science learning experiences that can have lasting impact beyond the original visit [3]. However, modeling museum visitor engagement poses significant challenges. Museum learning is a *free choice* experience: visitors are free to approach and leave museum exhibits at any time. This can lead to exhibit dwell times that are exceedingly short [4–6]. Leveraging multimodal trajectory analyses using computational models that incorporate temporal data holds considerable promise for measuring and predicting visitor engagement. Previous work has utilized multimodal data for tasks such as modeling learner knowledge [7, 8] and engagement [9, 10], but

© Springer Nature Switzerland AG 2021
I. Roll et al. (Eds.): AIED 2021, LNAI 12749, pp. 151–155, 2021.
https://doi.org/10.1007/978-3-030-78270-2_27

comparatively little work has explored the use of multimodal trajectories to inform models of visitor engagement in museums. It is important to develop computational models of visitor engagement that account for inherent differences in visitor characteristics and to understand which modalities provide the most predictive value to these models.

In this paper, we introduce a multimodal learning analytics framework that induces computational models of visitor engagement using multimodal trajectory analysis. We focus on modeling visitor dwell time, a measure of behavioral engagement, with an interactive exhibit about environmental sustainability, FUTURE WORLDS. The exhibit was instrumented with multiple sensors to capture visitor behavior at the exhibit, including their posture, facial expressions, exhibit interaction logs, and eye gaze. We construct a random effects linear model of visitor dwell time that accounts for different levels of visitors' *science fascination*, a construct that integrates science interest, curiosity, and mastery goals [11]. Leveraging temporal features extracted from each of the multimodal data channels, we investigate relationships among the modalities, visitors' science fascination levels, and dwell times.

2 Multimodal Trajectory Models in FUTURE WORLDS

FUTURE WORLDS is an interactive science museum exhibit that combines game-based learning and interactive tabletop technologies to support visitor explorations of environmental sustainability (Fig. 1). The exhibit enables learners to explore environmental sustainability problem scenarios by investigating the impacts of alternate environmental decisions on a 3D simulated environment [12]. The science content in FUTURE WORLDS is designed for learners aged 10–11.

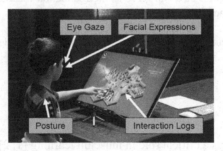

Fig. 1. Visitor interacting with the instrumented FUTURE WORLDS exhibit.

Multimodal visitor trajectory data was collected during a study with FUTURE WORLDS at a science museum [13]. There were 116 elementary school students aged 10–11 ($M = 10.4$, $SD = 0.57$) that participated in the study. The students completed a series of questionnaires and surveys before and after interacting with the exhibit. These included a demographics survey, sustainability content knowledge assessment, and the Fascination in Science scale [14] prior to exploring FUTURE WORLDS, and a sustainability content knowledge assessment and engagement survey afterward. Several participants were missing data, leaving 70 participants' data that were used for modeling. Each visitor interacted with FUTURE WORLDS individually until they successfully solved the exhibit's environmental problem or up to a maximum of approximately 12 min ($M = 4.13$, $SD = 2.38$).

This analysis specifically focuses on differences between visitors with low and high scores on the Fascination in Science scale [14]. The survey consists of eight 4-point Likert scale items. Responses are averaged to produce an overall science fascination score ($M = 3.23$, $SD = 0.58$). We split the 70 visitors into two groups (low and high) based on the median value of their overall science fascination scores.

Multimodal Features. FUTURE WORLDS was fully instrumented to track visitor behavior in real-time as described in [13]. To generate temporal representations of each visitor's interaction trajectory across their time spent exploring the exhibit, several features were generated by averaging or summing over the captured data from the start of the session to the current timestamp, using 10-s time intervals. Two posture-based features were distilled from four vertices tracked by the Microsoft Kinect, resulting in eight features. The facial expression data was used to generate 17 distinct features. Eight features were distilled from the interaction log data. Four gaze-based features were distilled to reflect the total duration of time spent fixating on specific areas of interest shown on the FUTURE WORLDS exhibit's display [13].

The distilled features were then converted into slope-based representations. Specifically, we fit an ordinary least squares (OLS) linear regression model to the series of points associated with each feature per visitor (i.e., one regression line per feature per visitor). Based upon these models, we derived a slope coefficient summarizing the trajectory of that feature for that visitor. For example, if a visitor interacted with FUTURE WORLDS for two minutes, they would have 12 data points (one data point per 10-s interval) per feature per modality. Upon fitting the OLS to the 12 points for each individual feature, we use the regression model's slope to represent the feature's temporal trajectory, thus capturing the rate at which the feature changes over time.

Multimodal Trajectory Models of Visitor Engagement. We induce predictive models of visitor dwell time using two linear models: (1) a baseline linear regression model that groups all museum visitors into a single group, and (2) a multilevel linear regression model that introduces random intercepts dependent on visitors' science fascination levels (low or high). We also conducted ablation analyses to investigate how the accuracy of the models of visitor dwell time vary with different sets of modalities, where we start with the full set of modalities: posture (P), facial expression (F), interaction logs (I), and eye gaze (E).

Predictive Performance. Each of the predictive linear models was trained and evaluated using visitor-level leave-one-out cross-validation. We report the R^2 for each type of model and ablation condition. Additionally, we report model performance across all visitors as well as for the high (HF) and low (LF) science fascination groups.

Within each predictive linear model, the features used were the slopes of the set of multimodal features for each ablation condition. We performed feature selection within each training fold of cross-validation by conducting a univariate linear regression test between the training set features and the target variable using a significance threshold of

0.3. In addition to feature selection, all input data were standardized within the training fold by using each individual feature's mean and standard deviation.

Table 1 shows the performance of each model predicting visitor dwell time in seconds. The random effects (RE) linear model outperforms the baseline linear model in three of the four ablation conditions. This suggests that differentiating between visitors with high and low science fascination scores enables linear models to better explain the variance in dwell time under different multimodal ablation conditions. RE achieves high predictive performance for both the PFI and PFIE ablation conditions, with the PFI condition outperforming all other conditions for the entire visitor population. The ablation analyses suggest that slope-based representations of facial expression and posture data provide limited benefit; therefore it may be helpful to investigate a hybrid approach that combines extracted features and slope-based representations for those two modalities.

Table 1. Predictive performance of the Baseline Model and Random Effects Linear Model. Bold values indicate the best performance for a specific ablation condition and group.

Model type	Group	PFIE R^2	PFI R^2	PF R^2	P R^2
Baseline model	*All Visitors*	0.464	0.614	−0.349	**−0.082**
	LF	**0.623**	**0.612**	−0.135	**−0.167**
	HF	0.359	0.604	−0.510	**−0.067**
Random effects linear model	*All Visitors*	**0.586**	**0.634**	**−0.090**	−0.106
	LF	0.580	0.602	**−0.003**	−0.236
	HF	**0.577**	**0.641**	**−0.171**	−0.068

3   Conclusion and Future Work

Engagement plays a critical role in museum-based learning. Modeling visitor engagement in science museums presents significant challenges, as visitor interaction with exhibits is often brief, and visitor engagement dynamics are affected by a wide range of factors. We have introduced a multimodal trajectory analysis framework for modeling visitor dwell time with interactive science museum exhibits. Results show random effects models that account for visitors' fascination in science yield more accurate models of visitor dwell time than baseline linear models. In addition, multimodal models incorporating visitor posture, facial expressions, and interaction logs outperform models with other modality configurations. Trajectory-based feature representations effectively incorporate temporal attributes of behavioral cues for modeling visitor engagement. Potential avenues for future work include investigating more temporal features within a single visitor's interactions, exploring additional factors that influence visitor engagement, and incorporating visitor engagement models into exhibits to operate at run-time to adaptively enhance visitors' learning experiences.

Acknowledgements. The authors would like to thank the staff and visitors of the North Carolina Museum of Natural Sciences. This research was supported by the National Science Foundation under Grant DRL-1713545. Any opinions, findings, and conclusions or recommendations expressed in this material are those of the authors and do not necessarily reflect the views of the National Science Foundation.

References

1. Hein, G.: Learning science in informal environments: People, places, and pursuits. Museums Soc. Issues **4**(1), 113–124 (2009)
2. Falk, J., Dierking, L.: Learning from Museums. Rowman & Littlefield (2018)
3. Allen, S.: Designs for learning: Studying science museum exhibits that do more than entertain. Sci. Educ. **88**(S1), S17–S33 (2004)
4. Diamond, J., Horn, M., Uttal, D.: Practical Evaluation Guide: Tools for Museums and Other Informal Educational Settings. Rowman & Littlefield (2016)
5. Lane, H., Noren, D., Auerbach, D., Birch, M., Swartout, W.: Intelligent tutoring goes to the museum in the big city: a pedagogical agent for informal science education. In: Proceedings of the 15th International Conference on Artificial Intelligence in Education, pp. 155–162. Springer, Berlin, Heidelberg (2011). https://doi.org/10.1007/978-3-642-21869-9_22
6. Long, D., McKlin, T., Weisling, A., Martin, W., Guthrie, H., Magerko, B.: Trajectories of physical engagement and expression in a co-creative museum installation. In: Proceedings of the 12th Annual ACM Conference on Creativity and Cognition, pp. 246–257 (2019)
7. Sharma, K., Papamitsiou, Z.: Giannakos, M: Building pipelines for educational data using AI and multimodal analytics: a "grey-box" approach. Br. J. Edu. Technol. **50**(6), 3004–3031 (2019)
8. Sharma, K., Papamitsiou, Z., Olsen, J., Giannakos, M.: Predicting learners' effortful behaviour in adaptive assessment using multimodal data. In: Proceedings of the Tenth International Conference on Learning Analytics & Knowledge, pp. 480–489 (2020)
9. Bosch, N., et al.: Detecting student emotions in computer-enabled classrooms. In: Proceedings of the 25th International Joint Conference on Artificial Intelligence, pp. 4125–4129 (2016)
10. Dhall, A., Kaur, A., Goecke, R., Gedeon, T.: Emotiw 2018: audio-video, student engagement and group-level affect prediction. In: Proceedings of the 20th ACM International Conference on Multimodal Interaction, pp. 653–656 (2018)
11. Bonnette, R.N., Crowley, K., Schunn, C.D.: Falling in love and staying in love with science: ongoing informal science experiences support fascination for all children. Int. J. Sci. Educ. **41**(12), 1626–1643 (2019)
12. Rowe, J., Lobene, E., Mott, B., Lester, J.: Play in the museum: Design and development of a game-based learning exhibit for informal science education. Int. J. Gaming Comput.-Mediated Simul. **9**(3), 96–113 (2017)
13. Emerson, A., et al.: Early prediction of visitor engagement in science museums with multimodal learning analytics. In: Proceedings of the Twenty-Second International Conference on Multimodal Interaction, pp. 107–116 (2020)
14. Chung, J., Cannady, M. A., Schunn, C., Dorph, R., Bathgate, M.: Measures Technical Brief: Fascination in Science. http://www.activationlab.org/wp-content/uploads/2016/02/Fascination-Report-3.2-20160331.pdf. Accessed 2016

Analytics of Emerging and Scripted Roles in Online Discussions: An Epistemic Network Analysis Approach

Máverick Ferreira[1], Rafael Ferreira Mello[2]([✉]), Rafael Dueire Lins[2],
and Dragan Gašević[3]

[1] Informatics Center, Universidade Federal de Pernambuco, Recife, Brazil
[2] Department of Computing, Universidade Federal Rural de Pernambuco,
Recife, Brazil
{rafael.mello,rafael.dueirelins}@ufrpe.br
[3] Faculty of Information Technology, Monash University, Melbourne, Australia
dragan.gasevic@monash.edu

Abstract. This paper investigates emerging roles in the context of the community of inquiry model. The paper reports the results of a study that demonstrated the application of epistemic network and clustering analyses to reveal the roles that different students assumed during an asynchronous course with online discussions. The proposed method highlights the differences and similarities between emerging and scripted roles based on the development of social and cognitive presences, two key constructs of the model of communities of inquiry.

Keywords: Emerging roles · Epistemic network analysis · CSCL

1 Introduction

Asynchronous online discussions are the most widely adopted resource to promote social interactions in computer-supported collaborative learning settings [4]. There is a rising need to understand asynchronous online discussions under different perspectives like engagement, knowledge (co-)construction and the roles of each learner within the discussion [3,19,24]. The Community of Inquiry (CoI) model highlights the social nature of interactions in learning by describing three different constructs [5,6,8]: the social, teaching, and cognitive presences. They measure the student social interactions and their cognitive development, unpack interactions in computer-supported collaborative scenarios [12], demonstrate the importance of teaching presence for the development of an effective community of learners [9,10], and measure the learners' participation in online communities [13,14].

Another relevant pedagogical aspect to observe in the asynchronous online communication process is the roles that learners assume during discussions. It could influence their contribution and interaction patterns with other group

© Springer Nature Switzerland AG 2021
I. Roll et al. (Eds.): AIED 2021, LNAI 12749, pp. 156–161, 2021.
https://doi.org/10.1007/978-3-030-78270-2_28

members [1]. In general, there are two categories of student roles [23]: i) emerging, when the student assumes a role intrinsically during the discussion and (ii) scripted, when the moderator pre-defines the role for each student. These role categories have a high impact on the social knowledge construction, group collaboration and cohesion during online discussions. While the predefined roles have been widely studied in the literature [17,18,23,25], fewer efforts have been devoted towards the analysis of the emerging roles [3]. The study reported in this paper unpacks the relationship between emerging and scripted roles and categories of social and cognitive presences in a CoI. To reach this goal, this study performed a cluster analysis in combination with an epistemic network analysis [21] to shed light on the scripted and emerging roles assumed by the students in asynchronous online discussions. More specifically, the aim of the stud was to answer the following research question: "*To what extent can a clustering algorithm using indicators of the social and cognitive presences predict emerging roles in an asynchronous online discussion?*"

1.1 Data and Course Design

The data used here encompass six offerings of a fully-online master level research-intensive course, with a total of 82 students that posted 1,747 messages, which accounted for 15% of the final mark. During the online discussions, there were two scripted roles defined by the instructors: (i) **experts**, the students that initiated the discussion by posting the video about the research paper they presented, (ii) **practicing researchers**, students who were requested to watch the video and interact asking questions and posting comments for the experts[1]. Not all students played both roles.

Two experts labeled the dataset using the coding scheme proposed by [7]. The coding unit of analysis was the entire message. However, each message could have more that one social presence indicator, but just one cognitive presence phase. The coders reached 84% and 98.1% of the agreement for social presence and cognitive presences, respectively. The differences were resolved through discussions among them. Table 1 presents the distribution of messages in the dataset.

Table 1. Distribution of social and cognitive presences.

Social messages	#	%	Cognitive messages	#	%
Affective positive	530	33.33%	*Other*	140	8.01%
Affective negative	1,217	66.67%	Triggering event	308	17.63%
Interactive positive	1,030	58.95%	Exploration	684	39.15%
Interactive negative	741	41.05%	Integration	508	29.08%
Cohesive negative	1,326	75.90%	Resolution	107	6.13%
Cohesive positive	421	24.10%	*Total*	1,747	100.00%

---

[1] For further details on the course design [10].

1.2   Clustering Algorithm

Clustering algorithms aim to create sets of high intra-group and low inter-group similarities between elements [22]. The messages posted by the students were clustered using the k-means algorithm [26] to analyze their emerging roles. The total number of instances evaluated was 163 equivalent to all pairs of student/role. The social presence indicators and the phases of cognitive presence were the features for the clustering algorithm. Such indicators from all student/role instance contributions were encompassed into a single line used as input to the k-means algorithm. The silhouette approach was used to identify the ideal number of clusters [11].

1.3   Epistemic Network Analysis (ENA)

ENA provides a mechanism to compare the differences between different groups of analysis units – such as between the emerging and scripted roles in the present study [20]. Each message was coded capturing the presence/absence of the social presence indicators and the cognitive presence phases. Both units of analysis and stanzas were students (i.e., all student messages) within different roles (emerging and scripted). The use of students as units of analysis here enabled us to see the connections between phases of cognitive presence and the indicators of social presence for each student. We removed the social presence indicators (Continuing a thread, Complementing, and Vocatives) that could generate a dominant code to enhance the ENA results [16].

2   Results

This study identified that the best number of groups was 3 (silhouette = 0.3644). The k-means algorithm was applied to identify the three clusters, considered the *emerging roles*. Table 2 shows the percentage of indicators for each cluster identified and the scripted roles (expert and practicing research). Cluster 0 presented the greater values for the social presence indicators, showing the concern in creating discussions that could reach a social climax, similar to the expert role. Cluster 1 had slightly higher values for the affective category indicators, but in general, the values for the social presence were similar to those of clusters 0 and 2. Both clusters 0 and 1 presented similar values of cognitive presence. Cluster 2 had a similar trend with the practicing researcher role when compared the social presence indicators; however, it also presented a lower development of the cognitive presence.

Figure 1 presents the projection of the average students' networks with relationships between the social and cognitive presences. The rectangles represent group-average networks (95% CI are also outlined) for students in expert (red), practicing researcher (blue), cluster 0 (green), cluster 1 (orange) and cluster 2 (purple) roles. The visualization was done using 1 and 2, which accounted for 18.8 and 12.2% of variability between students' network models, respectively.

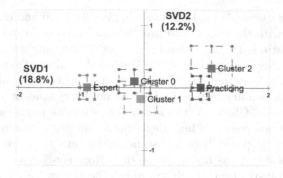

Fig. 1. ENA projections of the means values of group networks of the students who assumed different scripted and emerging roles. Each colored square (95% CI are outlined around the group means) represent a mean network of one role (emerging or scripted). (Color figure online)

Table 2. Indicators for emerging and scripted (Expert and Practicing Researcher)

		Indicator	Cluster 0	Cluster 1	Cluster 2	Expert	Practicing
Social Presence	Affective	Emotions	16.43%	17.20%	15.84%	22.12%	11.15%
		Humor	02.28%	02.42%	00.42%	02.94%	02.12%
		Self discosure	17.00%	17.74%	23.91%	17.06%	19.73%
	Interactive	Continuing thread	95.73%	90.32%	99.38%	90.24%	100.00%
		Quoting message	04.75%	03.23%	00.93%	06.47%	01.11%
		Referring message	05.32%	05.91%	04.04%	05.65%	04.79%
		Asking question	40.36%	37.37%	73.29%	14.47%	75.47%
		Complimenting	81.67%	64.25%	90.68%	70.94%	87.85%
		Agreement	14.81%	16.67%	07.76%	17.18%	10.81%
	Cohesive	Vocatives	84.14%	66.94%	92.55%	77.41%	86.40%
		Group	08.55%	08.87%	06.52%	09.41%	07.13%
		Salutations	75.78%	60.48%	80.12%	72.00%	74.58%
Cognitive presence		Others	07.41%	10.48%	07.14%	07.29%	08.70%
		Triggering event	16.14%	16.13%	24.22%	17.65%	17.61%
		Exploration	38.46%	37.10%	43.79%	39.76%	38.57%
		Integration	31.53%	29.30%	20.81%	29.18%	28.99%
		Resolution	06.46%	06.99%	04.04%	06.12%	06.13%
		Number of students/role	82	67	14	06.12%	06.13%
		Number of posts	1053	372	322	850	897

3   Conclusions and Lines for Further Work

The cluster analysis revealed that the social presence indicators and cognitive presence phases were effective in dividing the students into groups as the silhouette value reached 0.3644 [22]. Table 2 shows that cluster 1 and cluster 2 erre the most diverse groups in relation to the social and cognitive presences, respectively. The literature shows that when the instructor assigns students to scripted roles, the students play their scripted roles (especially when the participation in group work counts towards final marks) [15,23]. Cluster 0 and cluster 2 showed high similarities with the expert and practicing researcher, respectively, corroborating the literature [15,23]. However, cluster 2 had only 14 students in the practicing researcher role, and cluster 0, included students from both scripted roles. It is possible to say that the definition of the scripted roles impacted on the students' participation in group activities, but not every student acted according to their scripted role. Cluster 1, the less active students (an average of 5 posts per student), did not present a high degree of similarity with any scripted role.

The approach proposed here has some limitations that must be acknowledged. First, this study was based on the data from six-course offerings, from a single course and institution, which can affect the generalizability of the results obtained. Second, the findings of the present study may be limited, given the features of the course design and the scripted roles. The authors intend to apply the same analytic approach with other datasets form different course settings and with online discussions in a different language to address those problems. Finally, cluster analysis involves making many decisions, such as on the number of clusters and the features used. Different algorithms and parameters may yields different results. The authors intend to evaluate different clustering algorithms and incorporate other features as suggested in [2,3,18].

References

1. De Laat, M., Lally, V.: It's not so easy: researching the complexity of emergent participant roles and awareness in asynchronous networked learning discussions. J. Comput. Assist. Learn. **20**(3), 165–171 (2004)
2. De Wever, B., Van Keer, H., Schellens, T., Valcke, M.: Roles as a structuring tool in online discussion groups: the differential impact of different roles on social knowledge construction. Comput. Hum. Behav. **26**(4), 516–523 (2010)
3. Dowell, N.M., Poquet, O.: SCIP: combining group communication and interpersonal positioning to identify emergent roles in scaled digital environments. Comput. Hum. Behav. 106709 (2021)
4. Ferreira-Mello, R., André, M., Pinheiro, A., Costa, E., Romero, C.: Text mining in education. Wiley Interdiscip. Rev.: Data Mining Knowl. Disc. **9**(6), e1332(2019)
5. Garrison, D.R.: Thinking Collaboratively: Learning in a Community of Inquiry. Routledge, New York (2016)
6. Garrison, D.R., Anderson, T., Archer, W.: Critical inquiry in a text-based environment: computer conferencing in higher education. Internet High. Educ. **2**(2–3), 87–105 (2000). https://doi.org/10.1016/S1096-7516(00)00016-6
7. Garrison, D.R., Anderson, T., Archer, W.: Critical thinking, cognitive presence, and computer conferencing in distance education. Am. J. Distance Educ. **15**(1), 7–23 (2001). https://doi.org/10.1080/08923640109527071
8. Garrison, D.R., Anderson, T., Archer, W.: The first decade of the community of inquiry framework: a retrospective. Internet High. Educ. **13**(1–2), 5–9 (2010)
9. Garrison, D.R., Cleveland-Innes, M.: Facilitating cognitive presence in online learning: interaction is not enough. Am. J. Distance Educ. **19**(3), 133–148 (2005)
10. Gašević, D., Adesope, O., Joksimović, S., Kovanović, V.: Externally-facilitated regulation scaffolding and role assignment to develop cognitive presence in asynchronous online discussions. Internet High. Educ. **24**, 53–65 (2015)

11. Hamerly, G., Elkan, C.: Learning the k in k-means. Adv. Neural Inf. Process. Syst. **16**, 281–288 (2004)
12. Joksimović, S., Gašević, D., Kovanović, V., Adesope, O., Hatala, M.: Psychological characteristics in cognitive presence of communities of inquiry: a linguistic analysis of online discussions. Internet High. Educ. **22**, 1–10 (2014)
13. Kovanović, V., Gašević, D., Joksimović, S., Hatala, M., Adesope, O.: Analytics of communities of inquiry: effects of learning technology use on cognitive presence in asynchronous online discussions. Internet High. Educ. **27**, 74–89 (2015)
14. Kovanović, V., et al.: Examining communities of inquiry in massive open online courses: the role of study strategies. Internet High. Educ. **40**, 20–43 (2019)
15. Mayordomo, R.M., Onrubia, J.: Work coordination and collaborative knowledge construction in a small group collaborative virtual task. Internet High. Educ. **25**, 96–104 (2015)
16. Ferreira Mello, R., Gašević, D.: What is the effect of a dominant code in an epistemic network analysis? In: Eagan, B., Misfeldt, M., Siebert-Evenstone, A. (eds.) ICQE 2019. CCIS, vol. 1112, pp. 66–76. Springer, Cham (2019). https://doi.org/10.1007/978-3-030-33232-7_6
17. Näykki, P., Isohätälä, J., Järvelä, S., Pöysä-Tarhonen, J., Häkkinen, P.: Facilitating socio-cognitive and socio-emotional monitoring in collaborative learning with a regulation macro script-an exploratory study. Int. J. Comput.-Supported Collab. Learn. **12**(3), 251–279 (2017)
18. Pozzi, F.: The impact of scripted roles on online collaborative learning processes. Int. J. Comput.-Supported Collab. Learn. **6**(3), 471–484 (2011)
19. Rolim, V., Ferreira, R., Lins, R.D., Găsević, D.: A network-based analytic approach to uncovering the relationship between social and cognitive presences in communities of inquiry. Internet High. Educ. **42**, 53–65 (2019)
20. Shaffer, D.W.: Epistemic Frames and islands of expertise: learning from infusion experiences. In: Proceedings of the 6th International Conference on Learning Sciences, ICLS 2004, pp. 473–480. International Society of the Learning Sciences, Santa Monica, California (2004). http://dl.acm.org/citation.cfm?id=1149126.1149184
21. Shaffer, D.W., et al.: Epistemic network analysis: a prototype for 21st-century assessment of learning. Int. J. Learn. Med. **1**(2) (2009)
22. Starczewski, A., Krzyżak, A.: Performance evaluation of the silhouette index. In: Rutkowski, L., Korytkowski, M., Scherer, R., Tadeusiewicz, R., Zadeh, L.A., Zurada, J.M. (eds.) ICAISC 2015. LNCS (LNAI), vol. 9120, pp. 49–58. Springer, Cham (2015). https://doi.org/10.1007/978-3-319-19369-4_5
23. Strijbos, J.W., Weinberger, A.: Emerging and scripted roles in computer-supported collaborative learning. Comput. Hum. Behav. **26**(4), 491–494 (2010)
24. Thomas, J.: Exploring the use of asynchronous online discussion in health care education: a literature review. Comput. Educ. **69**, 199–215 (2013)
25. Wise, A.F., Chiu, M.M.: The impact of rotating summarizing roles in online discussions: effects on learners' listening behaviors during and subsequent to role assignment. Comput. Hum. Behav. **38**, 261–271 (2014)
26. Xu, R., Wunsch, D.: Clustering, vol. 10. John Wiley (2008)

Towards Automatic Content Analysis of Rhetorical Structure in Brazilian College Entrance Essays

Rafael Ferreira Mello[1]([✉]), Giuseppe Fiorentino[1], Péricles Miranda[1],
Hilário Oliveira[2], Mladen Raković[3], and Dragan Gašević[3]

[1] Department of computing, Universidade Federal Rural de Pernambuco,
Recife, Brazil
{rafael.mello,pericles.miranda}@ufrpe.br
[2] Instituto Federal do Espírito Santo, Vitória, Brazil
[3] Centre for Learning Analytics, Faculty of Information Technology,
Monash University, Melbourne, Australia
{mladen.rakovic,dragan.gasevic}@monash.edu

Abstract. Essay scorers manually look for the presence of required rhetorical categories to evaluate coherence, which is a time-consuming task. Several attempts in the literature have been reported to automate the identification of rhetorical categories in essays with machine learning. However, existing machine learning algorithms are mostly trained on content features which can lead to over-fitting and hindering model generalizability. Thus, this paper proposed a set of content-independent features to identify rhetorical categories. The best performing classifier, XGBoost, achieved performance comparable to human annotation and outperformed previous models.

Keywords: Essay analysis · Content analytics · Rhetoric structure

1 Introduction

Essays are short literary compositions that reflect an author's perspective on a particular topic [30]. At Brazilian universities, the essay writing exam is among the key components in the admission process. In 2020, 5.8 million students had applied for university admission and took essay writing exams[1]; this increased the challenge to provide a good quality assessment for every essay submitted.

Prospective students are required to write a dissertative-argumentative essay that needs to be coherent and cohesive and written in Portuguese using a formal academic style. Students are required to compose sentences with different rhetorical functions (e.g., thesis, argument) and connect them into a coherent essay with an introduction, argumentation, and conclusion [16]. Thus, automatic

---

[1] https://bit.ly/36LivBB.

© Springer Nature Switzerland AG 2021
I. Roll et al. (Eds.): AIED 2021, LNAI 12749, pp. 162–167, 2021.
https://doi.org/10.1007/978-3-030-78270-2_29

identification of rhetorical categories in Brazilian entrance essays can improve the essay scoring process's efficiency during the university admission process.

To date, researchers have proposed several computational solutions to automatic identification of rhetorical components in student essays, e.g., [4,24,25]; of specific relevance is the work by dos Santos et al. [26] who propose an approach for the context of entrance essays at Brazilian universities. To this end, researchers have developed classification models based on machine learning (ML) and natural language processing (NLP) methods. While those classifiers demonstrated empirical validity and attractive classification performance (73%–93%), they typically involved some form of content features, i.e., features determined by the vocabulary in an essay (e.g., key terms, noun phrases). Due to differences in vocabulary use among students, however, reliance on content features can diminish classification performance, leading to over-fitting and hindering model robustness and generalizability [18].

In this study, we looked at the automatic identification of rhetorical components in student essays a step forward, as we developed a supervised machine learning model that relies upon content-independent features. To accomplish this goal, we examined a set of features derived from the two empirically validated linguistic tools, the Linguistic Inquiry Word Count (LIWC, [28]) and Coh-Metrix [15], and also computed order and relational features. The features in our study measured text cohesion, readability, and semantic relations and are not affected by a student's vocabulary use. Further, we developed five machine learning classification models to identify rhetorical categories in the essays.

2 Method

2.1 Dataset

The structure of the Brazilian essay writing exam was detailed in [26]. This framework represents characteristics of genre observed in essays written by students on entrance exams. Since the students are required to write an opinion paper on an assigned topic, it is common for their write-ups to start with the title (class **s0**). Introductory sentences that follow contextualize and present the subject addressed in the text.

These sentences are classified as Theme (**t1**). Thereafter, students are required to present their viewpoints regarding the essay topic. These sentences are categorized as Thesis (**t2**). After the introduction, students must present factual and logically valid arguments to support their thesis. These sentences are classified as (**s2**). The argumentation section is generally most elaborate in the essay, as students need to discuss their viewpoints. Finally, students conclude their essay. Conclusions commonly involve sentences summarising the initial thesis (classified as Background, **t3**) and sentences presenting final arguments (classified as Conclusion **s3**) that may or may not offer solutions to the problem discussed in the essay.

In this study, we used the dataset initially created by [26]. It encompasses 271 essays, divided into 2,562 sentences, written by candidates that took entrance

exams applying for Brazilian universities in 2014 and 2016. Three human annotators with background in computer science and linguistics coded each sentence
in the essay corpus according to the following categories: Title, Theme, Thesis,
Argumentation, Background, Conclusion, and Author. The value of Fleiss's κ
agreement between the coders reached 0.78. The disagreements were resolved by
adopting the category the majority of coders elected for.

2.2   Feature Extraction

We examined the performance of predictive models that use features based on
linguistic resources for rhetorical structure identification. Those features have
been largely harnessed in other problems in educational research [7,13,23].

LIWC Features: The Linguistic Inquiry Word Count (LIWC) is a dictionary of
measures indicative of different psychological processes (e.g., affective, cognitive,
social, perceptual) [28]. In this study, we utilized the Portuguese version of LIWC
proposed in 2019 [6]. The Portuguese version contains 73 categories of word
counts that were used as features in this paper.

Coh-Metrix Features: Coh-Metrix is a computational linguistics tool that
provides measures of linguistic complexity, text coherence, text readability, and
lexical category [22]. It has been widely used in previous studies to analyze essay
coherence and structure (e.g., [1,11,21]). The Portuguese version of Coh-Metrix
used in this paper [5] has 98 different measures.

Ordering Features: In addition to the indicators implemented by LIWC and
Coh-Metrix, different theories indicated that capturing the flow of the ideas in
the document is essential to categorize text blocks into a rhetorical structure
model [29]. Therefore, we also incorporated two features capturing the order of
the sentences in the text: i) the position from the first sentence to the last; ii)
the position of the sentences from the last to the first.

Features Extracted from Adjacent Sentences: The initial feature space
used in this work had a total of 173 features. However, previous works [14,26]
indicated that the use of sequence-based machine learning models could reach
better results for this problem. We adopted the features extracted from the
actual sentence and the previous and following sentences for each sentence to
incorporate the notion of sequence into traditional machine learning algorithms.
Thus, the final feature space in our analysis contained 519 features.

2.3   Model Selection and Evaluation

We trained several machine learning classifiers, including Random Forest, Gaussian kernel SVM, AdaBoost, XGBoost, and CRF. Random Forest and Gaussian
kernel SVM were included based on the good performance of several previous
analyses [12]. Moreover, AdaBoost and XGBoost decision tree algorithms have
demonstrated better results when compared to Random Forest [8,9]. Finally, we
also evaluated the Conditional Random Fields (CRF) algorithm's performance.

This algorithm is largely adopted for sequence labeling problems such as the analysis of rhetorical structures [27].

We used the same evaluation process performed in the previous work [26] to compare the classification results. To measure the performance of supervised machine learning algorithms, we adopted Cohen's κ [10], a metrics commonly used in educational data mining and learning analytics [20,23], and precision, recall, and f-measure, which are widely used metrics in the field of machine learning [2]. We applied 10-fold stratified cross-validation to evaluate all the measures obtained.

3   Results

Table 1 shows the best results achieved by each algorithm using 10-fold cross-validation (as described in Sect. 2.3). The outcomes revealed that the XGBoost algorithm reached the best results in general for all metrics evaluated. The XGBoost outperformed, in terms of Cohen's κ, by 6.34% and 11.70% the CRF and Random Forrest classifiers, respectively. Adaboost and SVM achieved the worst results in this experiment.

Table 1. Results for the analysed algorithms in terms of precision, recall, F1-score, and Cohen's κ.

Algorithm	Precision	Recall	F1-score	κ
SVM	0.56	0.58	0.50	0.42
Random forest	0.71	0.71	0.71	0.60
AdaBoost	0.65	0.55	0.59	0.45
XGBoost	0.73	0.75	0.73	0.67
CRF	0.70	0.72	0.71	0.63

4   Discussion and Practical Implications

The best performing classifier, XGBoost, achieved κ of 0.67, the performance comparable to human annotation as discussed in [26]. Importantly, as a part of our modeling approach, we utilized the non-content features to improve the performance and generalizability of the classifier. As these features represent the structure (e.g., cohesiveness, legibility, semantic relationships, number of nouns and pronouns) of the text instead of the content itself [3,13], a considerably accurate classifier we developed based on those features promises robustness in predicting rhetorical categories across different writing styles and genres. Equally important, this approach to feature extraction reduces the total number of features in the model, decreasing the chances of over-fitting [18].

The study's practical implications include: (i) the classifier we developed may provide accurate automatic identification of rhetorical categories in entrance essays and reduce the time the assessors need to review and score each essay manually; (ii) in the context of writing assignments in university courses, the automatic analysis of rhetorical structures could generate valuable information in creating formative feedback to guide essay revisions [17]; and (iii) the results provide a foundation for the development of learning analytics tools for instructors and students based on the rhetorical structure theory [19].

References

1. Abba, K.A., Joshi, R.M., Ji, X.R.: Analyzing writing performance of l1, l2, and generation 1.5 community college students through coh-metrix. Written Lang. Literacy **22**(1), 67–94 (2019)
2. Aggarwal, C.C., Zhai, C.: A survey of text classification algorithms. In: Mining Text Data, pp. 163–222. Springer (2012). https://doi.org/10.1007/978-1-4614-3223-4_6
3. Barbosa, G., et al.: Towards automatic cross-language classification of cognitive presence in online discussions. In: Proceedings of the Tenth International Conference on Learning Analytics & Knowledge, pp. 605–614 (2020)
4. Burstein, J., Marcu, D., Knight, K.: Finding the write stuff: automatic identification of discourse structure in student essays. IEEE Intell. Syst. **18**(1), 32–39 (2003). https://doi.org/10.1109/MIS.2003.1179191
5. Camelo, R., Justino, S., de Mello, R.F.L.: Coh-metrix PT-BR: uma API web de análise textual para a educação. In: Anais dos Workshops do IX Congresso Brasileiro de Informática na Educação, pp. 179–186. SBC (2020)D
6. Carvalho, F., Rodrigues, R.G., Santos, G., Cruz, P., Ferrari, L., Guedes, G.P.: Evaluating the Brazilian Portuguese version of the 2015 LIWC lexicon with sentiment analysis in social networks. In: Anais do VIII Brazilian Workshop on Social Network Analysis and Mining, pp. 24–34. SBC (2019)
7. Cavalcanti, A.P., et al.: How good is my feedback? A content analysis of written feedback. In: Proceedings of the Tenth International Conference on Learning Analytics & Knowledge, pp. 428–437 (2020)
8. Chan, J.C.W., Paelinckx, D.: Evaluation of random forest and adaboost tree-based ensemble classification and spectral band selection for ecotope mapping using airborne hyperspectral imagery. Remote Sensing Environ. **112**(6), 2999–3011 (2008)
9. Chen, T., Guestrin, C.: Xgboost: a scalable tree boosting system. In: Proceedings of the 22nd ACM SIGKDD International Conference on Knowledge Discovery and Data Mining, pp. 785–794 (2016)
10. Cohen, J.: A coefficient of agreement for nominal scales. Educ. Psychol. Meas. **20**(1), 37–46 (1960)
11. Crossley, S.A., McNamara, D.S.: Understanding expert ratings of essay quality: Coh-metrix analyses of first and second language writing. Int. J. Continuing Eng. Educ. Life Long Learn. **21**(2–3), 170–191 (2011)
12. Fernández-Delgado, M., Cernadas, E., Barro, S., Amorim, D.: Do we need hundreds of classifiers to solve real world classification problems? J. Mach. Learn. Res. **15**(1), 3133–3181 (2014)
13. Ferreira, M., Rolim, V., Mello, R.F., Lins, R.D., Chen, G., Gašević, D.: Towards automatic content analysis of social presence in transcripts of online discussions. In: Proceedings of the Tenth International Conference on Learning Analytics & Knowledge, pp. 141–150 (2020)
14. Fiacco, J., Cotos, E., Rose, C.: Towards enabling feedback on rhetorical structure with neural sequence models. In: Proceedings of the 9th International Conference on Learning Analytics & Knowledge, pp. 310–319 (2019)

15. Graesser, A.C., McNamara, D.S., Kulikowich, J.M.: Coh-metrix: providing multi-level analyses of text characteristics. Educ. Res. **40**(5), 223–234 (2011)
16. Haendchen Filho, A., do Prado, H.A., Ferneda, E., Nau, J.: An approach to evaluate adherence to the theme and the argumentative structure of essays. Proc. Comput. Sci. **126**, 788–797 (2018)
17. Jiang, S., Yang, K., Suvarna, C., Casula, P., Zhang, M., Rose, C.: Applying rhetorical structure theory to student essays for providing automated writing feedback. In: Proceedings of the Workshop on Discourse Relation Parsing and Treebanking 2019, pp. 163–168 (2019)
18. Khalid, S., Khalil, T., Nasreen, S.: A survey of feature selection and feature extraction techniques in machine learning. In: 2014 Science and Information Conference, pp. 372–378. IEEE (2014)
19. Kiesel, D., Riehmann, P., Wachsmuth, H., Stein, B., Froehlich, B.: Visual analysis of argumentation in essays. IEEE Trans. Visual. Comput. Graph. **27**, 1139–1148 (2020)
20. Kovanovic, V., Joksimovic, S., Gasevic, D., Hatala, M.: What is the source of social capital? The association between social network position and social presence in communities of inquiry. In: Workshop at Educational Data Mining Conference. EDM (2014)
21. Latifi, S., Gierl, M.: Automated scoring of junior and senior high essays using coh-metrix features: implications for large-scale language testing. Lang. Test. 0265532220929918 (2020)
22. McNamara, D.S., Graesser, A.C., McCarthy, P.M., Cai, Z.: Automated Evaluation of Text and Discourse with Coh-Metrix. Cambridge University Press, Cambridge (2014)
23. Neto, V., Rolim, V., Ferreira, R., Kovanović, V., Gašević, D., Lins, R.D., Lins, R.: Automated analysis of cognitive presence in online discussions written in Portuguese. In: European Conference on Technology Enhanced Learning, pp. 245–261. Springer (2018). https://doi.org/10.1007/978-3-319-98572-5_19
24. Nguyen, H., Litman, D.: Context-aware argumentative relation mining. In: Proceedings of the 54th Annual Meeting of the Association for Computational Linguistics (Volume 1: Long Papers), pp. 1127–1137 (2016)
25. Rakovic, M., Winne, P., Marzouk, Z., Chang, D.: Automatic identification of knowledge transforming content in argument essays developed from multiple sources. J. Comput. Assist. Learn
26. dos Santos, K.S., Soder, M., Marques, B.S.B., Feltrim, V.D.: Analyzing the rhetorical structure of opinion articles in the context of a Brazilian college entrance examination. In: Villavicencio, A., et al. (eds.) PROPOR 2018. LNCS (LNAI), vol. 11122, pp. 3–12. Springer, Cham (2018). https://doi.org/10.1007/978-3-319-99722-3_1
27. Sutton, C., McCallum, A.: An introduction to conditional random fields for relational learning. In: Introduction to Statistical Relational Learning, vol. 2, pp. 93–128 (2006)
28. Tausczik, Y.R., Pennebaker, J.W.: The psychological meaning of words: LIWC and computerized text analysis methods. J. Lang. Soc. Psychol. **29**(1), 24–54 (2010). https://doi.org/10.1177/0261927X09351676
29. Van Dijk, T.A.: Macrostructures: An Interdisciplinary Study of Global Structures in Discourse, Interaction, and Cognition. Routledge (2019)
30. Zupanc, K., Bosnić, Z.: Automated essay evaluation with semantic analysis **120**(C), 118–132 (2017). https://doi.org/10.1016/j.knosys.2017.01.006

Contrasting Automatic and Manual Group Formation: A Case Study in a Software Engineering Postgraduate Course

Giuseppe Fiorentino[1], Péricles Miranda[1], André Nascimento[1],
Ana Paula Furtado[1], Henrik Bellhäuser[2], Dragan Gašević[3],
and Rafael Ferreira Mello[1(✉)]

[1] Department of computing, Universidade Federal Rural de Pernambuco,
Recife, Brazil
{pericles.miranda,andre.camara,anapaula.furtado,rafael.mello}@ufrpe.br
[2] Johannes Gutenberg-Universität Mainz, Mainz, Germany
bellhaeuser@uni-mainz.de
[3] Centre for Learning Analytics, Faculty of Information Technology,
Monash University, Melbourne, Australia
dragan.gasevic@monash.edu

Abstract. This paper proposes the comparison of a group formation approach based on an evolutionary algorithm with a manual approach performed by an instructor with ten years of experience on this task. The groups were created based on the professional, psychological, and experience profile of each student. The results obtained demonstrated the algorithm's potential, reaching an average similarity of 83.46% with the groups formed manually by the instructor.

Keywords: Group formation · Optimization · Collaborative learning

1 Introduction

Group formation is a complex problem that considers multiple personal characteristics and skills of individuals, and the complexity increases depending on the number of students enrolled in the course [9]. Moreover, the literature shows that the creation of diverse and cohesive groups of students enhances the learning process and the outcome of individual students [3,5,7,12,14].

The state-of-the-art algorithms define group formation as a combinatorial optimization problem, with one or more criteria [10]. Different studies [2,4,6, 8,11] have applied optimization methods to the problem of the automatic formation of groups, considering features such as the skills of those involved, their interactions, and performance in previous courses. Recently, Miranda et al., [10] proposed a group formation evolutionary optimization algorithm that considers different skills (i.e., ability in coding, problem interpretation, communication,

I. Roll et al. (Eds.): AIED 2021, LNAI 12749, pp. 168–172, 2021.
https://doi.org/10.1007/978-3-030-78270-2_30

and leadership) and aims to satisfy three objectives: intra-group heterogeneity (groups with people with complementary skills), inter-group homogeneity (overall similar profile between different groups), and empathy. This approach was validated on different scenarios reaching better results when compared to the literature.

In this context, the present work proposes an experimental study that applies the multi-objective algorithm proposed in [10] in forming groups of face-to-face classes of a Postgraduate course in Software Engineering. More specifically, this paper intends to answer the following research question: *To what extent can optimization algorithms automatically reproduce the results of an experienced instructor in the activity of forming groups in a learning context?*

2   Method

2.1   Data and Course Context

The data used in the present study consisted of four offerings of a professional master-level course in software engineering offered in a face-to-face setting at a Brazilian private university between 2019 and 2020. As part of the requirements for the master program, the students were divided into groups to develop a project in partnership with real companies. This project accounts for 60% of the credits to meet the requirements of the master's degree program. In those four offerings, a total of 93 students were divided into 14 groups. The different runs of the course had the following numbers:

- **Class 1**: Period: 2019.1; Number of Students: 30; Number of groups: 5.
- **Class 2**: Period: 2019.1; Number of Students: 21; Number of groups: 3.
- **Class 3**: Period: 2019.2; Number of Students: 14; Number of groups: 2.
- **Class 4**: Period: 2020.1; Number of Students: 28; Number of groups: 4.

The data collection took place during the first week of the course through an online questionnaire. The data required from students included the place where they live, information about the undergraduate course, professional profile, time availability, and MBTI classification. Figure 1 summarizes students' responses considering the professional experience and profile, and the Typological classification. It shows that the main profile is composed of developers with medium and high professional experience and who are mostly introverted.

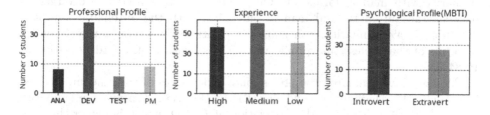

Fig. 1. Visualization of all numerical data and their distribution in all classes

2.2   Manual Methodology for Group Formation

Based on the information collected from the students, the instructor used three criteria to define the final groups: (i) Consider professional backgrounds of the students in order balance their skills across different groups. This criterion assures the heterogeneity of skills needed for the execution of the project; (ii) Consider years of professional experience and graduation year to create groups with mixed experiences and prevent a group from having an imbalance across professional experience levels; (iii) Adoption of the MBTI profile and availability for the project development so that groups can better understand each member's psychological profile and dedication throughout the project.

2.3   Automatic Methodology for Group Formation

The study adopted the multi-objective algorithm proposed in [10] for the automatic group formation of face-to-face classes in a postgraduate course in software engineering. The aim was to verify whether the GFMOA is able to form groups with features similar to those formed by the manual approach. The GFMOA optimizes two objectives: maximizing inter-homogeneity between groups and maximizing intra-heterogeneity for each group. This is done by taking into account the same aspects adopted by the instructor (see Sect. 2.2).

3   Results

The results present the percentage of features extracted with the automatic group formation algorithm that matched the manual approach; this is referred to as similarity. Initially, we evaluated the similarity concerning the professional profile. Regarding the similarity with the groups formed by the manual approach, in 2019.1a, the instructor allowed groups of different sizes and the GFMOA can only form groups of the same size. Thus, this variability led to a lower similarity between the manual and automatic approach in this course. For instance, in two groups the similarity achieved only 50.0% and 66.6%. However, the other groups showed similarities above 70%, reaching up to 85.7%. In the other courses analyzed (2019.1b and 2019.2) groups of the same size were formed. Consequently, GFMOA's performance reached 85.7% similarity across all groups formed. It means that the algorithm made the wrong choice in only one of the seven possible profiles per group. In 2020.1, of the four groups formed, three showed 85.7% similarity with manual formation. Only one group presented two divergences between the automatic and manual outcomes, with the automatic approach obtaining 71.4% similarity with manual formation. On average, the similarity in terms of the professional profile was 79.58%, with a standard deviation of 10.84%.

Moreover, we evaluated the professional experiences of the students divided into High (H), Medium(M), and Low(L). In this case, the average similarity reached 83.31% (with a standard deviation of 7.59%), which is higher than the professional profile result. In general, the algorithm made one wrong choice per group. The lower accurate group achieved 71.4% of similarity in the class of 2020.1. On the other hand, the GFMOA reached 85.7% of similarity for all groups in 2019.1b and 2019.2.

Finally, we also investigated the automatic and manual group matching in terms of the MBTI, which divides the students into Introvert (I) and Extrovert (E). In this case, the average similarity reached 87.57%, with a standard deviation of 11.06%. This result was expected as MBTI divides the students into only two groups, while the professional profile and experience divide the students into four and three groups, respectively.

Overall, the average similarity of the GFMOA in relation to the three aspects was 83.46%, with a standard deviation of 9.83%. It shows the potential in using GFMOA to reproduce the manual formation methodology. It is important to mention that we considered features such as professional, psychological, and experience profile in this study.

4 Practical Implications

Practical implications of the results presented includes three main findings. First, the use of an automatic group formation algorithm reduces the dependency on and the bias of the instructor [13]. For instance, the institution where we collected the data delegates to the same instructor the responsibility of creating the groups for an activity that accounts for 60% of all credits required for students. In short, it represents a weakness of this course, where the change of the instructor could be a threat to the students' success.

Second, the manual group formation process could be biased [1] as the process of group creation happens after the first class where the instructor met the entire cohort of students in a face-to-face activity [1]. Thus, the group formation could potentially lead to a decision based on the students' behavior in the class and not based on the predefined criteria.

Finally, the results of the study presented in the paper outcome demonstrate the potential of GFMOA to be adopted in practice in two contexts: i) to perform an automatic group formation process that could support collaborative learning activities; and ii) the use of GFMOA to provide recommendations for instructors, with less experience, to decide how to create a group of students efficiently [10]. It is important to highlight that GFMOA effectively reproduces instructors' choice (reaching more than 80% of concordance with the manual decision) and efficiently generates the groups in up to 120 s.

As future work, we intended to expand the characterization process and describe the students' profiles in greater detail. In terms of evaluation, we aim to use different approaches, including a) a questionnaire that will be administered after the group activity to measure the students' satisfaction and b) conduct a randomized control trail [3] to measure the effects of different group formation approaches on the individual and group performance. Finally, we aim to develop a recommendation system to support instructors' decisions on the best groups for specific activities.

References

1. Alqahtani, M., Gauch, S., Salman, O., Ibrahim, M., Al-Saffar, R.: Diverse group formation based on multiple demographic features. arXiv preprint arXiv:2008.03808 (2020)
2. Ani, Z.C., Yasin, A., Husin, M.Z., Hamid, Z.A.: A method for group formation using genetic algorithm. Int. J. Comput. Sci. Eng. **2**(9), 3060–3064 (2010)
3. Bellhäuser, H., Konert, J., Müller, A., Röpke, R.: Who is the perfect match? effects of algorithmic learning group formation using personality traits. J. Interact. Med. (i-com) **17**(1), 65–77 (2018). https://doi.org/10.1515/icom-2018-0004
4. Contreras, R., Salcedo, P., et al.: Genetic algorithms as a tool for structuring collaborative groups. Natural Comput. **16**(2), 1–9 (2016)
5. Dillenbourg, P.: Collaborative Learning: Cognitive and Computational Approaches. Advances in Learning and Instruction Series. ERIC (1999)
6. Graf, S., Bekele, R.: Forming heterogeneous groups for intelligent collaborative learning systems with ant colony optimization. In: Ikeda, M., Ashley, K.D., Chan, T.-W. (eds.) ITS 2006. LNCS, vol. 4053, pp. 217–226. Springer, Heidelberg (2006). https://doi.org/10.1007/11774303_22
7. Järvelä, S., et al.: Socially shared regulation of learning in CSCL: understanding and prompting individual-and group-level shared regulatory activities. Int. J. Comput.-Supported Collab. Learn. **11**(3), 263–280 (2016)
8. Lin, Y.T., Huang, Y.M., Cheng, S.C.: An automatic group composition system for composing collaborative learning groups using enhanced particle swarm optimization. Comput. Educ. **55**(4), 1483–1493 (2010)
9. Liu, C.C., Tsai, C.C.: An analysis of peer interaction patterns as discoursed by online small group problem-solving activity. Comput. Educ. **50**(3), 627–639 (2008)
10. Miranda, P.B., Mello, R.F., Nascimento, A.C.: A multi-objective optimization approach for the group formation problem. Exp. Syst. Appl. **162**, 113828 (2020)
11. Moreno, J., Ovalle, D.A., Vicari, R.M.: A genetic algorithm approach for group formation in collaborative learning considering multiple student characteristics. Comput. Educ. **58**(1), 560–569 (2012)
12. Müller, A., Bellhäuser, H., Konert, J., Röpke, R.: Effects of group formation on student satisfaction and performance: a field experiment. Small Group Res. 1046496420988592 (2021). https://doi.org/10.1177/1046496420988592
13. Odo, C., Masthoff, J., Beacham, N.: Group formation for collaborative learning. In: Isotani, S., Millán, E., Ogan, A., Hastings, P., McLaren, B., Luckin, R. (eds.) AIED 2019. LNCS (LNAI), vol. 11626, pp. 206–212. Springer, Cham (2019). https://doi.org/10.1007/978-3-030-23207-8_39
14. Zheng, B., Niiya, M., Warschauer, M.: Wikis and collaborative learning in higher education. Technol. Pedagogy Educ. **24**(3), 357–374 (2015)

Aligning Expectations About the Adoption of Learning Analytics in a Brazilian Higher Education Institution

Samantha Garcia[1], Elaine Cristina Moreira Marques[1],
Rafael Ferreira Mello[1(✉)], Dragan Gašević[2], and Taciana Pontual Falcão[1]

[1] Department of computing, Universidade Federal Rural de Pernambuco,
Recife, Brazil
`{rafael.mello,taciana.pontual}@ufrpe.br`
[2] Centre for Learning Analytics, Faculty of Information Technology,
Monash University, Melbourne, Australia
`dragan.gasevic@monash.edu`

Abstract. Stakeholders' buy-in is fundamental for the successful implementation of Learning Analytics (LA) in Higher Education. We present the results of a survey in a Brazilian HEI, to investigate the ideal and realistic expectations of students and instructors about the adoption of LA. Results indicate a high interest in using LA for improving the learning experience, but with ideal expectations higher than realistic expectations, and point out key challenges and opportunities for Latin American researchers to join efforts towards building solid evidence that can inform educational policy-makers and managers, and support the development of strategies for LA services in the region.

Keywords: Learning Analytics · Student and instructor expectations

1 Introduction

Learning Analytics (LA) has been increasingly used over the last years, as the analysis of large masses of data became more accessible and popular [8]. In higher education institutions (HEI), LA can optimize learning and its environments [16] with great potential to face educational challenges, such as student drop-out, failure, and personalized feedback at scale [5,12]. In Latin America (LATAM), LA adoption is still timid [1,6], but the amount of data collected over the last years indicates that LATAM countries can implement LA strategies to target challenges in the educational system [2], addressing known problems in the region like student dropout and program quality [7,11]. In Brazil, interest in LA is growing, particularly as online and blended learning expand, through the extensive use of Learning Management Systems (LMS) [1].

However, LA implementation is highly dependent on contextual factors [9, 15]. Projects like SHEILA (Supporting Higher Education to Integrate Learning

© Springer Nature Switzerland AG 2021
I. Roll et al. (Eds.): AIED 2021, LNAI 12749, pp. 173–177, 2021.
https://doi.org/10.1007/978-3-030-78270-2_31

Analytics) [14] and LALA (Learning Analytics in Latin America) [10] encourage diagnoses of HEI towards a successful adoption of LA, particularly identifying key stakeholders and the changes they desire. Stakeholder engagement and buy-in is one of the four main challenges for LA adoption, along with pedagogical grounding, resources, and ethics and privacy [15]. So far little is known about how stakeholders' opinions and behaviors impact LA adoption in the LATAM context [7]. We seek to address this gap by performing empirical research in a Brazilian HEI, generating evidence about stakeholders' opinions and perceptions that can guide LA implementation with maximized buy-in.

Qualitative findings from previous research indicate high interest of students and instructors in using LA for improving the learning experience, by providing personalized feedback, adapting teaching practices to students' needs, and making evidence-based pedagogical decisions [3,4]. In this paper, we complement qualitative evidence with quantitative data collected through SHEILA's survey instrument, which allows to investigate stakeholders' ideal and predicted expectations regarding the adoption of LA [9,17], considering the institutional context and the actual feasibility of LA successful implementation. While ideal expectations are desired outcomes based on the individual's hope, predicted expectations are realistic beliefs about what is perceived as feasible. Together, they provide deeper understanding of stakeholders' perspectives, and allow identifying main areas to focus, with realistically expected topics being considered priority in service planning [17].

2   Results

The survey items were formulated as statements with two 7-point Likert scales from total disagreement to total agreement, for both ideal and realistic expectations. The themes were data privacy, academic progress, feedback, decision-making, intervention and training. The survey had 241 participants from a Brazilian HEI (192 students and 49 instructors), from various areas of knowledge and courses (online and face-to-face).

Overall, the responses from instructors show high ideal expectations about the adoption of LA in their institution, with less optimistic views about the viability in their current context (median rating scores between 5 and 6) (Fig. 1). Items regarding access to students' progress (Q4-I and Q5-I), university support on data analysis (Q7-I), understanding of data (Q11-I), learning profile (Q12-I) and visualization of learning performance (Q16-I) showed almost unanimously high ideal expectations. Some of these items also had the highest median ratings of perceived feasibility in the present context (Q4-I, Q5-I, Q11-I, Q12-I and Q16-I). The university support on data analysis (Q7-I) showed the biggest interval, with answers between 3 and 7, oscillating between agreement and neutrality.

From students' perspective, the ideal expectations are also high on the use of LA, but slightly lower compared to the instructors' expectations (Fig. 2). For what students' expected as realistically applicable in their context, the median rating scores are between 5 and 7, i.e., higher values than those expressed through

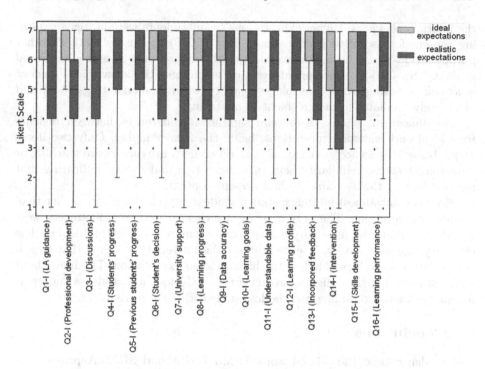

Fig. 1. Box plot of instructors' responses

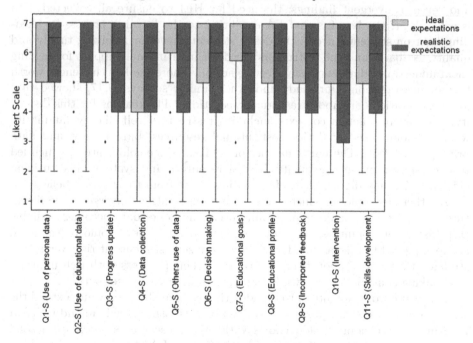

Fig. 2. Box plot of students' responses

the instructors' views. Higher expectations from students were in items regarding consent for use of their educational data (Q2-S) and use of data for other purposes (Q5-S); accessing their educational progress (Q3-S) and educational goals (Q7-S). Q10-S, regarding intervention when analytics show that a student is at-risk of failing, shows the biggest gap between median ratings (3–7) and most likely oscillates from agreement to neutrality.

For almost all items, there were significant differences between instructors' ideal and realistic expectations, being the former higher. Only two items about being able to access students' data on courses instructors are teaching or have taught showed similarity between expectation and reality, indicating that instructors find this is viable in their present context.

Similarly, for almost all items, students' ideal expectations were higher than realistic expectations. The only item for which no significant differences were found was about the university asking for consent to use identifiable data like ethnicity, age and gender. The item about the university ensuring that educational data will be kept safe had the highest ideal expectation. The item about instructors having the obligation to act if the analytics show students underperforming or at-risk of failing had the lowest realistic expectations rating.

3   Conclusions

Our evidences and the related work within LALA and SHEILA projects, in LATAM [7] and globally [9,17], reinforce the importance of stakeholder buy-in and reveal convergent findings: the need for HEI to ensure all collected data is safely kept; the benefits that LA can bring to the learning process by shedding light on students' needs; the desire students have for receiving timely and quality feedback; and the instructors' need for institutional support for helping them understand data and take effective action. There are key challenges worth further investigation. Our study and other similar surveys [9,17] showed that ideal expectations are above realistic expectations. The reasons for this disparity may vary in different contexts, including instructors' self-efficacy, familiarity with technology and analytics, institutional resources, bureaucracy, and data privacy legislation. The particularities of LATAM since colonization, which led to deep socioeconomic inequality, lack of resources and systemic institutional efficiency [7], may drive stakeholders' wishes apart from their actual beliefs.

Another key topic on which opinions and expectations diverge in the literature relates to whose main responsibility it is to act, once data becomes available [13]. Instructors' opinions vary on the extent to which they should be the main group expected to take action, for example to rescue students at-risk, versus the students themselves, upon being informed of their progress with rich information, taking control of their own learning, with instructors' support.

For future work, we intend to broaden the reach of the survey and extend the study to managers, with possible partnerships with other Brazilian and LATAM institutions. We hope to add efforts with other researchers to create a solid corpus of evidence that reflects the identity(ies) of LATAM [6,7], and leads to effective strategies that promote the adoption of LA in LATAM institutions.

References

1. Cechinel, C., Ochoa, X., Lemos dos Santos, H., Carvalho Nunes, J.B., Rodés, V., Marques Queiroga, E.: Mapping learning analytics initiatives in Latin America. Br. J. Educ. Technol. **51**(4), 892–914 (2020)
2. Cobo, C., Aguerrebere, C.: Building capacity for learning analytics in Latin America. Include us all! Directions for adoption of Learning Analytics in the global south p. 58 (2017)
3. Falcao, T.P., Ferreira, R., Rodrigues, R.L., Diniz, J., Gasevic, D.: Students' perceptions about learning analytics in a brazilian higher education institution. In: 2019 IEEE 19th International Conference on Advanced Learning Technologies (ICALT), vol. 2161, pp. 204–206. IEEE (2019)
4. Falcão, T.P., Mello, R.F., Rodrigues, R.L., Diniz, J.R.B., Tsai, Y.S., Gašević, D.: Perceptions and expectations about learning analytics from a brazilian higher education institution. In: Proceedings of the Tenth International Conference on Learning Analytics & Knowledge, pp. 240–249 (2020)
5. Ferguson, R.: Learning analytics: drivers, developments and challenges. Int. J. Technol. Enhanc. Learn. **4**(5–6), 304–317 (2012)
6. Hilliger, I., et al.: Towards learning analytics adoption: a mixed methods study of data-related practices and policies in Latin American universities. Br. J. Edu. Technol. **51**(4), 915–937 (2020)
7. Hilliger, I., et al.: Identifying needs for learning analytics adoption in Latin American universities: a mixed-methods approach. Int. High. Educ. **45**, 100726 (2020)
8. Joksimović, S., Kovanović, V., Dawson, S.: The journey of learning analytics. HERDSA Rev. High. Educ. **6**, 27–63 (2019)
9. Kollom, K., et al.: A four-country cross-case analysis of academic staff expectations about learning analytics in higher education. Int. High. Educ. **49**, 100788 (2021)
10. Maldonado-Mahauad, J., et al.: The LALA project: Building capacity to use learning analytics to improve higher education in Latin America. In: Companion Proceedings of the 8th International Learning Analytics & Knowledge conference, pp. 630–637 (2018)
11. Marta Ferreyra, M., Avitabile, C., Botero Álvarez, J., Haimovich Paz, F., Urzúa, S.: At a crossroads: higher education in Latin America and the Caribbean. The World Bank (2017)
12. Pardo, A., Jovanovic, J., Dawson, S., Gašević, D., Mirriahi, N.: Using learning analytics to scale the provision of personalised feedback. Br. J. Educ. Technol. **50**(1), 128–138 (2019)
13. Prinsloo, P., Slade, S.: An elephant in the learning analytics room: the obligation to act. In: Proceedings of the Seventh International Conference on Learning analytics & Knowledge, pp. 46–55. ACM, New York (2017)
14. Tsai, Y.S., et al.: The Sheila framework: informing institutional strategies and policy processes of learning analytics. J. Learn. Anal. **5**(3), 5–20 (2018)
15. Tsai, Y.S., et al.: Learning analytics in European higher education-trends and barriers. Comput. Educ. **155**, 103933 (2020)
16. Whitelock-Wainwright, A., Gašević, D., Tejeiro, R.: What do students want? Towards an instrument for students' evaluation of quality of learning analytics services. In: Proceedings of the Seventh International Conference on Learning Analytics & Knowledge, pp. 368–372 (2017)
17. Whitelock-Wainwright, A., Gašević, D., Tejeiro, R., Tsai, Y.S., Bennett, K.: The student expectations of learning analytics questionnaire. J. Comput. Assist. Learn. **35**(5), 633–666 (2019)

Interactive Teaching with Groups of Unknown Bayesian Learners

Carla Guerra[✉], Francisco S. Melo, and Manuel Lopes

INESC-ID & Instituto Superior Técnico, Universidade de Lisboa, Lisbon, Portugal
carla.guerra@gaips.inesc-id.pt, fmelo@inesc-id.pt,
manuel.lopes@tecnico.ulisboa.pt

Abstract. In this work we empirically explore the extension of an interactive approach for machine teaching from single learners to groups of learners. We use interactivity to overcome the common mismatch between the knowledge the teacher has about the students and the students themselves. With a multi-learner setting we also investigated the best way to consider the class—as a whole or divided in partitions accordingly to the students priors. The results of an user study where we teach a Bayesian estimation task have shown that, regardless of considering partitions or not, the interactive approaches significantly increase the learning performance of the class when compared to non-interactive alternatives.

Keywords: Machine teaching · Interactivity · Group-learning

1 Introduction

Machine teaching (MT) considers the problem of finding the smallest set of examples that allows a specific learner to acquire a given concept, explicitly considering a computational learning algorithm for the student [5,6,8]. Since a significant amount of teaching relies on providing examples, the learning efficiency can be greatly improved if the teacher selects the examples that are more informative for each particular learner using MT techniques. The main problem with MT is that it often assumes that the learner, or the learning algorithm, is completely known. This is a very strong assumption that does not hold in the general case, and certainly not in the case where the learner is a human. Melo et al. [4] explicitly address the unavoidable mismatch between what the machine teaching system assumes about the learner and the learner himself and propose interactivity as the means to overcome it. However, that work and most of MT research so far has focused on single learner scenarios. There are not yet many advances in machine teaching applied in a setting where the teacher must teach multiple learners although this is the reality in real-world classrooms. The students have different backgrounds and prior knowledges, which the teacher must take into consideration when delivering the same lecture to everyone.

© Springer Nature Switzerland AG 2021
I. Roll et al. (Eds.): AIED 2021, LNAI 12749, pp. 178–182, 2021.
https://doi.org/10.1007/978-3-030-78270-2_32

Our goal in this work is to empirically explore the impact of that imperfect knowledge in a classroom scenario with multiple learners and how the interactive approach proposed by Melo et al. [4] can help in overcoming it. Within this problem we formulated two hypotheses:

- Hypothesis 1: Considering interactivity in the teaching process outperforms other non-interactive approaches. This hypothesis is supported by numerous pedagogical studies found in the literature suggesting that interactivity enhances students' engagement and learning [1–3];
- Hypothesis 2: Dividing the class into partitions accordingly to the students priors and giving one sample per partition makes the learning process faster than considering the class as a whole. This hypothesis was inspired in the work by Zhu et al. [7] that investigated machine teaching with multiple learners but no interactivity.

To confirm these hypotheses we conducted an user study with classes of students. We found that interactivity can be the means to close the gap between the student and teacher parameters. Also, partitioning the whole group into smaller groups, although increasing the effort of the teacher in each run, revealed to improve the overall performance when combined with interactivity.

2   User Study

In this section we present an user study created to validate the hypotheses formulated related to our work. We want to test in a real-world scenario if the interactive approach proposed by Melo et al. [4] is still faster when we extend the discussion of interactivity from single-learner to multiple-learners settings. This raises several novel questions regarding the teaching process: How to interact with multiple students? How to deal with the individual differences between students? How much does the feedback from one student inform the teacher about the state of the class? We also explored different ways of considering the class—as a whole, or partitioned.

2.1   Experimental Design

We used the same artificial problem with a Bayesian estimation task proposed by Melo et al. [4], where each participant has to estimate the mean monthly rent of an 1-bedroom apartment in a city in the US. In this case, instead of teaching only one student, we considered groups of 10 students at the same time. In each run each student gives an answer (not shared with the rest of the group). After that, we give an example (said to be real) of an 1-bedroom apartment rented in that city to each student (which is, again, not shared with the others). This is repeated for 10 runs with every group.

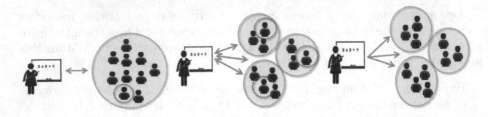

Fig. 1. Diagram with the 3 approaches considered: the interactive teaching approach with no partitions (left); the interactive teaching approach with partitions (middle); the non-interactive teaching approach with partitions (right).

To compute the teaching samples to show to the group we explored 3 different approaches:

- Condition 1: Interactive Teaching with No Partitions—Fig. 1 (left)—where we teach the whole group at the same time, considering in each iteration one randomly selected answer from the total group as the answer of the class. We then follow the algorithm of Melo et al. [4], where we take into account the considered answer of the class in each iteration and use it to calculate the sample to show. We show the same sample to every participant.
- Condition 2: Interactive Teaching with Partitions—Fig. 1 (middle)—instead of considering the whole-group, here we divide the group into smaller partitions. To do so we first ask each participant to select the interval where his estimate falls. We use that information to aggregate the participants into smaller groups (partitions) accordingly to this prior knowledge. In each iteration and for each partition we consider one randomly selected answer as the answer of that partition. If we have n partitions, we will consider n answers. And we will then show n samples, one per partition. Thus, not every participant sees the same sample as in condition 1—participants in different partitions see different samples.
- Condition 3: Non-Interactive Teaching with Partitions—Fig. 1 (right)—the group is partitioned as in Condition 2. However, the samples are presented to the students in each partition without taking into account the answers given by the students. Instead, the system assumes perfect knowledge about the student parameters and uses the mean value of the interval of prior estimates accepted in each partition as the estimate of that partition.

2.2 Results

The results confirmed our two hypothesis regarding the use of interactivity in the teaching process and addressing the class divided into partitions instead of as a whole. The study involved a total of 239 engineering students distributed uniformly among the three conditions. The average age was 23, with 71% males.

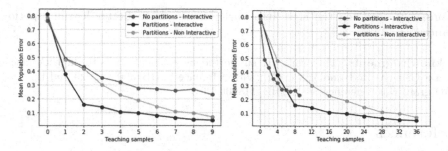

Fig. 2. Comparison of the learning performance when teaching multiple learners not considering the extra cost of teaching more partitions (left) and considering this extra effort (right).

Partitions vs. Whole Group. Figure 2 (left) shows that the conditions with partitions seem to have a better performance than the one with no partitions. However, when a teacher decides to partition the group, his teaching effort (the number of samples needed to teach) increases with the number of partitions considered. To take this into account, we multiplied the teaching samples of the conditions with partitions by the average number of partitions in all the groups acquired in each condition (around 4 in both of them)—Fig. 2 (left). Including this extra effort, it is not so obvious that the conditions with partitions have a better performance. When comparing the interactive cases with a Mann-Whitney U test, having partitions is indeed statistically better. However, the interactive condition with no partitions has significantly lower error rates when comparing with the non-interactive with partitions approach.

Interactivity vs. Non-interactivity. The previously mentioned results show that having partitions is not necessarily better and that one must also consider the factor of using interactivity in the teaching process. Indeed, with or without partitions, the interactive conditions showed significantly better performances when performing Mann-Whitney U tests, even when considering the extra teaching effort associated to the partitions.

3   Conclusions

In this work we empirically investigate the use of interactivity when teaching a Bayesian estimation task to groups of learners. However, we assume there is a mismatch between what the teacher knows about the learners and the learners themselves. We inspected two ways to consider the class: as a whole or partitioned accordingly to the learners priors. We could confirm that, by allowing the teacher to interactively assess the state of the class, the impact of the aforementioned mismatch is significantly mitigated. The results of an user study have shown that the interactive teaching approaches (with partitions or not) significantly outperform the non-interactive alternative. Between the interactive approaches, dividing the class into partitions leads to better learning performances.

Acknowledgments. This work was supported by national funds through Fundação para a Ciência e a Tecnologia (FCT) with reference UIDB/50021/2020 and the FCT PhD grant with reference SFRH/BD/118006/2016.

References

1. Alexander, R.: Developing dialogic teaching: genesis, process, trial. Res. Papers Educ. **33**(5), 561–598 (2018)
2. Beauchamp, G., Kennewell, S.: Interactivity in the classroom and its impact on learning. Comput. Educ. **54**(3), 759–766 (2010)
3. Kennewell, S., Beauchamp, G.: The features of interactive whiteboards and their influence on learning. Learn. Media Technol. **32**(3), 227–241 (2007)
4. Melo, F.S., Guerra, C., Lopes, M.: Interactive optimal teaching with unknown learners. In: Proceedings of the 27th International Joint Conference on Artificial Intelligence, pp. 2567–2573. Stockholm, Sweden (2018)
5. Zhu, X.: Machine teaching for bayesian learners in the exponential family. In: Advances in Neural Information Processing Systems, pp. 1905–1913 (2013)
6. Zhu, X.: Machine teaching: an inverse problem to machine learning and an approach toward optimal education. In: Twenty-Ninth AAAI Conference on Artificial Intelligence (2015)
7. Zhu, X., Liu, J., Lopes, M.: No learner left behind: on the complexity of teaching multiple learners simultaneously. In: Proceedings of the 26th International Joint Conference on Artificial Intelligence, pp. 3588–3594 (2017)
8. Zhu, X., Singla, A., Zilles, S., Rafferty, A.N.: An overview of machine teaching (2018). arXiv preprint, arXiv:1801.05927

Multi-task Learning Based Online Dialogic Instruction Detection with Pre-trained Language Models

Yang Hao[1], Hang Li[1], Wenbiao Ding[1], Zhongqin Wu[1], Jiliang Tang[2], Rose Luckin[3], and Zitao Liu[1(✉)]

[1] TAL Education Group, Beijing, China
{haoyang2,lihang4,dingwenbiao,wuzhongqin,liuzitao}@tal.com
[2] Data Science and Engineering Lab, Michigan State University, East Lansing, USA
tangjili@msu.edu
[3] UCL Knowledge Lab, London, UK
r.luckin@ucl.ac.uk

Abstract. In this work, we study computational approaches to detect online dialogic instructions, which are widely used to help students understand learning materials, and build effective study habits. This task is rather challenging due to the widely-varying quality and pedagogical styles of dialogic instructions. To address these challenges, we utilize pre-trained language models, and propose a multi-task paradigm which enhances the ability to distinguish instances of different classes by enlarging the margin between categories via contrastive loss. Furthermore, we design a strategy to fully exploit the misclassified examples during the training stage. Extensive experiments on a real-world online educational data set demonstrate that our approach achieves superior performance compared to representative baselines. To encourage reproducible results, we make our implementation online available at https://github.com/AIED2021/multitask-dialogic-instruction.

Keywords: Dialogic instruction · Multi-task learning · Pre-trained language model · Hard example mining

1 Introduction

Teaching online classes is a very challenging task for classroom instructors trained to work offline [12,24]. When sitting in front of a camera or a laptop, traditional classroom instructors lack effective pedagogical instructions to ensure the overall quality of their online classes [4,15]. In this paper, we develop a set of dialogic instructions for online classes aiming to encourage talks and discourses between teachers and students, in addition to teacher-presentation [7,10,11,14]. Furthermore, we study computational approaches to automatically detect these dialogic instructions from online class videos, which provides timely feedback to teachers and help them improve their online teaching skills.

© Springer Nature Switzerland AG 2021
I. Roll et al. (Eds.): AIED 2021, LNAI 12749, pp. 183–189, 2021.
https://doi.org/10.1007/978-3-030-78270-2_33

However, automatic dialogic instruction detection poses numerous challenges in real-life teaching scenarios. First, online teaching is not a standardized procedure. Even for the same learning content, different instructors may teach it in various ways according to their own pedagogical styles. Furthermore, the quality of dialogic instructions varies a lot from junior to senior instructors. The second challenge is that the model has to be robust enough to errors from automatic speech recognition (ASR) transcriptions. Publicly available ASR services may yield very high transcription errors, which lead to inferior performance in the noisy and dynamic classroom environments [3].

To address the above challenges, in this study, we propose an end-to-end multi-task framework for automatic dialogic instruction detection from online videos. Specifically, we (1) propose a contrastive loss based multi-task framework to distinguish instances by enlarging the distances between instances of different categories; (2) utilize the pre-trained neural language model to robustly handle errors from ASR transcriptions without the need for manual annotation efforts; and (3) propose a strategy to select and exploit hard instances in the training process to achieve higher performance.

2   The Dialogic Instruction Detection Framework

In this work, we aim to capture the following eight types of well-studied dialogic instructions that (1) motivate students and make them feel easy about the class: *greeting* [8,17] and *commending* [7,11], (2) help students understand learning materials and retain them: *guidance* [26], *example-giving* [20], *repeating* [2], and *reviewing* [1], and (3) build effective learning habits: *note-taking* [10,14] and *summarization* [18].

Our multi-task dialogic instruction detection framework has three key components: (1) a pre-trained language model, which serves as the base model in the classification task; (2) a multi-task learning module, which distinguishes effective instructions from similar but ineffective ones by pushing instances from different categories apart; and (3) a hard example mining strategy, which establishes a hard example set to select instances when constructing input pairs.

Pre-trained Language Model. To extract contextual information, in this study we utilize the Transformer-based pre-trained language model as our base model in our detection framework. To perform the instruction detection task on a sentence, similar to [6,16], we first add a special token $[CLS]$ in front of the sentence. After that, sentences are fed into multiple Transformer encoders sequentially. Finally the hidden state of the special token $[CLS]$ from the last layer of Transformer encoders is obtained as the representation of the sentence.

Multi-task Learning Module. The multi-task learning framework consists of two sub-tasks: (1) a multi-class classification task to decide which category a dialogic instruction belongs to, where the cross-entropy loss is used; and (2) an additional task with an objective to enlarge the distances between pairs of instructions from different categories by using contrastive loss. The total loss is a combination of the two parts above defined as follows:

$$L = \gamma \cdot \underbrace{\sum_{i=1}^{b} \sum_{c \in \mathbf{C}} -y_i^c \cdot \log(\hat{y}_i^c)}_{\text{cross-entropy loss}} + (1-\gamma) \cdot \underbrace{\sum_{i=1}^{b} \left(\max\left\{0, M - \|\mathcal{F}_\Theta(\mathbf{x}_i) - \mathcal{F}_\Theta(\mathbf{x}_j^{\tilde{g}})\|_2\right\}\right)^2}_{\text{contrastive loss}}$$

where $\mathbf{x}_i$ denotes the raw feature of the ith instance and y_i^c represents the indicator variable that is equal to 1 if and only if the ith instance belongs to the ground truth category g. $\hat{y}_i^c$ is the predicted probability that the ith instance belongs to category c and b is the batch size. $\mathcal{F}_\Theta(\cdot)$ denotes the pre-trained language model, which extracts representation of an input instance. γ and M are hyper-parameters. $\mathbf{x}_j^{\tilde{g}}$ denotes an arbitrary instance (indexed by j) that comes from a different category of $\mathbf{x}_i$, and $\tilde{g} = \mathbf{C}\backslash\{g\}$.

Hard Example Mining Strategy. Instances easily classified correctly by the model contribute little to the contrastive loss [19,22]. That is to say, a randomly selected instance $\mathbf{x}_j^{\tilde{g}}$ probably has been far away from an instance $\mathbf{x}_i$ after epochs of training. Therefore, instead of generating pairs by random sampling, we focus on hard examples, i.e., instances that are misclassified into a wrong category. Hence, the hard example set $\mathbf{H}$ is discovered by: $\mathbf{H} = \{\mathbf{x}_j | \arg\max y_j \neq \arg\max \hat{y}_j, j = 1, \cdots, b\}$. Pairs of training inputs are selected by first randomly choosing an instance $\mathbf{x}_i$ from the entire training set $\mathbf{X}$, and then randomly choosing $\mathbf{x}_j^{\tilde{g}}$ from the hard example set $\mathbf{H}$.

3   Experiments

We collected online-class video recordings from a third-party educational platform. Similar to [12,24], audio tracks are extracted from video recordings and then cut into utterances by a self-trained VAD model [21]. After that, utterances are transcribed into text using a self-trained ASR model [27] with a character error rate of 11.36% in classroom scenarios. The training and validation sets contains 16,174 and 4,088 instances respectively. Performance on each category (except *others*) is separately evaluated on a binary test set containing 2000 positive instances that belong to this category, and 2000 negative ones from the other categories (other seven categories of instructions, or *others*). We select a series of widely-used baselines, including BiLSTM [9], TextRCNN [13], and pre-trained language models: BERT [6], ELECTRA [5], NEZHA [23], RoBERTa [16], and XLNet [25]. Moreover, we compare different strategies of negative example selection in our multi-task framework: (1) random selection from all the instances of other categories, i.e., *M-RoBERTa-All*; and (2) hard example mining, i.e., *M-RoBERTa-Hard*.

3.1   Results Discussion

From Table 1, we can find that pre-trained language models such as ELECTRA, NEZHA, and RoBERTa achieve better performance than classic approaches, i.e.,

Table 1. Performance of different pre-trained language models.

Instruction	Model	Accuracy	F1	Instruction	Model	Accuracy	F1
macro-average	BiLSTM	0.781	0.783	micro-average	BiLSTM	0.781	0.791
	TextRCNN	0.785	0.788		TextRCNN	0.785	0.789
	BERT	0.781	0.787		BERT	0.781	0.778
	ELECTRA	0.791	0.790		ELECTRA	0.791	0.794
	NEZHA	0.797	0.803		NEZHA	0.797	**0.797**
	XLNet	0.770	0.775		XLNet	0.770	0.764
	RoBERTa	**0.799**	**0.812**		RoBERTa	**0.799**	0.795

Table 2. Performance of the proposed method and its variants.

Instruction	Model	Accuracy	F1	Instruction	Model	Accuracy	F1
commending	RoBERTa	0.828	0.831	guidance	RoBERTa	0.809	0.829
	M-RoBERTa-All	0.831	0.844		M-RoBERTa-All	0.847	0.850
	M-RoBERTa-Hard	**0.842**	**0.855**		M-RoBERTa-Hard	**0.868**	**0.872**
summarization	RoBERTa	0.803	0.829	greeting	RoBERTa	0.788	0.803
	M-RoBERTa-All	0.862	0.875		M-RoBERTa-All	0.791	0.810
	M-RoBERTa-Hard	**0.876**	**0.886**		M-RoBERTa-Hard	**0.802**	**0.830**
note-taking	RoBERTa	0.814	0.830	repeating	RoBERTa	0.690	0.725
	M-RoBERTa-All	0.735	0.771		M-RoBERTa-All	0.749	0.774
	M-RoBERTa-Hard	**0.880**	**0.889**		M-RoBERTa-Hard	**0.750**	**0.776**
reviewing	RoBERTa	0.796	0.787	example-giving	RoBERTa	0.868	0.859
	M-RoBERTa-All	**0.824**	**0.811**		M-RoBERTa-All	0.861	0.854
	M-RoBERTa-Hard	0.822	**0.811**		M-RoBERTa-Hard	**0.929**	**0.893**
macro-average	RoBERTa	0.799	0.812	micro-average	RoBERTa	0.799	0.795
	M-RoBERTa-All	0.812	0.824		M-RoBERTa-All	0.812	0.804
	M-RoBERTa-Hard	**0.847**	**0.852**		M-RoBERTa-Hard	**0.847**	**0.823**

BiLSTM and TextRCNN, which indicates their stronger capacity to model dialogic instructions by utilizing contextual information. ELECTRA, RoBERTa, and NEZHA have a better overall performance than BERT, which is not surprising since they are pre-trained with improved training objectives and larger corpus.

We demonstrate the effectiveness of our multi-task framework by comparing with standard RoBERTa model (in Table 1). Table 2 shows that: (1) by adding a contrastive loss to enlarge the margin between different categories, *M-RoBERTa-All* outperforms the original RoBERTa model in 6 out of 8 types of dialogic instructions and the overall performance; and (2) by fully utilizing instances misclassified by the model, *M-RoBERTa-Hard* outperforms *M-RoBERTa-All* and achieves the best prediction performance compared with other methods in terms of accuracy, macro- and micro-F1 scores.

4   Conclusion

We present a multi-task dialogic instruction detection framework using pre-trained language models. Furthermore, we design a strategy to select hard instances and exploit them during training. Experiments conducted on a real-world data set show that our framework outperforms both classic methods and pre-trained language models fine-tuned solely with the classification objective.

Acknowledgment. This work was supported in part by National Key R&D Program of China, under Grant No. 2020AAA0104500 and in part by Beijing Nova Program (Z201100006820068) from Beijing Municipal Science & Technology Commission.

References

1. AN, S.: Capturing the Chinese way of teaching: the learning-questioning and learning-reviewing instructional model. In: How Chinese learn mathematics: Perspectives from insiders, pp. 462–482. World Scientific (2004)
2. Anthony, G., Hunter, J., Hunter, R.: Supporting prospective teachers to notice students' mathematical thinking through rehearsal activities. Math. Teach. Educ. Dev. **17**(2), 7–24 (2015)
3. Blanchard, N., Brady, M., Olney, A.M., Glaus, M., Sun, X., Nystrand, M., Samei, B., Kelly, S., D'Mello, S.: A study of automatic speech recognition in noisy classroom environments for automated dialog analysis. In: Conati, C., Heffernan, N., Mitrovic, A., Verdejo, M.F. (eds.) AIED 2015. LNCS (LNAI), vol. 9112, pp. 23–33. Springer, Cham (2015). https://doi.org/10.1007/978-3-319-19773-9_3
4. Chen, J., Li, H., Wang, W., Ding, W., Huang, G.Y., Liu, Z.: A multimodal alerting system for online class quality assurance. In: Isotani, S., Millán, E., Ogan, A., Hastings, P., McLaren, B., Luckin, R. (eds.) AIED 2019. LNCS (LNAI), vol. 11626, pp. 381–385. Springer, Cham (2019). https://doi.org/10.1007/978-3-030-23207-8_70
5. Clark, K., Luong, M., Le, Q.V., Manning, C.D.: ELECTRA: pre-training text encoders as discriminators rather than generators. In: 8th International Conference on Learning Representations, ICLR 2020, Addis Ababa, Ethiopia, 26–30 April 2020. OpenReview.net (2020). https://openreview.net/forum?id=r1xMH1BtvB
6. Devlin, J., Chang, M.W., Lee, K., Toutanova, K.: BERT: pre-training of deep bidirectional transformers for language understanding. In: Proceedings of the 2019 Conference of the North American Chapter of the Association for Computational Linguistics: Human Language Technologies, Volume 1 (Long and Short Papers). pp. 4171–4186. Association for Computational Linguistics, Minneapolis, Minnesota (2019). https://doi.org/10.18653/v1/N19-1423, https://www.aclweb.org/anthology/N19-1423
7. Dweck, C.S.: Boosting achievement with messages that motivate. Educ. Can. **47**(2), 6–10 (2007)
8. Goodenow, C.: The psychological sense of school membership among adolescents: scale development and educational correlates. Psychol. Sch. **30**(1), 79–90 (1993)
9. Graves, A., Mohamed, A.r., Hinton, G.: Speech recognition with deep recurrent neural networks. In: 2013 IEEE international conference on acoustics, speech and signal processing, pp. 6645–6649. IEEE (2013)
10. HAGHVERDİ, H., Biria, R., Karimi, L.: Note-taking strategies and academic achievement. Dil ve Dilbilimi Çalışmaları Dergisi **6**(1) (2010)
11. Henderlong, J., Lepper, M.R.: The effects of praise on children's intrinsic motivation: a review and synthesis. Psychol. Bull. **128**(5), 774 (2002)

12. Huang, G.Y., Chen, J., Liu, H., Fu, W., Ding, W., Tang, J., Yang, S., Li, G., Liu, Z.: Neural multi-task learning for teacher question detection in online classrooms. In: Bittencourt, I.I., Cukurova, M., Muldner, K., Luckin, R., Millán, E. (eds.) AIED 2020. LNCS (LNAI), vol. 12163, pp. 269–281. Springer, Cham (2020). https://doi.org/10.1007/978-3-030-52237-7_22

13. Lai, S., Xu, L., Liu, K., Zhao, J.: Recurrent convolutional neural networks for text classification. In: Bonet, B., Koenig, S. (eds.) Proceedings of the Twenty-Ninth AAAI Conference on Artificial Intelligence, 25–30 January 2015, Austin, Texas, USA, pp. 2267–2273. AAAI Press (2015). http://www.aaai.org/ocs/index.php/AAAI/AAAI15/paper/view/9745

14. Lee, P.L., Lan, W., Hamman, D., Hendricks, B.: The effects of teaching notetaking strategies on elementary students' science learning. Instr. Sci. **36**(3), 191–201 (2008)

15. Li, H., Kang, Y., Ding, W., Yang, S., Yang, S., Huang, G.Y., Liu, Z.: Multimodal learning for classroom activity detection. In: 2020 IEEE International Conference on Acoustics, Speech and Signal Processing, ICASSP 2020, Barcelona, Spain, 4–8 May 2020, pp. 9234–9238. IEEE (2020). https://doi.org/10.1109/ICASSP40776.2020.9054407

16. Liu, Y., et al.: Roberta: A robustly optimized bert pretraining approach (2019). arXiv preprint arXiv:1907.11692

17. Osterman K.F.: Teacher practice and students' sense of belonging. In: Lovat T., Toomey R., Clement N. (eds) International Research Handbook on Values Education and Student Wellbeing. Springer, Dordrecht (2010). https://doi.org/10.1007/978-90-481-8675-4_15

18. Rinehart, S.D., Stahl, S.A., Erickson, L.G.: Some effects of summarization training on reading and studying. Read. Res. Q. 422–438 (1986)

19. Schroff, F., Kalenichenko, D., Philbin, J.: Facenet: A unified embedding for face recognition and clustering. In: IEEE Conference on Computer Vision and Pattern Recognition, CVPR 2015, Boston, MA, USA, 7–12 June 2015, pp. 815–823. IEEE Computer Society (2015). https://doi.org/10.1109/CVPR.2015.7298682

20. Shafto, P., Goodman, N.D., Griffiths, T.L.: A rational account of pedagogical reasoning: teaching by, and learning from, examples. Cogn. Psychol. **71**, 55–89 (2014)

21. Tashev, I., Mirsamadi, S.: DNN-based causal voice activity detector. In: Information Theory and Applications Workshop (2016)

22. Wang, W., Xu, G., Ding, W., Huang, Y., Li, G., Tang, J., Liu, Z.: Representation learning from limited educational data with crowdsourced labels. In: TKDE (2020)

23. Wei, J., et al.: Nezha: Neural contextualized representation for Chinese language understanding (2019). arXiv preprint arXiv:1909.00204

24. Xu, S., Ding, W., Liu, Z.: Automatic dialogic instruction detection for k-12 online one-on-one classes. In: Bittencourt, I.I., Cukurova, M., Muldner, K., Luckin, R., Millán, E. (eds.) AIED 2020. LNCS (LNAI), vol. 12164, pp. 340–345. Springer, Cham (2020). https://doi.org/10.1007/978-3-030-52240-7_62

25. Yang, Z., Dai, Z., Yang, Y., Carbonell, J.G., Salakhutdinov, R., Le, Q.V.: Xlnet: Generalized autoregressive pretraining for language understanding. In: Wallach, H.M., Larochelle, H., Beygelzimer, A., d'Alché-Buc, F., Fox, E.B., Garnett, R. (eds.) Advances in Neural Information Processing Systems 32: Annual Conference on Neural Information Processing Systems 2019, NeurIPS 2019, December 8–14, 2019, Vancouver, BC, Canada, pp. 5754–5764 (2019). https://proceedings.neurips.cc/paper/2019/hash/dc6a7e655d7e5840e66733e9ee67cc69-Abstract.html

26. Yelland, N., Masters, J.: Rethinking scaffolding in the information age. Comput. Educ. **48**(3), 362–382 (2007)

27. Zhang, S., Lei, M., Yan, Z., Dai, L.: Deep-FSMN for large vocabulary continuous speech recognition. In: 2018 IEEE International Conference on Acoustics, Speech and Signal Processing, ICASSP 2018, Calgary, AB, Canada, 15–20 April 2018, pp. 5869–5873. IEEE (2018). https://doi.org/10.1109/ICASSP.2018.8461404

Impact of Predictive Learning Analytics on Course Awarding Gap of Disadvantaged Students in STEM

Martin Hlosta(✉) ⓘ, Christothea Herodotou ⓘ, Vaclav Bayer ⓘ,
and Miriam Fernandez ⓘ

The Open University, Milton Keynes, UK
martin.hlosta@open.ac.uk

Abstract. In this work, we investigate the degree-awarding gap in distance higher education by studying the impact of a Predictive Learning Analytics system, when applying it to 3 STEM (Science, Technology, Engineering and Mathematics) courses with over 1,500 students. We focus on Black, Asian and Minority Ethnicity (BAME) students and students from areas with high deprivation, a proxy for low socio-economic status. Nineteen teachers used the system to obtain predictions of which students were at risk of failing and got in touch with them to support them (intervention group). The learning outcomes of these students were compared with students whose teachers did not use the system (comparison group). Our results show that students in the intervention group had 7% higher chances of passing the course, when controlling for other potential factors of success, with the actual pass rates being 64% vs 61%. When disaggregated: 1) BAME students had 10% higher pass rates (55 %vs 45%) than BAME students in the comparison group and 2) students from the most deprived areas had 4% higher pass rates (58% vs 54%) in the intervention group compared to the comparison group.

Keywords: Predictive analytics · Course awarding gap · BAME · SES

1 Introduction

Historically, the performance of some demographic groups of students has been persistently worse than others. The impact of low socio-economic status (SES) on learning has increased over the last 50 years across countries, including the UK [3]. The attainment of ethnic minorities is consistently worse than White students. In the UK, in the past decade, 57% of Black students gained an upper second or first in their undergraduate degree, compared with 81% of White students [10]. There may be a significant overlap between Black, Asian and Minority Ethnic (BAME) students and low SES students. Recent post-pandemic statistics show that nearly half of BAME households (46%) live in poverty as opposed to 20% of White households [11].

© Springer Nature Switzerland AG 2021
I. Roll et al. (Eds.): AIED 2021, LNAI 12749, pp. 190–195, 2021.
https://doi.org/10.1007/978-3-030-78270-2_34

Predictive Learning Analytics (PLA) focuses on forecasting the future students' outcomes using Machine Learning (ML) models and provide actionable feedback to students or teachers, leading to improved student outcomes [6]. Growing evidence suggest that using PLA to trigger interventions leads to improved student outcomes in some studies [7,14] but not in others [2,5]. This suggests that further fine-grained analysis is needed to understand who of the students may benefit the most from PLA interventions. Previous studies reported the importance of a teacher in improving student outcomes and closing the attainment gap [4,13].

Research Questions. To the best of authors' knowledge, there are no studies directly investigating the impact of PLA on different demographic subgroups. To fill this research gap, we examined the impact of PLA on the course awarding gap of BAME students and low SES students. We formulated two research questions (RQs): **RQ1:** What is the impact of PLA on student pass rates and their final score when deployed by teachers? **RQ2:** What is the impact of PLA when disaggregating the results by ethnicity and by SES?

2   Methods

Three STEM courses were selected based on their historically low retention and because they have not used Predictive technology before. Nineteen out of the 59 course teachers took part in the study (Intervention group). The remaining teachers (N = 37) were treated as a Control group. Teachers were asked to log in before the first three assignments; 1, 2 and 3 weeks before the assignment's submission deadline. For each access, they were asked to consider contacting students that were identified as at-risk of 1) not submitting or 2) predicted as Fail or achieve low grade (50–60). Teachers were compensated to complete this research activity.

The predictions, generated weekly, estimate each student's likelihood to submit their next assignment and a likely banded score in the assignment. To generate these predictions the model utilises data from the previous run of the same course, i.e. 1) demographics, workload and prev. results, 2) student engagement in VLE, and 3) previous assignment performance. Gradient Boosting Machines (GBM) has been selected in the previous years as the best performing model [8].

Evaluation. For each RQ, we focus on students completion, passing and their overall score. Completion means that a student satisfied the course requirements and sat the exam; passing means that they were successful in the exam. Logistic regression models were applied for binary outcomes (completion and pass) and linear regression was used for the overall score. The unit of analysis were students ($N = 1,412$). The factors entered into the regression analysis included: (1) **Student** (age, gender, an indicator of linked qualification, declared disability, caring responsibility, new/continuing, highest previous education, avg. previous score, no. of other credits studied, no. of previous attempts of the course, IMD[1]

---

[1] In the UK, the SES gap can be expressed as a difference between students from low and high deprived areas, measured by Index of Multiple Deprivation (IMD) [9,12].

and whether the student is identified as BAME), (2) **Teacher** (no. of students the teacher is responsible for, avg. student pass rate in the previous years they have been teaching), (3) **Course** - dummy encoded as variables Course 1, 2 and 3.

Similarly, as [12], IMD was discretised into quintiles - Q1 representing the most deprived areas and Q5 the least deprived areas. The check for homogeneity of variances, multicollinearity and normality were conducted to ensure no assumption violation. Except for the number of students in the teachers' group, the continuous variables did not follow a normal distribution and were discretised. IMD, previous student score, and teacher previous pass rates contained missing values, and we encoded them as a special category. The previous score was discretised for each course separately.

To answer RQ2, we created separate regression models for each demographic group - i.e. for BAME/non-BAME and each IMD quintile $Q1 - Q5$. BAME students encompassed 57 Asian, 46 Black, 39 Mixed and 18 Minor Ethnicity students (11% of all students). This was conducted again for completion, pass and overall score. For each regression model, we investigated the coefficient indicating any differences between the Intervention and the Control group.

3   Results

The accuracy of the model for predicting completion was $Acc = 0.71$ and for pass $Acc = 0.69$, for the continuous target overall score $R^2 = 0.22$. Table 1 shows the coefficients of the regression for pass, completion and score, with their statistical significance and standard errors for all students, regardless of the demographic group.[2] The results show that students in the Intervention group (factor group_INT) were much more likely to pass the module ($\beta = 0.36, p < 0.01$) and also obtain higher overall score ($\beta = 5.07, p < 0.01$). The positive coefficient for completion was however not statistically significant ($\beta = 0.11, p >= 0.1$). This might suggest that students in the Intervention group were better prepared to be successful in the exam. The pass rate beta $\beta = 0.36$ can be converted to an Average Marginal Effect 0.07, which means that keeping all attributes constant, students in the Intervention group have 7% higher chances of passing the course and obtaining 5.07 more points in the overall score.

Disaggregation by BAME and IMD. Overall, the pass rates (61% vs 65%) and overall score (44 vs 46.5) were higher in the Intervention group. The positive differences were higher for BAME students for passing (52% in the Control vs 62% in the Intervention) and lower IMD quintiles, IMD1-3. Regression models were created only for the specific demographic group, controlling for potential confounding variables. Fig. 1 shows the β regression coefficients for the Intervention group extracted from these models. Except for completion, the lower SES groups have higher coefficients, with statistical significant results measured for

---

[2] The results only include attributes where at least one of the factors had $p < 0.05$. The full analysis can be found at https://doi.org/10.21954/ou.rd.14414774.

Table 1. Regression table for completion, pass and overall score

	Completion		Pass		Overall Score	
	β	SE	β	SE	β	SE
prev_sc_VERY_HIGH	0.41	0.75	1.33**	0.69	23.22**	10.08
disability	−0.34*	0.28	−0.44**	0.26	−8.12**	3.83
is_new	0.29	0.73	0.74*	0.67	13.66**	10.00
course_2	0.26	0.37	0.46**	0.35	2.19	4.88
group_INT	0.11	0.29	0.36**	0.27	5.07**	3.82
credits_other_[1−60]	−0.67**	0.34	−0.70**	0.30	−11.80**	4.14
credits_other_>=61	−1.09**	0.44	−1.04**	0.41	−17.12**	5.79
stud_in_group	−0.01*	0.01	−0.02**	0.01	−0.23**	0.16

*p<0.05 **p<0.01 ***p<0.001

IMD Q1, $\beta = 0.78, p < 0.05$. For BAME, the most significant factor related to passing the course was teachers' previous low pass rates $\beta = -3.08, p < 0.05$. This factor was not present for non-BAME students. This suggests that teachers who had students with consistently low pass rates in the past are more likely to have lower pass rates for BAME students but not for non-BAME students. The same attribute was significant also for the second most deprived areas IMD_Q2 $\beta = -1.78, p < 0.01$. Overall, a great concentration of BAME has been observed in low SES and conditions of poverty [1,11], suggesting that any intervention that tackles students in low SES would be particularly beneficial for BAME students alongside other ethnicities found in low SES.

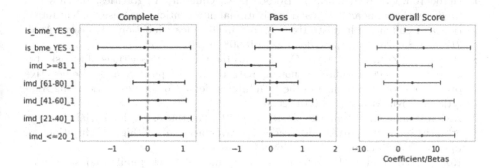

Fig. 1. Outcomes beta coefficients for being in the Intervention group

4   Conclusions

The results demonstrated a positive impact on students' performance, particularly those who were coming from low SES, as measured by the Index of Multiple

Deprivation (IMD). This suggests that students found in rather disadvantaged contexts such as poverty are more likely to benefit from PLA systems. BAME are shown to have the greatest representation in low SES (32% as opposed to 10% non-BAME students), stressing the significance of early PLA support for BAME students in particular. Because our study was conducted only on 3 STEM courses and less than 1,500 students, the scaled experiment should try to replicate the study across more courses, examine separately specific student groups within BAME such as Black or Asian students and investigate the context, i.e. whether some conditions need to be met to observe the same or similar effect.

References

1. American Psychology Association and others: Ethnic and racial minorities & socioeconomic status (2016). https://www.apa.org/pi/ses/resources/publications/minorities
2. Borrella, I., Caballero-Caballero, S., Ponce-Cueto, E.: Predict and intervene: addressing the dropout problem in a MOOC-based program. In: Proceedings of the Sixth (2019) ACM Conference on Learning@ Scale, pp. 1–9 (2019)
3. Chmielewski, A.K.: The global increase in the socioeconomic achievement gap, 1964 to 2015. Am. Sociol. Rev. **84**(3), 517–544 (2019)
4. Crenna-Jennings, W.: Key drivers of the disadvantage gap: literature review (2018)
5. Dawson, S., Jovanovic, J., Gašević, D., Pardo, A.: From prediction to impact: evaluation of a learning analytics retention program. In: Proceedings of the seventh international learning analytics & knowledge conference, pp. 474–478 (2017)
6. Herodotou, C., Hlosta, M., Boroowa, A., Rienties, B., Zdrahal, Z., Mangafa, C.: Empowering online teachers through predictive learning analytics. Br. J. Educ. Technol. **50**(6), 3064–3079 (2019). https://doi.org/10.1111/bjet.12853
7. Herodotou, C., Naydenova, G., Boroowa, A., Gilmour, A., Rienties, B.: How can predictive learning analytics and motivational interventions increase student retention and enhance administrative support in distance education? J. Learn. Anal. **7**(2), 72–83 (2020). https://doi.org/10.18608/jla.2020.72.4
8. Hlosta, M., Zdrahal, Z., Bayer, V., Herodotou, C.: Why predictions of at-risk students are not 100% accurate? showing patterns in false positive and false negative predictions. In: Proceedings of the 10th International Conference on Learning Analytics and Knowledge (LAK20) (2020)
9. Richardson, J.T., Mittelmeier, J., Rienties, B.: The role of gender, social class and ethnicity in participation and academic attainment in UK higher education: an update. Oxf. Rev.Educ. **46**(3), 346–362 (2020)
10. Roberts, N., Bolton, P.: Educational outcomes of black pupils and students - research briefing, October 2020. https://commonslibrary.parliament.uk/research-briefings/cbp-9023/
11. Stroud, P.: Measuring poverty 2020, a report of the social metrics commission, July 2020. https://socialmetricscommission.org.uk/wp-content/uploads/2020/06/Measuring-Poverty-2020-Web.pdf. Accessed 08 February 2021
12. Thiele, T., Pope, D., Singleton, A., Stanistreet, D.: Role of students' context in predicting academic performance at a medical school: a retrospective cohort study. BMJ Open **6**(3), e010169 (2016)

13. Warschauer, M., Matuchniak, T.: New technology and digital worlds: analyzing evidence of equity in access, use, and outcomes. Rev. Res. Educ. **34**(1), 179–225 (2010)
14. Wong, B.T.M., Li, K.C.: Learning analytics intervention: a review of case studies. In: 2018 International Symposium on Educational Technology (ISET), pp. 178–182 (2018). https://doi.org/10.1109/ISET.2018.00047

Evaluation of Automated Image Descriptions for Visually Impaired Students

Anett Hoppe[1]([✉]) [iD], David Morris[1] [iD], and Ralph Ewerth[1,2] [iD]

[1] TIB–Leibniz Information Centre for Science and Technology, Hannover, Germany
{anett.hoppe,david.morris,ralph.ewerth}@tib.eu
[2] L3S Research Center, Leibniz University Hannover, Hannover, Germany

Abstract. Illustrations are widely used in education, and sometimes, alternatives are not available for visually impaired students. Therefore, those students would benefit greatly from an automatic illustration description system, but only if those descriptions were complete, correct, and easily understandable using a screenreader. In this paper, we report on a study for the assessment of automated image descriptions. We interviewed experts to establish evaluation criteria, which we then used to create an evaluation questionnaire for sighted non-expert raters, and description templates. We used this questionnaire to evaluate the quality of descriptions which could be generated with a template-based automatic image describer. We present evidence that these templates have the potential to generate useful descriptions, and that the questionnaire identifies problems with description templates.

Keywords: Accessibility · Blind and visually impaired · Automatic image description · Educational resources

1 Introduction

Images are widely used in educational resources, but their usefulness is reduced for visually impaired learners. In specialised professional settings, image descriptions are provided by experts based on their experience and accredited guidelines [9,11]. This is, however, not applicable to informal learning settings, especially on the Web: Open Educational Resources are available for a multitude of topics, but with only limited accessibility for students with visual impairments due to missing alternative texts (alt-texts) and image descriptions. Relying on experts to generate descriptions does not scale.

Computer vision systems have made considerable progress in recent years on topics such as image captioning [10,12] and object detection [5]. However,

This work is financially supported by the German Federal Ministry of Education and Research (BMBF) and the European Social Fund (ESF) (Project InclusiveOCW, no. 01PE17004).

© Springer Nature Switzerland AG 2021
I. Roll et al. (Eds.): AIED 2021, LNAI 12749, pp. 196–201, 2021.
https://doi.org/10.1007/978-3-030-78270-2_35

results of automatic description algorithms do not yet provide results with sufficient structure and reliable completeness, especially for figures typically used in
educational materials such as slidesets. They are, however, promising in specific
information extraction tasks, such as formula recognition [4,13], diagram structure identification [3], and the recognition of text in scientific figures [2,7]. In con-
sequence, we let ourselves be inspired by prior work on accessibility technology:
studies used requirements analysis to identify needs of the future users [1]; and
explored templates to gather image descriptions by untrained volunteers [6]. Other
previous work suggested HTML to structure screenreader-friendly documents and
used Likert scales for the evaluation of descriptions [8].

In this paper, we build on this related work and examine scalable ways to
provide high-quality descriptions for educational image types, such as bar or
pie charts. The objective is to generate these descriptions automatically based
on state-of-the-art computer vision techniques – using the structured templates
as a guide, and filling in the blanks using specialised Deep Learning techniques.
Consequently, we explore (1) a simplification of the description task by the use of
structured templates derived from expert knowledge; and (2) scalable evaluation
of the descriptions based on structured questionnaires. In doing so, we setup
the context necessary to use current computer vision methods for automatic
description of visual educational resources.

Section 2 introduces our method to acquire the necessary expertise for the
creation of structured descriptions and the resulting templates; the structured
evaluation procedure is described in Sect. 3. The developed materials are available here: http://go.lu-h.de/gbxfC.

2   Structured Image Description

2.1   User Needs Analysis

We consulted three college-educated congenitally blind people with different levels of expertise in image description, who all agreed to be identified: Anja Winkler
(TU Dresden) trains sighted people to write image descriptions. Anja Pfaffen-
zeller is a teacher at a school for blind students, and has previously worked
teaching the use of assistive technology. Hunter Jozwiak, is a technology professional with experience using different computer environments. Our experts
were selected by reference. Interview transcripts are available upon request. The
interviews served as the basis to develop guiding principles for the design of the
description templates (Sect. 2.2) and the evaluation questionnaire (Sect. 3):

Alt-text: Alt-texts, or short descriptions should be as **concise** as possible. They
should let the screenreader user know as quickly as possible whether the image
is relevant for their purposes. It should further make **use of available information** – if the author provided a title, for instance, it should be used here.

Long description: **Complete and correct information** needs to be contained. Preferably, the user gets to navigate varying levels of detail. Moreover, **context** is key: Used symbols should be described with their semantic meaning, not only their shape. Similarly, **tables** need to be carefully formatted, as they are difficult to navigate with a screenreader. They need to be understandable when reading one row at a time, without referring back to the column titles.

2.2 Description Templates

We drafted template structures for four commonly used types of illustrations: line and scatter plots, bar charts, node-link diagrams, and pie charts. They use an HTML format which is easily navigated with a screenreader and present the information of the image in an ascending level of granularity. They are designed to cover all information possibly contained in a plot and are thus quite detailed. The objective is to provide access to all discernible information, and they might appear overwhelming to a seeing user. But, by a clear and known structure, a screenreader user is enabled to decide when sufficient information has been consumed. This is best exemplified by scatter plots, which might contain hundreds of data points. Instead of listing each point (and potentially confusing the user), the area of the diagram is be grouped in sectors, and conflated information displayed for each sector.

3 Structured Evaluation

Besides the generation of image descriptions, their evaluation is a bottleneck. We explore the possibility of description evaluation with untrained volunteers, enabled by structured questionnaires and thus, reducing the need for expert knowledge. The questions have been developed based on the expert interviews (Sect. 2.1), and iteratively refined . The result is a two-stage evaluation . In the first stage, the evaluator only sees the description and answers questions on its perceived comprehensibility and its capacity to evoke a mental image. Then, the evaluator judges completeness and correctness using the displayed image.

3.1 Experimental Setup

We performed a comparative evaluation against a set of best-case descriptions. As a reference, we used example descriptions supplied as part of the image description guidelines developed by the National Center for Accessible Media (NCAM) and the Benetech DIAGRAM (Digital Image And Graphic Resources for Accessible Materials) center [9]. Nine images were used for evaluation. For each, the study participants rate the description from the guidelines, and the template-based one, allowing us to compare both scores.

All raters evaluate two descriptions per image without knowing the respective source. Order effects are counterbalanced by randomising the sequence of evaluated descriptions. For each image, the evaluators finish the first stage of evaluation (without seeing the image) for both available descriptions, then proceed with stage two (with the image as a reference for correctness and completeness).

Table 1. Table of description scores: TB is the score of the template-based descriptions, DIA the score of the control descriptions from the DIAGRAM center guidelines.

Image Filename	Image Type	Score (TB)	Score (DIA)	Difference
image030	Bar chart	64.1	65.2	−1.1
image031	Bar chart	67.1	65.4	1.7
image032	Bar chart	72.5	68.1	4.4
image024	Node-link diagram	54.5	63.8	−9.3
image028	Node-link diagram	57.7	68.3	−10.6
Flow-Chart-1	Node-link diagram	52.9	65.1	−12.2
image033	Line/scatter plot	67.7	69.0	−1.4
scatter-plot-3	Line/scatter plot	63.6	55.7	7.9
image034	Pie chart	67.7	66.0	1.7

We report combined scores to assess our descriptions. All 15 rating items (there were three other items recording metadata) used a scale from one to five, where five was the best score possible. Thus, the maximum score was 75. For each description, we average the responses to each question, add the averages together, and report this as "description score" in Table 1. Nine untrained evaluators were recruited using social media. The analysis is limited to the six of them who finished rating at least 14 of the 18 descriptions (two descriptions are rated for each of the nine example images, each is evaluated with 15 questions).

3.2   Results

Table 1 shows the average description scores for the template-based descriptions (TB) and the control descriptions drawn from the DIAGRAM center guidelines (DIA). Four out of nine descriptions scored within two points of the control. In two cases, our descriptions even scored higher than the control. However, our descriptions of node-link diagrams scored worse than the controls, indicating further need to improve those.

4   Discussion and Conclusion

Online educational resources often contain images of informative nature. Due to missing or incomplete alt-texts and descriptions, those are often inaccessible for learners with visual impairments. For this reason, we investigated the use of structured templates to simplify the task of automatically generating high-quality descriptions and propose a procedure to evaluate the resulting descriptions *without expert involvement*. The results indicate that the current structured templates successfully capture the information in simpler diagram types, such as bar and pie charts; but need further refinement for complex schemata such as node-link diagrams. While our study is a first pointer to necessary next steps, the results of the volunteer evaluation need to be complemented by an assessment involving both, visually impaired users and description experts to confirm the procedure. Other future work includes the addition of other image types to the template repertoire and the development of adapted computer vision methods to automatically fill the templates with diagram information.

References

1. Ferres, L., Parush, A., Roberts, S., Lindgaard, G.: Helping people with visual impairments gain access to graphical information through natural language: the iGraph system. In: Miesenberger, K., Klaus, J., Zagler, W.L., Karshmer, A.I. (eds.) ICCHP 2006. LNCS, vol. 4061, pp. 1122–1130. Springer, Heidelberg (2006). https://doi.org/10.1007/11788713_163
2. Jessen, M., Böschen, F., Scherp, A.: Text localization in scientific figures using fully convolutional neural networks on limited training data. In: Schimmler, S., Borghoff, U.M. (eds.) Proceedings of the CM Symposium on Document Engineering 2019, Berlin, Germany, 23–26 September 2019, pp. 13:1–13:10. ACM (2019). https://doi.org/10.1145/3342558.3345396
3. Kembhavi, A., Salvato, M., Kolve, E., Seo, M., Hajishirzi, H., Farhadi, A.: A diagram is worth a dozen images. In: Leibe, B., Matas, J., Sebe, N., Welling, M. (eds.) ECCV 2016. LNCS, vol. 9908, pp. 235–251. Springer, Cham (2016). https://doi.org/10.1007/978-3-319-46493-0_15
4. Le, A.D., Indurkhya, B., Nakagawa, M.: Pattern generation strategies for improving recognition of handwritten mathematical expressions. Pattern Recogn. Lett. **128**, 255–262 (2019). https://doi.org/10.1016/j.patrec.2019.09.002
5. Liu, L., Ouyang, W., Wang, X., Fieguth, P.W., Chen, J., Liu, X., Pietikäinen, M.: Deep learning for generic object detection: a survey. Int. J. Comput. Vis. **128**(2), 261–318 (2020). https://doi.org/10.1007/s11263-019-01247-4
6. Morash, V.S., Siu, Y., Miele, J.A., Hasty, L., Landau, S.: Guiding novice web workers in making image descriptions using templates. TACCESS **7**(4), 12:1–12:21 (2015). https://doi.org/10.1145/2764916
7. Morris, D., Tang, P., Ewerth, R.: A neural approach for text extraction from scholarly figures. In: 2019 International Conference on Document Analysis and Recognition, ICDAR 2019, Sydney, Australia, 20–25 September 2019, pp. 1438–1443. IEEE (2019). https://doi.org/10.1109/ICDAR.2019.00231
8. Morris, M.R., Johnson, J., Bennett, C.L., Cutrell, E.: Rich representations of visual content for screen reader users. In: Proceedings of the 2018 CHI Conference on Human Factors in Computing Systems, CHI 2018, Montreal, QC, Canada, 21–26 April 2018. p. 59 (2018). https://doi.org/10.1145/3173574.3173633
9. NCAM, DIAGRAM: Image description guidelines. http://diagramcenter.org/table-of-contents-2.html
10. Park, C.C., Kim, B., Kim, G.: Attend to you: Personalized image captioning with context sequence memory networks. In: 2017 IEEE Conference on Computer Vision and Pattern Recognition, CVPR 2017, Honolulu, HI, USA, 21–26 July 2017, pp. 6432–6440. IEEE Computer Society (2017). https://doi.org/10.1109/CVPR.2017.681
11. Reid, L.G., Snow-Weaver, A.: WCAG 2.0: a web accessibility standard for the evolving web. In: Yesilada, Y., Sloan, D. (eds.) Proceedings of the International Cross-Disciplinary Conference on Web Accessibility, W4A 2008, Beijing, China, 21–22 April 2008, pp. 109–115. ACM International Conference Proceeding Series, ACM (2008).https://doi.org/10.1145/1368044.1368069

12. Shuster, K., Humeau, S., Hu, H., Bordes, A., Weston, J.: Engaging image captioning via personality. In: IEEE Conference on Computer Vision and Pattern Recognition, CVPR 2019, Long Beach, CA, USA, 16–20 June 2019, pp. 12516–12526. Computer Vision Foundation / IEEE (2019). https://doi.org/10.1109/CVPR.2019.01280

13. Zhang, J., Du, J., Dai, L.: Multi-scale attention with dense encoder for handwritten mathematical expression recognition. In: 24th International Conference on Pattern Recognition, ICPR 2018, Beijing, China, 20–24 August 2018, pp. 2245–2250. IEEE Computer Society (2018). https://doi.org/10.1109/ICPR.2018.8546031

Way to Go! Effects of Motivational Support and Agents on Reducing Foreign Language Anxiety

Daneih Ismail[✉][iD] and Peter Hastings[✉][iD]

DePaul University, Chicago, IL 60614, USA
dismail1@depaul.edu, peterh@cdm.depaul.edu

Abstract. Using a tutoring system for English as a foreign language, we studied the impact on students' anxiety levels of an animated agent that provides motivational, supportive feedback. We compared two types of feedback — explanatory and motivational supportive feedback — presented in three ways: by text, by voice, or by a character agent. Results showed that using an agent that gives motivational, supportive feedback decreases the learners' anxiety levels overall. We also found that performance and gender interact with the effectiveness of the treatment for reducing foreign language anxiety (FLA). Our findings have implications for promoting equity and determining how best to improve positive emotions and reduce anxiety for all students.

Keywords: Feedback · Motivation · Agent · Foreign language · Anxiety · Support

1 Introduction

Foreign language anxiety (FLA) is a feeling of tension, stress, or worry when learning a new language [12–14,18]. Decreasing FLA can significantly improve learning achievement [21]. Researchers have investigated ways of reducing FLA in general, like providing supportive, empathetic feedback, either by teachers, peers, or animated agents [2,4,5,16,17,25]. Others have studied the use of animated agents to improve learning [1,7,8,26] and support emotions [19]. Researchers used multiple forms of animated agents such as voice assistants or characters with bodies and voices [1,7]. Conversational agents that provide empathetic support increased the willingness to communicate in the foreign language, which presumably alleviated anxiety and enhanced self-confidence [4,5].

Different types of feedback have also been studied, for example, sandwich feedback [23], explanatory feedback [8], and corrective feedback [8,20,24]. Sandwich feedback is providing an explanation or correction between two positive comments [22,23]. Explanatory feedback is explaining the right answer instead of focusing on evaluating the learner. Corrective feedback informs the learner if their answers were correct or incorrect without any explanation [8].

© Springer Nature Switzerland AG 2021
I. Roll et al. (Eds.): AIED 2021, LNAI 12749, pp. 202–207, 2021.
https://doi.org/10.1007/978-3-030-78270-2_36

Factors such as learner's achievement [9] and gender differences [3] affect how the learner benefits from motivational, supportive feedback and animated agents. Equity in education implies that all students should be empowered to succeed in learning based on their own needs. An adaptive learning environment can aim to ensure success no matter the learner's gender or performance level [9,12]. Research on FLA has shown that struggling learners feel more anxious than successful students [11]. The research has shown mixed results about FLA differences according to gender [10,11,17,27].

2   Methods and Experimental Design

We built an e-learning system for teaching English as a foreign language and for researching FLA. We performed an experiment using a 2×3 factorial design where the factors were *feedback type* (Explanatory vs Motivational Supportive) and *feedback modality* (Text vs Voice vs Agent). The 56 non-native English speaking participants were randomly assigned to one of the six conditions. In all conditions, textual feedback was shown on the screen. In the voice modality condition, the text was accompanied by narration. The agent modality used an animated agent[1]. After the learner answered a question, the system evaluated their answer and provided its feedback depending on the condition. In every case, an *explanation* like this one for a vocabulary exercise was given:

Decreased is the right answer because we need a word that means fewer.

In the explanatory feedback condition, if the learner's answer was correct, the feedback was, "Yes", followed by the explanation. If it was incorrect or partially correct, then only the explanation was given. The motivational supportive feedback conditions used a sandwich feedback model, which put the comment between two positive statements [23]. Figure 1 shows how the explanation was embedded in the motivational, supportive feedback, depending on the evaluation of the learner's answer. After each exercise, the learner answered a question about their level of anxiety during that exercise [14,15].

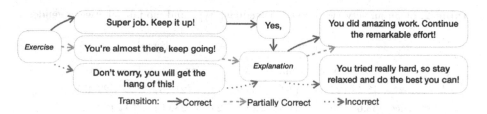

Fig. 1. Example of motivational supportive feedback. Correct: green straight line, Partially Correct: yellow dashed line, Incorrect: red dotted line (Color figure online)

---

[1] Media Semantics (https://www.mediasemantics.com) provided us with a free license for educational use.

3   Results

We did an ANOVA with feedback type and feedback modality as factors and the self-reported FLA as the dependent variable. The results revealed no main effect. There was, however, a crossover interaction between feedback type and modality, $F(2, 1114) = 7.163, p < .001$ (see Table 1).

Table 1. Mean anxiety (with SD) for feedback modality and type

	Explanatory	Supportive
Text	24.13 (23.44)	33.52 (28.62)
Voice	30.35 (34.08)	29.57 (22.99)
Agent	29.01 (28.27)	22.62 (21.31)

To test the effects of performance, we grouped answers as correct, partially correct, or incorrect. For each group we did a t-test with feedback type as the independent variable and level of anxiety as the dependent variable. We found no significant differences for feedback type within the incorrect group $t(304) = 1.744, p = .082$ and partially correct group $t(187) = 0.684, p = 0.495$, but there was a significant difference for the correct group, $t(623) = -3.308, p < .001$. When receiving explanatory feedback, anxiety was lower(M = 17.36, SD = 25.52), than it was with motivational supportive feedback (M = 23.67, SD = 21.85).

For gender, we did an ANOVA with FLA as dependent variable and gender, feedback type and feedback modality as the factors. The results revealed a main effect of gender, $F(1, 1088) = 7.519, p = .006$. This was qualified by interactions between gender and feedback modality, $F(2, 1088) = 3.305, p = .037$. There were no interactions between gender and feedback type $F(1, 1088) = 2.543, p = .111$. The interaction among gender, feedback type, and feedback modality was significant $F(2, 1088) = 13.098, p < .001$ (see Table 2).

Table 2. Mean anxiety (with SD) for feedback type and modality between gender

	Male		Female	
	Explanatory	Supportive	Explanatory	Supportive
Text	16.48 (22.52)	44.17 (23.09)	31.77 (21.95)	29.97 (29.45)
Voice	33.66 (36.51)	34.82 (22.99)	24.53 (28.66)	23.14 (22.32)
Agent	34.24 (29.39)	20.38 (20.41)	22.05 (25.29)	23.46 (21.65)

To further investigate the interaction between gender and feedback type and modality, we did separate ANOVAs for males and females. For females, there was a main effect of feedback modality $F(2, 634) = 5.353, p = 0.005$. Feedback from the animated agent produced the lowest level of FLA (M = 23.08, SD = 22.65), followed by voice (M = 23.76, SD = 25.27), then text (M = 30.42, SD = 27.73). There was no significant interaction between feedback type and modality $F(2, 634) = 0.208, p = 0.812$. For males, we found a statistically significant interaction between the effects of feedback type and modality on FLA $F(2, 454) = 17.202, p < 0.001$. There were no other significant effects.

4   Discussion and Conclusion

Focusing first on feedback type alone, we did not find a main effect on FLA. Both explanatory and motivational supportive feedback types included explanations which focused on the right answers. Because the explanations did not dwell on incorrect answers, learners with incorrect answers should not have been overly threatened by the feedback. We also found that the modality for providing the feedback did not have an overall effect on FLA. As discussed below, there may be other factors that affect the overall impact of feedback modality. Learners who received supportive feedback from animated agents reported the lowest anxiety levels. This result echoes [5] which found that a conversational agent that gave empathetic support effectively reduced FLA.

The highest level of anxiety was reported by learners who gave incorrect answers and received explanatory feedback, but the difference between that and the supportive feedback did not reach the level of significance. This differs from the findings of [9], but it should be noted that they were based on a median pre-test split, and we analyzed the data on an exercise-by-exercise basis. We found that the lowest anxiety level was reported by learners who answered correctly and received explanatory feedback, and this was significantly lower than the level of anxiety for correct answers which received supportive feedback. The highest anxiety level was reported by learners answering incorrectly and receiving explanatory feedback. This suggests that motivational support should be applied judiciously. It can reduce anxiety when the learner gives an incorrect answer. It may, however, increase anxiety when the learner has answered correctly, perhaps by implying that they're not doing as well as they thought. This is in line with [9] which indicated the importance of being supportive *only* when needed.

To advance gender equity in foreign language, researchers recommend understanding how gender influences which aspects of a learning environment are most effective for both learning and for anxiety [3,6]. We did not find gender-based differences for different feedback types. We did, however, find gender differences based on the feedback modality and the combination of feedback type and modality. For women, feedback from the agent produced significantly lower anxiety than from the other modalities, with the lowest levels coming from agent-based explanatory feedback. Males' anxiety levels were lowest when they received text-based explanatory feedback but they were highest when they received text-based supportive feedback. Future studies will focus on understanding the effectiveness of the interaction between feedback type, gender and performance within an adaptive system.

References

1. Al-Kaisi, A., Arkhangelskaya, A., Rudenko-Morgun, O., Lopanova, E.: Pedagogical agents in teaching language: types and implementation opportunities. Int. E-J. Adv. Educ. **5**(15), 275–285 (2020)
2. Ansari, M.S.: Speaking anxiety in ESL/EFL classrooms: a holistic approach and practical study. Int. J. Educ. Invest. **2**(4), 38–46 (2015)
3. Arroyo, I., Woolf, B.P., Cooper, D.G., Burleson, W., Muldner, K.: The impact of animated pedagogical agents on girls' and boys' emotions, attitudes, behaviors and learning. In: 2011 IEEE 11th International Conference on Advanced Learning Technologies, pp. 506–510. IEEE (2011)
4. Ayedoun, E., Hayashi, Y., Seta, K.: A conversational agent to encourage willingness to communicate in the context of English as a foreign language. Procedia Comput. Sci. **60**, 1433–1442 (2015)
5. Ayedoun, E., Hayashi, Y., Seta, K.: Adding communicative and affective strategies to an embodied conversational agent to enhance second language learners' willingness to communicate. Int. J. Artif. Intell. Educ. **29**(1), 29–57 (2019). https://doi.org/10.1007/s40593-018-0171-6
6. Brantmeier, C., Schueller, J., Wilde, J.A., Kinginger, C.: Gender equity in foreign and second language learning. Handb. Achiev. Gend. Equity Educ. **2**, 305–333 (2007)
7. Carlotto, T., Jaques, P.A.: The effects of animated pedagogical agents in an English-as-a-foreign-language learning environment. Int. J. Hum. Comput. Stud. **95**, 15–26 (2016)
8. Clark, R.C., Mayer, R.E.: E-learning and the Science of Instruction: Proven Guidelines for Consumers and Designers of Multimedia Learning. Wiley, Hoboken (2016)
9. D'Mello, S., Graesser, A.: AutoTutor and Affective AutoTutor: Learning by talking with cognitively and emotionally intelligent computers that talk back. ACM Trans. Interact. Intell. Syst. **2**(4), 1–39 (2013)
10. Fariadian, E., Azizifar, A., Gowhary, H.: Gender contribution in anxiety in speaking EFL among Iranian learners. Int. Res. J. Appl. Basic Sci. **8**(11), 2095–2099 (2014)
11. Genç, G.: Can ambiguity tolerance, success in reading, and gender predict the foreign language reading anxiety? J. Lang. Linguist. Stud. **12**(2), 135–151 (2016)
12. Hasan, D.C., Fatimah, S.: Foreign language anxiety in relation to gender equity in foreign language learning. In: Zhang, H., Chan, P.W.K., Boyle, C. (eds.) Equality in Education, pp. 183–193. SensePublishers, Rotterdam (2014). https://doi.org/10.1007/978-94-6209-692-9_14
13. Horwitz, E.K., Horwitz, M.B., Cope, J.: Foreign language classroom anxiety. Mod. Lang. J. **70**(2), 125–132 (1986). https://doi.org/10.2307/327317
14. Ismail, D., Hastings, P.: Identifying anxiety when learning a second language using e-learning system. In: Proceedings of the 2019 Conference on Interfaces and Human Computer Interaction, pp. 131–140 (2019). http://www.iadisportal.org/digital-library/identifying-foreign-language-anxiety-when-using-an-e-learning-system
15. Ismail, D., Hastings, P.: A sensor-lite anxiety detector for foreign language learning. In: Proceedings of the 2020 Conference on Interfaces and Human Computer Interaction, pp. 19–26 (2020)
16. Jin, Y.X., Dewaele, J.M.: The effect of positive orientation and perceived social support on foreign language classroom anxiety. System **74**, 149–157 (2018)
17. Kralova, Z., Petrova, G.: Causes and consequences of foreign language anxiety. XLinguae **10**(3), 110–122 (2017)

18. MacIntyre, P.D., Gardner, R.C.: The subtle effects of language anxiety on cognitive processing in the second language. Lang. Learn. **44**(2), 283–305 (1994). https://doi.org/10.1111/j.1467-1770.1994.tb01103.x

19. Van der Meij, H., Van der Meij, J., Harmsen, R.: Animated pedagogical agents effects on enhancing student motivation and learning in a science inquiry learning environment. Educ. Technol. Res. Dev. **63**(3), 381–403 (2015). https://doi.org/10.1007/s11423-015-9378-5

20. Moreno, R.: Decreasing cognitive load for novice students: Effects of explanatory versus corrective feedback in discovery-based multimedia. Instr. Sci. **32**(1–2), 99–113 (2004)

21. Onwuegbuzie, A.J., Bailey, P., Daley, C.E.: Cognitive, affective, personality, and demographic predictors of foreign-language achievement. J. Educ. Res. **94**(1), 3–15 (2000)

22. Parkes, J., Abercrombie, S., McCarty, T.: Feedback sandwiches affect perceptions but not performance. Adv. Health Sci. Educ. **18**(3), 397–407 (2013)

23. Prochazka, J., Ovcari, M., Durinik, M.: Sandwich feedback: The empirical evidence of its effectiveness. Learn. Motiv. **71**, 101649 (2020)

24. Qutob, M.M., Madini, A.A.: Saudi EFL learners' preferences of the corrective feedback on written assignment. Engl. Lang. Teach. **13**(2), 16–27 (2020)

25. Rafada, S.H., Madini, A.A., et al.: Effective solutions for reducing Saudi learners' speaking anxiety in EFL classrooms. Arab World Engl. J. **8**(2), 15 (2017)

26. Romero-Hall, E.: Animated pedagogical agents and emotion. In: Emotions, technology, design, and learning, pp. 225–237. Elsevier Academic Press, Cambridge (2016)

27. Taghinezhad, A., Abdollahzadeh, P., Dastpak, M., Rezaei, Z.: Investigating the impact of gender on foreign language learning anxiety of Iranian EFL learners. Mod. J. Lang. Teach. Methods **6**(5), 417–426 (2016)

"I didn't copy his code": Code Plagiarism Detection with Visual Proof

Samuel John[1(✉)] and George Boateng[2(✉)]

[1] FUOYE, Oye-Ekiti, Nigeria
[2] ETH Zurich, Zürich, Switzerland
`gboateng@ethz.ch`

Abstract. Code plagiarism in online courses gives a false idea of the performance of students. In 2020, we run a smartphone-based online coding course, SuaCode Africa 2.0 in which 27% of plagiarism cases was found in the final assignment submissions. Hence, a need arose to develop software that detects plagiarism among source code. The software described in this paper detects plagiarized source code containing English and French texts. Also, the code examples provided by the instructors is taken into consideration. In other words, code blocks present in the examples can be reused by any student. We trained machine learning models on three cosine similarity based metric extracted from the TF-IDF feature vector of the code files. The system provides proof of plagiarism on a GUI tool that visualizes the similar sections of the flagged files. This software will contribute to having a sincere evaluation of the impact of SuaCode on the students, thereby preventing the production of incompetent programmers.

Keywords: Code plagiarism · Machine learning · Online course · TD-IDF · NLP · Coding · Introductory programming · Processing · Africa

1 Introduction

Source-code plagiarism can be defined as trying to pass off (parts of) source code written by someone else as one's own (i.e., without indicating which parts are copied from which author) [6]. In an academic environment, source-code plagiarism arises in programming assignments. Students having the intention of achieving good grades with less or almost no effort, often try to copy the assignments from their friends. The instructor of a course can receive false feedback about the difficulty level of the course and the performance of the students. This situation makes the problem of code plagiarism detection an important task. It is hard to manually inspect and decide whether a submission is genuine or plagiarized since the number of unique pairs that could contain plagiarism grows quadratically.

© Springer Nature Switzerland AG 2021
I. Roll et al. (Eds.): AIED 2021, LNAI 12749, pp. 208–212, 2021.
https://doi.org/10.1007/978-3-030-78270-2_37

In 2020, a smartphone-based online coding course, SuaCode Africa 2.0 was run in both English and French [3] building upon past works to leverage smartphones to teach Africans to code [1,4]. Over 2,000 applications from 69 countries were received, 740 students were accepted and trained. Out of that, 431 students submitted the final assignment in Processing (a Java-based programming language) and 27% were found culpable of the offense of plagiarism. It was laborious and time-consuming for the instructors to manually detect dishonest students. This process is not scalable with the program growing. Hence, the need for software that automates code plagiarism detection and provides interpretable proof arose. Some students in the course were Francophone students and hence parts of their submitted code were written in French. Students were not allowed to share their code files with students or use code written by other students. One challenge that had to be addressed was that code examples were provided to students which they could freely use in their code. The situation could be a basis for wrongly flagging students for plagiarism since the same example code blocks could be found in several students' code. Hence the system needs to be robust in such cases. Additionally, given the penalty of being implicated – not receiving a certificate of completion – the system should provide proof to students if they are accused of plagiarism to enable transparency and give an opportunity to challenge the accusation.

In comparison to other works on plagiarism detection, the authors of a work called InfiniteMonkey developed a GUI-based software trained on a synthetic dataset [7]. In another work, the compiled-state features of a source code were extracted to create a machine learning powered system [8] In [11], an algorithm was specifically created to detect plagiarism using three low-level similarity based metrics. Domin et al. [5] created a system that eliminates the common ground code block before checking the pairwise similarity used to train a random forest algorithm that classifies if a code is plagiarised or not. Additionally, early solutions such as JPlag and MOSS described in [9] and [2] respectively are dependent on the structure of code pairs in the detection of plagiarism. JPlag implements a 'Greedy String Tiling' algorithm to tokenize texts while MOSS implements a 'winnowing' algorithm for its detection and also, it accepts code templates to be ignored if found in the files under inspection. Our work is the first that builds a complete end-to-end system using machine learning algorithms in detecting plagiarized source code containing English and French texts while taking the code examples provided by the instructors into consideration, and providing visual proofs for the implicated cases.

2 Data and Features

We compiled 432 source codes (the instructor code example inclusive) submitted for the last assignment. Then, 5 instructors manually inspected and annotated 230 suspected files (from the 92655 unique pairwise combinations of all files) as plagiarised (1) or non-plagiarised (0). This resulted in 117 confirmed cases. The suspected files were those that were above a threshold of the cosine similarity scores between TF-IDF n-grams of all code file pairs.

We computed the TF-IDF vectors of the code files using an n-gram range of (2, 6). Also, the keywords peculiar to the language syntax were used as stopwords. The n-gram range was found experimentally to limit flagging of files containing French texts. Next, we grouped all student code files in pairs. Our features were extracted from the resultant 92665 unique pairwise combinations. We computed the cosine similarity of the TF-IDF vector of the code file pairs. The resultant array is of shape (432, 432). This process resulted in 3 features for each pair of students' code which we standardized by scaling the numerical values between -1 and 1. Our feature set consists of the cosine similarity score of the following: (1) student 1 code and student 2 code, (2) student 1 code and example code and (3) student 2 code and example code.

3   Experiments and Evaluation

We set out to have a model with a high balanced accuracy ($\geq 80\%$) and less than 1% false positives in order reduce the chances of wrongly accusing students. For our first experiment, we trained five (5) models: logistic regression (LR), decision trees (DT), linear support vector machine (SVM), k-nearest neighbor (KNN), and random forest (RF) classifier. The data has a very high class imbalance (117 of 92665 cases of plagiarism) and so we used the 'balanced' parameter for the models and used the balanced accuracy and false positives as the metrics. The data was split into a 70:30 train test stratified split. Additionally, we explored if using only the comments in the code would be enough to detect plagiarism.

Table 1. Classification results with different models

Model	Balanced accuracy	False positives
Logistic regression	81%	5
Support vector machine	81%	5
Decision tree	82%	6
K nearest neighbor	86%	6
Random forest	**84%**	**1**

4   Results and Discussion

We present the results of five models in Table 1 obtained from our experiments here using the validation set. Given our goal has a low number of false positives, we select RF to be the model that the system would make use of in production. We next used the selected RF model to perform classification with only the code comments. That classification produced an 80% balanced accuracy with 1 false positive. This result is not far off from the result using the whole code and shows that comments alone can be used to detect code plagiarism.

Furthermore, we built software that provides visual proof of plagiarism, which is interpretable (Fig. 1) using a GUI tool named Pydiff [10], an open-source minimalistic GUI for python's difflib module shown. It compares two files and then highlights the textual difference found. Pydiff does not highlight textual "similarities". So, we modified its algorithm to make it suitable for this work. Color red and green (darker) are used to highlight the similar sections while the dissimilar characters are highlighted in the transparent mode of these colors. The dissimilar sections in both file pair are not highlighted at all.

Given the system is trained using Processing-based code files, it may not effectively detect plagiarized cases in codes written in other languages like Python, Ruby, etc. Also, the model has not been deployed for use in a new course yet which will be done in future cohort of SuaCode. Doing this would give a true evaluation of its performance in the real world.

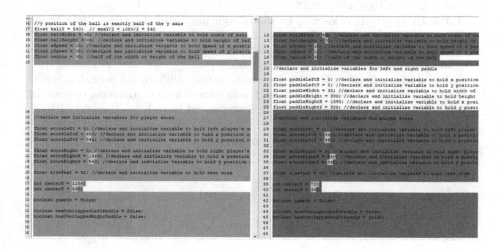

Fig. 1. GUI tool highlighting plagiarized code sections in two files

5   Conclusion

We developed a machine-learning powered software that detects plagiarism in the assignments submitted for a coding course and it also provides interpretable visual proofs. The system contributes to having a sincere evaluation of the competency of students in SuaCode courses, thereby preventing the production of incompetent programmers.

212   S. John and G. Boateng

References

1. Keep calm and code on your phone: a pilot of suacode, an online smartphone-based coding course, pp. 1173–1179 (2019). https://doi.org/10.1145/2851581.2892512
2. Aiken, A.: Moss, a system for detecting software plagiarism (2002)
3. Boateng, G.: Kwame: a bilingual ai teaching assistant for online suacode courses. arXiv preprint arXiv:2010.11387 (2020)
4. Boateng, G., Kumbol, V.: Project iSWEST: Promoting a culture of innovation in Africa through STEM. In: 2018 IEEE Integrated STEM Education Conference (ISEC), pp. 104–111, March 2018. https://doi.org/10.1109/ISECon.2018.8340459
5. Domin, C., Pohl, H., Krause, M.: Improving plagiarism detection in coding assignments by dynamic removal of common ground, pp. 9–14 (2016)
6. Hage, J., Rademaker, P., van Vugt, N.: A comparison of plagiarism detection tools. Utrecht University. Utrecht, The Netherlands 28(1) (2010)
7. Heres, D.: Source Code Plagiarism Detection using Machine Learning. Master's thesis, Utrecht University (2017)
8. Katta, J.Y.B.: Machine learning for source-code plagiarism detection. Ph.D. thesis, International Institute of Information Technology Hyderabad (2018)
9. Prechelt, L., Guido, M., Philippsen, M.: Finding plagiarisms among a set of programs with JPlag. J. UCS 8(11), 1016 (2002)
10. pydiff - a minimalistic difflib GUI, December. https://github.com/yebrahim/pydiff
11. Inoue, U., Wada, S.: Detecting plagiarisms in elementary programming courses. In: 9th International Conference on Fuzzy Systems and Knowledge Discovery (2012)

An Epistemic Model-Based Tutor for Imperative Programming

Amruth N. Kumar[✉]

Ramapo College of New Jersey, Mahwah, NJ 07430, USA
amruth@ramapo.edu

Abstract. We developed a tutor for imperative programming in C++. It covers algorithm formulation, program design and coding – all three stages involved in writing a program to solve a problem. The design of the tutor is epistemic, i.e., true to real-life programming practice. The student works through all the three stages of programming in interleaved fashion, and within the context of a single code canvas. The student has the sole agency to compose the program and write the code. The tutor uses goals and plans as prompts to scaffold the student through the programming process designed by an expert. It provides drill-down immediate feedback at the abstract, concrete and bottom-out levels at each step. So, by the end of the session, the student is guaranteed to write the complete and correct program for a given problem. We used model-based architecture to implement the tutor because of the ease with which it facilitates adding problems to the tutor. In a preliminary study, we found that practicing with the tutor helped students solve problems with fewer erroneous actions and less time.

Keywords: Programming tutor · Imperative programming · Model-based architecture

1 Introduction

Numerous tutors have been developed to help students learn to write code in high-level languages [15–17] such as LISP [5], Haskell [27] and Prolog [21], imperative languages such as Pascal [6], Java [24–26] and C# [23], and scripting languages such as Python [22, 28]. We built a tutor for imperative programming in the popular language C++ for use by introductory programming students.

When learning to program, students need explicit instruction on formulating the algorithm [14]. Several flowchart-based programming environments have been developed to help improve the algorithm formulation skills of students (e.g., [1, 2, 13, 19, 20, 29]). ProPL [8] uses natural language to help students write pseudocode.

Several systems have been reported that integrate all three stages of programming, viz., algorithm formulation, program design and coding, including LISP Tutor for LISP [5] and PROUST [6], BRIDGE [7] and GPCEditor [9] for Pascal and Guided-Planning and Assisted-Coding tutor [10, 11] and J-Latte [25] for Java. Typically, these tutors deal with algorithm formulation in terms of goals and plans - the student identifies goals

© Springer Nature Switzerland AG 2021
I. Roll et al. (Eds.): AIED 2021, LNAI 12749, pp. 213–218, 2021.
https://doi.org/10.1007/978-3-030-78270-2_38

(what should be done next) and plans (how it should be done), before writing code for the plans.

LISP Tutor for LISP [5] and PROUST for Pascal [6] use goals and plans to diagnose the code written by the student. Instead, we use goals and plans as prompts to scaffold the problem-solving process of the student. BRIDGE [7] uses a visual intermediate language to represent the algorithm, whereas J-Latte [25] uses a visual representation. Instead, we use pseudocode as comments, which naturally belong in a program. Guided-planning and assisted coding tutor [11] provides feedback on demand during coding. It places each line of code in the program instead of asking the student to do so. In contrast, we provide immediate feedback, which has been shown to be more efficient for programming instruction [12]. In addition, we make placing each new statement in its correct location in the program the responsibility of the student. Goal-Plan-Code Editor (GPCEditor) [9] translates the plans of students into code. Instead, we have the student write the code.

Epistemic Design. Our design of the tutor is epistemic, i.e., true to real-life problem-solving and programing practice because of the following design choices:

Actions. The tutor facilitates three operations: selecting, locating and coding. In select-ing operation, the student selects the appropriate step in the algorithm (e.g., which input to process next) that is translated into pseudocode in the program. The student also uses selecting operation for program design, e.g., to identify the type of control construct to use for a step in the algorithm. The student uses locating operation to compose the program, i.e., the location in the program where the next step in the algorithm should be coded. Thereafter, the student proceeds to write code for the step. The tutor does not use affordances such as drag-and-drop tiles (e.g., [25]) or flowcharts (e.g., [1, 2, 13, 19, 20, 29]) or intermediate languages [7] to design the algorithm – affordances not found in real-life programming environments.

Agency. The student is responsible for identifying the location of each step in the algo-rithm and program – it is not automatically determined by the tutor for the student (e.g., [11]). When coding, the student is expected to enter the frame [18] of each control construct by hand – it is not provided to the student (e.g., [25]).

Temporal Order. Trying to end algorithm design stage before going to coding (e.g., [7]) forces the novice programmer to either design based on assumptions about the code (e.g., assumptions about the control statement that will be used in a section) or code based on decisions taken too far in advance for the novice to properly appreciate (e.g., why statements must appear in a certain order in the program). In real life, programmers go back and forth between algorithm design and coding: each informing the other. In our tutor, the student goes through algorithm design, program design and coding steps in an interleaved fashion for each step.

Code Canvas. In the tutor, the student takes all the actions in the context of a single code canvas. The algorithm as pseudocode is embedded as comments within the program. So, the novice can conceptually connect each step in the algorithm with the corresponding statement(s) in the program.

Scaffolding. The tutor uses goals and plans as prompts to scaffold the student through algorithm formulation and program design instead of using them to diagnose the student's program (e.g., [5, 6]). Every time the student selects an incorrect step in the algorithm or an incorrect choice for program design, the tutor provides immediate drill-down feedback that steers the student towards the correct step/choice. So, the student gets the opportunity to practice the process of problem-solving and programming as designed by an expert every time the student works with the tutor.

Reified Steps. During coding, the tutor provides feedback at the level of statements and expressions. When control statements such as `if-else` and `while` loop are involved, the tutor uses a script to step the student through the various components of the control statement, e.g., frame, initialization, condition, body and update for `while` loop. Such reification of steps not only makes diagnosing and providing feedback more tractable, it also trains the student to use a pedagogically effective algorithm to compose each control statement in terms of its components.

Non-deterministic. The tutor admits equivalent answers. For example, the student can select any equivalent data type for a variable (`short`, `int`, or `long`), can locate inputs in any order in the program and write a commutative expression in any order. This design acknowledges the fact that a program can be written in a multitude of ways for a problem.

Model Based Architecture. A problem is represented using a problem specification and a reference solution template. A problem specification is an annotated problem statement wherein, input and output data elements are identified, and other attributes are specified such as the expected data type for the input/output data, preferred name, etc. The reference solution template is a complete solution (i.e., program) written in BNF notation, with meta-variables for variable names (e.g., <V1>), data types (e.g., <T1>), and other program elements.

The tutor model, user interface and domain model of the tutor are all problem-independent. The tutor model includes scripts for the steps in the problem-solving process. For example, the script for input data object is: 1) Locate where the data will be input; 2) Declare the variable and 3) Input the variable. Each step in the above script may itself generate additional scripts. A declarative representation is used for these scripts so that they can be swapped to test various problem-solving processes with the same tutor. The user interface of the tutor translates each atomic step in the script into one of the following three user inputs:

1. **Select** from a drop-down menu of options. A built-in feedback server for each select action encodes drill-down feedback at abstract, concrete and bottom-out levels for each incorrect menu option selected by the student.
2. **Locate** in code by clicking in it. If the student correctly locates a statement, the tutor inserts pseudocode as a comment at that location and presents a dialog box for the student to enter the code. If the location is incorrect, the tutor provides drill-down feedback to steer the student towards the correct location.
3. Write the **code** for the next statement or expression. The domain model (described next) provides drill-down explanation for incorrect code.

The tutor determines the correctness of select actions by comparing them with the annotations in the problem statement. It determines the correctness of locate and code actions by comparing them with the reference solution provided for the problem.

The Domain Model is a model of the programming domain built using Model-Based Reasoning principles [4]: each programming construct is modeled as a component with a text representation and behavior [3]. The tutor uses the Domain Model to build a model of the student's solution, called the Program Model. The Program Model is used to generate the text representation of the student's program. It is also used to provide feedback for coding and locating actions. Each component in the Program Model is associated with a bug library relevant to that programming construct. The component generates drill-down feedback for coding errors by using the bug library to diagnose the error in the student's code. Each component is also associated with a catalog of program transformations. It uses this catalog to approve semantically equivalent code alternatives (e.g., count++ is equivalent to count+ = 1). This ability of the Program Model to provide drill-down feedback for locating and coding actions is a significant advantage of using model-based reasoning instead of some of the other AI techniques used for modeling the domain in programming tutors such as rule-based (e.g., [5]) and constraint-based (e.g., [25]) reasoning: the drill-down feedback need not be individually specified for each problem added to the tutor. So, adding a new problem takes minimal effort – only the problem specification and reference solution template need to be specified for each new problem.

Whether it is select, locate or code action, the student cannot proceed to the next step in the algorithm until the student answers that step correctly and completely. So, by the end of the session, the student is guaranteed to have written the correct program for a given problem. For each action, the tutor provides immediate drill-down feedback at abstract, concrete and bottom-out levels, thereby ensuring that the student is never stranded at a dead-end. Given this design, the proficiency of a student is measured not in terms of the correctness of the final program, but in terms of the number of actions needed by the student to arrive at the correct program: the more actions the student takes, the less proficient the student.

The tutor is not a novel intervention for introductory programming as much as a technological facilitator of a pedagogy well understood to help introductory students learn programming – the pedagogy of practice. The more programs a student writes, the better the student becomes at the process of problem-solving and programming. The role the tutor plays is of a facilitator – it provides one-on-one scaffolding and feedback throughout the process of programming. The alternative to using the tutor in an introductory programming class would be to assign multiple programming projects on each topic, which is untenable because of the workload it entails for the instructor, not to mention the reluctance of students to engage in such labor, especially without the one-on-one feedback facilitated by the tutor. Given this, we evaluated the tutor to see whether the benefits of practice would accrue to students who use it. Preliminary results show that practicing with the tutor indeed helped students solve subsequent problems with fewer erroneous actions and in less time.

Acknowledgments. Partial support for this work was provided by the National Science Foundation under grant DUE-1432190.

References

1. Gomes, A., Mendes, A.J.: SICAS: interactive system for algorithm development and simulation. In: Computers and Education - Towards an Interconnected Society, pp. 159–166 (2001)
2. Carlisle, M., Wilson, T., Humphries, J., Hadfield, S.: RAPTOR: a visual programming environment for teaching algorithmic problem solving. ACM SIGCSE Bull. **37**(1), 176–180 (2005)
3. Kumar, A.N.: Model-based reasoning for domain modeling in a web-based intelligent tutoring system to help students learn to debug C++ programs. In: Cerri, S.A., Gouardères, G., Paraguaçu, F. (eds.) ITS 2002. LNCS, vol. 2363, pp. 792–801. Springer, Heidelberg (2002). https://doi.org/10.1007/3-540-47987-2_79
4. Davis, R.: Diagnostic reasoning based on structure and behavior. Artif. Intell. **24**, 347–410 (1984)
5. Reiser, B., Anderson, J., Farrell, R.: Dynamic student modelling in an intelligent tutor for lisp programming. In: Proceedings of the Ninth International Joint Conference on Artificial Intelligence, pp. 8–14 (1985)
6. Johnson, W.L.: Intention-Based Diagnosis of Errors in Novice Programs. Morgan Kaufman, Palo Alto (1986)
7. Bonar, J., Cunningham, R.: BRIDGE: tutoring the programming process, in intelligent tutoring systems: lessons learned. In: Psotka, J., Massey, L., Mutter, S. (eds.) Lawrence Erlbaum Associates, Hillsdale (1988)
8. Lane, C., VanLehn, K.: Teaching the tacit knowledge of programming to novices with natural language tutoring. Comput. Sci. Educ. **15**(3), 183–201 (2005)
9. Guzdial, M., Hohmann, L., Konneman, M., Walton, C., Soloway, E.: Supporting programming and learning-to-program with an integrated CAD and scaffolding workbench. Interact. Learn. Environ. **6**(1&2), 143–179 (1998)
10. Jin, W.: Pre-programming analysis tutors help students learn basic programming concepts. In: Proceedings of the 39th SIGCSE Technical Symposium on Computer Science Education (SIGCSE 2008), New York, NY, USA, pp. 276–280. Association for Computing Machinery (2008)
11. Jin, W., Corbett, A., Lloyd, W., Baumstark, L., Rolka, C.: Evaluation of guided-planning and assisted-coding with task relevant dynamic hinting. In: Trausan-Matu, S., Boyer, K.E., Crosby, M., Panourgia, K. (eds.) ITS 2014. LNCS, vol. 8474, pp. 318–328. Springer, Cham (2014). https://doi.org/10.1007/978-3-319-07221-0_40
12. Barr, V., Trytten, D.: Using Turing's craft Codelab to support CS1 students as they learn to program. ACM Inroads **7**(2), 67–75 (2016)
13. Hu, M., Winikoff, M., Cranefield, S.: A process for novice programming using goals and plans. In: Proceedings of the Fifteenth Australasian Computing Education Conference (ACE 2013), vol. 136, pp. 3–12. Australian Computer Society, Inc. (2013)
14. Soloway, E.: Learning to program = learning to construct mechanisms and explanations. Commun. ACM **29**(9), 850–858 (1986)
15. Crow, T., Luxton-Reilly, A., Wuensche, B.: Intelligent tutoring systems for programming education: a systematic review. In: Proceedings of the 20th Australasian Computing Education Conference (ACE 2018), New York, NY, USA, pp. 53–62. Association for Computing Machinery (2018). https://doi.org/10.1145/3160489.3160492
16. Keuning, H., Jeuring, J., Heeren, B.: A systematic literature review of automated feedback generation for programming exercises. ACM Trans. Comput. Educ. **19**(1), 1–43 (2018)

17. Le, N.T., Strickroth, S., Gross, S., Pinkwart, N.: A review of AI-supported tutoring approaches for learning programming. In: Nguyen, N., van Do, T., le Thi, H. (eds.) Advanced Computational Methods for Knowledge Engineering, pp. 267–279. Springer, Heidelberg (2013). https://doi.org/10.1007/978-3-319-00293-4_20

18. Price, T.W., Brown, N.C., Lipovac, D., Barnes, T., Kölling, M.: Evaluation of a frame-based programming editor. In: Proceedings of the 2016 ACM Conference on International Computing Education Research (ICER 2016), New York, NY, USA, pp. 33–42. Association for Computing Machinery (2016). https://doi.org/10.1145/2960310.2960319

19. Hooshyar, D., Ahmad, R.B., Yousefi, M., Yusop, F.D., Horng, S.-J.: A flowchart-based intelligent tutoring system for improving problem-solving skills of novice programmers. J. Comput. Assist. Learn. **31**(4), 345–361 (2015). https://doi.org/10.1111/jcal.12099

20. Scott, A., Watkins, M., McPhee, D.: E-learning for novice programmers – a dynamic visualization and problem solving tool. In: 3rd International Conference Information and Communication Technologies: From Theory to Applications, ICTTA, 7–11 April, Damascus, Syria, pp. 1–6 (2008)

21. Gegg-Harrison, T.S.: Exploiting program schemata in a prolog tutoring system. Ph.d. thesis, Duke University, Durham (1993)

22. Weragama, D., Reye, J.: Analysing student programs in the PHP intelligent tutoring system. Int. J. Artific. Intell. Edu. **24**(2), 162–188 (2014)

23. Hartanto, B., Reye, J.: CSTutor: an intelligent tutoring system that supports natural learning. In: Proceedings of the Conference on Computer Science Education Innovation and Technology, pp. 19–26 (2013)

24. Sykes, E.: Design, development and evaluation of the java intelligent tutoring system. Technol. Instr. Cogn. Learn. **8**(1), 25–65 (2010)

25. Holland, J., Mitrovic, A., Martin, B.: J-LATTE: a constraint-based tutor for Java. In: Proceedings of the Conference on Computers in Education, pp. 142–146 (2009)

26. Gross, S., Pinkwart, N.: Towards an integrative learning environment for java programming. In: Proceedings of the IEEE Conference on Advanced Learning Technologies, pp. 24–28 (2015)

27. Gerdes, A., Heeren, B., Jeuring, J., Thomas van Binsbergen. , L.: Ask-Elle: an adaptable programming tutor for Haskell giving automated feedback. Int. J. Artif. Intell. Educ. **2016**, 1–36 (2016)

28. Brusilovsky, P., Malmi, L., Hosseini, R., Guerra, J., Sirkiä, T., Pollari-Malmi, K.: An integrated practice system for learning programming in python: design and evaluation. Res. Pract. Technol. Enhanc. Learn. **13**(1), 1–40 (2018). https://doi.org/10.1186/s41039-018-0085-9

29. Chen, S., Morris, S.: Iconic programming for flowcharts. In: Java, Turing, ETC', Conference on Innovation and Teaching Computer Science Education (ITiCSE), Caparica, Portugal, ACM, pp. 104–107 (2005)

Long Term Retention of Programming Concepts Learned Using Tracing Versus Debugging Tutors

Amruth N. Kumar[✉] [iD]

Ramapo College of New Jersey, Mahwah, NJ 07430, USA
amruth@ramapo.edu

Abstract. We studied long-term retention of the concepts that introductory programming students learned using two software tutors on tracing the behavior of functions and debugging functions. Whereas the concepts covered by the tutor on the behavior of functions were interdependent, the concepts covered by debugging tutor were independent. We analyzed the data of the students who had used the tutors more than once, hours to weeks apart. Our objective was to find whether students retained what they had learned during the first session till the second session. We found that the more the problems students solved during the first session, the greater the retention. Knowledge and retention varied between debugging and behavior tutors, even though they both dealt with functions, possibly because debugging tutor covered independent concepts whereas behavior tutor covered interdependent concepts.

Keywords: Retention of learning · Programming · Code-tracing · Debugging

1 Introduction

Researchers have studied interventions for improving long term retention of learning, such as data-driven examples (e.g., [1]), game-based environments (e.g., [2]), task interleaving (e.g., [9]), spacing (e.g., [3]) and active construction of digital artifacts (e.g., [4]). They have also attempted to incorporate retention into student models (e.g., [10]) in order to be able to predict the performance of a student on the next problem on a concept, when the problem is attempted a few hours, days or weeks later.

In order to find out if students retained the concepts learned using a tutor over the long term, in this observational study, we analyzed the data collected by two tutors when students used them more than once, a few days or weeks apart, of their own volition and on their own time. One tutor was on function behavior wherein students were asked to identify the output of a program and the other was on debugging functions wherein students were asked to identify bugs in a program. Both had reified interface [11], making it hard to guess the correct answer. The tutor on function behavior covered ten concepts: four on function call, two on function definition and four on parameter passing. The concepts are **interdependent**, i.e., a student who learns one parameter passing concept is likely to be able to solve problems on other parameter-passing concepts correctly. The tutor on debugging functions covered nine concepts: three on function call, four

© Springer Nature Switzerland AG 2021
I. Roll et al. (Eds.): AIED 2021, LNAI 12749, pp. 219–223, 2021.
https://doi.org/10.1007/978-3-030-78270-2_39

on function definition and two on parameter passing. The bugs are **independent**, i.e., knowledge of one bug is unlikely to help a student solve problems on another bug correctly. The tutors presented isomorphic problems generated as randomized instances of parameterized templates which are still challenging for novices [5, 6]. So, students saw different problems each time they used the tutors.

The tutors administered pre-test-adaptive practice-post-test protocol every time they were used [7]. The pre-test was used to prime the student model. Practice was provided on only the concepts on which the student solved a pretest problem incorrectly. Practice was provided on a concept until the student had mastered the concept by solving a minimum number and percentage of problems correctly. Post-test was presented on only the concepts mastered during practice. Pretest, practice and post-test were administered by the tutors back-to-back, all online and without any interruptions. The entire protocol was limited to 30 min. Each concept covered by the tutors can be classified as known, tested, practiced or learned for each student, as summarized in Table 1.

Table 1. Types of learning experience with the tutors.

Pretest	Practice	Posttest	Type of Learning
Correct			Known
Incorrect	None		Tested
Incorrect	Some		Practiced
Incorrect	Mastered	Incorrect	Practiced
Incorrect	Mastered	Correct	Learned

If a student who returns to use the tutor a second time at a later date or time solves the pretest problem on a concept correctly, the student has **retained** the concept from the previous session. If the student solves the pretest problem incorrectly, the student has **forgotten** the concept from the previous session. Based on the student's learning experience during the first tutoring session and pretest performance in the second tutoring session, the eight possible retention behaviors of a student on a concept are: known-retained, known-forgotten, tested-retained, tested-forgotten, practiced-retained, practiced-forgotten, learned-retained and learned-forgotten. Neither known-retained nor known-forgotten concepts are affected by the use of the tutor. These served as the comparison group in the study. On the other hand, tested-retained, practiced-retained and learned-retained all provide evidence in support of long-term retention of what was learned using the tutor, the hypothesis of this study, whereas tested-forgotten, practiced-forgotten and learned-forgotten all provide evidence disproving retention. These served as experimental data points in the study.

We used the data collected by the tutors over 14 semesters: Fall 2012 – Spring 2019. The tutors were used by introductory programming students in high schools and colleges as after-class assignments. The students could use the tutors as often as they pleased. We used data only from the students who had used the tutors at least twice and gave us permission to use their data for research purposes.

Function Behavior Tutor Results: 513 students used the tutor more than once. They solved problems at least twice on 3918 concepts, representing an average of 7.64 concepts per repeat user. Table 2 lists the number of student concepts N in each type of retention behavior, the percentage of the total student concepts represented by that retention behavior O%, the percentage of retained and forgotten concepts within the learning category L%, the mean pretest score on the first and second pretests, and the mean time between the two sessions in hours. The score on each problem was normalized to the range 0 → 1.0.

Table 2. Functions behavior tutor - types of retention behavior

Retention behavior type	N	O%	L%	Pretest1	Pretest2	Time (hours)
Known-Retained	2226	56.81	94.56	1.0	1.0	402.15 ± 107.0
Known-Forgotten	128	3.27	5.44	1.0	0.08	872.22 ± 446.4
Tested-Retained	760	19.40	67.86	0.12	1.0	27.07 ± 183.2
Tested-Forgotten	360	9.19	32.14	0.10	0.13	87.16 ± 266.2
Practiced-Retained	167	4.26	70.17	0.14	1.0	688.39 ± 390.8
Practiced-Forgotten	71	1.81	29.83	0.15	0.14	549.51 ± 599.4
Learned-Retained	164	4.19	79.61	0.17	1.0	588.52 ± 394.4
Learned-Forgotten	42	1.07	20.39	0.24	0.20	1622.82 ± 779.3

Known-forgotten concepts represent transience, the deterioration of learning over time. The student concepts in this category were 5.44% of all known student concepts. Based on the column titled L%, *students retained over 67% of the concepts covered by the tutor on function behavior.* Conversely, tested-forgotten, practiced-forgotten and learned-forgotten figures were all greater than known-forgotten percentage (5.44%) attributable to transience of learning. So, although students retained over 67% of the concepts, *there is room for improvement of the tutor to promote retention of learning.* We note two additional patterns in the descriptive statistics: in the column L%, learned-retained was greater than both practiced-retained and tested-retained. Since students solved more problems on learned concepts than practiced concepts and on practiced concepts than on tested concepts, this supports the observation that *the more the practice problems solved during the first session, the more likely students retained the concept till the second session.* From Table 2, we also note that the mean time between sessions is 2–3 times greater for forgotten concepts in each learning category compared to retained concepts, except in practiced category. It is possible that this observational study captured retained and forgotten student concepts in each category at different points in time, and eventually, more retained student concepts will convert to forgotten concepts without additional reinforcement of learning.

Debugging Tutor Results: 642 students used the tutor more than once. They solved problems at least twice on 5489 concepts, representing an average of 8.55 concepts per repeat user. Table 3 lists the retention behavior figures for debugging tutor. Since students either correctly identified a bug or did not, the score on a problem was either 0 or 1.

Table 3. Debugging functions tutor - types of retention behavior

Retention Behavior Type	N	O%	L%	Pretest1	Pretest2	Time (hours)
Known-Retained	2224	40.52	88.68	1.0	1.0	63.16 ± 21.52
Known-Forgotten	284	5.17	11.32	1.0	0.0	210.43 ± 60.23
Tested-Retained	1616	29.44	72.53	0.0	1.0	12.46 ± 25.25
Tested-Forgotten	612	11.15	27.47	0.0	0.0	93.43 ± 41.03
Practiced-Retained	402	7.32	76.14	0.0	1.0	52.55 ± 50.62
Practiced-Forgotten	126	2.30	23.86	0.0	0.0	125.41 ± 90.42
Learned-Retained	205	3.73	91.11	0.0	1.0	56.72 ± 70.89
Learned-Forgotten	20	0.36	8.89	0.0	0.0	137.89 ± 226.96

Known-retained is far smaller than 56.81% for function behavior tutor. Known-forgotten as a percentage of known concepts, which accounts for transience of learning, is larger (11.32%) than that for function behavior tutor. *So, knowledge and retention of learning varied between debugging and tracing skills, even though they both pertained to functions.* This confirms the results from our earlier study conducted using selection tutor [8]. Tested-retained and practiced-retained percentages (L%) on the other hand were greater for debugging than behavior of functions. One explanation is that each bug is unique and **independent**, and the short explanation provided for it clarifies the genesis of the bug. On the other hand, students must synthesize a lot of **interdependent** concepts to understand and predict the behavior of functions, making the behavior of functions harder to learn and retain.

Based on the column L%, *students retained over 72% of the concepts covered by the tutor on debugging functions.* Learned-forgotten (8.89%) was less than transience of learning. So, the mastery criterion used by debugging tutor during practice stage is robust. Here again, we found that *the more the practice problems solved during the first session, the more likely students retained the concept till the second session:* in column L%, learned-retained was greater than practiced-retained and tested-retained. Just as in the case of behavior tutor, we note that the mean time between sessions is at least twice as much for forgotten concepts in each category compared to retained concepts suggesting that we captured retained and forgotten student concepts in each category at different points in time.

In this study, we did not consider guesses and slips: the reified user interface makes it hard to guess the correct answer and error-flagging feedback provided by the tutors offers the opportunity for students to recover from slips. On the other hand, students who use a tutor repeatedly of their own volition are typically self-motivated. They are also likely to have had extraneous opportunities to practice the tutored concepts between the two tutoring sessions, which could have affected retention. These are confounding factors in terms of being able to generalize the results of this study.

Acknowledgments. Partial support for this work was provided by the National Science Foundation under grant DUE-1432190.

References

1. Mostafavi, B., Zhou, G., Lynch, C., Chi, M., Barnes, T.: Data-driven worked examples improve retention and completion in a logic tutor. In: Conati, C., Heffernan, N., Mitrovic, A., Verdejo, M.F. (eds.) AIED 2015. LNCS, vol. 9112, pp. 726–729. Springer, Heidelberg (2015). https://doi.org/10.1007/978-3-319-19773-9_102
2. Wouters, P., et al.: A meta-analysis of the cognitive and motivational effects of serious games. J. Educ. Psychol. **105**, 249–265 (2013)
3. Cepeda, N.J., Vul, E., Rohrer, D., Wixted, J.T., Pashler, H.: Spacing effects in learning, a temporal ridgeline of optimal retention. Psychol. Sci. **19**(11), 1095–1102 (2008)
4. Federici, S., Medas, C. and Gola, E.: Who learns better - achieving long-term knowledge retention by programming-based learning. In: Proceedings of the 10th International Conference on Computer Supported Education (CSEDU 2018) – vol. 2, pp. 124–133 (2018)
5. Bassok, M., Novick, L.R.: Problem solving. In: Holyoak, K.J., Morrison, R.G. (Eds.), Oxford handbook of thinking and reasoning, pp. 413-432. New York: Oxford University Press, Oxford (2012)
6. Martin, S.A., Bassok, M.: Effects of semantic cues on mathematical modeling: evidence from word-problem solving and equation construction tasks. Mem. Cogn. **33**(3), 471–478 (2005)
7. Kumar, A.N.: A model for deploying software tutors. In: IEEE 6th International Conference on Technology for Education (T4E), Amritapuri, India, pp. 3–9, 18-21 December 2014
8. Kumar, A.N.: Long Term Retention of Programming Concepts Learned Using Software Tutors. In: Kumar V., Troussas C. (eds) Proceedings of Intelligent Tutoring Systems (ITS 2020). Athens, Greece, LNCS 12149, pp. 382–387, 8-12 June 2020 https://doi.org/10.1007/978-3-030-49663-0_46
9. LeBlanc, K., Simon, D.: Mixed practice enhances retention and JOL accuracy for mathematical skills. In: 49th Annual Meeting of the Psychonomic Society, Chicago, IL (2008)
10. Wang, Y., Beck, J.E.: Using student modeling to estimate student knowledge retention. In: Proceedings of 5th International Conference on Educational Data Mining, pp. 200–203 (2012)
11. Kumar, A.N.: A reified interface for a tutor on program debugging. In: Proceedings of Third IEEE International Conference on Advanced Learning Technologies (ICALT 2003), Athens, Greece, pp. 190–194, 9–11 July 2003

Facilitating the Implementation of AI-Based Assistive Technologies for Persons with Disabilities in Vocational Rehabilitation: A Practical Design Thinking Approach

Marco Kähler[✉], Rolf Feichtenbeiner, and Susan Beudt

Deutsches Forschungszentrum für Künstliche Intelligenz (DFKI), Berlin, Germany
marco.kaehler@dfki.de

Abstract. Digital and AI-based assistive technologies (AI-AT) are becoming more important for the inclusion of persons with disabilities (PWD). One challenge in providing PWD with AI-AT is to meet their requirements and needs. At the same time, they are often embedded in organizational contexts and thus need to be cost-effective and easy to learn and handle. This short paper introduces a systematic approach to match the individual needs and organizational context with AI-AT that support working and learning of PWD. The approach combines Design Thinking (DT) methods, participatory elements, and online collaboration tools in a cycle of three workshops. The aim is to understand the target group better, identify, evaluate and choose appropriate AI-AT and develop innovation spaces that help introduce and test AI-AT. The approach was developed for a vocational rehabilitation setting but can also be easily adapted for various settings (e.g., educational technology or corporate AI projects).

Keywords: Design thinking · AI-based assistive technology · Inclusion · Participatory design · Innovation spaces

1 Matching of AI-AT and the Needs of PWD

Due to scientific progress, for example, AI-based recognition or recommendation systems, AI-based assistive technologies (AI-AT) are becoming more important to overcome the barriers PWD face every day [1, 2]. One challenge in providing PWD with suitable AI-AT is matching them with the individual's needs and requirements. Innovation research offers various approaches (e.g., design sprints, experience design, business model canvas) to introduce technology up to socio-technical transformation [3]. One particularly person-centered approach is the Design Thinking (DT) methodology [4] which is of particular importance for the inclusion of PWD [5, 6]. This paper presents an adapted and applied DT approach, which was developed in the context of innovation spaces in vocational rehabilitation and can be used to provide PWD with existing AI-AT. Innovation spaces help organizations cope with new challenges (e.g., in the field

© Springer Nature Switzerland AG 2021
I. Roll et al. (Eds.): AIED 2021, LNAI 12749, pp. 224–228, 2021.
https://doi.org/10.1007/978-3-030-78270-2_40

of digitalization) by experimenting, learning, and finding innovative solutions in a participatory way. As a process-based method, DT can help facilitate the different phases of introducing and developing new technologies and the necessary adaptations through a participatory process with relevant stakeholders [7]. The goal of the DT process can be to develop person-centered innovations in the form of different products or services from the very beginning and throughout the entire process [8, 9]. The most common DT approach describes five different phases in which a set of different creative and analytical methods (e.g., empathy cards, personas) can be applied [10–12]: (1) *empathy phase* (better understanding of the target group), (2) *define* (target group is defined), (3) *ideate* (develop and choose ideas on how the target group can be supported and meeting its needs), (4) *prototyping* (developing prototypes based on the ideas) and (5) *testing* (testing prototypes with the target group). The traditional DT process aims for the development of user-centered prototypes to kick off the technology development process. If new technology development is not desired (e.g., due to limited resources or time constraints), an adaptation of the classic DT process is necessary to match existing technologies with the specific requirements of the target group. Therefore, a DT approach with three successive workshops and participative phases was developed to assess the matching of AI-AT with the needs of PWD and create innovation spaces that help test the selected AI-AT by the target group. Due to the Covid 19 pandemic, the workshops were designed as digital workshops with online collaboration tools.

2 Selecting AI-AT with a Design Thinking Approach

To assess, select and test existing AI-AT and design innovation spaces for PWD, the target group must be determined, the AI-AT needs to be selected, and spaces, times, people, and approaches must be chosen for the innovation space. These must be in accordance with the conditions and capabilities of the organization, in which the AI-AT will be tested. Three coordinated DT workshops were developed with participative phases in between (see Fig. 1) to realize this process and open it up for PWD. The chosen procedure deviates from the traditional DT approach, where the user group is often only involved again in the testing phase. The necessary changes were applied in phases 2 (define) till 5 (testing) - in particular, the prototyping phase is mainly affected.

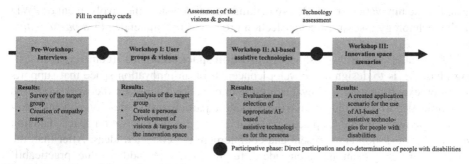

Fig. 1. Depiction of the design thinking approach to create a persona, select a suitable AI-based assistive technology, and create a scenario for an innovation space.

The first workshop, **"User groups & visions"** (which includes phase 2 and partly phase 3), aims to identify the desires, challenges, and needs of the potential user groups and develop a vision for the innovation spaces. For this purpose, interviews with the target group are conducted in advance, and the most crucial findings are recorded in so-called empathy maps (1st *participative phase*). Empathy maps are a common method in the DT process to structure answers from the interviews into certain clusters like "what does the interviewed person say about their problems" [13]. The interview results are the basis for the development of a persona. A persona is also a commonly used DT method which is the design of a fictional person with their own story [14, 15]. The persona templates for the workshops were expanded with some essential attributes (including problems in daily life due to the disability, degree of assistance needed, limitations and barriers in the world of work) to consider the special needs of PWD. This approach can help identify initial indications of potential assistance through AI. After that, a contrived daily routine of the persona is developed, which helps identify and specify the persona's central challenges and needs. Based on the steps before, visions are derived by using "How might we"-questions, which guide the assessment of AI-AT (see workshop II) and the development of the innovation space (see workshop III). After the workshop, assessing the vision and the goals together with PWD is recommended (*2nd participative phase*).

Building on the first workshop's results, the goal of the second workshop, **"AI-based assistive technologies"** (including the changed prototyping phase 4), is to assess and select suitable AI-AT for PWD in a participative manner. Before the workshop, a diverse set of six to eight suitable AI-AT should be identified and prepared for the workshop. Based on the results of the first workshop, the second workshop starts with defining a persona profile containing its main working activities, commonly used technologies, wishes and goals, and problems to be addressed by AI-AT. After that, the workshop's central part is assessing the user-technology-matching of each of the presented AI-AT. Therefore, each technology is assessed regarding criteria such as support of work activities, wishes and goals, and potential to overcome identified problems of the persona. The assessment process results in a total score (calculated from a set of the criteria mentioned above) for each technology representing its user-technology potential from the persona perspective (each score is evaluated on a scale of one to ten, added together to determine the total score). At the end of the workshop, the total scores of each AI-AT are used to create a ranking that helps to discuss and decide for one or more AI-AT to be tested in the innovation space. Before making a final decision, the involvement of PWD in the technology assessment and selection process is recommended (*3rd participative phase*).

The goal of the third workshop, **"Innovation space scenarios"** (comparable to testing phase 5), is to design and develop concepts for an innovation space that supports learning, experimenting, and evaluating the chosen AI-AT. Before the workshop, it is recommended to collect ideas from the workshop participants to identify relevant stakeholders, activities, methods, places, and time frames for the innovation space. The first step in the workshop is to present and complement the collected ideas. After that, the ideas are critically examined alongside different filters (e.g., added value, practicability). The interim result is a preliminary concept for the innovation space, which is then verified in terms of inner consistency and goal attainment (see workshop I). As a final step, the persona's typical day in the innovation space is envisioned to verify and specify the innovation space concept (regarding responsibilities, activities, places, and times). It is recommended to involve PWD after the workshop, for example, by giving feedback to the innovation space concept before implementation.

3 First Results and Outlook

This short paper presents a DT approach and workshop process conducted and evaluated in three vocational rehabilitation institutions. The first results of the formative evaluation (involving feedback surveys) show that the person-centered DT approach led to a strong emphasis on individual needs and challenges (e.g., by interviews with the target group before the first workshop and continually focusing on the persona), a more participatory process (e.g., by using democratic instruments like voting tools for decisions in the process), and a critical examination of the assistive potential of AI-AT for learning (e.g., job application trainer supporting emotional regulation) and working (e.g., voice assistance in care work via smart glasses) for PWD. This approach also revealed a significant gap in the availability of AI-AT on the market that really fit the individual needs and challenges of persons with specific disabilities (e.g., mental disabilities). This situation presents challenges for technology assessment and matching (see workshop II). Due to its high flexibility, the developed DT approach allows addressing such challenges with adaptions in the process, for example, by widening the scope of problems and needs of the persona or by using the developed personas (see workshop I) to design prototypes and develop new AI-AT. Also, the workshop participants showed very high expectations concerning the effectiveness, adaption possibilities, and availability of AI-AT. Potential frustrations and disappointed expectations, which can be detrimental to the implementation of AI-AT, could be resolved through transparent communication about the current state of AI-AT and active expectation management. Unfortunately, the available information basis for AI-AT is often insufficient [16], especially in (research or development) projects in which AI-AT are still under development. Due to the complexity of AI, it can also be challenging to explain AI-AT in a simple and concise form in a workshop setting. The experience so far has also shown that online creative DT workshops can be successfully facilitated in a similar quality as in face-to-face settings. They can even be advantageous regarding efficiency (easy and quick access to the digital space) and flexibility (digital workshops can easily be adapted for face-to-face settings, but not vice versa). Online workshop settings can be either more accessible (e.g., physical disabilities) or create new barriers depending on the individual disabilities. The complexity and dynamics of the DT process and disparate competencies and hierarchies between participants can be challenging for PWD and the facilitation of the workshops. The results from the DT workshops in overall nine institutions based on a summative evaluation will be presented in a separate paper.

References

1. Marzin, C.: Plug and Pray? A Disability Perspective on artificial intelligence, automated decision-making and Emerging Technologies. European Disability Forum, Brussels (2018)

2. Mark, B.G., Hofmayer, S., Rauch, E., Matt, D.T.: Inclusion of workers with disabilities in production 4.0: legal foundations in europe and potentials through worker assistance systems. Sustainability **11**(21), 5978 (2019)
3. Keijzer-Broers, W.J.W., de Reuver, M.: Applying agile design sprint methods in action design research: prototyping a health and wellbeing platform. In: Parsons, J., Tuunanen, T., Venable, J., Donnellan, B., Helfert, M., Kenneally, J. (eds.) DESRIST 2016. LNCS, vol. 9661, pp. 68–80. Springer, Cham (2016). https://doi.org/10.1007/978-3-319-39294-3_5
4. Plattner, H., Meinel, C., Leifer, L.: Design Thinking. Understand – improve – apply, Springer, Heidelberg (2011) https://doi.org/10.1007/978-3-642-21643-5_1
5. Chasanidou, D., Gasparini, A.A., Lee, E.: Design thinking methods and tools for innovation. In: Marcus, A. (ed.) DUXU 2015. LNCS, vol. 9186, pp. 12–23. Springer, Cham (2015). https://doi.org/10.1007/978-3-319-20886-2_2
6. Vechakul, J., Shrimali, B.P., Sandhu, J.S.: Human-centered design as an approach for place-based innovation in public health: a case study from Oakland. California. Matern. Child Health J. **19**, 2552–2559 (2015)
7. Carlgren, L.: Design Thinking as an Enabler of Innovation: Exploring the concept and its relation to building innovation capabilities. Chamlers University of technology, Gothenburg (2013)
8. Royalty, A., Shepard, S.: Mapping and measuring design thinking in organizational environments. In: Plattner, H., Meinel, C., Leifer, L. (eds.) Design Thinking Research. UI, pp. 301–312. Springer, Cham (2018). https://doi.org/10.1007/978-3-319-60967-6_15
9. von Thienen, J.P.A., Clancey, W.J., Corazza, G.E., Meinel, C.: Theoretical foundations of design thinking. In: Plattner, H., Meinel, C., Leifer, L. (eds.) Design Thinking Research. UI, pp. 13–40. Springer, Cham (2018). https://doi.org/10.1007/978-3-319-60967-6_2
10. Hasso Plattner Institute of Design at Stanford: An Introduction to D PROCESS Guide. https://web.stanford.edu/~mshanks/MichaelShanks/files/509554.pdf. Accessed 21 Jan 2021
11. Plattner, H., Meinel, C., Leifer, L.: Design Thinking Research: Studying Co-creation in Practice. Springer, Berlin (2012)
12. Brown, T.: Change by Design: How Design Thinking Transforms Organizations and Inspires Innovation. Harper Business, New York (2009)
13. Ferreira, B.M., Barbosa, S.D.J., Conte, T.: PATHY: using empathy with personas to design applications that meet the users' needs. In: Kurosu, M. (ed.) HCI 2016. LNCS, vol. 9731, pp. 153–165. Springer, Cham (2016). https://doi.org/10.1007/978-3-319-39510-4_15
14. Junior, P.T.A., Filgueiras, L.V.L.: User modeling with personas. In: Proceedings of CLIHC Latin American Conference on Human Computer Interaction, pp. 277–282 (2005)
15. Negru, S, Buraga, S: Towards a conceptual model for describing the personas methodology. In: Proceedings of the ICCP 2012. IEEE (2012)
16. Beudt, S., Blanc, B., Feichtenbeiner, R., Kähler, M.: Critical reflection of AI applications for persons with disabilities in vocational rehabilitation. In: Proceedings of DELFI Workshops. Bonn: Gesellschaft für Informatik e.V.z (2020)

Quantifying the Impact of Severe Weather Conditions on Online Learning During the COVID-19 Pandemic

Ezekiel Adriel Lagmay[✉] [iD] and Ma. Mercedes T. Rodrigo[✉] [iD]

Ateneo de Manila University, Quezon City, Metro Manila, Philippines
ezekiel.lagmay@obf.ateneo.edu, mrodrigo@ateneo.edu

Abstract. From October to November 2020 the Philippines was struck by eight typhoons, two of which caused widespread flooding, utilities interruptions, property destruction, and loss of life. How did these severe weather conditions affect online learning participation of students pursuing their undergraduate and graduate studies in the midst of the COVID-19 pandemic? We used CausalImpact analysis to explore September 2020 to January 2021 data collected from the Moodle Learning Management System data of one university in the Philippines. We found that overall student online participation was significantly negatively affected by typhoons. However, the effect on participation in Assignments and Quizzes were not significant. These findings suggested that students continued to invest their time and energy on activities that have a direct bearing on their final grades.

Keywords: CausalImpact · COVID-19 · Learning management system · Typhoon · Philippines

The Philippines is an archipelago in South East Asia with a population of 106 million people, 55 million of whom are less than 25 years old [10]. Even prior to the COVID-19 pandemic, the Philippine educational system was already in crisis. Philippines students' achievement levels in Math, Science, and English were among the poorest if not the poorest among countries in least three international achievement tests [5, 8, 11].

During the COVID-19 pandemic, students' and teachers' struggle to transition to online learning worsened when eight typhoons entered the Philippine Area of Responsibility from October 11 to November 12, 2020 [6]. Two of them—Typhoon Goni and Typhoon Vamco—caused widespread property destruction, utilities disruptions, and loss of life. Schools responded to this crisis with a post-Vamco class suspension, after which online learning resumed.

To what extent were students affected by these severe weather events? Were they able to return to normal levels of online participation or did these calamities dampen the performance for the rest of the post-typhoon period? In this paper, we use CausalImpact analysis [2] to quantify the extent to which student participation in an online learning environment was affected by Typhoons Goni and Vamco.

© Springer Nature Switzerland AG 2021
I. Roll et al. (Eds.): AIED 2021, LNAI 12749, pp. 229–233, 2021.
https://doi.org/10.1007/978-3-030-78270-2_41

1   CausalImpact Analysis

CausalImpact is a method of estimating the impact of an intervention such as an ad campaign on an outcome variable such as additional clicks [1, 2]. Given time series data, we first identify predictor variables, the outcome variable, and the pre- and post-intervention time segments. CausalImpact uses the pre-intervention data to model the relationship between the predictor variables and the outcome variable. It then uses the model to estimate the post-intervention counterfactual. The impact of the intervention is the difference between the counterfactual and the observed post-intervention data.

For this analysis, we used a time series of log data from the Moodle Learning Management System (LMS) of a university in Metro Manila, Philippines, collected from September 9, 2020 to January 9, 2021. This time period represented two distinct academic terms: the first quarter (September 9 to October 24) and second quarter (October 28 to January 9). The dataset contained a total of 2,641,461 logs from 12,699 users.

We used transaction log volume, i.e. counts, as the indicator of participation. We did not consider the actual content of the transactions.

Moodle subcategorizes each transaction into components. We performed a CausalImpact analysis for four outcome variables: overall student LMS activity, the System component (all transactions related to course communications and management), the Assignment component (all transactions related to editing, viewing, completion, and grading of assignments), and the Quiz component (all transactions related to quiz attempts, submissions, creation, and grading).

The users of Moodle fell into three categories: teachers, non-editing teachers (e.g. teaching assistants), and students. The data was first tallied according to User Type, Component, and Date. Within each user type we normalized the data by dividing each User Type-Component-Date row with the maximum possible value of User Type-Component. The normalized values ranged from 0 to 1.

We opted to use teacher and non-editing teacher components as our predictor variables. Our theoretical grounding for this choice is the teacher expectancy effect which asserts that teacher expectations have an impact on students' academic progress. These effects have been observed at the individual and class level for both achievement outcomes and self-concept [see 3, 9] and have been shown to persist over time [see 9]. We used Dynamic Time Warping (DTW) to arrive at a parsimonious set of predictor variables [7].

We defined our pre-intervention period as September 9 to October 28, our intervention period as October 29 to November 13, and our post-intervention period as November 14 to December 23.

2   Results

Figure 1 shows the CausalImpact graph of all LMS activity. The topmost graph labeled "original" shows a solid line representing the actual observed data. The broken line represents the prediction. The light blue band represents the confidence interval of the prediction. The middle graph labeled "pointwise" shows the difference between the prediction and the actual values. Finally, the cumulative graph at the bottom shows the accumulated difference between the prediction and the actual. The gap in the pointwise and cumulative graphs is the intervention period. There is no accumulated difference during the pre-intervention period. The differences are accumulated post-intervention. Note that the cumulative graph shows a downward trend during the post-intervention period and that there was indeed a slump in the week or so following the typhoons.

All LMS activity decreased significantly ($p = 0.01$) after the typhoons (see Fig. 1). The response variable had an average value of 0.18 in contrast to the counterfactual prediction of 0.23 ($SD = 0.018$). The typhoons therefore had an estimated effect of -0.045 with a 95% interval of $[−0.082, −0.010]$. When the data points during the intervention period are summed, the response variable had an overall value of 7.41. The counterfactual prediction was 9.25 ($SD = 0.743$) with a 95% confidence interval of $[7.83, 10.77]$.

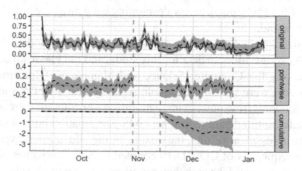

Fig. 1. CausalImpact graph for all LMS activity.

System activity during this period also significantly decreased ($p = 0.01$). The response variable averaged 0.11 as opposed to a counterfactual prediction of 0.15 ($SD = 0.017$) with a 95% interval of $[0.12, 0.19]$. The sum of the response variable data points during the post-intervention period was 4.43 in contrast to a predicted 6.30 ($SD = 0.710$) with a 95% interval of $[4.96, 7.71]$. The effects of the typhoons on student behavior on Assignments and Quizzes was not statistically significant.

3   Discussion

The purpose of this paper was to determine the extent to which severe weather affected student participation in online classes. All LMS Logs and System components significantly decreased.

It was interesting to see that actual participation in the Assignments and Quizzes components were not significantly different from their predicted behavior. This suggests that students continued to comply with academic assessments as assignments and quizzes make measurable contributions to their grades. System behavior, on the other hand, refer to actions such as checking the course for announcements. These activities are generally not graded.

Limitations. The generalizability of these findings is subject to at least four limitations. First, CausalImpact analysis requires that the predictor variables should not be affected by the same intervention as the response variable [4]. In this case, it was the likely case that the teachers and non-editing teachers were affected by the typhoons, just as their students were. To this point, we offer two counterarguments: First, we used DTW to find the teacher and non-editing teacher features that were most predictive of student behaviors. The algorithm eliminated the features with no predictive power, leaving only those that could give us a reasonable estimate of student behavior. Second, we return to our theoretical framework regarding the teacher expectancy effects [see 9]. Teacher expectations have an impact on student achievement and self-concept, so it is arguable that students will take their cues from the pace and requirements that the teachers set.

Our second limitation has to do with the population from which the data was taken. The students in this sample were among the best in the country. They generally came from well-to-do socio-economic backgrounds. Hence, their resilience is not indicative of the resilience of the Philippines or any developing country as a whole. It may, at best, serve as an upper bound.

Third, the university had two LMSs working in parallel, Moodle and Canvas. We were only able to capture Moodle data for this study, and the classes using the Moodle server were generally the Computer Science and Management Information Systems classes. The students were therefore technology-savvy and adept at online modes of communication. Students from other courses might have encountered greater challenges.

Finally, the data captured here represents LMS participation but not other important outcomes such as assessment results, the quality of the educational experience, or the mental health consequences of COVID-19 coupled by severe weather. While students and faculty evidently powered through their requirements, it would be best to triangulate these results with findings and observations from other constituency checks, for a more complete reading of our community.

Contributions. Despite these limitations, this paper contributes to the literature by applying CausalImpact analysis on LMS data from the Philippines to determine the effects of severe weather on students. To our knowledge and as of the time of this writing, this is the first study of this kind. It also contributes to what is quantitatively known about how Philippine students cope with online learning. In the context of COVID-19, quantitative research on this subject is still scarce.

Acknowledgements. We would like to thank Hiroyuki Kuromiya and Hiroaki Ogata of Kyoto University, the Ateneo Research Institute for Science and Engineering (ARISE), and the Ateneo Laboratory for the Learning Sciences for their support in this research.

References

1. Brodersen, K.: CausalImpact: a new open-source package for estimating causal effects in time series. https://opensource.googleblog.com/2014/09/causalimpact-new-open-source-pac kage.html. Accessed 25 Jan 2021
2. Brodersen, K.H., Gallusser, F., Koehler, J., Remy, N., Scott, S.L.: Inferring causal impact using Bayesian structural time-series models. Ann. Appl. Stat. **9**(1), 247–274 (2015)

3. Friedrich, A., Flunger, B., Nagengast, B., Jonkmann, K., Trautwein, U.: Pygmalion effects in the classroom: teacher expectancy effects on students' math achievement. Contemp. Educ. Psychol. **41**, 1–12 (2015)
4. Google. CausalImpact. http://google.github.io/CausalImpact/CausalImpact.html. Accessed 29 Jan 2021
5. IEA, TIMSS 2019 International Results in Mathematics and Science. https://timssandpirls.bc.edu/timss2019/international-results/wp-content/themes/timssandpirls/download-center/TIMSS-2019-International-Results-in-Mathematics-and-Science.pdf. Accessed 09 Jan 2021
6. Lalu, G.P., Student group wants academic freeze until floods clear, internet fixed. https://newsinfo.inquirer.net/1361470/student-group-wants-academic-freeze-until-floods-clear-internet-is-fixed. Accessed 25 Jan 2021
7. Larsen, K.: MarketMatching Package Vignette. https://cran.r-project.org/web/packages/MarketMatching/vignettes/MarketMatching-Vignette.html. Accessed 29 Jan 2021
8. Philippines Department of Education, PISA 2018 National Report of the Philippines. https://www.deped.gov.ph/wp-content/uploads/2019/12/PISA-2018-Philippine-National-Report.pdf. Accessed 03 Dec 2020
9. Szumski, G., Karwowski, M.: Exploring the Pygmalion effect: The role of teacher expectations, academic self-concept, and class context in students' math achievement. Contemp. Educ. Psychol. **59**, 101787 (2019)
10. UNESCO. Philippines: Education and Literacy. http://uis.unesco.org/en/country/ph. Accessed 25 Jan 2021
11. UNICEF & SEAMEO, SEA-PLM 2019 Main Regional Report, Children's Learning in 6 Southeast Asian Countries. https://www.seaplm.org/index.php?option=com_content&view=article&id=44&Itemid=332. Accessed 03 Dec 2020

I-Mouse: A Framework for Player Assistance in Adaptive Serious Games

Riya Lalwani[1], Ashish Chouhan[1(✉)], Varun John[1], Prashant Sonar[1],
Aakash Mahajan[1], Naresh Pendyala[1], Alexander Streicher[2],
and Ajinkya Prabhune[1]

[1] SRH Hochschule Heidelberg, Heidelberg, Germany
{Riya.Lalwani,Varun.John,Sonar.Prashant,Aakash.Mahajan,
NareshKumar.Pendyala}@stud.hochschule-heidelberg.de,
{Ashish.Chouhan,Ajinkya.Prabhune}@srh.de
[2] Fraunhofer IOSB, Karlsruhe, Germany
Alexander.Streicher@iosb.fraunhofer.de

Abstract. A serious game is an educational digital game created to entertain and achieve characterizing goal to promote learning. However, a serious game's major challenge is capturing and sustaining player attention and motivation, thus restricting learning abilities. Adaptive frameworks in serious games (Adaptive serious games) tackle the challenge by automatically assisting players in balancing boredom and frustration. The current state-of-the-art in Adaptive serious games targets modeling a player's cognitive states by considering eye-tracking characteristics like gaze, fixation, pupil diameter, or mouse tracking characteristics such as mouse positions. However, a combination of eye and mouse tracking characteristics has seldom been used. Hence, we present I-Mouse, a framework for predicting the need for player assistance in educational serious games through a combination of eye and mouse-tracking data. I-Mouse framework comprises four steps: (a) Feature generation for identifying cognitive states, (b) Partition clustering for player state modeling, (c) Data balancing of the clustered data, and (d) Classification to predict the need for assistance. We evaluate the framework using a real game data set to predict the need for assistance, and Random Forest is the best performing model with an accuracy of 99% amongst the trained classification models.

Keywords: Serious games · Adaptivity · Eye and mouse tracking

1 Introduction

Serious Game (SG) is an entertaining tool for education. The main goal of SG is to promote learning besides entertainment by cultivating knowledge in players and allowing them to practice their skills through overcoming numerous obstacles in the game [13]. It is essential to maintain an efficient balance between motivation and boredom in SG. Adaptivity in SG is used to capture and process

© Springer Nature Switzerland AG 2021
I. Roll et al. (Eds.): AIED 2021, LNAI 12749, pp. 234–238, 2021.
https://doi.org/10.1007/978-3-030-78270-2_42

data to aid a player. These games are termed Adaptive Serious Games (ASG). According to Streicher and Smeddinck [11], personalization and adaptivity can promote motivated usage, increased user acceptance, and user identification in serious games. However, not assisting the player at the appropriate moment may lead to repetitive attempts by the player resulting in frustration, loss of interest, and hampering the player's progress. The current state-of-the-art utilizes various physical and behavioural biometrics like the player's eye-tracking data such as fixations, gaze, pupil size, or mouse-tracking data. Eye-tracking data can be used for cognitive load analysis [3,8,12], however, the effectiveness of the use of eye-tracking in computer games as a direct control input is questioned [1]. Khedher et al. [5] conclude that eye-tracking data is not the only indicator for cognitive load analysis. In the domain of cognitive state modeling using behavioural biometrics, Grimes et al. [4] show that mouse movement is also useful for cognitive load analysis.

In comparison with these adaptive studies and solutions, the I-Mouse framework incorporates eye-tracking and mouse-tracking to predict the need for assistance in SG. I-Mouse framework is equipped with (a) Data Preparation service to process the data (b) Feature creation service to create features from the eye and mouse tracking data (c) Partition clustering service to cluster different cognitive states (d) Data balancing service to balance the game data (e) Classification model creation service to predict instances when the player needs assistance.

2   I-Mouse Framework

As shown in Fig. 1, the I-Mouse framework comprises different services executed sequentially to perform a specific task. The orchestration of each service is coordinated by creating workflows using Apache Airflow[1].

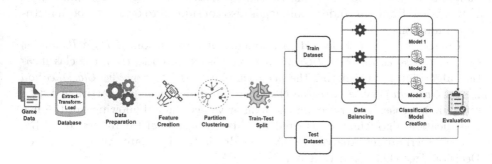

Fig. 1. I-Mouse framework

I-Mouse framework uses the SaFIRa (Seek and Find for Image Reconnaissance adaptive) game data set collected by Streicher et al. [10]. The SaFIRa

---

[1] https://airflow.apache.org/.

game data set comprises eye-tracking and mouse-tracking logs of twenty-four players with information about each player asking for assistance while playing the game. Assistance provided to the player is in the form of hints comprising of information regarding the distance remaining to reach the target and the direction in which the target lies. SaFIRa data set is divided into two classes, i.e., "assistance required" and "assistance not required", with a high bias towards the "assistance not required" class. *Data Preparation* service's main functionality is to create a document-based database using SaFIRa dataset by executing Extract, Transform, and Load (ETL) jobs.

Grimes et al. [4] show that mouse position on screen does not have any relation to cognitive load. However, the frequency of mouse direction change is a good indicator of cognitive load. As the SaFIRa dataset records only mouse position, the *Feature Creation* service creates an additional mouse feature, i.e. Mouse Click Direction Change. The Mouse Click Direction Change is the total number of significant direction changes in 20 consecutive moves, where the direction changes are significant when direction changes by an angle greater than 90 degrees. This newly created feature, along with pupil size, fixation duration [3,8,12] helps to predict the player cognitive load.

After obtaining the new feature from the *Feature Creation* service, the *Partition Clustering* service is executed to form clusters of player cognitive states [7]. The *Partition Clustering* service uses the K-means clustering algorithm due to its scalability and time-efficiency compared to the K-Medoids clustering algorithm. The optimum k value representing the number of clusters obtained from the data set is determined using the elbow curve technique.

Due to the imbalanced distribution of classes in the SaFIRa data set, each cluster obtained from the *Partition Clustering* service is individually balanced for model training with the help of *Data Balancing* service. *Synthetic Minority Over-sampling Technique* (SMOTE) [2] for Over-sampling and *Random Under Sampling* [6] for Under-sampling is used for data balancing. A combination of SMOTE and Random Under Sampling is also considered to overcome both methods' limitations.

Considering the balanced data obtained after execution of *Data Balancing* service, *Classification Model Creation* service is executed to train the classification algorithm for predicting the need for player assistance. The *Classification Model Creation* service creates a classification model for every partitioned data set as it increases the framework's cumulative accuracy. Following are the five classification algorithms considered by this service: Logistic Regression (LR), Linear Discriminant Analysis (LDA), Quadratic Discriminant Analysis (QDA), Decision Tree (DT), and Random Forest (RF).

3   Evaluation

We evaluate the I-Mouse framework by combining high, low, and normal cognitive load data sets. Data from the clustering service is split into train and

Table 1. Evaluation of I-Mouse framework

Sr. No	Data partition	Data balancing technique	Model	Accuracy (%)	Precision (%)	Recall (%)	F1 score
1			LR	89	89	99	0.93
2			LDA	97	88	73	0.80
3		Under sampling	QDA	80	80	76	0.75
4			DT	97	93	92	0.90
5			RF	98	99	86	0.92
6			LR	98	89	99	0.93
7	No		LDA	93	88	73	0.80
8		Over sampling	QDA	78	83	76	0.76
9			DT	97	93	92	0.90
10			RF	98	99	86	0.92
11			LR	98	88	99	0.94
12			LDA	96	88	73	0.83
13			QDA	82	89	74	0.85
14			DT	96	98	83	0.88
15		Combination	RF	98	99	86	0.92
16			LR	93	61	93	0.65
17			LDA	87	66	93	0.68
18	Yes		QDA	93	61	90	0.75
19			DT	99	93	92	0.90
20			**RF**	**99**	**99**	**99**	**0.99**

test set using cross-validation technique to avoid over-fitting classification models [9]. Table 1 shows evaluation metric scores for different combinations of components present in the framework. The *Data Partition* column denotes whether the trained data is partitioned into different data sets based on the *Partitioning Clustering* service results, and *Data Balancing Technique* column denotes the data balancing technique used for the respective combination. The *Model* column denotes the classification algorithm used for the evaluation of the framework, and *Accuracy, Precision, Recall,* and *F1 Score* represents the evaluation metric scores for respective combination. The test data set contains 100,000 records with 91,646 records of the "assistance not required" class and 8,354 records of "assistance required" class. Test data are not passed through the data balancing service as test data emulates the real-world game data that is always imbalanced. The reason for high accuracy for most combinations is the data balancing techniques integrated during the model training process. Out of different combinations, the Random Forest classification model combined with data partition and combination of data balancing techniques is the best performing with an accuracy of 99%. This high accuracy value and the majority class's influence will be the subject of future in-detail studies.

4 Conclusion and Future Work

In this paper, we presented the I-Mouse framework that uses a combination of eye and mouse tracking data to predict the need for player assistance. I-Mouse framework is evaluated with different combinations of data sampling, data balancing, and classification algorithms. Out of different combinations, the Random Forest classification model combined with data partition and combination of data balancing techniques is the best performing with an accuracy of 99%. In our future work, we plan to replace cognitive state modeling with player state modeling leveraging a player's behavioural states and actions by employing the Hidden Markov Model (HMM) and Reinforcement Learning.

References

1. Antunes, J., Santana, P.: A study on the use of eye tracking to adapt gameplay and procedural content generation in first-person shooter games. Multimodal Technol. Interact. **2**(2), 23 (2018)
2. Chawla, N.V., Bowyer, K.W., Hall, L.O., Kegelmeyer, W.P.: Smote: synthetic minority over-sampling technique. J. Artif. Intell. Res. **16**, 321–357 (2002)
3. Eckstein, M.K., Guerra-Carrillo, B., Singley, A.T.M., Bunge, S.A.: Beyond eye gaze: What else can eye tracking reveal about cognition and cognitive development? Dev. Cogn. Neurosci. **25**, 69–91 (2017)
4. Grimes, M., Valacich, J.: Mind over mouse: the effect of cognitive load on mouse movement behavior. In: Proceedings of Thirty Sixth International Conference on Information Systems (2015)
5. Khedher, A.B., Jraidi, I., Frasson, C.: Exploring students' eye movements to assess learning performance in a serious game. In: Proceedings of EdMedia+ Innovate Learning, pp. 394–401. Association for the Advancement of Computing in Education (AACE) (2018)
6. Prusa, J., Khoshgoftaar, T.M., Dittman, D.J., Napolitano, A.: Using random undersampling to alleviate class imbalance on tweet sentiment data. In: Proceedings of IEEE International Conference on Information Reuse and Integration. pp. 197–202. IEEE (2015)
7. Ramirez-Cano, D., Colton, S., Baumgarten, R.: Player classification using a meta-clustering approach. In: Proceedings of 3rd Annual International Conference Computer Games, Multimedia & Allied Technology, pp. 297–304 (2010)
8. Rodden, K., Fu, X., Aula, A., Spiro, I.: Eye-mouse coordination patterns on web search results pages. In: Proceedings of CHI 2008 Extended Abstracts on Human Factors in Computing Systems, pp. 2997–3002. ACM (2008)
9. Stone, M.: Cross-validation: A review. Stat.: J. Theor. Appl. Stat. **9**(1), 127–139 (1978)
10. Streicher, A., Leidig, S., Roller, W.: Eye-tracking for user attention evaluation in adaptive serious games. In: Pammer-Schindler, V., Pérez-Sanagustín, M., Drachsler, H., Elferink, R., Scheffel, M. (eds.) EC-TEL 2018. LNCS, vol. 11082, pp. 583–586. Springer, Cham (2018). https://doi.org/10.1007/978-3-319-98572-5_50
11. Streicher, A., Smeddinck, J.D.: Personalized and adaptive serious games. In: Dörner, R., Göbel, S., Kickmeier-Rust, M., Masuch, M., Zweig, K. (eds.) Entertainment Computing and Serious Games. LNCS, vol. 9970, pp. 332–377. Springer, Cham (2016). https://doi.org/10.1007/978-3-319-46152-6_14
12. Van der Wel, P., Van Steenbergen, H.: Pupil dilation as an index of effort in cognitive control tasks: A review. Psychon. Bull. Rev. **25**(6), 2005–2015 (2018)
13. Zhonggen, Y.: A meta-analysis of use of serious games in education over a decade. Int. J. Comput. Game. Technol. **2019** (2019)

Parent-EMBRACE: An Adaptive Dialogic Reading Intervention

Arun Balajiee Lekshmi Narayanan[1]([✉]) [iD], Ju Eun Lim[1] [iD], Tri Nguyen[3],
Ligia E. Gomez[2] [iD], M. Adelaida Restrepo[3] [iD], Chris Blais[3], Arthur M. Glenberg[3] [iD],
and Erin Walker[1] [iD]

[1] University of Pittsburgh, Pittsburgh, PA 15260, USA
{arl122,jul118,eawalker}@pitt.edu
[2] Ball State University, 2000 W. University Avenue, Muncie, IN 47306, USA
legomezfranco@bsu.edu
[3] Arizona State University, Tempe, AZ 85281, USA
{Tri.D.Nguyen,Laida.Restrepo,chris.blais,
Arthur.Glenberg}@asu.edu

Abstract. Dialogic reading is a practice where adults and children engage in a dialogue as they read together to improve children's language strategies and comprehension. These dialogues are often initiated by parent questioning behaviors, but parents do not always engage in this behavior spontaneously. In this paper, we describe an adaptive intervention for dialogic reading, Parent-EMBRACE, built into an iPad application that uses an embodied cognition approach and is designed specifically for Latino dual language learners in the US. The intervention: 1) Models parent question asking, 2) Provides parents with on-demand hints on questions that can be asked at particular moments during the story, 3) Prompts parents to ask questions at appropriate times, 4) Includes a dashboard that presents parents with data on their question-asking behaviors, 5) Provides all support in both English and Spanish. We discuss the implications of this intervention as an intelligent tutoring system for parent-child interactions, plans to extend and evaluate the system.

Keywords: Dialogic reading · Embodied cognition · Parent-child interactions · Intelligent tutoring systems

1 Introduction and Related Work

Dialogic reading (DR) is a practice where adults and children engage in dialogue as they read together. It has been demonstrated to improve children's language skills, such as vocabulary and syntax development, and inference-making skills [1, 2]. Parents can be trained dialogic reading strategies (e.g., asking questions or recasting children's verbal contributions) to facilitate these outcomes [3]. For example, Schwanenflugel and colleagues [4] used a DR model with three types of questions, denoted using the acronym CAR. C stands for competence questions such as, "What are the ingredients in the bowl?". A stands for abstract questions such as, "Why do they need the bowl of chilis?".

I. Roll et al. (Eds.): AIED 2021, LNAI 12749, pp. 239–244, 2021.
https://doi.org/10.1007/978-3-030-78270-2_43

R stands for questions that relate to the child such as, "Have you seen a bowl of chilis in our kitchen?". In a recent parent-child question-asking study [5], Parents were trained in CAR questions asked an increased number of questions while reading with their child, even five weeks later.

However, there are some limitations to this approach. First, training is expensive in terms of cost and person-hours. Therefore, it does not scale well beyond a handful of families. Second, parents must implement the training after it is over. But, parent implementation of the target practices not only varies with existing family literacy practices and family income, but also with cultural background and possibly the language of training (e.g., [6–8]). Often programs assume parental literacy skills that parents from low-income homes may not have [7, 9–11]. In contrast, Mesa and Restrepo [12] found that modeling and coaching in the native language changed parents' practices and attitudes towards reading with their children, while also affecting the children's language use. There may be promise in embedding this type of training into digital environments, and, in fact Troseth and colleagues [13] modelled good question-asking within an e-book by embedding a character that presented example DR prompts on each page of a story. This approach led to strong gains in both parent and child book-related talk, although there was no improvement in story comprehension compared to a condition without DR.

This paper describes Parent-EMBRACE (Enhanced Moved By Reading to Accelerate Comprehension in English), an adaptive system for scaffolding parents in dialogic reading practices as they read an interactive storybook with their children. Parent-EMBRACE extends an iPad application that uses an embodied cognition approach to reading and is designed specifically for parents from Latino communities in the US with children between the ages of 5–10 [14]. Beyond this, our intervention builds on the literature described above by adapting question prompts to parent behaviors, including a dashboard that presents parents with data on their question-asking behaviors, and providing all support in both English and Spanish. There are a limited number of intelligent systems that support parent-child learning, and thus Parent-EMBRACE provides one blueprint for such systems.

2 Parent-EMBRACE System

EMBRACE. In previous work, we developed EMBRACE, an app that leverages theories of embodied cognition, dual language learning, and intelligent tutoring systems to promote reading comprehension in Latino dual language learners [14]. The app follows principles of embodied cognition by engaging the reader in physical and cognitive simulation. The reader uses an iPad that presents texts and pictures much like in a child's picture book. However, after reading key sentences, the app prompts the reader to move the pictures to correspond to the sentence, an approach that yields improved reading comprehension outcomes over typical reading practice [15]. This system was developed specifically for Latino populations, and consistent with research on bilingual education, provides support in Spanish (e.g., vocabulary help is presented in English and Spanish; [16]). In addition, the system functions as an intelligent tutor by using how children move the pictures within the application to make inferences about the child's vocabulary and syntactic knowledge. This information is then used to provide the child with tailored feedback and vocabulary practice. The system is implemented in Objective-C, with storybooks and related metadata for the ITS encoded in xml.

Fig. 1. Main interface to Parent-EMBRACE. Parents indicate the type of question they ask and to receive example questions in the large pane at the bottom of the screen.

Assessment of Question-Asking. Parent-EMBRACE currently defines a few basic rules for how questions should be asked: Parents should not go more than 3 pages of the story without asking a question, and parents should ask roughly equal numbers of C, A, and R questions. The app includes a parent-facing interface, shown in Fig. 1, asks parents to indicate when they are asking their child questions and whether they are asking a Concrete (competence), Abstract, or Relational question by clicking on the "C", the "A", or the "R". When they click on the relevant button, the question counter below the button increments, and we maintain a simple parent model by storing the total number of C, A, and R questions asked per book.

On-Demand Hints and Adaptive Prompts. There are two mechanisms built into the application to adaptively encourage parents to engage in DR, following principles of ITS design [17]. First, they can use the parent interface to request a hint from the application. Hints consist of three example questions that can be asked. Examples are presented one at a time, and parents can view the next example by swiping right or left. Second, if parents have not asked a question after they have read 3 story pages, an example question is revealed in the interface without any hint request being necessary. Examples are drawn from a stored question bank and are always relevant to the current page. The questions are ordered based on question type, where the types that parents have asked the least are put first. Assistance is faded as parents and children reread the books, where on the third reading of the same text, the system only indicates a good time to ask a question of a particular type but does not provide examples, and on the fourth reading there is no support given. Across all readings, this approach depends on parents' accuracy in indicating and labeling the questions they are asking.

Fig. 2. Both sides of a book card in the parent dashboard for a book. One side (left) indicates information about who has read and in what language. The other side (right) indicates information related to number of questions asked and time spent reading.

Parent Dashboard. A dashboard for parents in Parent-EMBRACE with two-sided book cards (see Fig. 2) encourages them to ask questions and reflect on their progress. The dashboard displays the number of times a book has been read, the language of the reading (Spanish or English), the reader (Parent or Child), the total number of questions asked by the Parent along with a chart that displays that number by question types and the time spent reading the book.

Language. All aspects of our application are implemented in both Spanish and English to facilitate dual language learning. For example, parents can toggle the question interface between Spanish (ES) and English (EN). This allows a parent who may not be comfortable enough in one language to switch to the other available language. Low reading proficiency in the second language is not a barrier to use this application.

3   Discussion and Future Work

We present an implementation of Parent-EMBRACE, which has several complementary features that support parent-child dialogic reading, including a parent interface that adaptively displays example questions and a dashboard that displays parent progress. Our next step is to test Parent-EMBRACE in a controlled study to determine its effects on interaction during DR and children's reading comprehension. In addition, there are many ways to extend the current implementation. Whereas we cannot currently ensure that the parents accurately self-label the questions they ask, as future work, using Automated Speech Recognition (ASR), we can build a model for dynamic classification of parent-child dialogue into questions and question types. We can extend the model of parent reading to include questions and question types that are particularly appropriate for certain parts of the book or phases of reading, and even use question generation methods to automatically generate example questions for each page. Currently, the original ITS and parent-ITS operate separately from each other, but in the future, we could use information provided by the original ITS to inform the questions suggested by the parent ITS. Overall, this project represents a framework for supporting parent-child interactions using ITS-based approaches.

Acknowledgments. This work was supported by the National Science Foundation Award Nos. CISE-IIS-1917625 and CISE-IIS-1917636. We would like to thank Sarah M. Fialko, Purav Patel and Nora Carrillo for the project resource management.

References

1. Lonigan, C.J., Anthony, J.L., Bloomfield, B.G., Dyer, S.M., Samwel, C.S.: Effects of two shared-reading interventions on emergent literacy skills of at-risk preschoolers. J. Early Interv. **22**(4), 306–322 (1999)
2. WWC: U.S. Department of Education, Institute of Education Sciences, National Center for Education Evaluation and Regional Assistance, What Works Clearinghouse (2007). https://ies.ed.gov/ncee/wwc/Intervention/271
3. Whitehurst, G.J., Falco, F.L., Lonigan, C.J., Fischel, J.E., DeBaryshe, B.D., Valdez-Menchaca, M.C., Caulfield, M.: Accelerating language development through picture book reading. Dev. Psychol. **24**(4), 552 (1988)
4. Schwanenflugel, P.J., Hamilton, C.E., Neuharth-Pritchett, S., Restrepo, M.A., Bradley, B.A., Ruston, H.P.: PAVEd for Success: An evaluation of a program to improve the preliteracy skills of 4-year old children. J. Literacy Res. (2010)
5. Gómez, L.E., Restrepo, M.A., Glenberg, A.M, Walker, E.: Enhancing Latino parent question-asking during shared reading. J. Latinos Educ. (Under Review)
6. Shanahan, T., Mulhern, M., Rodriguez-Brown, F.: Project FLAME: lessons learned from a family literacy program for linguistic minority families. Read. Teach. **48**(7), 586–593 (1995)
7. Hockenberger, E.H., Goldstein, H., Sirianni Haas, L.: Effects of commenting during joint book reading by mothers with low SES. Top. Early Child. Spec. Educ. **19**(1), 15–27 (1999)
8. Tsybina, I., Eriks-Brophy, A.: Bilingual dialogic book-reading intervention for preschoolers with slow expressive vocabulary development. J. Commun. Disord. **43**(6), 538–556 (2010)
9. Hoff, E.: Interpreting the early language trajectories of children from low-SES and language minority homes: implications for closing achievement gaps. Dev. Psychol. **49**(1), 4 (2013)
10. Raikes, H., Green, B.L., Atwater, J., Kisker, E., Constantine, J., Chazan-Cohen, R.: Involvement in early Head Start home visiting services: demographic predictors and relations to child and parent outcomes. Early Child. Res. Q. **21**(1), 2–24 (2006)
11. Van Steensel, R., McElvany, N., Kurvers, J., Herppich, S.: How effective are family literacy programs? Results of a meta-analysis. Rev. Educ. Res. **81**(1), 69–96 (2011)
12. Mesa, C., Restrepo, M.A.: Effects of a family literacy program for Latino parents: evidence from a single-subject design. Lang. Speech Hear. Serv. Sch. **50**(3), 356–372 (2019)
13. Troseth, G.L., Strouse, G.A., Flores, I., Stuckelman, Z.D., Johnson, C.R.: An enhanced eBook facilitates parent–child talk during shared reading by families of low socioeconomic status. Early Child. Res. Q. **50**, 45–58 (2020)
14. Walker, E., Wong, A., Fialko, S., Restrepo, M.A., Glenberg, A.M.: EMBRACE: applying cognitive tutor principles to reading comprehension. In: André, E., Baker, R., Hu, X., Rodrigo, M.M.T., du Boulay, B. (eds.) AIED 2017. LNCS (LNAI), vol. 10331, pp. 578–581. Springer, Cham (2017). https://doi.org/10.1007/978-3-319-61425-0_68

15. Walker, E., Adams, A., Restrepo, M.A., Fialko, S., Glenberg, A.M.: When (and how) interacting with technology-enhanced storybooks helps dual language learners. Transl. Issues Psychol. Sci. **3**(1), 66 (2017)
16. Restrepo, M.A., Morgan, G., Thompson, M.: The efficacy of a vocabulary intervention for dual language learners with language impairment. J. Speech Lang. Hear. Res. **56**(2), 748–765 (2013). https://doi.org/10.1044/1092-4388(2012/11-0173
17. Koedinger, K.R. Corbett, A.: Cognitive tutors: Technology bringing learning sciences to the classroom (2006)

Using Fair AI with Debiased Network Embeddings to Support Help Seeking in an Online Math Learning Platform

Chenglu Li, Wanli Xing[✉], and Walter Leite

University of Florida, Gainesville, FL, USA
li.chenglu@ufl.edu, {wanli.xing,walter.leite}@coe.ufl.edu

Abstract. There has been a long-standing issue of sparse discussion forums participation in online learning, which can impede students' help seeking practices. Researchers have examined AI techniques such as link prediction with network analysis to connect help seekers with help providers. However, little is known whether these AI systems will treat students fairly. In this study, we aim to start a foundation work to build a recommender system that can (1) fairly suggest peers who are likely to answer a question and (2) predict the response quality of students.

Keywords: Fair AI · Link prediction · Recommender system

1 Introduction and Related Work

The discussion forum in online learning has been demonstrated to be an important learning tool given its collaborative nature that enhances learning through knowledge exchange [1,2]. However, there has been a long-standing issue of discussion forums that peer interactions are sparse [6]. The inactive use of discussion forums in online settings can impede students' help seeking practices [13]. Help seeking is an important skill in self-regulated learning and can positively affect students' learning outcomes [13,17,20]. To support students' help seeking in online discussion forums at a large scale, researchers have examined AI techniques such as link prediction with network analysis to connect help seekers with help providers [10,12,16]. Other than using network analysis such as structural similarity for link prediction, network embedding has recently shown to be a strong candidate [24]. Network embedding represents nodes in a graph with latent vectors such that neighboring nodes would have high similarity scores [19]. Studies have shown that network embedding can outperform prior link prediction algorithms [9,19,24].

While promising results on predictive accuracy have been presented in prior studies on automatically supporting help seeking in discussion forums, little is known whether these AI systems will treat students fairly. Algorithmically, studies have shown that AI can reflect humans' hidden values due to the existing

© Springer Nature Switzerland AG 2021
I. Roll et al. (Eds.): AIED 2021, LNAI 12749, pp. 245–250, 2021.
https://doi.org/10.1007/978-3-030-78270-2_44

bias in training datasets. For example, Caliskan et al. [5] found word embedding algorithms can perpetuated cultural stereotypes (e.g., females are highly correlated with family-oriented careers). Empirically, biases in AI have been found in domains such as education, hiring, and finance, where participants with specific demographics can be favored by predictive models [3, 8, 22]. In the case of link prediction, students might form communities of specific demographics. For example, white students dominantly interact with other white students because they come from the same school where minority students are scarce. Trained with such a dataset, models can reinforce the status quo and not give students opportunities to establish diversified connections that can be equally helpful. Therefore, to make AI in education sustainable, researchers need to purposefully address fairness issues [21, 23]. In this study, we aim to start a foundation work to build a recommender system that can (1) fairly suggest peers who are likely to answer a question and (2) predict the response quality of students.

2   Methods

2.1   Research Context and Dataset

This study uses students' discussion forum, demographics, and log data on Algebra I from Algebra Nation (AN), an online math learning platform originated in Florida. The dataset consists of 17,794 post-reply pairs by 3,726 students with over 6 million logs in the academic year of 2018–2019. Post-reply pairs include contents of post and reply, poster IDs, and replier IDs. The log data captured students' interactions with AN (e.g., lecturing videos, reviewing videos, and discussion board).

2.2   Model Procedure

Link Prediction with Network Embeddings. Link prediction models are trained with network embeddings to predict if two students will be connected. For the network embeddings, we have examined Node2Vec [9] and DeBayes [4]. Node2Vec is inspired by the widely-applied algorithm Word2Vec [18]. In Node2Vec, nodes are analogous to words in Word2Vec, and the random walks algorithm is used to construct sequences of nodes. Latent vectors (embeddings) of nodes are then extracted from a neural network's hidden layer trained with the sequences. Node2Vec is selected because previous studies have achieved desired link prediction results with it. However, Node2Vec is fairness-unaware. To ensure the fairness of link prediction, we have also examined DeBayes modified based on Conditional Network Embeddings (CNE) [14] to learn fair representations. Conceptually, CNE solves for

$$P(G|X) = \frac{P(X|G)P(G)}{P(X)} \tag{1}$$

by finding an embedding X using Maximum Likelihood estimation, where G is the given network. Thus, the embedding will only need to capture information that is NOT represented by the prior $P(G)$. DeBayes utilizes this property to get debiased embeddings by introducing a biased prior, where sensitive information related to protected groups is retained in the biased prior so that embeddings are not aware of such information.

Representation Bias and Equalized Odds. To evaluate fairness, we have examined representation bias (RB) [25] of network embeddings and equalized odds (EO) [11] in terms of **gender** and **races**. RB is the weighted average AUC scores of using embeddings to predict sensitive attributes (e.g., gender). Conceptually, embeddings are fair when RB is close to 0.5 since an AUC of 0.5 suggests a random classifier and we cannot infer students' sensitive information from embeddings. EO is defined as

$$P(\hat{Y} = 1 | A = 0, Y = 1) = P(\hat{Y} = 1 | A = 1, Y = 1) \tag{2}$$

$$P(\hat{Y} = 1 | A = 0, Y = 0) = P(\hat{Y} = 1 | A = 1, Y = 0) \tag{3}$$

, where $\hat{Y}$ is the predicted outcome of the model, Y is the binary outcome from the dataset (e.g., connected or not), and A is the comparison group (e.g., female vs. male). EO is satisfied when Eqs. 2 and 3 are met.

Response Quality Prediction. We have conducted a multiple linear regression analysis to understand what contributes to response quality. There were 27 predictors, which were repliers' standardized frequencies of interactions on Algebra Nation. Variance inflation factors (VIF) were calculated to avoid multicollinearity. Response quality of a reply is calculated based on its linguistic features (number of words and number of named entities), reputations (number of up-votes), readability (Flesch reading ease [15]), and coherence (cosine similarity between post and reply using BERT embeddings [7]). Log-transformation was applied to linguistic features and readability as we think the contribution of them decays as their values increase.

3   Results

Link Prediction. We evaluated models' predictive accuracy with AUC. The results show that Node2Vec has an AUC of 0.88 and DeBayes achieves that of 0.94. For the fairness evaluation (see Fig. 1), the representation bias of Node2Vec's embedding is 0.54 for gender and 0.53 for the race and that of DeBayes is 0.49 for gender and 0.5 for the race. In terms of equalized odds (EO), lower is fairer. Node2Vec has an EO of 0.037 for gender and 0.038 for race, while DeBayes has an EO of 0.002 for gender and 0.005 for race. The results suggested that DeBayes greatly outperformed Node2Vec in predictive accuracy and fairness.

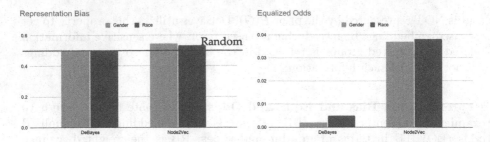

Fig. 1. Fairness evaluation of network embeddings and link prediction.

Table 1. Regression analysis results of the significant predictors

	Coef	P-value	Definition
Load discussions	.0528	<.000	Load the discussion forum page
Answer assessment	.0300	.002	Answer an assessment item
Finish assessment	−.1357	.016	Finish a whole assessment
Review incorrect assessment	−.0223	.006	Review the solution of an incorrect item
Create post	−.3183	<.000	Create a post in the discussion forum
Search discussions	.1714	.044	Search within the discussion forum
View document	.3191	.001	View learning resource files

Response Quality Prediction. The regression model demonstrates that 7 behaviors in discussion forums, video watching, and assessment taking are significant predictors of response quality. Table 1 illustrates the regression results and the significant predictors' definitions.

4 Conclusion

This paper has shown the possibility of conducting link prediction fairly while producing desirable accuracy. Although the fairness evaluation of Node2Vec does not suggest that the model is highly biased in our context, unlike DeBayes, Node2Vec is fairness-unaware and potential equity issues can arise without careful handling. Meanwhile, the regression analysis sheds light on the factors contributing to response quality. From a learning perspective, these significant predictors' effects on response quality are reasonable, indicating the validity of the computed response quality. In the future, we intend to triangulate the reliability and validity of the computed response quality with qualitative approaches.

Funding. The research reported here was supported by the Institute of Education Sciences, U.S. Department of Education, through Grant R305C160004 to the University of Florida and University of Florida AI Catalyst Grant. The opinions expressed are those of the authors and do not represent views of the Institute or the U.S. Department of Education.

References

1. Almatrafi, O., Johri, A., Rangwala, H.: Needle in a haystack: identifying learner posts that require urgent response in mooc discussion forums. Comput. Educ. **118**, 1–9 (2018)
2. Biasutti, M.: A comparative analysis of forums and wikis as tools for online collaborative learning. Comput. Educ. **111**, 158–171 (2017)
3. Binns, R.: Fairness in machine learning: lessons from political philosophy. In: Conference on Fairness, Accountability and Transparency, pp. 149–159 (2018)
4. Buyl, M., De Bie, T.: Debayes: a bayesian method for debiasing network embeddings. In: International Conference on Machine Learning, pp. 1220–1229. PMLR (2020)
5. Caliskan, A., Bryson, J.J., Narayanan, A.: Semantics derived automatically from language corpora contain human-like biases. Science **356**(6334), 183–186 (2017)
6. Chiu, T.K., Hew, T.K.: Factors influencing peer learning and performance in mooc asynchronous online discussion forum. Australas. J. Educ. Technol. **34**(4) (2018)
7. Devlin, J., Chang, M.W., Lee, K., Toutanova, K.: Bert: pre-training of deep bidirectional transformers for language understanding. In: Proceedings of the 2019 Conference of the North American Chapter of the Association for Computational Linguistics: Human Language Technologies, Volume 1 (Long and Short Papers), pp. 4171–4186 (2019)
8. Giang, V.: The potential hidden bias in automated hiring systems. The Future of Work. Fast Company (2018)
9. Grover, A., Leskovec, J.: node2vec: Scalable feature learning for networks. In: Proceedings of the 22nd ACM SIGKDD International Conference on Knowledge Discovery and Data Mining, pp. 855–864 (2016)
10. Hansen, P., et al.: Predicting the timing and quality of responses in online discussion forums. In: 2019 IEEE 39th International Conference on Distributed Computing Systems (ICDCS), pp. 1931–1940. IEEE (2019)
11. Hardt, M., Price, E., Srebro, N.: Equality of opportunity in supervised learning. In: Proceedings of the 30th International Conference on Neural Information Processing Systems, pp. 3323–3331 (2016)
12. Howley, I., Tomar, G., Yang, D., Ferschke, O., Rosé, C.P.: Alleviating the negative effect of up and downvoting on help seeking in MOOC discussion forums. In: Conati, C., Heffernan, N., Mitrovic, A., Verdejo, M.F. (eds.) AIED 2015. LNCS (LNAI), vol. 9112, pp. 629–632. Springer, Cham (2015). https://doi.org/10.1007/978-3-319-19773-9_78
13. Howley, I., Tomar, G.S., Ferschke, O., Rosé, C.P.: Reputation systems impact on help seeking in mooc discussion forums. IEEE Trans. Learn. Technol. (2017)
14. Kang, B., Lijffijt, J., De Bie, T.: Conditional network embeddings. In: 7th International Conference on Learning Representations, ICLR 2019, p. 16 (2019). https://openreview.net/forum?id=ryepUj0qtX
15. Kincaid, J.P., Fishburne Jr., R.P., Rogers, R.L., Chissom, B.S.: Derivation of New Readability Formulas (Automated Readability Index, Fog Count and Flesch Reading Ease Formula) for Navy Enlisted Personnel. Tech. rep, Naval Technical Training Command Millington TN Research Branch (1975)
16. Lan, A.S., Spencer, J.C., Chen, Z., Brinton, C.G., Chiang, M.: Personalized thread recommendation for MOOC discussion forums. In: Berlingerio, M., Bonchi, F., Gärtner, T., Hurley, N., Ifrim, G. (eds.) ECML PKDD 2018. LNCS (LNAI), vol. 11052, pp. 725–740. Springer, Cham (2019). https://doi.org/10.1007/978-3-030-10928-8_43

17. Magnusson, J.L., Perry, R.P.: Academic help-seeking in the university setting: the effects of motivational set, attributional style, and help source characteristics. Res. High. Educ. **33**(2), 227–245 (1992)
18. Mikolov, T., Chen, K., Corrado, G., Dean, J.: Efficient estimation of word representations in vector space. arXiv preprint arXiv:1301.3781 (2013)
19. Nelson, W., Zitnik, M., Wang, B., Leskovec, J., Goldenberg, A., Sharan, R.: To embed or not: network embedding as a paradigm in computational biology. Front. Genet. **10**, 381 (2019)
20. Newman, R.S.: How self-regulated learners cope with academic difficulty: the role of adaptive help seeking. Theory Pract. **41**(2), 132–138 (2002)
21. Pedro, F., Subosa, M., Rivas, A., Valverde, P.: Artificial intelligence in education: Challenges and opportunities for sustainable development (2019)
22. Riazy, S., Simbeck, K.: Predictive algorithms in learning analytics and their fairness. DELFI **2019** (2019)
23. Vincent-Lancrin, S., Van der Vlies, R.: Trustworthy artificial intelligence (ai) in education: Promises and challenges (2020)
24. Xu, Z., Ou, Z., Su, Q., Yu, J., Quan, X., Lin, Z.: Embedding dynamic attributed networks by modeling the evolution processes. In: Proceedings of the 28th International Conference on Computational Linguistics, pp. 6809–6819 (2020)
25. Zemel, R., Wu, Y., Swersky, K., Pitassi, T., Dwork, C.: Learning fair representations. In: International Conference on Machine Learning, pp. 325–333. PMLR (2013)

A Multimodal Machine Learning Framework for Teacher Vocal Delivery Evaluation

Hang Li, Yu Kang, Yang Hao, Wenbiao Ding, Zhongqin Wu, and Zitao Liu[✉]

TAL Education Group, Beijing, China
{lihang4,kangyu,haoyang2,dingwenbiao,wuzhongqin,liuzitao}@tal.com

Abstract. The quality of vocal delivery is one of the key indicators for evaluating teacher enthusiasm, which has been widely accepted to be connected to the overall course qualities. However, existing evaluation for vocal delivery is mainly conducted with manual ratings, which faces two core challenges: subjectivity and time-consuming. In this paper, we present a novel machine learning approach that utilizes pairwise comparisons and a multimodal orthogonal fusing algorithm to generate large-scale objective evaluation results of the teacher vocal delivery in terms of fluency and passion. We collect two datasets from real-world education scenarios and the experiment results demonstrate the effectiveness of our algorithm. To encourage reproducible results, we make our code public available at https://github.com/tal-ai/ML4VocalDelivery.git.

Keywords: Vocal delivery · Multimodal machine learning · Pairwise comparison

1 Introduction

Teacher enthusiasm has been widely accepted by recent researches that is highly correlated with the high-quality instructions, which provides students with learning opportunity and fosters their learning and achievement [6,11,13,17]. To evaluate teacher enthusiasm, multiple statistical algorithms focusing on counting and scoring different aspects of instruction behaviors, i.e., vocal delivery, facial expressions, have been employed as the basic indicators of enthusiastic teaching in their own systems [1,3,4,9,14]. Among these studies, vocal delivery is one of the most commonly accepted indicators due to its irreplaceability in student-teacher communication. Therefore, we focus on improving the existing vocal delivery evaluation (VDE) via the advanced machine learning algorithms.

Traditionally, VDE is conducted by human observers and the evaluation results face two challenges: (1) *subjectivity*: human annotators may have different understandings about evaluation rules; and (2) *time-consuming*: vocal delivery manual evaluation requires annotators to examine the vocal samples multiple times. To solve these two challenges, we propose a multimodal machine learning framework to conduct objective VDE in terms of both fluency and passion. The

I. Roll et al. (Eds.): AIED 2021, LNAI 12749, pp. 251–255, 2021.
https://doi.org/10.1007/978-3-030-78270-2_45

fluency indicator is designed to detect poor articulations between the words and topics, and the passion indicator is utilized to evaluate the variations of pitch, volume, and speed.

In summary, the contributions of this work are: (1) we alleviate the subjectivity problem in current VDE by utilizing the pairwise comparisons; (2) we propose a multimodal orthogonal fusing algorithm, which helps embeddings from different unimodal pre-trained models fuse in an informative way; (3) we demonstrate that our proposed method is able to provide accurate and objective evaluation results.

2   Label Generation via Pairwise Comparison

In our framework, in order to obtain reliable training labels for VDE, we design a two-step label generation algorithm via pairwise comparison to eliminate the discrepancies caused by the ambiguous descriptions to the anchors of some subjective perceptions such as passion [2,12,15].

Anchor Selection. We collect a moderate-size unlabeled dataset $S = \{s_i\}_{i=1}^{N}$ via uniform sampling. After that, for each paired samples (s_i, s_j), we ask human annotators to judge which sample is better fitting the requirements (e.g., "Is sample A more passionate than sample B?"). After collecting plenty of these comparing results, we model the probability of choosing s_i over s_j by utilizing the Bradley-Terry model [2], i.e., $P(s_i > s_j) = f(a_i - a_j)$ where $f(u) = \frac{1}{1+\exp(-u/\sigma)}$, a_i is the estimated ranking score and σ is standard deviation of $A = \{a_i\}_{i=1}^{N}$. Following the prior work by Tsukida and Gupta [16], the ranking scores A is obtained through maximum a posteriori estimation. After that, we carefully choose L anchor samples $\mathcal{G} = \{s_{g_1}, \cdots, s_{g_L}\}$ by percentiles that represent the ranking score distribution.

Comparison Labeling. Once we obtain the anchor samples, we label the remaining samples based on their comparing results with selected anchors. More specifically, for a new sample s^*, we first conduct its pairwise comparisons with each anchor in $\mathcal{G}$. Then, similar to the ranking score generation process in the anchor selection step, we learn the Bradley-Terry model from these comparison results and obtain the corresponding ranking score a^*. Finally, we compare a^* with the ranking scores, i.e., $\{a_{g_1}, \cdots, a_{g_L}\}$ of our select anchors in $\mathcal{G}$ and the final label y^* is obtained by computing number of anchors ordered after s^*, i.e., $y^* = \sum_{l=1}^{L} \mathbb{1}_{a^* > a_{g_l}}$, where $\mathbb{1}_{(.)}$ is an indicator function.

3   Multimodal Learning

The traditional evaluation of vocal delivery usually involves complicated considerations on multiple facets of speeches [1,3,4]. To make full use of these information in each speech sample, we propose a multimodal learning framework with three modules: a language encoder, an audio encoder, and a multimodal fusion block. The overall framework architecture is shown in Fig. 1.

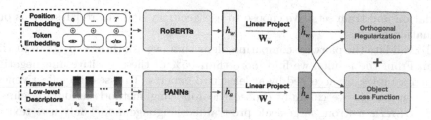

Fig. 1. The proposed multimodal learning framework.

Language Encoder. The pre-trained language models like BERT [5], RoBERTa [10], BART [8] have been demonstrated to have strong capabilities in capturing semantic information. In our framework, we choose to use RoBERTa as our backbone model which accepts text token embeddings combining with their corresponding position embeddings as inputs. Following prior researches [10], we use the first token's output representation $\mathbf{h}_w$ as the extracted semantic sentence embedding.

Audio Encoder. Similar to language encoder, we use a pretrained audio neural networks (PAANs) [7] as our backbone module to extract acoustic features. The inputs of the audio encoder are the frame-level low-level descriptors and the output is a single vector $\mathbf{h}_a$, which summarizes the acoustic features of the entire utterance.

Orthogonal Fusion. Multimodal learning aims to exhibit and capture information from different modalities and therefore, we propose an orthogonal fusion method to enforce representations from different modalities to be dissimilar. Specifically, we design an additional orthogonal regularization penalty as follows: $\mathcal{L}_{\mathrm{Orth}} = \frac{|(\mathbf{W}_a \cdot \mathbf{h}_a)^\top (\mathbf{W}_w \cdot \mathbf{h}_w)|}{\|\mathbf{W}_a \cdot \mathbf{h}_a\| \|\mathbf{W}_w \cdot \mathbf{h}_w\|}$, where $\mathbf{W}_w$ and $\mathbf{W}_a$ are trainable parameters that project $\mathbf{h}_w$ and $\mathbf{h}_a$ to the same hidden space respectively. In the final objective functions, we use the fused representation $\mathbf{h}^{fuse}$, i.e., $\mathbf{h}^{fuse} = \mathbf{W}_w \cdot \mathbf{h}_w \oplus \mathbf{W}_a \cdot \mathbf{h}_a$ to optimize the VDE loss together with the regularization term $\mathcal{L}_{\mathrm{Orth}}$.

4   Experiments

We evaluate teacher vocal delivery in two aspects: fluency and passion. We collect two datasets from real-world K-12 education scenarios: (1) the *Passion* dataset contains 18,000 teacher speech samples extracted from a third-party online class platform; and (2) the *Fluency* dataset includes 15,000 utterances and each sample is labeled based on its fluency level. The sample labels for these two datasets are obtained through pairwise comparisons discussed in Sect. 2. We choose two anchors, i.e., set $L = 2$, which represent the 25% and 75% percentiles. Hence, samples are split into three groups: high, medium and low. In terms of model training, we exclude samples of medium group to reduce the ambiguity. 1,000 utterances are randomly sampled from each dataset and used as test data. Additionally, we perform a 20%/80% split over the remaining dataset to generate validation and train sets. We choose to use accuracy and macro F1-score as our evaluation metrics.

To validate the pairwise-comparing algorithm, we ask three experts to justify them. From the results, we find more than 95% of these positive and negative labeled samples are accepted by at least two experts. To assess the effectiveness of our approach, we carefully choose the following methods as our baselines: (1) *RoBERTa*: a strong large-scale pre-trained language model only uses text as input. (2) *PANNs*: a uni-modal pre-trained model that only uses audio signals as input. (3) *Concat*: a multimodal model which uses both pre-trained RoBERTa and PANNs to extract features and simply concatenates the representations of different modalities for classification. The detailed results for both fluency and passion datasets are shown in Table 1.

From Table 1, we have several observations: (1) by comparing *RoBERTa* and *PANNs* on Fluency dataset, we find language information is more important than audio for fluency evaluation; (2) we observe *PANNs* outperforms *RoBERTa* on Passion dataset, which is consistent with our expectation that acoustic features should be better in evaluating the passion of the utterance; (3) when comparing *Concat* with prior two unimodal models, we find it outperforms the two unimodal baselines by a great margin, which indicates the effectiveness of multimodal learning; (4) by comparing *Ours* to *Concat*, we find the model's performance is further improved.

Table 1. Model performances on two datasets. Acc and $F1_{macro}$ indicate the accuracy and macro F1-score respectively.

Task	PAANs		RoBERTa		Concat		Ours	
	Acc	$F1_{macro}$	Acc	$F1_{macro}$	Acc	$F1_{macro}$	Acc	$F1_{macro}$
Passion	0.775	0.723	0.763	0.714	0.808	0.758	0.846	0.805
Fluency	0.654	0.628	0.788	0.777	0.838	0.828	0.872	0.862

5   Conclusion

In this work, we present an efficient machine learning approach to evaluate teacher vocal delivery for online classes. Experiments demonstrate that our framework achieves accurate evaluations in terms of both fluency and passion aspects. In the future, we would like to conduct further researches to the other facets of the teacher enthusiasm.

Acknowledgments. This work was supported in part by National Key R&D Program of China, under Grant No. 2020AAA0104500 and in part by Beijing Nova Program (Z201100006820068) from Beijing Municipal Science & Technology Commission.

References

1. Bettencourt, E.M., Gillett, M.H., Gall, M.D., Hull, R.E.: Effects of teacher enthusiasm training on student on-task behavior and achievement. Am. Educ. Res. J. **20**(3), 435–450 (1983)
2. Bradley, R.A., Terry, M.E.: Rank analysis of incomplete block designs: i. the method of paired comparisons. Biometrika **39**(3/4), 324–345 (1952)
3. Brigham, F.J., Scruggs, T.E., Mastropieri, M.A.: Teacher enthusiasm in learning disabilities classrooms: effects on learning and behavior. Learn. Disabil. Res. Pract. (1992)
4. Collins, M.L.: Effects of enthusiasm training on preservice elementary teachers. J. Teach. Educ. **29**(1), 53–57 (1978)
5. Devlin, J., Chang, M.W., Lee, K., Toutanova, K.: Bert: pre-training of deep bidirectional transformers for language understanding. In: Proceedings of the 2019 Conference of the North American Chapter of the Association for Computational Linguistics: Human Language Technologies, pp. 4171–4186. Association for Computational Linguistics (2019)
6. Feldman, K.A.: Identifying exemplary teachers and teaching: evidence from student ratings. In: The scholarship of teaching and learning in higher education: an evidence-based perspective, pp. 93–143. Springer (2007). https://doi.org/10.1007/1-4020-5742-3_5
7. Kong, Q., Cao, Y., Iqbal, T., Wang, Y., Wang, W., Plumbley, M.D.: Panns: large-scale pretrained audio neural networks for audio pattern recognition. IEEE/ACM Trans. Audio, Speech, Lang. Process. **28**, 2880–2894 (2020)
8. Lewis, M., et al.: Bart: Denoising sequence-to-sequence pre-training for natural language generation, translation, and comprehension. In: Proceedings of the 58th Annual Meeting of the Association for Computational Linguistics, pp. 7871–7880. Association for Computational Linguistics (2020)
9. Li, H., Wang, Z., Tang, J., Ding, W., Liu, Z.: Siamese neural networks for class activity detection. In: Bittencourt, I.I., Cukurova, M., Muldner, K., Luckin, R., Millán, E. (eds.) AIED 2020. LNCS (LNAI), vol. 12164, pp. 162–167. Springer, Cham (2020). https://doi.org/10.1007/978-3-030-52240-7_30
10. Liu, Y., et al.: Roberta: a robustly optimized bert pretraining approach. CoRR abs/1907.11692 (2019)
11. Liu, Z., et al.: Dolphin: a spoken language proficiency assessment system for elementary education. Proc. Web Conf. **2020**, 2641–2647 (2020)
12. Maystre, L., Grossglauser, M.: Fast and accurate inference of plackett-luce models. Tech. rep. (2015)
13. Moulding, N.T.: Intelligent design: student perceptions of teaching and learning in large social work classes. High. Educ. Res. Dev. **29**(2), 151–165 (2010)
14. Murray, H.G.: Low-inference classroom teaching behaviors and student ratings of college teaching effectiveness. J. Educ. Psychol. **75**(1), 138 (1983)
15. Plackett, R.L.: The analysis of permutations. J. Roy. Stat. Soc.: Ser. C (Appl. Stat.) **24**(2), 193–202 (1975)
16. Tsukida, K., Gupta, M.R.: How to Analyze Paired Comparison Data. WASHINGTON UNIV SEATTLE DEPT OF ELECTRICAL ENGINEERING, Tech. rep. (2011)
17. Zeidner, M.: Test anxiety in educational contexts: Concepts, findings, and future directions. In: Emotion in education, pp. 165–184. Elsevier (2007)

Solving ESL Sentence Completion Questions via Pre-trained Neural Language Models

Qiongqiong Liu[1], Tianqiao Liu[1], Jiafu Zhao[1], Qiang Fang[1], Wenbiao Ding[1], Zhongqin Wu[1], Feng Xia[3], Jiliang Tang[2], and Zitao Liu[1](✉)

[1] TAL Education Group, Beijing, China
{liuqiongqiong1,liutianqiao,zhaojiafu,fangqiang,dingwenbiao,
wuzhongqin,liuzitao}@tal.com
[2] Data Science and Engineering Lab, Michigan State University, East Lansing, USA
tangjili@msu.edu
[3] Federation University Australia, Ballarat, Australia
f.xia@ieee.org

Abstract. Sentence completion (SC) questions present a sentence with one or more blanks that need to be filled in, three to five possible words or phrases as options. SC questions are widely used for students learning English as a Second Language (ESL) and building computational approaches to automatically solve such questions is beneficial to language learners. In this work, we propose a neural framework to solve SC questions in English examinations by utilizing pre-trained language models. We conduct extensive experiments on a real-world K-12 ESL SC question dataset and the results demonstrate the superiority of our model in terms of prediction accuracy. Furthermore, we run precision-recall trade-off analysis to discuss the practical issues when deploying it in real-life scenarios. To encourage reproducible results, we make our code publicly available at https://github.com/AIED2021/ESL-SentenceCompletion.

Keywords: Sentence completion · Pre-trained language model · Neural networks

1 Introduction

Sentence completion (SC) questions present a sentence with one or more blanks that need to be filled in. Three to five possible words (or short phrases) are given as options for each blank and only one of the options yields to a reasonable sentence. SC questions have been proven a necessary source of evaluation data for investigating and diagnosing the situations that the English as a Second Language (ESL) learners grasp the essential language knowledge [1,3,6,9,13]. An example of SC question is shown in Table 1.

In this work, we study computational approaches to automatically solve such ESL SC questions. They are valuable for many reasons: (1) they are able to provide instant feedback to students and help students learn and practice ESL

© Springer Nature Switzerland AG 2021
I. Roll et al. (Eds.): AIED 2021, LNAI 12749, pp. 256–261, 2021.
https://doi.org/10.1007/978-3-030-78270-2_46

Table 1. An illustrative example of SC questions.

—That T-shirt with Yao Ming's picture on it ____ belong to John. He likes him a lot
—No, it ____ be his. He hates black color
(A) can; can't (B) may; needn't (C) must; mustn't (D) must; can't

questions anytime anywhere; (2) they provide feasible solutions to evaluate distractors in SC questions and help teachers revise and improve the overall qualities of SC questions; and (3) they shed light on the opposite tasks like automatically generating questions for language proficiency evaluation and provide as many as possible training samples for building effective question-answering systems or intelligent tutoring systems.

Various approaches have been proposed to automatically solve the ESL SC questions. For example, Zweig et al. chose to use a trigram language model (LM) for solving the SC questions in Scholastic Aptitude Test (SAT) where the trigram LM is trained on 1.1B words from newspaper data [13]. Shen et al. proposed a blank LM to iteratively determine which word to place in a blank and whether to insert new blanks, until no blanks need to be filled [10]. Donahue et al. trained the LM by using the concatenation of artificially-masked texts and the texts which are masked as input [5].

However, automatically solving ESL SC questions still presents numerous challenges that come from special characteristics of real-world educational scenarios as follows: (1) confusing distractors: the ESL SC questions are created by English teaching professionals and the corresponding distractors are very similar; (2) detailed linguistic knowledge: due to the evaluation propose, SC questions always embed detailed linguistic knowledge including grammar, syntax, and semantics; and (3) arbitrary number of blanks and tokens: the ESL SC questions may have one or more missing blanks to be filled and each of which may require an arbitrary unknown number of tokens.

To overcome the above challenges, we propose to utilize a large-scale neural LM to automatically solve the ESL SC questions in students' real-life scenarios. Our approach is based on the standard Transformer-based neural machine translation architecture and utilizes a denoising autoencoder for pre-training sequence-to-sequence models. Our approach shows a powerful generalization capability for automatically solving ESL SC questions of various types from real-world scenarios. Experiments conducted on a real-world online education dataset demonstrate the superiority of our proposed framework compared with competitive baseline models.

2 Our Approach

The SC question is composed of (1) a question, i.e., **q**, formed in natural language with one or more blanks, and (2) m candidate options, i.e., $o_1, \cdots, o_m$. Solving the SC question is to find the option that leads to the highest correct probability after completing the to-be-filled sentence with the selected option, i.e., $\arg\max_{i=1,\cdots,m} \Pr(\mathbf{o}_i|\mathbf{q})$.

In this work, we first fill candidate options into the corresponding blanks to get complete sentences. Then we treat sentences that contain the correct options as positive examples and the rest as negative examples. After that, we build a neural LM model to extract the semantically meaningful information within each sentence and make final SC question predictions via a multilayer perceptron (MLP).

We choose to use a denoising autoencoder for pretraining sequence-to-sequence models, i.e., BART, [7] as our neural LM model. BART adapts standard Transformer [11] as its backbone model and is pre-trained to map corrupted document to their original. We apply pre-trained BART model to our SC questions task with simple modifications on the output layers and loss function. Specifically, given a complete sentence $\mathbf{q} = (w_1, w_2, \cdots, w_n)$, we first convert it into token embeddings $\mathbf{E} = (\mathbf{e}_1, \mathbf{e}_2, \cdots, \mathbf{e}_n)$, where $\mathbf{E} \in \mathbb{R}^{n \times d}$, and d is the embedding size. Then we pass $\mathbf{E}$ through multiple Transformer encoder layers to obtain the contextualized token representations $\mathbf{H} = (\mathbf{h}_1, \mathbf{h}_2, \cdots, \mathbf{h}_n)$. The input of the decoder is the same as the encoder and we pass $\mathbf{E}$ to a stack of Transformer decoder layers. Different from the encoder, masked self-attention is applied to ensure that the predictions can depend only on the information at prior positions in the decoder. Additionally decoder performs cross-attention over the final hidden representations of the encoder, i.e., $\mathbf{H}$. Finally, we obtain the final hidden states $(\mathbf{t}_1, \mathbf{t}_2, \cdots, \mathbf{t}_n)$, where $\mathbf{t}_i \in \mathbb{R}^{d \times 1}$. We utilize the final hidden state $\mathbf{t}_n$ as the aggregated sentence representation. We introduce two additional fully-connected layers to perform the binary classification task, i.e., $\mathbf{x} = \mathrm{softmax}(\mathbf{W}_1 \tanh(\mathbf{W}_0 \mathbf{t}_n + \mathbf{b}_0) + \mathbf{b}_1))$, where $\mathbf{W}_0 \in \mathbb{R}^{1024 \times d}$, $\mathbf{b}_0 \in \mathbb{R}^{1024}$, $\mathbf{W}_1 \in \mathbb{R}^{2 \times 1024}$ and $\mathbf{b}_1 \in \mathbb{R}^2$. The first entry of $\mathbf{x}$ gives the probability of wrong option while the second entry gives right option probability. The objective is to minimize the cross entropy of the right or wrong option labels.

3   Experiments

We collect real-world K-12 English SC exam questions from a thirty-party educational company. After data cleaning and random shuffling, we end up with 250,918 and 48,686 SC questions as our training and testing datasets. Due to the fact that the difficulty of a particular SC question heavily depends on the number of to-be-filled blanks and the number of tokens in the candidate options. Therefore, we divide the SC questions into the following four categories: C1: one-blank and one-token; C2: one-blank and many-token; C3: many-blank and one-token; and C4: many-blank and many-token. Specifically, we have 114,547, 138,392, 28,738 and 17,927 SC questions in each category.

We carefully choose the following state-of-the-art pre-trained LM approaches as our baselines (1) BERT [4]: a pre-trained natural language understanding model with transformer encoder blocks; (2) XLNet [12]: an autoregressive based pre-training method with transformer decoder blocks; (3) ELECTRA [2]: a more sample-efficient pre-training framework which adapts a generator to perform the textualized masked language modeling task and a discriminator to perform token-level "real-fake" binary classification task; (4) RoBERTa [8]: improves BERT by replacing static masking with dynamic masking, pre-training more epochs with larger batch size, and removing the next sentence prediction task.

3.1   Results

As we can see from Table 2, our model outperforms all other methods in terms of prediction accuracy on all SC question categories. Specifically, when comparing the prediction performance of all the methods on C1 to C2, C3 and C4, we can see that the increase of either the number of blanks or the length of options does not hurt the accuracy of ESL SC question solvers. The pre-trained large-scaled LMs are very robust and insensitive to SC questions in different categories.

Table 2. Results on different categories of SC question datasets in terms of accuracy.

	C1	C2	C3	C4
BERT	0.8840	0.8894	0.9221	0.9166
XLNet	0.9128	0.9165	0.9290	0.9264
ELECTRA	0.9212	0.9186	0.9346	0.9236
RoBERTa	0.9171	0.9321	0.9380	0.9304
BART	**0.9381**	**0.9428**	**0.9475**	**0.9445**

Furthermore, we conduct a precision-recall trade-off analysis on the results. When deploying the model in practice, a wrong answer may give bad guidance to students. In order to reduce such problem, we may refuse to solve some difficult questions and improve the precision of more solvable questions. We set a threshold to the correct probability of the model's selected option and accept the above-the-threshold questions as our solvable questions. The recall is computed as (the number of solvable questions)/(the number of all test questions), and the precision is calculated as (the number of both solvable and correct-answered questions)/(the number of solvable questions). Finally, we find that when the threshold is 0.95, the precision reaches 97.22% and the recall is 88.17% which can be used in practice.

4   Conclusion

In this paper, we present a neural framework for automatically solving the ESL sentence completion questions. Experimental results based on the real-world English examinations indicate that our proposed model works well in different kinds of sentence completion questions. Furthermore, we conduct fine-grained performance analysis on ESL SC questions from different categories and a trade-off analysis between precision and recall, which reveals insights of applying the proposed approach in the real-world production system.

Acknowledgment. This work was supported in part by National Key R&D Program of China, under Grant No. 2020AAA0104500 and in part by Beijing Nova Program (Z201100006820068) from Beijing Municipal Science & Technology Commission.

References

1. Beinborn, L., Zesch, T., Gurevych, I.: Candidate evaluation strategies for improved difficulty prediction of language tests. In: Proceedings of the Tenth Workshop on Innovative Use of NLP for Building Educational Applications, pp. 1–11. Association for Computational Linguistics, Denver, Colorado (2015). https://doi.org/10.3115/v1/W15-0601, https://www.aclweb.org/anthology/W15-0601
2. Clark, K., Luong, M., Le, Q.V., Manning, C.D.: ELECTRA: pre-training text encoders as discriminators rather than generators. In: 8th International Conference on Learning Representations, ICLR 2020, Addis Ababa, Ethiopia, April 26–30, 2020. OpenReview.net (2020). https://openreview.net/forum?id=r1xMH1BtvB
3. Davey, G., De Lian, C., Higgins, L.: The university entrance examination system in china. J. Furth. High. Educ. **31**(4), 385–396 (2007)
4. Devlin, J., Chang, M.W., Lee, K., Toutanova, K.: BERT: pre-training of deep bidirectional transformers for language understanding. In: Proceedings of the 2019 Conference of the North American Chapter of the Association for Computational Linguistics: Human Language Technologies, Volume 1 (Long and Short Papers), pp. 4171–4186. Association for Computational Linguistics, Minneapolis, Minnesota (2019). https://doi.org/10.18653/v1/N19-1423, https://www.aclweb.org/anthology/N19-1423
5. Donahue, C., Lee, M., Liang, P.: Enabling language models to fill in the blanks. In: Proceedings of the 58th Annual Meeting of the Association for Computational Linguistics, pp. 2492–2501. Association for Computational Linguistics, Online (2020). https://doi.org/10.18653/v1/2020.acl-main.225, https://www.aclweb.org/anthology/2020.acl-main.225
6. Franke, W.: The reform and abolition of the traditional Chinese examination system, vol. 10. Harvard Univ Asia Center (1960)
7. Lewis, M., et al.: BART: denoising sequence-to-sequence pre-training for natural language generation, translation, and comprehension. In: Proceedings of the 58th Annual Meeting of the Association for Computational Linguistics, pp. 7871–7880. Association for Computational Linguistics, Online (2020). https://doi.org/10.18653/v1/2020.acl-main.703, https://www.aclweb.org/anthology/2020.acl-main.703
8. Liu, Y., et al.: RoBERTa: a robustly optimized BERT pretraining approach. arXiv preprint arXiv:1907.11692 (2019)
9. Madaus, G.F.: The effects of important tests on students: implications for a national examination system. Phi Delta Kappan **73**(3), 226–231 (1991)
10. Shen, T., Quach, V., Barzilay, R., Jaakkola, T.: Blank language models. In: Proceedings of the 2020 Conference on Empirical Methods in Natural Language Processing (EMNLP), pp. 5186–5198. Association for Computational Linguistics, Online (2020). https://doi.org/10.18653/v1/2020.emnlp-main.420, https://www.aclweb.org/anthology/2020.emnlp-main.420

11. Vaswani, A., et al.: Attention is all you need. In: Guyon, I., et al. (eds.) Advances in Neural Information Processing Systems 30: Annual Conference on Neural Information Processing Systems 2017, December 4–9, 2017, Long Beach, CA, USA, pp. 5998–6008 (2017). https://proceedings.neurips.cc/paper/2017/hash/3f5ee243547dee91fbd053c1c4a845aa-Abstract.html

12. Yang, Z., Dai, Z., Yang, Y., Carbonell, J.G., Salakhutdinov, R., Le, Q.V.: Xlnet: generalized autoregressive pretraining for language understanding. In: Wallach, H.M., Larochelle, H., Beygelzimer, A., d'Alché-Buc, F., Fox, E.B., Garnett, R. (eds.) Advances in Neural Information Processing Systems 32: Annual Conference on Neural Information Processing Systems 2019, NeurIPS 2019, December 8–14, 2019, Vancouver, BC, Canada, pp. 5754–5764 (2019). https://proceedings.neurips.cc/paper/2019/hash/dc6a7e655d7e5840e66733e9ee67cc69-Abstract.html

13. Zweig, G., Platt, J.C., Meek, C., Burges, C.J., Yessenalina, A., Liu, Q.: Computational approaches to sentence completion. In: Proceedings of the 50th Annual Meeting of the Association for Computational Linguistics (Volume 1: Long Papers), pp. 601–610. Association for Computational Linguistics, Jeju Island, Korea (2012). https://www.aclweb.org/anthology/P12-1063

DanceTutor: An ITS for Coaching Novice Ballet Dancers Using Pose Recognition of Whole-Body Movements

Lurlynn Maharaj-Pariagsingh and Phaedra S. Mohammed(✉)

Department of Computing and Information Technology, The University of the West Indies, St. Augustine, Trinidad and Tobago

Abstract. This paper presents the design, development and evaluation of a prototype intelligent dance tutoring system, DanceTutor, for coaching students in low-resource settings. The system evaluates seventeen core body points on a dancer using video footage captured from a mobile phone or web camera using a combination of simple algorithms and 2D pose estimation software. Detailed feedback is provided on the quality and correctness of the dancer's pose for the first five static dance positions in Ballet, and then for intermediate to advanced exercises with permutations of the five basic Ballet positions. Evaluation of the prototype revealed the highly subjective nature and cultural biases of evaluating the quality of a dancer's technique. Three experienced dance teachers, trained in different countries, evaluated 165 video recordings of 11 candidate dancers. The system was only able to achieve 47% consensus overall with the feedback and grading results produced by the dance teachers, who each evaluated tension and height differently. There was however a 60% agreement between DanceTutor and one teacher who used the most granular evaluation strategy matching DanceTutor's baseline and assessment features.

Keywords: Intelligent tutoring system · Ballet · Gesture recognition · Student modelling · Whole-body movement · Feedback

1 Introduction

Dance education is a subject area that was late to adopt any application of technology to its teaching strategies [1]. Dance studios are traditionally built with walls of mirrors that allow dancers the opportunity to view their movements. However, these reflections provide little to no knowledge of whether or not movements are correct for novice dancers, how to adjust correctly and safely, and by extension if certain actions may be injurious [2]. Researchers have studied the idea of automated feedback and have implemented electronic systems using pose recognition [3], virtual reality [4], visual and verbal responses [5] to assist teachers with correcting and providing feedback to Ballet dancers in anticipation of better quality learning experiences and skill set improvement. The use of cameras, augmented mirrors and virtual dance teachers, motion-capture systems, extended feedback [3–6] have been featured in software systems for the purpose

© Springer Nature Switzerland AG 2021
I. Roll et al. (Eds.): AIED 2021, LNAI 12749, pp. 262–266, 2021.
https://doi.org/10.1007/978-3-030-78270-2_47

of identifying faults and the application of corrective measures. However, a thorough analysis of a dancer's full body with respect to the basic five positions of Ballet has not been pursued. These are deemed foundational to any other movement in the style. The Super Mirror prototype [5] for example focused only on the lower body parts, namely the knee and hip joints. Another issue is handling variations in body shape and type. In [5], Marquardt et al. evaluated the feedback quality produced by Super Mirror in comparison to a human dance teacher's feedback. Results showed that the teacher evaluated aspects that were not considered by the system because the data model did not cater for the height of the dancers. The YouMove system [6] dealt with this challenge by tracking 20 joints via a skeleton framing algorithm using Microsoft Kinect however feedback was limited. Huang et al. [4] showed that a virtual dance instructor with real-time feedback for ballroom dance successfully assessed a dancer's performance by interrupting the execution and repeating the explanation or demonstration until the dancer's performance improved or by praising correct movements.

The previous examples highlight the value of pose recognition [7] and automated feedback and these were therefore the core focus of this research. However, a simpler yet still effective approach was key since access to augmented mirrors, movement tracking hardware and live-streaming virtual avatars are far from the norm for many dance classes in developing countries. Consequently, the DanceTutor system presented in this paper targets full body movement for the first five foundational Ballet positions using open-source pose recognition software that works on any body type. Automated corrective feedback is however based on video footage from mobile phones using a simple system architecture that relies on inexpensive hardware devices and lightweight processing. The aim was to generate customized audio feedback that assists a dancer (novice or trained) in perfecting the various Ballet positions through corrective advice on how to adjust misaligned parts of his/her body. The research also aimed to discover standardized aspects of Ballet dance evaluation and correction, and to assess the efficacy and accuracy of DanceTutor in evaluating a dancer's Ballet pose.

2 The DanceTutor System

The DanceTutor architecture follows the Model-View-Controller (MVC) design pattern alongside the traditional ITS structure. The components of the system comprise four (4) modules shown in Fig. 1, a database, and a workstation for accepting input and delivering output. The Pose Detection Module recognises, tracks and produces annotated data identifying seventeen core body points as a dancer moves using the 2D OpenPose software [8]. An overlay on the dancer's body, shown in Fig. 1, is produced from video footage using JSON files with x and y coordinates for each data point. This produces a dynamic skeletal model of the dancer used by the Pose Evaluation Module to continuously evaluate the alignment of the dancer's body parts for a given Ballet position. This was done using several custom algorithms that operated on the JSON data that determine if a dancer deviates from the correct position at specific body points based on threshold values for angles that would indicate correct and incorrect poses. The baselines were mined from data gathered from video recordings of both advanced and novice dancers gesturing the five Ballet positions. The Feedback module provides audio feedback for

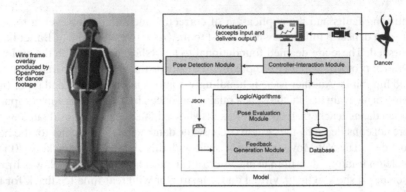

Fig. 1. High level sketch of the overall architecture of DanceTutor.

guiding self-corrective adjustment of the dancer's body in response to the analysis produced by the Pose Evaluation Module. The Interaction Module displays a Graphical User Interface and coordinates the interactions between the Pose Detection, Pose Evaluation and Feedback Generation Modules. Personal data of the dancer, feedback statements, progress reports, achievements and status logs are stored in simple student models in the database.

3   Experimental Evaluation and Results

Three experiments were conducted to evaluate the effectiveness and accuracy of the DanceTutor system. Three dance teachers evaluated 165 videos of 11 Ballet dancers and produced 609 reviews containing feedback and ratings for all of the core body parts for each Ballet position in the videos. There was no observed agreement amongst the three dance teachers, T1, T2 and T3 (Fleiss $\kappa < 0$). 56 out of the 609 reviews were generic where the dance teachers gave an overall statement on the position, for example, "Perfect" or "You look tense". 289 responses out of the 609 reviews were similar to the feedback generated by DanceTutor. They were not exactly the same words but implied the same misalignment was detected. For example:

Dance Teacher: "Heels - right heel needs to move to the left to be directly in front of left toes. I am not sure if dancer is showing an open fourth where the heels line up."

DanceTutor: "Move your right heel line with your left toes"

There was a 60% similarity between DanceTutor feedback and T2, and 47% similarity to the feedback statements collated from all three dance teachers. Grades awarded showed that the DanceTutor had a 60% similarity to T2 and 40% similarity to the grades given by all three dance teachers (Table 1). The variations in the teacher evaluations could have affected the thresholds built into the DanceTutor algorithms resulting in a lack of standardization in the evaluation criteria for each of the body parts. This is a known challenge. The results also revealed gaps between the evaluation processes used by DanceTutor and the dance teachers. Tension and roundedness in the arms as well as turn out from the hips were identified by all three teachers but predominantly by T2.

Table 1. Measure of the grading matches between T1, T2, T3 and DanceTutor for 11 Dancers

Teacher	D1	D2	D3	D4	D5	D6	D7	D8	D9	D10	D11	Avg
T1- F1 score	0.57	0.00	0.00	0.42	0.67	0.92	0.75	0.80	0.13	0.00	0.97	0.48
T2- F1 score	0.00	0.00	0.67	0.70	0.75	0.89	0.75	0.75	0.70	0.13	0.50	0.53
T3- F1 score	0.00	0.00	0.29	0.24	0.13	0.64	0.42	0.57	0.70	0.00	0.50	0.32
Avg.	0.13	0.00	0.00	0.50	0.35	0.89	0.70	0.70	0.64	0.00	0.50	0.40

Incidentally, T2 matched the DanceTutor system's evaluation the closest owing to more granular evaluations in her ratings and feedback, resulting in a mix of good and poor ratings of the body parts for a given video.

4   Conclusion

The COVID-19 pandemic highlighted the significance of DanceTutor since teachers currently struggle to examine a student's entire body movements based on the orientation of their cameras in remote online meeting environments. DanceTutor can assist with remote dance training since it caters for low resource settings and offers individualised attention and feedback on full body movements. For future research, an improvement in the evaluation of the tension, arms and turnout and the transition between each position of the combinations can be investigated.

References

1. Daniaa, A., Hatziharistosa, D., Koutsoubaa, D., Tyrovolaa, V.: The Use of Technology in Movement and Dance Education: Recent Practices and Future Perspectives, pp. 3–4. Athens, Greece: Elsevier Ltd (2011). https://doi.org/10.1016/j.sbspro.2011.04.299
2. Trajkova, M., Francesco, C.: Takes tutu to ballet: designing visual and verbal feedback for augmented mirrors. Proc. ACM Interact. Mobile, Wearable Ubiquit. Technol. 2(1), 1–30 (2018). https://doi.org/10.1145/3191770
3. Cheng, D., Hartmann, B., Chi, P., Wright, P., Kwak, T.: Body-Tracking Camera Control for Demonstration Videos, Paris, France: CHI 2013 Extended Abstracts on Human Factors in Computing Systems, pp. 1–6 (2013). https://doi.org/10.1145/2468356.2468568
4. Huang, H., Uejo, M., Seki, Y., Lee J., Kawagoe, K.: Realizing real-time feedback on learners' practice for a virtual ballroom dance instructor, pp. 5–6. Workshop on Real-time Conversations with Virtual Agents (RCVA), IVA 2012, Santa Cruz, USA (2012)
5. Marquardt, Z., Beira, J., Paiva, I., Em, N., Kox, S.: Super Mirror: a kinect interface for ballet dancers, pp. 1–6. CHI EA 2012 (2012). https://doi.org/10.1145/2212776.2223682
6. Anderson, F., Grossman, T., Matejka, J., Fitzmaurice, G.: YouMove: enhancing movement training with an augmented reality mirror, pp. 311–319. In: Proceedings of the 26th ACM Symposium on User Interface Software and Technology (2013). https://doi.org/10.1145/2501988.2502045

7. Shotton, J., Fitzgibbon, A.: Real-time human pose recognition in parts from single depth images, pp. 1–8. In Proceedings of the 24th IEEE Conference on Computer Vision and Pattern Recognition, CVPR 2011, Colorado Springs, USA, pp. 20–25 (2011). https://doi.org/10.1109/CVPR.2011.5995316

8. Zhe Cao, G., Hidalgo, S., Wei, S., Sheikh, Y.: OpenPose: realtime multi-person 2D Pose Estimation using part affinity fields, pp. 1–14. IEEE Transactions on Pattern Analysis and Machine Intelligence (2019). https://doi.org/10.1109/TPAMI.2019.2929257

Tracing Embodied Narratives of Critical Thinking

Shitanshu Mishra[1], Rwitajit Majumdar[2]($\boxtimes$), Aditi Kothiyal[3], Prajakt Pande[4],
and Jayakrishnan Madathil Warriem[5]

[1] IIT Bombay, Mumbai, India
[2] Kyoto University, Kyoto, Japan
[3] EPFL, Lausanne, Switzerland
[4] RUC, Roskilde, Denmark
[5] IIT Madras, Chennai, India

Abstract. Critical Thinking (CrT) is generally characterized as an abstract thinking process, detached from the (bodily) actions one engages in during the process. Though recent cognitive theories assert that all thinking is action-based, the embodied and distributed cognitive processes underlying CrT have not been identified. We present preliminary findings from the first iteration of a design-based research project which involves probing possible connections between CrT and one's (bodily) action sequences. We performed sequential pattern mining and qualitative analysis on the study participants' actions logs to find differences in participants CrT processes. Our analysis showed that only a subset of participants contextualized their assumptions, inferences, and implications in the different information resources available in the environment. A majority of participants' actions performed within the interface were incoherent. These results have implications for automated analyses of the CrT process, and for the design of AI-based scaffolds to support CrT development.

Keywords: Critical thinking · Embodied cognition · ENaCT framework · Cognitive strategies · Learning analytics · Sequential pattern mining

1 Introduction

Critical thinking (CrT) is one of the most important 21st-century skills [1]. It has been defined in multiple ways, but for the purpose of this work, we consider the following definition: "Critical thinking involves analyzing and evaluating thought processes with a view to improve them" [2]. We adopt Paul and Elder's (2019) [2] framework, which prescribes that intellectual standards, such as clarity, accuracy, and precision must be applied to different elements of thought while reasoning in order to develop CrT. These elements of thought include purpose, questions, points of view, information, concepts, inferences, implications, and assumptions.

With the advent of computer-based task environments, sensors, and automatic data logging, it has become easier to track complex thinking and problem-solving processes.

© Springer Nature Switzerland AG 2021
I. Roll et al. (Eds.): AIED 2021, LNAI 12749, pp. 267–272, 2021.
https://doi.org/10.1007/978-3-030-78270-2_48

For instance, the processes of designing, programming, and problem-solving have been analyzed and documented in great detail [3–6]. Similar analyses of CrT in computer environments have found that the prevalence of use of certain actions and words among people is indicative of CrT [7–9].

Newer approaches to cognition such as 4E cognition [10], suggest that the features of task environments, and one's actions on them are constitutive of thinking [11]. For instance, actions performed within a problem-solving environment can help offload parts of thinking to the environment, thus distributing thinking to the environment and making it more efficient and effective [12, 13]. This theoretical stance motivates us to identify which actions and sequences of actions constitute CrT within a computer environment, thus grounding our analysis in theory, and improving its validity [14]. Specifically, we ask the following research question, *What (interaction and thinking) processes do participants follow while executing a CrT task?* Here, we present a scalable analysis methodology to obtain a characterization of the CrT processes.

2 Research Methods

Our computer environment (called 'ENaCT') is based on a framework that aligns the design and analytics of CrT environments [15]. ENaCT system supports the following four main problem-solving affordances (Fig. 1 [16]): (i) Task description (ii) Information resources presenting task-related data and concepts. (iii) Expression affordances, as per the "Elements of Thought" [2] (iv) Summary panel with all the text generated by the solver presented together.

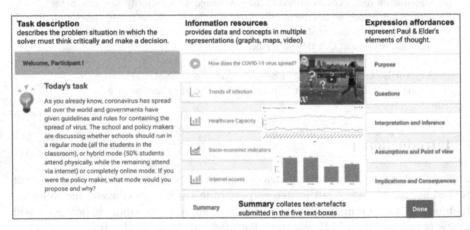

Fig. 1. Computer task environment ENaCT [16]

Twenty five students participated in an online synchronous session (conducted in an undergraduate human-computer interaction course). The participants were given 35 min to complete the CrT task individually on ENaCT. We collected time-stamped logs of students' activities in the ENaCT environment that captured the click-interactions with each element of the environment (e.g. name of the UI element, and the submitted artefacts) along with the timestamp of each interaction. 522 click-interaction logs and 46 text-artefacts were collected from the 25 participants. For our detailed analysis, we consider the data from only twelve participants who completed all parts of the required task and created text-artefacts.

3 Data Analysis and Findings

The aim of the data analysis was to identify participants' CrT processes. First, we extracted all the counts of each interaction namely: 1) opening of information panels (*info*) 2) opening of expression panels (*touch*) 3) submissions in the expression panels (*submit*) 4) opening of the summary panel (*summary*). *Info* was the most frequently performed interaction, while *summary* was the least performed interaction. On an average, a student submitted texts in the expression panel 3.83 times. One out of twelve participants did not use *summary* interaction.

At the first level of analysis, we performed Sequential Pattern Mining (SPM; [17, 18]) to identify the sequences of these four interactions with the following parameters: MinSupport = 0.3; MinGap = 0; MaxGap = 0. MinSupport of 0.3 corresponds to all the sequential patterns that were evident in the sequences of at least 30% of participants. I-Support or mean of I-frequency represents the number of times a pattern occurred per participant averaged across all participants. S-Frequency represents the number of participants who executed the pattern at least once. An example of an extracted pattern is: *"summary- > info- > submit"*, *i.e.* a participant accessed the summary panel, then looked at one of the info panels, and finally submitted a text-input in one of the expression panels. This pattern occurred on an average 0.41 times per participant.

In total, there were 124 patterns. Of these, 4 patterns were of unit length, 76 were of length more than five, and the remaining 44 were of lengths between two and five. There were patterns that contained only one type of interaction, i.e., info-only, touch-only, submit-only, summary-only. For example, *"info- > info- > info- > info"* is an *info-only* pattern that represents that participant(s) consecutively accessed different information panels. Also, there were groups of patterns with mixed types of interactions (e.g. *"summary- > info- > submit"*). Out of the 44 patterns of length between 2 and 5, 14 consisted of a single type of interaction, while 30 (68%) patterns were composed of more than one type of interaction. The SPM statistics show that *info-only* and *touch-only* pattern groups had maximum I-support (7.08 and 3.17, respectively), while *Submit-only, Summary-only,* and *interleaved patterns* had I-support values of 1.6, 0.92 and 1.0, respectively. To develop a more detailed understanding of the CrT, we focus on the interleaved interaction patterns, particularly those involving at least one *submit* interaction. We discuss one such pattern below.

Participants Submitting Text-Artefacts after Referring to the Provided Information: Examples of such patterns are: *"info- > submit- > submit"*, *"info- > submit"*, *"info- > submit- > submit- > submit"*, *"info- > info- > submit"*. We conjectured that while engaging in a CrT process, participants would first refer to the information panels to build an understanding of the task and then write the "elements of thought" in the expression panels. Interleaving between the *info* and *submit* interactions confirms such a process. However, of all participant sequences, one participant adopted a completely opposite course. They first submitted the text and then accessed the information panels. Similarly, another interesting and desirable pattern is where participants looked at the *summary*, followed by *info* and then submitted text in the expression panels.

Next, we wanted to understand how participants' usage of information panels was related to their text submissions in the expression panels. So, we qualitatively compared the contents of the information panels accessed by participants immediately before submitting a text, with the submitted content. Three researchers independently carried out this analysis, and disagreements were resolved through discussion. We discovered variations in the quality of participant responses along the dimensions of validity and coherence. Coherence was examined in the following two ways: (i) coherence between the submitted content and the previously accessed content in the information panel, and (ii) coherence between the submitted artefact in one expression panel with that in another expression panel. Qualitative categories of the submitted text expressions are provided in Fig. 2.

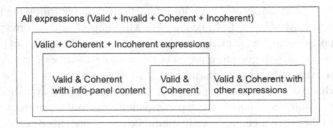

Fig. 2. Different types of submitted text in the expression panels

In summary, we find evidence for coherent as well as incoherent submissions, thus leading to the emergence of the following types of CrT processes. (1) **Type-1:** Participant doesn't make any submission in the expression panels; (2) **Type-2:** Participant submits text in the expression panel(s) without accessing any available resources (information panels); (3) **Type-3:** Participant submits after accessing one or more information panels. (3.1) **Type-3a:** Participant submissions were not coherent with the accessed information resources; (3.2) **Type-3b:** Participant submissions were coherent with the accessed information resources.

4 Discussion and Conclusion

The goal of this work was to develop an analysis methodology that can be scaled up in order to obtain a detailed characterization of CrT processes. We began with a 4E cognition [10] theoretical stance on the basis of which we conjectured that a rich interaction between a problem solver and a computer-based task environment would be desirable for good CrT. In concrete terms, this means that participants may interact with the environment affordances (information and expression) in an interleaved manner (i.e. transiting from one to the other, and/or back and forth). We performed sequential pattern mining in order to identify prominent patterns of interaction-sequences present in the data. We found that 68% of these interaction patterns were of 'interleaved' nature. Further detailed examination of CrT processes through a qualitative analysis of participant responses (text-artefacts) in terms of "elements of thought" highlighted that the interleaving of actions alone does not mark good CrT. An additional level of coherence, where submission in each "element of thought" is coherent with related information as well as with contents in other "elements of thought", may be required. This highlights the need for automating this coherence analysis using techniques such as semantic similarity and text overlap [19, 20] analysis of participant responses.

Acknowledgement. This collaborative research is partially funded by the SPIRITS 2020 grant of Kyoto University.

References

1. Larson, L.C., Miller, T.N.: 21st century skills: prepare students for the future. Kappa Delta Pi Rec. **47**(3), 121–123 (2011)
2. Paul, R., Elder, L.: The Miniature Guide to Critical Thinking Concepts and Tools. Rowman & Littlefield (2019)
3. Worsley, M., Blikstein, P.: A multimodal analysis of making. Int. J. Artif. Intell. Educ. **28**(3), 385–419 (2017). https://doi.org/10.1007/s40593-017-0160-1
4. Segedy, J.R., Biswas, G., Sulcer, B.: A model-based behavior analysis approach for Open-Ended Environments. J. Educ. Technol. Soc. **17**(1), 272–282 (2014)
5. Blikstein, P., Worsley, M., Piech, C., Sahami, M., Cooper, S., Koller, D.: programming pluralism: using learning analytics to detect patterns in the learning of computer programming. J. Learn. Sci. **23**(4), 561–599 (2014). https://doi.org/10.1080/10508406.2014.954750
6. Vieira, C., Hathaway Goldstein, M., Purzer, Ş., Magana, A.J.: Using learning analytics to characterize student experimentation strategies in engineering design. J. Learn. Analytics **3**(3), 291–317 (2016). https://doi.org/10.18608/jla.2016.33.14
7. Kovanović, V., et al.: Understand students' self-reflections through learning analytics. In: Proceedings Of The 8th International Conference on Learning Analytics and Knowledge, pp. 389–398 (2018)
8. Koh, E., Jonathan, C., Tan, J.P.L.: Exploring conditions for enhancing critical thinking in networked learning: findings from a secondary school learning analytics environment. Educ. Sci. **9**(4), 287 (2019)
9. Buckingham Shum, S., Crick, R.D.: Learning analytics for 21st century competencies. J. Learn. Analytics **3**(2), 6–21 (2016)
10. Newen, A., De Bruin, L., Gallagher, S. (eds.): The Oxford Handbook of 4E Cognition. Oxford University Press, Oxford (2018)
11. Landy, D., Allen, C., Zednik, C.: A perceptual account of symbolic reasoning. Front. Psychol. **5**, 275 (2014)
12. Kirsh, D.: Thinking with external representations. AI & Soc. **25**(4), 441–454 (2010). https://doi.org/10.1007/s00146-010-0272-8
13. Aurigemma, J., Chandrasekharan, S., Nersessian, N.J., Newstetter, W.: Turning experiments into objects: the cognitive processes involved in the design of a lab-on-a-chip device. J. Eng. Educ. **102**(1), 117–140 (2013)
14. Winne, P.H.: Construct and consequential validity for learning analytics based on trace data. Comput. Hum. Behav. **112**, 106457 (2020). https://doi.org/10.1016/j.chb.2020.106457

15. Mishra, S., Majumdar, R., Kothiyal, A., Pande, P., Warriem, J.M.: ENaCT: An Action-based Framework for the Learning and Analytics of Critical Thinking. in the Proceedings of ICCE 2020, pp.144–153 (2020)
16. Majumdar, R., et al.: Design of a Critical Thinking Task Environment based on ENaCT framework. In: The Proceedings of ICALT 2021 (2021)
17. Agrawal, R., Srikant, R.: Mining sequential patterns. In: Proceedings of the Eleventh IEEE International Conference on Data Engineering (ICDE), pp. 3–14 (1995)
18. Kinnebrew, J.S., Loretz, K.M., Biswas, G.: A contextualized, differential sequence mining method to derive students' learning behavior patterns. 5(1), 190–219 (2013)
19. Crossley, S.A., Kyle, K., Dascalu, M.: The tool for the automatic analysis of cohesion 2.0: integrating semantic similarity and text overlap. Behav. Res. Methods 51(1), 14–27 (2018). https://doi.org/10.3758/s13428-018-1142-4
20. Segedy, J.R., Kinnebrew, J.S., Biswas, G.: Using coherence analysis to characterize self-regulated learning behaviours in open-ended learning environments. J. Learn. Analytics 2(1), 13–48 (2015)

Multi-armed Bandit Algorithms for Adaptive Learning: A Survey

John Mui, Fuhua Lin[✉], and M. Ali Akber Dewan

Athabasca University, Athabasca, Canada
jmui1@athabasca.edu, {oscarl,adewan}@athabascau.ca

Abstract. Adaptive learning aims to provide each student individual tasks specifically tailed to his/her strengths and weaknesses. However, it is challenging to realize it, overcoming the complexity issue in online learning. There are many unsolved problems such as knowledge component sequencing, activity sequencing, exercise sequencing, question sequencing, and pedagogical strategy, to realize adaptive learning. Bandit algorithms are particularly suitable to model the process of planning and using feedback on the outcome of that decision to inform future decisions. They are finding their way into practical applications in various areas especially in online platforms where data is readily available, and automation is the only way to scale. This paper presents a survey on bandit algorithms for facilitating adaptive learning in different settings. The findings indicate that the various bandit algorithms have great potential to solve the above problems. Also, we discuss issues and challenges of developing and using adaptive learning systems based on the multi-armed bandit framework.

Keywords: Bandit algorithms · Multi-armed bandit algorithm · Adaptive learning · Exploration and exploitation · Personalized learning

1 Introduction

Adaptive learning aims to provide efficient, effective, and customized learning paths to engage each student. However, it is challenging to develop adaptive learning systems. There are many unsolved problems such as knowledge component sequencing, activity sequencing, exercise sequencing, question sequencing, and pedagogical strategy, to realize adaptive learning. Recent research in machine learning on bandit algorithms, also called multi-armed bandits (MAB) [1], for education shows that MAB algorithms is a more practical and scalable solution [2–5]. MAB algorithms provide an efficient way to balance between exploitation and exploration. They have been used in many areas including many areas such as healthcare, and recommendation systems [6].

This paper reviews the state of the art in MAB approaches for enabling adaptive learning. The rest of the paper is organized as follows. Section 2 explains the methodology we used for the survey. Section 3 discusses the existing bandit algorithms for solving various problems in adaptive learning. Section 4 concludes the paper.

© Springer Nature Switzerland AG 2021
I. Roll et al. (Eds.): AIED 2021, LNAI 12749, pp. 273–278, 2021.
https://doi.org/10.1007/978-3-030-78270-2_49

2 Methodology

The search strategy utilized for this survey employs a mixture of techniques. Google Scholar was the search engine utilized. For currency, the years 2014–2020 were chosen as the publishing dates, the search was performed in June 2020. To identify relevant sources of high-impact journal papers and conference proceedings, the metrics tool within Google Scholar was used, specifically from the categories of Educational Technology under Engineering and Computer Science. For each publication entry, an advanced search was done to find relevant articles. As these publications are already in the field of educational technology, the desire was to learn about the MAB usage within this domain using the keywords 'multi-armed bandits', 'bandit algorithms' combined with keywords such as 'adaptive exercises', 'personalized sequences', 'automated assignments', 'adaptive learning' along with its individual words, joined compound components, and various alternative synonyms. Similarly, a general search within Google Scholar for MAB algorithms in sequencing adaptive tasks for adaptive learning was done. This resulted in the selection of several papers, theses, and dissertations for further examination. Also, an ancillary search involving selected conferences and journals was conducted.

3 The Problems of Adaptive Learning

Adaptive learning is challenging because an instructor agent needs to deal with a large decision space that grows combinatorially with the number of knowledge components (KCs), the number of learning objectives, the number of pedagogical strategies, and the number of learning activities of a course. Therefore, through a tasks analysis of adaptive learning systems and based on the framework by [7], we divided the sequencing tasks into three types which correspond to the framework:

- Adaptive knowledge component sequencing for personalizing learning paths.
- Adaptive learning activity sequencing for increasing proficiency levels.
- Adaptive question sequencing for adaptive assessments.

And then we identified and grouped literature by their MAB algorithm techniques for each type, which we believe helps to predict future trends.

4 Motivation for Using Bandit Algorithms

The MAB family of algorithms (*aka*, bandit algorithms) is named after the problem for a gambler who must decide which arm of a "multi-armed bandit" slot machine to pull to maximize the total reward in a series of trials [1]. The MAB algorithms are data-driven and can balance exploration and exploitation and make sequential decisions under uncertainty. Thus, they are particularly relevant to decision-making about alternative pedagogies and lend themselves quite naturally to the problem of determining adaptive sequences in online learning environments. More importantly, the relative simplicity of the MAB framework (and its requirements for no or far less training data) makes it more effective than rule-based methods and more practical than the POMDP framework

[8]. Due to the complexity of adaptive learning, standard bandit algorithms can not be applied directly, and they have been extended to model the different settings in online adaptive learning. Table 1 of Sect. 5 shows the overview of the algorithms used in the reviewed papers and for the various problems in adaptive learning.

5 Bandit Algorithms for Adaptive Learning

For *adaptive assessment question sequencing*, the main approaches used to date are based on Item Response Theory (IRT) [9] or multidimensional IRT [10]. The main limitation of these approaches is high complexity in calibrating the item difficulties in the question bank. To overcome these limitations, Melesko and Novickij (2019) [11] proposed a method based on one of the standard MAB algorithms, Upper-Confidence Bound [1], to solve the quiz sequence generation problem and shows that the method can significantly decrease the quiz length without a decrease in accuracy. However, many research questions remain to be answered before their results can be applied in the wild [12].

Table 1. Overview of the algorithms used in the reviewed papers and for the various problems in adaptive learning.

Algorithms	Problems			
	Assessment question sequencing	Question/exercise sequencing	Action/task sequencing	KC sequencing
UCB	[11]		[17]	
Adversarial bandits		[2–4, 12, 13]		
Contextual bandits			[5, 14–16, 19]	[25]
Gittins		[20]		
Epsilon-greedy			[21]	
Softmax		[22]		
Stochastic bandits		[23]		
Recovering bandits		[24]		

For question or exercise sequencing, Clement et al. (2014) [1] used the EXP4 (Exponential weighting for Exploration and Exploitation with Experts) algorithm in MAB-based adaptive activity sequencing algorithms and developed two algorithms: Right Activity at Right Time (RiARiT) and Zone of Proximal Development and Empirical Success (ZPDES) [2, 3, 12]. The results indicate that their algorithm could lead to a faster learning speed, is better at adapting to the student, and could help the student progress to a higher and more complex skill level. Segal et al. (2018) [3] focused on the difficulty levels of practice questions as a central part for their algorithm to sequence questions for learning and knowledge tracking. They used an exploration policy like the one that is used by the EXP4 algorithm [1]. The paper did not address how to incorporate a dynamic or increasing pool of questions. Mu et al. (2017) [4] uses ZPDES to adapt and help students progress faster and reduce the number of problems needed to be completed to reach mastery [13] and incorporates the multiscale context model of forgetting to model the effect of forgetting via memory trace that decreases over time.

Several studies utilize contextual bandits to solve the problem of adaptive activity or task sequencing. Contextual bandit algorithms introduce the concept of side information, which is called context, to the bandit algorithm. The contexts are mapped and used to help with action selection in the decision process. Lan and Baraniuk (2016) aimed to provide personalized learning actions (PLAs) to the students between each assessment using a contextual bandit approach [5, 14]. Following [5], Manickam et al. (2017) [15] developed two new Bayesian-based contextual bandit PLA selection algorithms. Wan (2017) [16] attempted to use context to improve personalized tutorial strategies for each student in a web based ITS to avoid over-tutoring or under-tutoring, by developing bandits with decision tree algorithm. Lakhani (2016) [17] used both topic and student contextual information to personalize a learning path (content items) for the student by extending the most cited contextual bandit learning algorithm, LinUCB (Linear UCB) [18]. Cai et al. (2020) [19] illustrated how contextual bandits can fit into their overall personalized conversational agent for learning math to determine a pace that is suitable for the student. Nguyen (2014) [20] associated exercise and experience via the learning curve theory with the development of two new algorithms: Parametric Gittins Index (PGI) and Parametric Thompson Sampling (PTS). However, Gittins-based techniques are typically hard to implement in practice. Andersen et al. (2016) [21] developed a skill-based task selection that is used to determine customized tasks which should be put forward to the student to maximize the learning of the student by progressively providing tasks for their level. They created a knowledge matrix and incorporated the system with an epsilon-greedy approach and a dynamic-epsilon approach. Zhang and Goh (2019) [22] tackle the difficulty adaption problem with the aim of providing students with suitably challenging tasks by utilizing a softmax approach [1] with the system to select questions that aim to keep the student's performance close to a target grade to provide the optimal performance. Pike-Burke et al. (2018) [23] studied a problem that when a student answers a question, the benefit to their learning from doing so may not be evident immediately. They studied a variant of the stochastic bandit problem with delayed, aggregated anonymous feedback and presented a rarely switching algorithm which can learn from this kind of feedback and achieve almost the same performance as a state-of-the-art algorithm for the simpler delayed feedback bandit problem. Also, Pike-Burke and Grünewälder (2019) identified a factor that would have a clear effect on the student's ability to answer a question correctly based on the length of time since they have seen similar (or the same) questions. They developed a recovering bandit algorithm where recovery represents the appropriate elapsed period in which the benefit of reward by asking the same question has recovered [24].

Finally, we found that none of the above MAB algorithms have dealt with the adaptive KC problem even though Xu et al. (2016) [25] used contextual bandits to identify which sequences of courses lead students to obtain maximal GPAs.

6 Conclusions

We have presented a review on MAB algorithms for tackling the challenges of adaptive content sequencing in adaptative learning systems. While there are diverse variations of solutions illustrated in the literature, it was noticed that MAB studies for adaptive learning are not prevalent. This indicates that this is an area under development and has not yet been fully noticed by educational technology practitioners. At this time, simulations are extensively utilized in most of the papers. Many researchers highlighted the importance of real trials in the evaluation and testing of a system [16, 26]. Also, we found that none of the studies fully demonstrated coverage that included the adaptation and usage of their approaches to span the length of a whole course in sufficient detail. Thus, there are many exciting possibilities and questions such as adaptive pedagogical strategies [27] remaining to be answered through adapting the MAB algorithms into adaptive learning systems.

References

1. Lattimore, T., Szepesvári, C.: Bandit Algorithms. Cambridge University Press, Cambridge (2020)
2. Clément, B., Roy, D., Oudeyer, P.-Y., Lopes, M.: Online optimization of teaching sequences with multi-armed bandits. In: 7th International Conference on Educational Data Mining, London, United Kingdom (2014). https://hal.inria.fr/hal-01016428.
3. Segal, A., Ben David, Y., Williams, J.J., Gal, K., Shalom, Y.: Combining difficulty ranking with multi-armed bandits to sequence educational content. In: Penstein Rosé, C., et al. (eds.) AIED 2018. LNCS (LNAI), vol. 10948, pp. 317–321. Springer, Cham (2018). https://doi.org/10.1007/978-3-319-93846-2_59
4. Mu, T., Wang, S., Andersen, E., Brunskill, E.: Combining adaptivity with progression ordering for intelligent tutoring systems. In: Proceedings of the Fifth Annual ACM Conference on Learning at Scale, pp. 1–4 (2018). https://doi.org/10.1145/3231644.3231672
5. Lan, A.S., Baraniuk, R.G.: A contextual bandits framework for personalized learning action selection. In: International Conference on Educational Data Mining (2016)
6. Bouneffouf, D., Rish, I.: A survey on practical applications of multi-armed and contextual bandits. In: Congress on Evolutionary Computation, Glasgow, United Kingdom (2019)
7. Essa, A.: A possible future for next generation adaptive learning systems. Smart Learn. Environ. 3(1), 1–24 (2016). https://doi.org/10.1186/s40561-016-0038-y
8. Rafferty, A.N., Brunskill, E., Griffiths, T.L., Shafto, P.: Faster teaching by POMDP planning. In: Biswas, G., Bull, S., Kay, J., Mitrovic, A. (eds.) AIED 2011. LNCS (LNAI), vol. 6738, pp. 280–287. Springer, Heidelberg (2011). https://doi.org/10.1007/978-3-642-21869-9_37
9. Lord, F.M.: Applications of Item Response Theory to Practical Testing Problems. Routledge (1980)
10. Reckase, M.D.: Multidimensional item response theory models. In: Reckase, M.D. (ed.) Multidimensional Item Response Theory, pp. 79–112. Springer New York, New York (2009). https://doi.org/10.1007/978-0-387-89976-3_4

11. Melesko, J., Novickij, V.: Computer adaptive testing using upper-confidence bound algorithm for formative assessment. Appl. Sci. **9**(20), 4303 (2019)
12. Clement, B., Roy, D., Oudeyer, P.Y., Lopes, M.: Multi-armed bandits for intelligent tutoring systems. J. Educ. Data Min. **7**(2), 20–48 (2015)
13. Mu, T., Goel, K. Brunskill, E.: Program2Tutor: combining automatic curriculum generation with multi-armed bandits for intelligent tutoring systems. In: Conference on Neural Information Processing Systems (2017)
14. Lan, A.S.: Machine learning techniques for personalized learning, Doctoral dissertation, Rice University (2016)
15. Manickam, I., Lan, A.S., Baraniuk, R.G.: Contextual multi-armed bandit algorithms for personalized learning action selection. In: 2017 IEEE International Conference on Acoustics, Speech and Signal Processing (ICASSP), pp. 6344–6348 (2017). https://doi.org/10.1109/ICASSP.2017.7953377,
16. Wan, H.: Tutoring students with adaptive strategies, Doctoral dissertation, Worcester Polytechnic Institute (2017)
17. Lakhani, A.: Adaptive teaching: learning to teach, Master's thesis, University of Victoria (2018)
18. Li, L., Chu, W., Langford, J., Schapire, R.E.: A contextual-bandit approach to personalized news article recommendation. In: Proceedings of the 19th International Conference on World Wide Web - WWW 2010, pp. 661 (2010). https://doi.org/10.1145/1772690.1772758
19. Cai, W., et al.: MathBot: a personalized conversational agent for learning math, Cai2020MathBotAP (2020)
20. Nguyen, M.Q.: Multi-armed bandit problem and its applications in intelligent tutoring systems, Master's thesis. École Polytechnique (2014)
21. Andersen, P.-A., Kråkevik, C., Goodwin, M., Yazidi, A.: Adaptive task assignment in online learning environments. In: Proceedings of the 6th International Conference on Web Intelligence, Mining and Semantics, pp. 1–10 (2016). https://doi.org/10.1145/2912845.2912854
22. Zhang, Y., Goh, W.B.: Bootstrapped policy gradient for difficulty adaptation in intelligent tutoring systems. In: International Conference on Autonomous Agents and Multiagent Systems, pp. 711–719 (2019)
23. Pike-Burke, C., Agrawal, S., Szepesvari, C., Grunewalder, S.: Bandits with Delayed, Aggregated Anonymous Feedback. [Cs, Stat] (2018). http://arxiv.org/abs/1709.06853
24. Pike-Burke, C., Grünewälder, S.: Recovering Bandits. [Cs, Stat] (2019). http://arxiv.org/abs/1910.14354
25. Xu, J., Xing, T., Van Der Schaar, M.: Personalized course sequence recommendations. IEEE Trans. Sig. Process. **64**(20), 5340–5352 (2016)
26. Clement, B.: Adaptive personalization of pedagogical sequences using machine learning, Doctoral dissertation, Bordeaux (2018)
27. Shen, S., Ausin, M.S., Mostafavi, B., Chi, M.: 2018. Improving learning & reducing time: a constrained action-based reinforcement learning approach. In: UMAP 2018: 26th Conference on User Modeling, Adaptation and Personalization, 8–11 July 2018, Singapore, Singapore. ACM, New York, NY, USA, p. 9 (2018). https://doi.org/10.1145/3209219.3209232

Paraphrasing Academic Text: A Study of Back-Translating Anatomy and Physiology with Transformers

Andrew M. Olney[✉][iD]

University of Memphis, Memphis, TN 38152, USA
aolney@memphis.edu
https://olney.ai

Abstract. This paper explores a general approach to paraphrase generation using a pre-trained seq2seq model fine-tuned using a back-translated anatomy and physiology textbook. Human ratings indicate that the paraphrase model generally preserved meaning and grammaticality/fluency: 70% of meaning ratings were above 75, and 40% of paraphrases were considered more grammatical/fluent than the originals. An error analysis suggests potential avenues for future work.

Keywords: Paraphrase · Deep learning · Natural language generation

1 Introduction

Paraphrasing is a core task in natural language processing (NLP) and has multiple educational applications, like essay grading [5], short answer assessment [11], text simplification [4] and plagiarism detection [1]. Recent developments in automated paraphrase have largely tracked advances in machine translation using neural networks, i.e., neural machine translation (NMT), primarily using the LSTM [8,13,17] and Transformer [10,12,14,20] architectures. One approach to generating paraphrases is back-translation, by which a sentence is translated from a source language to a *pivot* language and back to the source language.

Paraphrasing academic text has its own challenges because it differs from normal text both in vocabulary and syntax, particularly in scientific domains [6,16] and it is usually copyright-restricted and therefore difficult to obtain in quantities necessary for machine learning models. The present study addresses these problems through NMT back-translation and fine-tuning a recent Transformer variant called T5 [18]. Our primary research questions are therefore (1) how well the paraphrases preserve the meaning of the source text and (2) how grammatical and fluent are the paraphrases with respect to the source text.

2 Model and Human Evaluation

We conducted a small pilot study to determine the best pivot languages for paraphrasing anatomy and physiology. Randomly selected sentences (N=24) from

© Springer Nature Switzerland AG 2021
I. Roll et al. (Eds.): AIED 2021, LNAI 12749, pp. 279–284, 2021.
https://doi.org/10.1007/978-3-030-78270-2_50

a textbook [19] were back-translated with different pivot languages using the Google Translate API. The paraphrases were evaluated by an expert judge on (1) the degree of change as none, word, or phrase (a measure of diversity) and (2) whether the paraphrase was disfluent or incorrect (a measure of acceptability). Results are presented in Table 1. Values are sentence counts except for weighted change, which weights word change counts by 1 and phrase change counts by 2.

An ideal pivot language would result in low unacceptability and high diversity. Our analysis suggests Czech introduces more changes at the word choice level, and Russian introduces marginally more changes at the phrasal level. On the intuition these properties may be additive, we conducted an additional evaluation using Czech and Russian as pivot languages together (English-Czech-Russian-English). As indicated by the results in the table, the combination appears to increase the weighted change above Czech and Russian individually without noticeably increasing error. Furthermore, the weighted change is comparable to most of the non-European pivot languages, which created substantially more unacceptable paraphrases. Based on these results, we back-translated the complete textbook (12,062 sentences) both with Czech as a pivot and with Czech-Russian as a double pivot, producing 24,124 source-paraphrase pairs.

Training and testing sets were prepared by aligning the two back-translations with the corresponding source and randomly selecting 90% of the 3-tuples for training and the remainder for test. These datasets were then augmented by permuting the 3-tuples to create combinations of all pairs in all orders. Pairs differing by less than 3 characters and sentences with less than 11 characters were excluded as noisy data. Augmentation resulted in 34,094 pairs in the training and 3,836 pairs in the test sets. The T5-BASE pre-trained model from the HuggingFace library [21] and fine-tuned using Pytorch with the training set for 8 epochs, though test set loss did not improve past epoch 4. The training process completed in approximately 3.5 h using an NVIDIA 1080Ti GPU.

A human evaluation was conducted to determine the quality of the model-generated paraphrases, specifically (1) how well the paraphrases preserve the meaning of the source text and (2) how grammatical and fluent the paraphrases are with respect to the source text. Raters ($N = 29$) were recruited through the Amazon Mechanical Turk (AMT) marketplace between January and February of 2021, using the CloudResearch platform [15]. In this study, raters were required to be native English speakers and be employed as a nurse or physician. Raters were further required to have completed at least 100 previous AMT tasks with at least a 95% approval rating. Raters were paid $7.

A separate textbook on anatomy and physiology [2] from OpenStax was used as a source for sentences to paraphrase. The book was downloaded and preprocessed by splitting main body text into sentences, removing sentences that refer to figure and tables, removing parenthetical elements, performing Unicode to ASCII translation, and performing spelling correction. The final sentences contained ranges, slashes, formulas, and chemical symbols. Paraphrases of these sentences were then generated using the model.

Six surveys were created on Qualtrics, an online survey tool, using randomly selected source-paraphrase pairs, each containing 100 pairs, as is common for this type of evaluation [3,7,9]. Each pair was formatted on a single survey page where the source text was formatted above the paraphrase, followed by two questions with slider-format response on a 0–100 scale. The first was a meaning-assessment question, "The paraphrase conveys the same meaning of the original," and was anchored by "not at all" on the left and "perfectly" on the right. The second was a fluency-assessment question, "Which is more grammatical and fluent?", with "original" on the left and "paraphrase" on the right. The sliders had no numeric indicators and were initialized at the midpoint. Following the direct assessment methodology [7,9], 12 of each 100 were control pairs were created by copying an existing item (a survey page) and then degrading the paraphrase on that page by deleting a random span of words, where $span_{length} = 0.21696 * word_{count} + 0.78698$, rounded down, which linearizes existing rules [9]. Twelve pairs are sufficient to detect a large (.8 SD) effect using a Wilcoxon signed-ranks test for matched pairs at $\alpha = .05$ and .80 power with a one-tailed test. If we do not detect a large effect between ratings of distinct items and their degraded versions, we infer the rater is not reliable. The degraded items were randomly positioned based on the position of their matched item, modulo 44.

3   Results and Discussion

Subsets of raters passed control checks for meaning ($n = 35$) and fluency ($n = 23$), with $p < .05$ on the signed-ranks test, except for a fluency check on the 2nd survey, $p = 0.06$, which was allowed because its control items were more difficult to distinguish. Cronbach's alpha for passing raters was high ($\alpha > .85$), except survey 6, $\alpha = .66$, until two raters were dropped to obtain high agreement, $\alpha = .77$. The mean meaning rating was high ($M = 78.78$, $SD = 16.89$, $CI_{95} = [77.33, 80.22]$), and the mean grammaticality/fluency rating was less than the midpoint of 50 ($M = 43.97$, $SD = 20.75$, $CI_{95} = [42.19, 45.74]$). The distribution of each rating may be examined in Fig. 1. The distribution for meaning illustrates that most paraphrases are rated as highly meaning preserving. The meaning distribution peaks at the most frequent rating of 89, and approximately 70% of all meaning ratings are above 75. The distribution for fluency reflects its anchoring at 50, at which point both the original (0) and paraphrase (100) are considered equally fluent. The grammaticality/fluency distribution is symmetric and peaks at a rating of 38, and approximately 40% of all grammaticality/fluency ratings are above 50, indicating that the paraphrase was considered more grammatical/fluent than the original sentence approximately 40% of the time.

Paraphrases associated with the lowest 5% of ratings for meaning and grammaticality/fluency were examined to determine common error types, four of which accounted for 76% of errors. Most common was the substitution of a near neighbor for the target, e.g. "membrane" for "diaphragm," and it more negatively impacted grammaticality/fluency than meaning. Second was the use of the wrong word sense for the target, e.g. "adults' volumes" for "volumes in adults,"

Table 1. Paraphrase change (None, Word, Phrase, Weighted) and error across pivot languages.

Language	Change				Err
	No	Wd	Ph	Wt	
Czech	3	15	6	27	4
Russian	7	9	8	25	2
Cz-Ru	3	11	10	31	4
Chinese	2	11	11	33	9
Persian	2	14	8	30	9
Arabic	2	13	9	31	11
Hindi	5	11	8	27	9
Turkish	0	8	16	40	8
Welsh	5	10	9	28	10

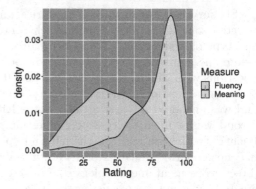

Fig. 1. Density plot for paraphrase ratings with indicated medians.

and more evenly affected both metrics. The third arose when the text contained an acronym, chemical formula, time range, or malformed Unicode, e.g. "Rh-abundant" for "Rh+," and adversely impacted meaning more than grammaticality/fluency. Forth was the replacement of a word with its antonym, e.g. "more mature" for "immature," and primarily impacted meaning. The other error types were approximately evenly represented and included pronoun insertion/deletion, replacement with a foreign word/phrase, insertion of a random word, and correct paraphrases that were misclassified. While some of these errors might be resolved with better or larger language models, we speculate that acronyms and chemical formulas may require a specialized approach.

4   Conclusion

Results from this study indicate that relatively high-quality paraphrases may be generated using a Transformer-based model fine-tuned with back-translated academic text. By leveraging a pre-trained Transformer like T5, researchers can construct a paraphrase model for a new domain in about a day, given available text in electronic format. An important limitation of these results is that only one domain was investigated, anatomy and physiology, raising the question of whether these results will generalize to other domains. Furthermore, while our results seem promising, we did not have a dataset to allow direct comparison to human performance, as is often the case in machine translation. Two important targets for future research are to replicate these findings in other domains and to conduct an evaluation directly comparing model-generated paraphrases with paraphrases generated by humans on the same source sentences.

Acknowledgements. This material is based upon work supported by the National Science Foundation (1918751, 1934745) the Institute of Education Sciences (R305A190448).

References

1. Barrón-Cedeño, A., Vila, M., Martí, M.A., Rosso, P.: Plagiarism meets paraphrasing: insights for the next generation in automatic plagiarism detection. Comput. Linguist. **39**(4), 917–947 (2013). https://doi.org/10.1162/COLI_a_00153
2. Betts, J.G., et al.: Anatomy and Physiology. OpenStax (2017)
3. Bojar, O., et al.: Findings of the 2018 conference on machine translation (WMT18). In: Proceedings of the Third Conference on Machine Translation: Shared Task Papers, pp. 272–303. Association for Computational Linguistics, Belgium, Brussels October (2018). https://doi.org/10.18653/v1/W18-6401
4. Botarleanu, R.-M., Dascalu, M., Crossley, S.A., McNamara, D.S.: Sequence-to-sequence models for automated text simplification. In: Bittencourt, I.I., Cukurova, M., Muldner, K., Luckin, R., Millán, E. (eds.) AIED 2020. LNCS (LNAI), vol. 12164, pp. 31–36. Springer, Cham (2020). https://doi.org/10.1007/978-3-030-52240-7_6
5. Burstein, J., Flor, M., Tetreault, J., Madnani, N., Holtzman, S.: Examining linguistic characteristics of paraphrase in test-taker summaries. ETS Res. Rep. Ser. **2012**(2), 1–46 (2012)
6. Fang, Z.: The language demands of science reading in middle school. Int. J. Sci. Educ. **28**(5), 491–520 (2006). https://doi.org/10.1080/09500690500339092
7. Federmann, C., Elachqar, O., Quirk, C.: Multilingual whispers: Generating paraphrases with translation. In: Proceedings of the 5th Workshop on Noisy User-Generated Text, pp. 17–26. Association for Computational Linguistics, Hong Kong, China, November 2019. https://doi.org/10.18653/v1/D19-5503
8. Fu, Y., Feng, Y., Cunningham, J.P.: Paraphrase generation with latent bag of words. In: Wallach, H.M., Larochelle, H., Beygelzimer, A., d'Alché-Buc, F., Fox, E.B., Garnett, R. (eds.) Proceedings of the Thirty-third Annual Conference on Neural Information Processing Systems, pp. 13623–13634 (2019)
9. Graham, Y., Baldwin, T., Moffat, A., Zobel, J.: Is machine translation getting better over time? In: Proceedings of the 14th Conference of the European Chapter of the Association for Computational Linguistics, pp. 443–451. Association for Computational Linguistics, Gothenburg, Sweden, April 2014. https://doi.org/10.3115/v1/E14-1047
10. Hu, J.E., et al.: Improved lexically constrained decoding for translation and monolingual rewriting. In: Proceedings of the 2019 Conference of the North American Chapter of the Association for Computational Linguistics: Human Language Technologies, Volume 1 (Long and Short Papers), pp. 839–850. Association for Computational Linguistics, Minneapolis, Minnesota, June 2019. https://doi.org/10.18653/v1/N19-1090
11. Koleva, N., Horbach, A., Palmer, A., Ostermann, S., Pinkal, M.: Paraphrase detection for short answer scoring. In: Proceedings of the Third Workshop on NLP for Computer-Assisted Language Learning, pp. 59–73. LiU Electronic Press, Uppsala, Sweden, November 2014
12. Krishna, K., Wieting, J., Iyyer, M.: Reformulating unsupervised style transfer as paraphrase generation. In: Proceedings of the 2020 Conference on Empirical Methods in Natural Language Processing (EMNLP), pp. 737–762. Association for Computational Linguistics, Online, November 2020. https://doi.org/10.18653/v1/2020.emnlp-main.55

13. Li, Z., Jiang, X., Shang, L., Li, H.: Paraphrase generation with deep reinforcement learning. In: Proceedings of the 2018 Conference on Empirical Methods in Natural Language Processing, pp. 3865–3878. Association for Computational Linguistics, Brussels, Belgium, October-November 2018. https://doi.org/10.18653/v1/D18-1421

14. Li, Z., Jiang, X., Shang, L., Liu, Q.: Decomposable neural paraphrase generation. In: Proceedings of the 57th Annual Meeting of the Association for Computational Linguistics, pp. 3403–3414. Association for Computational Linguistics, Florence, Italy, July 2019. https://doi.org/10.18653/v1/P19-1332

15. Litman, L., Robinson, J., Abberbock, T.: TurkPrime.com: a versatile crowdsourcing data acquisition platform for the behavioral sciences. Behav. Res. Methods **49**(2), 433–442 (2016). https://doi.org/10.3758/s13428-016-0727-z

16. Nagy, W., Townsend, D.: Words as tools: learning academic vocabulary as language acquisition. Read. Res. Q. **47**(1), 91–108 (2012). https://doi.org/10.1002/RRQ.011

17. Qian, L., Qiu, L., Zhang, W., Jiang, X., Yu, Y.: Exploring diverse expressions for paraphrase generation. In: Proceedings of the 2019 Conference on Empirical Methods in Natural Language Processing and the 9th International Joint Conference on Natural Language Processing (EMNLP-IJCNLP), pp. 3173–3182. Association for Computational Linguistics, Hong Kong, China, November 2019. https://doi.org/10.18653/v1/D19-1313

18. Raffel, C., Shazeer, N., Roberts, A., Lee, K., Narang, S., Matena, M., Zhou, Y., Li, W., Liu, P.J.: Exploring the limits of transfer learning with a unified text-to-text transformer. J. Mach. Learn. Res. **21**(140), 1–67 (2020)

19. Shier, D., Butler, J., Lewis, R.: Hole's Human Anatomy & Physiology. McGraw-Hill Education, 15th edn. (2019)

20. Witteveen, S., Andrews, M.: Paraphrasing with large language models. In: Proceedings of the 3rd Workshop on Neural Generation and Translation, pp. 215–220. Association for Computational Linguistics, Hong Kong, November 2019. https://doi.org/10.18653/v1/D19-5623

21. Wolf, T., et al.: Transformers: state-of-the-art natural language processing. In: Proceedings of the 2020 Conference on Empirical Methods in Natural Language Processing: System Demonstrations, pp. 38–45. Association for Computational Linguistics, Online, October 2020. https://doi.org/10.18653/v1/2020.emnlp-demos.6

PAKT: A Position-Aware Self-attentive Approach for Knowledge Tracing

Yuanxin Ouyang, Yucong Zhou[✉], Hongbo Zhang, Wenge Rong,
and Zhang Xiong

Engineering Research Center of Advanced Computer Application Technology,
Ministry of Education, Beihang University, Beijing, China
{oyyx,zhouyucong,jarviszhb,w.rong,xiongz}@buaa.edu.cn

Abstract. Knowledge Tracing aims to model a student's knowledge state from her past learning interactions and predict her performance in future. Although structures such as positional encoding or forgetting gate have already been used in Knowledge Tracing models, positional information with great potential is not fully utilized. In this paper, we propose a *Position-aware Self-Attentive Knowledge Tracing* (PAKT) model with a position supervision mechanism. Massive experimental results show that PAKT outperforms other benchmarks on several popular datasets.

Keywords: Knowledge tracing · Educational data mining · Student performance prediction

1 Introduction

Knowledge Tracing task, whose objective is modeling a student's learning process and tracing her knowledge state from past learning interactions, is formally regarded as a sequential supervised problem that given a student's chronological sequence $\mathbf{x} = \{x_1, x_2, ..., x_i\}$ including exercises q_i she tried and corresponding results r_i, predict the probability $p(r_{i+1} = 1|q_{i+1}, \mathbf{x})$ indicating whether she can answer another new exercise q_{i+1} correctly [4]. Thereby KT can support many downstream applications such as exercise recommendation [12] and diagnosis report [2]. Such an effective model for predicting student performance can also save lots of time for both students and teachers [3,10].

Since *Deep Knowledge Tracing* (DKT) combined RNN and KT task together [9], deep models have become the most popular approaches for KT. *Dynamic Key-Value Memory Network* (DKVMN) uses extra memory matrix to expand memory space for knowledge concepts [1,11]. *Self-Attentive Knowledge Tracing* (SAKT) exploits *Transformer* to model student knowledge state [7].

In real-world education scenarios, similar knowledge concepts are always taught in same chapters and students would also focus on relevant concepts during this period. However, most contributions merely model positional factor implicitly and thus may not fully leverage it. In this paper, we propose

© Springer Nature Switzerland AG 2021
I. Roll et al. (Eds.): AIED 2021, LNAI 12749, pp. 285–289, 2021.
https://doi.org/10.1007/978-3-030-78270-2_51

a position supervision mechanism and design a *Position-aware Self-Attentive Knowledge Tracing* (PAKT) model by exploiting positional feature both implicitly and explicitly. Experimental results on several public datasets show PAKT outperformed other baselines.

2 Methodology

2.1 Model Architecture

The overall architecture of our PAKT model is presented in Fig. 1. Our main contribution, position supervision mechanism in Knowledge Tracing, is reflected in *Positional Scaling* (PS) and *Local Feature Extraction* (LE) layer.

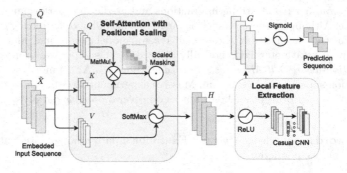

Fig. 1. Overall structure of PAKT. Two main parts are colored in green and yellow. (Color figure online)

Input and Embedding Layer. Following [7,9], an embedding matrix is used to transform each original one-hot input token into a dense question representation $\tilde{q}_i$. Furthermore, an all-zero vector is padded to the beginning or end indicating whether student answered this question correctly or not. Finally, question vector $\tilde{q}_i \in \mathbb{R}^d$ and interaction vector $\tilde{x}_i \in \mathbb{R}^{2d}$ are fed to the following layers.

Masking with Positional Scaling. Although positional encoding is used in SAKT [7], it needs to learn the representations of positions and their relationships on its own. Similar to [5], a multiplier β^{i-j} is applied on each entry s_{ij} $(i \geq j)$ in attention score matrix, where β is a hyperparameter less than 1. The output vector of attention layer h_i is calculated as follows.

$$s_{ij} = \frac{q_i^\mathsf{T} k_j}{\sqrt{d}} \cdot \beta^{i-j} \tag{1}$$

$$h_i = \sum_{j \leq i} \text{SoftMax}(s_{ij}) v_j \tag{2}$$

where q_i, k_j, v_j are query, key and value vectors of attention layer. By incorporating such an exponential weight decay mechanism with base β into attention, we can assign relatively higher attention scores to interactions with shorter distance explicitly. Meanwhile, distant interactions do not get fully vanished because their attention weights are small but non-zero.

Local Feature Extraction. Compared to fully-connected layer, *Local Connectivity* property of convolutional layer assures the locality of output features, and feature pattern is learned implicitly during training phase. Kernel size K controls the neighbor range, where only interactions within certain positional distance could be taken in to calculcation. From a practical point of view, *Causal CNN* is adopted to keep causality of input interaction sequence as shown in Eq. 3:

$$g_i = \sum_{k=1}^{K} h_{i-k+1}^{\mathsf{T}} W_k^C \tag{3}$$

where g_i represents the output vector of LE and W^C is convolution kernel.

2.2 Objective Function

Output layer converts latent vector into output probabilistic vector $o_i \in \mathbb{R}^N$:

$$o_i = \sigma(g_i^{\mathsf{T}} W^O) \tag{4}$$

where $W^O \in \mathbb{R}^{d*N}$, N is total questions number and σ is sigmoid function. Therefore we can obtain the correct probability of next step $\hat{p}_{i+1} = o_{t,q_{i+1}}$. We optimize model by minimize cross-entropy loss function defined as Eq. 5.

$$L = -\sum_{i}^{T} (r_i \log \hat{p}_i + (1 - r_i)(\log (1 - \hat{p}_i))) \tag{5}$$

3 Experiments

Datasets and Implementation Details. We performed experiments on three popular public datasets. Their statistics are shown in Table 1. 5-fold cross validation is performed for all non-simulated datasets.

The proposed model is implemented with PyTorch [8]. Source code has been released on GitHub[1]. To train model efficiently, we split or pad records into sequences with fixed length of 50. Other hyperparameters are chosen by validation set performance: hidden dimensionality $d = 128$, CNN kernel size $K = 7$, scaling rate $\beta = 0.6$. We use Adam optimizer [6] to train model in all experiments.

---

[1] https://github.com/EnrigleZ/pakt.

Table 1. Data statistics of public datasets.

Dataset	Students	Exercise tags	Interactions	Avg. interactions
ASSISTments2009[a]	4417	124	328K	78
ASSISTments2015[b]	19917	100	709K	34
Simulated-5 [9]	4000	50	200K	50

[a]https://sites.google.com/site/assistmentsdata/home/assistment-2009-2010-data/skill-builder-data-2009-2010
[b]https://sites.google.com/site/assistmentsdata/home/2015-assistments-skill-builder-data

Main Results. We compared our PAKT model with some genres of benchmark models on all three datasets by measuring area under the curve(AUC). Results are listed in Table 2.

Table 2. AUC of all the models on the three datasets

Model	ASSIST2009	ASSIST2015	Simulated-5
DKT(RNN)	83.90	68.87	82.49
DKT(LSTM)	84.31	69.15	82.30
DKVMN	83.87	68.47	82.03
SAKT	86.33	82.98	91.83
PAKT	**88.32**	**86.59**	**92.32**

The proposed model outperforms other models on all three datasets. For instance, for the ASSISTments2009, it yields an AUC at 88.32%, whereas SAKT and DKT(LSTM) yield only 86.33% and 84.31%. Compared with those RNN-based models, attention-based SAKT and PAKT have obvious advantage on three datasets. PAKT enjoys the computational efficiency inherited from SAKT since it only brings in limited extra parameters.

4   Conclusion

In this work, we proposed a novel attention-based model PAKT. We introduced a position supervision mechanism which is able to leverage positional information both explicitly and implicitly. Massive experimental results verify the effectiveness of PAKT.

Acknowledgements. This research was partially supported by the National Key R&D Program of China (No. 2018YFB2100800) and the National Natural Science Foundation of China (No. 61772132).

References

1. Abdelrahman, G., Wang, Q.: Knowledge tracing with sequential key-value memory networks. In: Proceedings of the 42nd International ACM SIGIR Conference on Research and Development in Information Retrieval, pp. 175–184 (2019)
2. Burns, H., Luckhardt, C.A., Parlett, J.W., Redfield, C.L.: Intelligent Tutoring Systems: Evolutions in Design. Psychology Press (2014)
3. Cen, H., Koedinger, K., Junker, B.: Learning factors analysis – a general method for cognitive model evaluation and improvement. In: Ikeda, M., Ashley, K.D., Chan, T.-W. (eds.) ITS 2006. LNCS, vol. 4053, pp. 164–175. Springer, Heidelberg (2006). https://doi.org/10.1007/11774303_17
4. Corbett, A.T., Anderson, J.R.: Knowledge tracing: modeling the acquisition of procedural knowledge. User Modeling User-Adapted Interact. 4(4), 253–278 (1994)
5. Ghosh, A., Heffernan, N., Lan, A.S.: Context-aware attentive knowledge tracing. In: Proceedings of the 26th ACM SIGKDD International Conference on Knowledge Discovery & Data Mining, pp. 2330–2339 (2020)
6. Kingma, D.P., Ba, J.: Adam: a method for stochastic optimization. In: Bengio, Y., LeCun, Y. (eds.) 3rd International Conference on Learning Representations, ICLR 2015, San Diego, CA, USA, 7–9 May 2015, Conference Track Proceedings (2015)
7. Pandey, S., Karypis, G.: A self attentive model for knowledge tracing. In: Desmarais, M.C., Lynch, C.F., Merceron, A., Nkambou, R. (eds.) Proceedings of the 12th International Conference on Educational Data Mining, EDM 2019, Montréal, Canada, 2–5 July 2019. International Educational Data Mining Society (IEDMS) (2019)
8. Paszke, A., et al.: Automatic differentiation in pytorch (2017)
9. Piech, C., Bassen, J., Huang, J., Ganguli, S., Sahami, M., Guibas, L.J., Sohl-Dickstein, J.: Deep knowledge tracing. Advances in Neural Information Processing Systems 28, 505–513 (2015)
10. Thai-Nghe, N., Horváth, T., Schmidt-Thieme, L.: Factorization models for forecasting student performance. In: Educational Data Mining 2011 (2010)
11. Zhang, J., Shi, X., King, I., Yeung, D.Y.: Dynamic key-value memory networks for knowledge tracing. In: Proceedings of the 26th International Conference on World Wide Web, WWW 2017, pp. 765–774. International World Wide Web Conferences Steering Committee, Republic and Canton of Geneva, CHE (2017). https://doi.org/10.1145/3038912.3052580
12. Zhao, J., Bhatt, S., Thille, C., Zimmaro, D., Gattani, N.: Interpretable personalized knowledge tracing and next learning activity recommendation. In: Proceedings of the Seventh ACM Conference on Learning @ Scale, pp. 325–328. L@S 2020. Association for Computing Machinery, New York (2020). https://doi.org/10.1145/3386527.3406739

Identifying Struggling Students by Comparing Online Tutor Clickstreams

Ethan Prihar[✉], Alexander Moore, and Neil Heffernan

Worcester Polytechnic Institute, Worcester, MA 01609, USA
ebprihar@wpi.edu

Abstract. New ways to identify students in need of assistance are imperative to the evolution of online tutoring platforms. Currently implemented models to identify struggling students use costly and tedious classroom observation paired with student's platform usage, and are often suitable for only a subset of students. With the recent influx of new students to online tutoring platforms due to COVID-19, a simple method to quickly identify struggling students could help facilitate effective remote learning. To this end, we created an anomaly detection algorithm that models the normal behavior of students during remote learning and recognizes when students deviate from this behavior. We demonstrated how anomalous behavior revealed which students needed additional assistance and predicted student learning outcomes.

Keywords: Online learning · Tutoring · Unsupervised learning · Anomaly detection · Outlier detection

1 Introduction

Finding patterns in student behavior that correlate negatively with learning is often costly, requiring professional observers to watch students as they complete assignments [1,8,10,13]. Algorithms created to identify these behaviors can be biased toward correctly identifying patterns in select populations [3] and can provide too specific or too great a quantity of information to be practically deployed by an instructor to help their students [8]. Furthermore, a model that requires expensive labeled data is unlikely to be updated often, which introduces model bias as populations and use cases change over time.

We would like to thank multiple NSF grants (e.g., 1917808, 1931523, 1940236, 1917713, 1903304, 1822830, 1759229, 1724889, 1636782, 1535428, 1440753, 1316736, 1252297, 1109483, & DRL-1031398), as well as the US Department of Education for three different funding lines; a) the Institute for Education Sciences (e.g., IES R305A170137, R305A170243, R305A180401, R305A120125, R305A180401, & R305C100024), b) the Graduate Assistance in Areas of National Need program (e.g., P200A180088 & P200A150306), and c) the EIR. We also thank the Office of Naval Research (N00014-18-1-2768), Schmidt Futures, and anonymous philanthropy.

© Springer Nature Switzerland AG 2021
I. Roll et al. (Eds.): AIED 2021, LNAI 12749, pp. 290–295, 2021.
https://doi.org/10.1007/978-3-030-78270-2_52

These common problems have been exacerbated by recent events. COVID-19 has led to an unprecedented demand for remote learning [17] and within the online learning platform ASSISTments [7] the number of users has grown tenfold since schools have switched to teaching remotely. Many students and teachers who have made the transition to remote learning have not previously used an online tutoring platform. This can cause inequity in students' quality of learning due to a lack of available resources and access to technology in lower income districts, exacerbating the achievement gap [5,11,12].

Unsupervised anomaly detection algorithms are a quickly trainable and deployable method to support instructors during this transition. Anomaly detection can identify unusual student clickstream patterns without needing a labelled dataset. This mitigates the time, expense, and subjectivity associated with manual classroom observation. Once trained, the model can be used to alert instructors when students are behaving abnormally and allow the instructor to assist the students as they see fit.

We define our objectives as follows:

1. Train a model capable of predicting student behavior using only students' clickstream data.
2. Use the student behavior model to identify abnormally behaving students.
3. Investigate the extent to which our measure of anomalous behavior correlates with learning outcomes and engagement.

2 Methodology

In order to identify anomalous students, we first trained a model to predict typical student behavior and then used the error in the model's predictions to identify students behaving anomalously. In the following sections we provide details on the data available for model training and evaluation, the structure of the models, and the model's training and validation process.

2.1 Data Processing

The data used in this work comes from ASSISTments, an online learning platform [7]. Within ASSISTments every action a student takes is recorded. The action records consist of action-timestamp pairs grouped by student and assignment. Working with this clickstream data is an extremely low-level interpretation of students' interactions with ASSISTments; it does not contain additional information such as features of the student, classroom, learning material, or past performance.

Only actions from Skill Builder assignments were used to train the model. Skill Builders are assignments in ASSISTments in which students answer a sequence of problems addressing a single math skill until they answer three problems in a row correctly. Skill Builders were used for training because they have a consistent format and are unlikely to cause divergences in typical student behavior. The distribution of the number of actions taken in Skill Builders is a highly-skewed exponential distribution: almost all students took less than 50 actions to complete each of their assignments, but outlying observations show some students taking 100 to 400 actions.

2.2   The Behavior Prediction Model

For our anomaly detection algorithm to be successful, the behavior prediction model had to be complex enough to capture trends in student behavior, but not so complex that it became capable of predicting the behavior of abnormally behaving students as well. To find a suitable model, we trained a logistic regression [9], neural network [16], decision tree [15], and Bernoulli naïve Bayes classifier [18] to predict a student's next action, given only their previous action and the time since taking an action.

To prepare the clickstream data for model training, we formatted the data into previous-action next-action pairs. To prepare the time data for model training, the time since taking an action was binned into 10 discrete ranges of increasing length. The ranges of the time bins grow to parallel the distribution of time between actions. The models therefore had 21 binary inputs (11 one-hot encoded actions and 10 time bins) and 11 binary outputs (11 one-hot encoded next actions).

To evaluate model quality, 985,000 actions from 7,300 students were used in 5-fold cross validation. The average accuracy, ROC AUC [6], and Cohen's kappa [4] for each model was calculated and used to select the model used to identify anomalous students in the following evaluation.

2.3   Identification of Anomalous Students

The best model from the previous section, which was a logistic regression, was trained on all the data used in the 5-fold cross validation and was then used to predict the next action of 985,000 actions from 7,300 different students the model had never seen data from before. The average absolute error of the model's predictions across each student's actions became their "anomaly score". To determine if anomaly scores correlated with student performance, we calculated Spearman correlations [14] between the students' anomaly scores and their average correctness and time on task for all the problems the students completed in ASSISTments, excluding the assignments used to calculate their anomaly scores. Additionally, we investigated differences between students in the 95th percentile of anomaly scores, which we labeled "anomalous students", and the rest of the students, which we labeled "normal students". We investigated differences in the frequency of actions taken and the time spent waiting before and after taking actions.

3   Results

3.1   Behavior Prediction Model Evaluation

The four models trained to predict students' next actions all performed relatively well. Each model obtained at least an accuracy of 65%, ROC AUC of 0.94, and Cohen's kappa of 0.66. The best performing model was the logistic regression, with an accuracy of 71%, ROC AUC of 0.96, and Cohen's kappa of 0.67. For this reason, logistic regression was the model of choice to evaluate the relationship between anomaly score and student behavior, discussed in the following section.

3.2   The Behavior of Anomalous Students

The students' anomaly scores were reliably negatively correlated with average correctness ($r = -0.21$, $p < 0.001$) and time on task ($r = -0.04$, $p < 0.001$). Students with higher anomaly scores took only slightly less time than students with lower anomaly scores, but got significantly more problems wrong. These results could indicate that students with high anomaly scores have more difficulty learning the material, or exhibit more gaming behavior [2]. This is an encouraging implication as it indicates that anomaly score could be used to inform teachers of struggling students in their classes.

Additionally, wrong answers occurred 60% more frequently and correct responses occurred 32% less frequently in anomalous students' action sequences. The time a student waited before and after they submitted a wrong answer or received tutoring was also significantly different ($p < 0.05$) between normal and anomalous students. Anomalous students spent about 20 s less looking at a problem before requesting tutoring or submitting a wrong answer. Then, anomalous students spent about 30 s less looking at the provided tutoring and about 50 s less thinking about their wrong responses before performing another action. These statistics paint the picture of a student that rushes to answer a problem, frequently submits wrong responses, and quickly requests tutoring. Then, without taking the time to rethink their answer, submits more wrong answers until they are able to move on. This behavior is essentially gaming [2], and would certainly be of interest to teachers as it is counterproductive to learning and should be corrected. Students' anomaly scores could therefore be a useful tool for identifying students in need of instructional intervention without having to define, or even be aware of, the specific kinds of negative behaviors of the students.

4   Conclusion

Students' anomaly scores, calculated only by comparing their clickstreams, negatively correlated with their average correctness and time on task. Additionally, anomalous students spent significantly less time thinking about a problem before getting the answer wrong or requesting tutoring, and once they were told they got the answer wrong or shown tutoring, they spent significantly less time before attempting the problem again. Using ASSISTments data, the anomaly detection algorithm was able to identify a common mode in unusual student behavior: rushing to complete assignments without trying to learn, i.e., gaming [2]. While this algorithm has the potential to be used to inform teachers in real time if their students need assistance, the behaviors identified as anomalous must be examined before choosing how to address them, lest students receive irrelevant interventions because of an incorrect assumption of what it means to be anomalous.

References

1. Baker, R.S., Corbett, A.T., Koedinger, K.R., Wagner, A.Z.: Off-task behavior in the cognitive tutor classroom: when students "game the system". In: Proceedings of the SIGCHI Conference on Human Factors in Computing Systems, pp. 383–390 (2004)
2. Baker, R.S.J., Mitrović, A., Mathews, M.: Detecting gaming the system in constraint-based tutors. In: De Bra, P., Kobsa, A., Chin, D. (eds.) UMAP 2010. LNCS, vol. 6075, pp. 267–278. Springer, Heidelberg (2010). https://doi.org/10.1007/978-3-642-13470-8_25
3. Botelho, A.F., Baker, R.S., Heffernan, N.T.: Improving sensor-free affect detection using deep learning. In: André, E., Baker, R., Hu, X., Rodrigo, M.M.T., du Boulay, B. (eds.) AIED 2017. LNCS (LNAI), vol. 10331, pp. 40–51. Springer, Cham (2017). https://doi.org/10.1007/978-3-319-61425-0_4
4. Cohen, J.: A coefficient of agreement for nominal scales. Educ. Psychol. Measur. **20**(1), 37–46 (1960)
5. DeWitt, P.: Teachers work two hours less per day during Covid-19: 8 key edweek survey findings. Education Week (2020)
6. Fawcett, T.: An introduction to ROC analysis. Pattern Recogn. Lett. **27**(8), 861–874 (2006)
7. Heffernan, N.T., Heffernan, C.L.: The assistments ecosystem: building a platform that brings scientists and teachers together for minimally invasive research on human learning and teaching. Int. J. Artif. Intell. Educ. **24**(4), 470–497 (2014)
8. Holstein, K., McLaren, B.M., Aleven, V.: Student learning benefits of a mixed-reality teacher awareness tool in AI-enhanced classrooms. In: Penstein Rosé, C., et al. (eds.) AIED 2018. LNCS (LNAI), vol. 10947, pp. 154–168. Springer, Cham (2018). https://doi.org/10.1007/978-3-319-93843-1_12
9. Hosmer Jr., D.W., Lemeshow, S., Sturdivant, R.X.: Applied Logistic Regression, vol. 398. Wiley, Hoboken (2013)
10. Lehman, B., Matthews, M., D'Mello, S., Person, N.: What are you feeling? Investigating student affective states during expert human tutoring sessions. In: Woolf, B.P., Aimeur, E., Nkambou, R., Lajoie, S. (eds.) International Conference on Intelligent Tutoring Systems, pp. 50–59. Springer, Heidelberg (2008). https://doi.org/10.1007/978-3-540-69132-7_10
11. Levinson, M., Cevik, M., Lipsitch, M.: Reopening primary schools during the pandemic (2020)
12. Middleton, K.V.: The longer-term impact of Covid-19 on k-12 student learning and assessment. Issues and Practice, Educational Measurement (2020)

13. Pardos, Z.A., Baker, R.S., San Pedro, M.O., Gowda, S.M., Gowda, S.M.: Affective states and state tests: investigating how affect throughout the school year predicts end of year learning outcomes. In: Proceedings of the Third International Conference on Learning Analytics and Knowledge, pp. 117–124 (2013)
14. Schober, P., Boer, C., Schwarte, L.A.: Correlation coefficients: appropriate use and interpretation. Anesthesia Analgesia **126**(5), 1763–1768 (2018)
15. Steinberg, D., Colla, P.: Cart: classification and regression trees. Top Ten Algorithms Data Mining **9**, 179 (2009)
16. Svozil, D., Kvasnicka, V., Pospichal, J.: Introduction to multi-layer feed-forward neural networks. Chemom. Intell. Lab. Syst. **39**(1), 43–62 (1997)
17. UNESCO: 290 million students out of school due to Covid-19: Unesco releases first global numbers and mobilizes response. UNESCO (2020)
18. Zhang, H.: Exploring conditions for the optimality of Naive Bayes. Int. J. Pattern Recognit. Artif. Intell. **19**(02), 183–198 (2005)

Exploring Dialogism Using Language Models

Stefan Ruseti[1], Maria-Dorinela Dascalu[1], Dragos-Georgian Corlatescu[1],
Mihai Dascalu[1,2(✉)], Stefan Trausan-Matu[1,2], and Danielle S. McNamara[3]

[1] University Politehnica of Bucharest, 313 Splaiul Independentei,
060042 Bucharest, Romania
{stefan.ruseti,dorinela.dascalu,dragos.corlatescu,
mihai.dascalu,stefan.trausan}@upb.ro
[2] Academy of Romanian Scientists, Str. Ilfov, Nr. 3, 050044 Bucharest, Romania
[3] Department of Psychology, Arizona State University, PO Box 871104,
Tempe, AZ, USA
dsmcnama@asu.edu

Abstract. Dialogism is a philosophical theory centered on the idea that
life involves a dialogue among multiple voices in a continuous exchange
and interaction. Considering human language, different ideas or points
of view take the form of voices, which spread throughout any discourse
and influence it. From a computational point of view, voices can be
operationlized as semantic chains that contain related words. This study
introduces and evaluates a novel method of identifying semantic chains
using BERT, a state-of-the-art language model for computational linguistics. The resulting model generalizes to multiple relations including
repetitions, semantically related concepts from WordNet (i.e., synonyms,
hypernyms, hyponyms, and siblings), as well as pronominal resolutions.
By combining the attention scores between words, word pairs are merged
into connected components that denote emerging voices from the discourse. The introduced visualization argues for a more dense capturing
of inner semantic links between words and even compound words in contrast to classical methods of building lexical chains.

Keywords: Dialogism · Semantic chains · Language models

1 Introduction

Dialogism is a philosophical theory introduced by Mikhail Bakhtin [1,2], centered
on the idea that everything, even life, is dialogic, a continual exchange and
interaction between voices: "Life by its very nature is dialogic... when dialogue
ends, everything ends" [2]. Trausan-Matu et al. [3] extended the concept of
voice for analyzing discourse, in general, and collaborative learning, in particular.
They consider voices to be generalized representations of different points of view
or ideas, which spread throughout the discourse, and influence it. Voices were
subsequently operationalized by Dascalu et al. [4] as semantic chains that were

I. Roll et al. (Eds.): AIED 2021, LNAI 12749, pp. 296–301, 2021.
https://doi.org/10.1007/978-3-030-78270-2_53

obtained by combining lexical chains, i.e., sequences of repeated or related words, including synonyms or hypernyms [5]. Semantic chains propagate along sentences and help create narrative threads throughout the text.

Recent studies on building lexical chains consider word repetitions, synonyms, and semantic relationships between nouns [6]. Mukherjee et al. [6] used lexical chains to distinguish easy from difficult medical texts. Identifying lexical chains that signal a difficult sentence helps in the simplification process. Olena [7] proposed a method for identifying lexical chains based on graphs, in which the nodes represent the terms in the document, and the edges the semantic relations between them. More recently, Ruas et al. [8] combined lexical chains with word embeddings to classify documents.

We introduce and evaluate a novel operationalization of voices using BERT [9], a state-of-the-art language model. This model enhances even further the Cohesion Network Analysis graph from the ReaderBench framework [10,11] by integrating semantic links of related concepts, indicative of semantic flow [12].

2 Method

A specific dataset with examples of links was required to identify the attention heads from BERT capable of detecting semantic links between words that belong to the same chain. A set of simple heuristics were used to extract links from sample texts, for all pairs of words tagged as noun, verb, or pronoun that fulfil one of the following conditions: a) repetitions of words having the same lemma; b) synonyms, hypernyms, or siblings in the WordNet taxonomy [13]; and c) coreferences identified using spaCy[1]. The TASA corpus[2] was selected as reference due to its diversity and covered complexity levels. The "correct" pairs were extracted from the entire dataset using the previous rules, while the "incorrect" ones were randomly sampled with 10% probability from all pairs of words that were not selected (i.e., otherwise, the number of negative samples would have been one order of magnitude larger than "correct" semantic associations). In total, 49 million word pairs were extracted, out of which around 20 million were positive examples.

Transformer-based models, in particular BERT [9], build contextual representations of words by stacking multi-head attention layers. Besides state-of-the-art results obtained on a vast range of tasks in Natural Language Processing, these models also provide insights regarding the importance of words and the relations between them by looking at the attention values. Clark et al. [14] explored the interpretability of different attention heads from different layers of the BERT model. The authors show that attention heads can be used to identify specific syntactic functions or perform coreference resolution.

No single attention head is accurate enough to predict these kinds of semantic relationships between words. Therefore, a prediction model that learns to combine the attention values from all the attention heads between two words was

---

[1] https://www.spacy.io, Retrieved April 15th, 2021.

[2] http://lsa.colorado.edu/spaces.html, Retrieved April 15th, 2021.

trained on the dataset constructed based on TASA. By considering both directions of the attention heads, 288 scores were used in total, similar to the approach
used by Clark et al. [14]. An issue to be tackled was the limited sequence length
accepted by the pretrained BERT model (i.e., 512 tokens). Texts in the TASA
dataset, but also in general, can be longer; thus, a sliding window was used to
compute the attention weights for all pairs of words. The sliding window had
a length of 256 for efficiency reasons, but also because semantic chains usually
do not contain links that are too far apart. An overlap of 128 tokens was used
so that words on different sides of the window could still be connected; if two
different attention values are computed between the same two words (because
of this overlap), the maximum value was used as the weight.

The previously described prediction model was used to score all pairs of words
that are within a given distance in the text. The next step consisted of grouping
these pairs of words into sets of semantically related words, i.e., semantic chains.
In order to filter the links based on the predicted weight, a fixed threshold
was experimentally set at 0.90. The semantic chains are selected in the form of
connected components from the resulting graph.

3   Results

Different architectures for identifying semantic links were trained and evaluated:
a linear model that only computes one weight for each attention head, and Multi-
Layer Perceptron (MLP) with one or two hidden layers. All models return one
number passed through a Sigmoid activation (see Table 1).

Table 1. Link prediction results.

Model	Hidden layer size	Accuracy (%)
Linear	–	79.75
MLP	16	85.67
MLP	32	86.24
MLP	64	86.65
MLP	64, 64	87.43
MLP	128, 64	87.99

An interactive view developed using Angular 6 (https://angular.io) was introduced to display the semantic chains - see Fig. 1 for a text selected from the
dataset described in McNamara et al. [15]. Each sentence is represented in a row,
while rows are grouped in their corresponding paragraph. Words and links from
a semantic chain share the same color. A higher density of the chains extracted
with our method can be observed in contrast to classical lexical chains. Surprising relations not present in the constructed dataset can be seen in the generated

chains. The linear model found connections between "colonists" and "Boston", or between "help" and "supplies", while the MLP model identified connections between "British" and "Great Britain" as a compound word. This example also shows that choosing the best model between linear and MLP is not straightforward, despite the substantial performance improvement of the latter on the word pairs dataset. Even though the linear model cannot perfectly learn the simple heuristics used to build the initial dataset, it can retrieve new insightful connections between words.

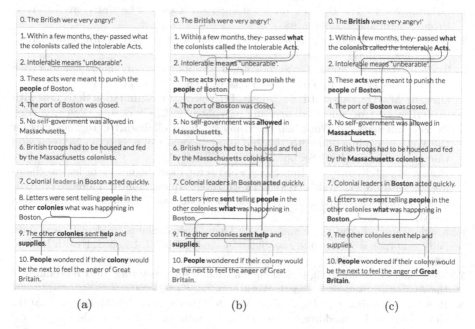

(a)   (b)   (c)

Fig. 1. Visualizations of a) lexical chains [5], b) semantic chains using the linear model, and c) semantic chains using the MLP model.

4 Conclusions

A novel method for identifying semantic links is introduced using only the attention scores computed by BERT, a core task for operationalizing dialogism as a discourse model. Choosing which attention heads are relevant for this task and how to combine them was achieved by building a dataset with pairs of words with simple rules. The introduced visualization argues for a more dense capturing of inner semantic links between words and even compound words, which are quite sparse when considering manually defined synsets from WordNet. Our aim is to further extend this model with sentiment analysis features derived from local contexts captured by BERT, thus further enriching the analysis with the identification of convergent and divergent points of view.

Acknowledgments. This research was supported by a grant of the Romanian National Authority for Scientific Research and Innovation, CNCS – UEFISCDI, project number TE 70 PN-III-P1-1.1-TE-2019-2209, ATES – "Automated Text Evaluation and Simplification", the Institute of Education Sciences (R305A180144 and R305A180261), and the Office of Naval Research (N00014-17-1-2300; N00014-20-1-2623). The opinions expressed are those of the authors and do not represent views of the IES or ONR.

References

1. Bakhtin, M.M.: The Dialogic Imagination: Four Essays. The University of Texas Press, Austin and London (1981)
2. Bakhtin, M.M.: Problems of Dostoevsky's Poetics. University of Minnesota Press, Minneapolis (1984)
3. Trausan-Matu, S., Stahl, G., Sarmiento, J.: Supporting polyphonic collaborative learning. E-service J. Indiana Univ. Press **6**(1), 58–74 (2007)
4. Dascalu, M., Trausan-Matu, S., Dessus, P.: Voices' inter-animation detection with readerbench - modelling and assessing polyphony in CSCL chats as voice synergy. In: 2nd International Workshop on Semantic and Collaborative Technologies for the Web, in conjunction with the 2nd International Conference on Systems and Computer Science (ICSCS), pp. 280–285. IEEE (2013)
5. Galley, M., McKeown, K.: Improving word sense disambiguation in lexical chaining. In: Gottlob, G., Walsh, T. (eds.) 18th International Joint Conference on Artificial Intelligence (IJCAI 2003), pp. 1486–1488. Morgan Kaufmann Publishers, Inc. (2003)
6. Mukherjee, P., Leroy, G., Kauchak, D.: Using lexical chains to identify text difficulty: a corpus statistics and classification study. IEEE J. Biomed. Health Inform. **23**(5), 2164–2173 (2018)
7. Medelyan, O.: Computing lexical chains with graph clustering. In: Proceedings of the ACL 2007 Student Research Workshop, pp. 85–90 (2007)
8. Ruas, T., Ferreira, C.H.P., Grosky, W., de França, F.O., de Medeiros, D.M.R.: Enhanced word embeddings using multi-semantic representation through lexical chains. Inf. Sci. (2020)
9. Devlin, J., Chang, M.W., Lee, K., Toutanova, K.: Bert: Pre-training of deep bidirectional transformers for language understanding. arXiv preprint arXiv:1810.04805 (2018)
10. Dascălu, M.: Analyzing Discourse and Text Complexity for Learning and Collaborating. SCI, vol. 534. Springer, Cham (2014). https://doi.org/10.1007/978-3-319-03419-5
11. Dascalu, M., Trausan-Matu, S., McNamara, D., Dessus, P.: Readerbench - automated evaluation of collaboration based on cohesion and dialogism. Int. J. Comput.-Support. Collab. Learn. **10**(4), 395–423 (2015)
12. O'Rourke, S., Calvo, R.: Analysing semantic flow in academic writing. In: Dimitrova, V., Mizoguchi, R., du Boulay, B., Graesser, A. (eds.) Artificial Intelligence in Education. Building Learning Systems That Care: From Knowledge Representation to Affective Modelling (AIED 2009), pp. 173–180. IOS Press, Amsterdam, The Netherlands (2009)
13. Miller, G.A.: Wordnet: a lexical database for English. Commun. ACM **38**(11), 39–41 (1995)

14. Clark, K., Khandelwal, U., Levy, O., Manning, C.D.: What does BERT look at? An analysis of BERT's attention. In: Proceedings of the 2019 ACL Workshop BlackboxNLP: Analyzing and Interpreting Neural Networks for NLP, pp. 276–286 (2019)
15. McNamara, D.S., Louwerse, M.M., McCarthy, P.M., Graesser, A.C.: Coh-metrix: capturing linguistic features of cohesion. Discourse Process. **47**(4), 292–330 (2010)

EduPal Leaves No Professor Behind: Supporting Faculty via a Peer-Powered Recommender System

Nourhan Sakr$^{(\boxtimes)}$, Aya Salama , Nadeen Tameesh , and Gihan Osman

The American University in Cairo, Cairo, Egypt
n.sakr@columbia.edu, {aya_salama,nadeentameesh14,gosman}@aucegypt.edu

Abstract. The swift transitions in higher education after the COVID-19 outbreak identified a gap in the pedagogical support available to faculty. We propose a smart, knowledge-based chatbot that addresses issues of knowledge distillation and provides faculty with personalized recommendations. Our collaborative system crowdsources useful pedagogical practices and continuously filters recommendations based on theory and user feedback, thus enhancing the experiences of subsequent peers. We build a prototype for our local STEM faculty as a proof concept and receive favorable feedback that encourages us to extend our development and outreach, especially to underresourced faculty.

Keywords: AI Chatbots · Knowledge-based recommender system · User-centric design · Personalization · Crowdsourcing · Collaborative filtering

1 Background and Related Work

The COVID-19 lock-down forced many higher education institutions globally to continue instruction via online modalities at an unprecedented pace and scale [5,10]. With many faculty scantily trained in teaching strategies or with little support on best online practices [4,10,22], instruction was maintained at the cost of education quality, equity and sound pedagogy [3,5,18]. Online education requires deliberate design and development [5,10,16,22], yet, the pandemic forced the adoption of *emergency remote learning*, regardless of any obstacles.

In non-emergency times, faculty in resourced institutions are often supported by instructional designers who provide personalized guidance on making sound design and technology decisions for the faculty's particular context [1]. However, given the sheer number of "overnight" transitions, individualized help became rather challenging [10]. The pandemic revealed the lacking capacities for support and infrastructure in institutions [3,11,14,21], thereby questioning readiness for the digital era. Looking further into under-resourced institutions, general capacity building and high-quality instructional guidance are considered a luxury.

In light of this extreme global test, we identify a gap in the pedagogical support available to educators. Social media and online webinars attempted closing

I. Roll et al. (Eds.): AIED 2021, LNAI 12749, pp. 302–307, 2021.
https://doi.org/10.1007/978-3-030-78270-2_54

this gap by providing platforms for sharing experiences and sound tips online. However, we see three issues with such channels: They are (1) less personalized, (2) suffer from information overflow and (3) are not guaranteed to continue after the pandemic. These issues make it hard for some faculty to find relevant resources or apply what they find to their personalized contexts. Based on our survey ($n = 103$), 86% believed that being able to readily access relevant experiences shared by peers would be beneficial to them, even in the long run.

Within this framework, we propose, EDUPAL, a virtual educational consultant. Our user-centric design provides a crowd-sourcing platform augmented with collaborative filtration to automate experience sharing and knowledge distillation. We wrap these within a recommendation system that provides personalized, context-aware guidance on best-fit pedagogical practices, as supported by research theory and faculty practice. As a proof of concept, EDUPAL was customized for our local STEM community at the American University in Cairo (AUC), classifies as an M1 university[1]. In this paper, we present our pilot's data collection methodology, system design and show positive user feedback declaring the system as promising for extension and generalization.

2   Data-Driven Modeling

Our data-driven design builds on a taxonomy that is the product of an elaborate data collection process, outlined briefly in this section. Despite our focus on supporting STEM faculty at AUC (for the pilot study), we consider various populations to build our data. We apply maximum variation sampling in recruiting participants and conclude any stage when the research team agrees on information saturation. Our findings heavily rely on qualitative analysis. Our secondary research builds on education and psychology literature, as well as social media narratives from all stakeholders, i.e. faculty, students and instructional designers.

Community Feedback. The first stage distills knowledge from 50 hours of semi-structured interviewing of faculty at AUC, spanning all schools. Faculty reflected on their teaching challenges, need for pedagogical support, and practices most effective for their specific class types . Results were augmented with secondary research to determine sets of: 1) features that identify profiles of instructors/classes and 2)pedagogical practices and technology tools that best fit each profile.

Filtration and Validation for STEM. The second stage refines the features and recommendations to those most applicable to STEM courses via a semi-open survey aimed at STEM faculty at AUC ($n = 100$). We, then, ran two seminars for STEM faculty at AUC ($n = 50$) and two seminars for instructional designers ($n = 14$). Our recommendations were presented for discussions regarding their viability and feasibility. This process seeds our recommendations bank.

---

[1] According to the Carnegie Classification of Institutions of Higher Education.

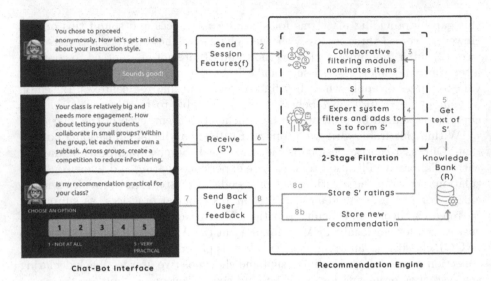

Fig. 1. System components and flow. (1) EDUPAL collects session features, f (2) f are sent to the recommendation engine (3) Collaborative filtering module selects S, from the knowledge bank(R) based on f (4) Expert system refines S, forming S' (5) S' text retrieved from R (6) Items in S' presented sequentially to the user (7) Ratings are assigned to S' by the user (8a) Ratings and (8b) new recommendations are stored

Learning from Experts. Finally, we conducted semi-structured interviews with global educational consultants ($n = 9$, 60 mins each). The interviews simulated a pedagogical consultation followed by a discussion on the usual process. The goal was to observe and model the thought process of an instructional designer during a consultation and identify the features they look for to formulate a recommendation. This model sets the question flow in EDUPAL.

3   Recommender System Design

EDUPAL is an instance of knowledge-based recommender systems [2,6,7,12,15, 19,20]. Via a chatbot interface, the user (assumed to be an instructor) interacts with the system as if talking to an instructional designer. The conversational session collects *session features* (f) about the user and their course. The design encompasses a two-staged pipeline that allows users to benefit from the perspectives of practicing peers and pedagogical experts, as well as rate and suggest recommendations, thus automating experience sharing. The proposed recommendations are meant to enhance the three class room interaction modes defined in [17]. Communication with the server is secure and confidential. An "anonymous" mode is added to maintain privacy of users who do not want to share their information. Figure 1 depicts the system components and flow.

Collaborative Filtering Based on Educators Feedback. A user-based collaborative filtering approach [20] is used to compile an initial set of recommendations. When a new session starts, f is collected and cosine similarity [8] is used to identify the most similar users and extract the set of recommendations S, top rated by those users. This approach is suitable only for recommendations that have received ratings, while unrated recommendations are sent to the *expert system*, which is authorized to update the knowledge bank R.

Expert System. This module mimics the decision making processes of an instructional designer and addresses the cold start problem[13]. It can be considered as a symbolic AI system where rules are constructed based on feedback from subject matter experts [9]. It ensures that the recommendations are not only based on popularity but are also pedagogically sound with support from research and practice. For each recommendation $r \in R$, experts identify the factors defined in f that should match r with a user, based on the learning sciences. Those rules are then translated into system logic, where, each element $s \in S$ is either accepted as part of the final set, S', or rejected. This logic is also used to add recommendations, deemed as best fit by experts, to S'.

Feedback. Each recommendation in S' is presented in a conversational format and rated by the user. The ratings are fed-back to the system to inform future selections[2]. Finally, the user may share other effective pedagogical practices, that are considered for extending R and fully streamlining experience sharing.

4   Prototype Evaluation and Conclusion

AUC faculty[3] ($n = 10$) evaluated the prototype via 30-minute usability tests. We first learned about (1) their means and frequency of seeking help with pedagogical matters and (2) their experience with chatbots. They, then used EDUPAL and provided ratings on the overall quality of the experience and the received recommendations. On a Likert scale out of 5, the mean responses to those questions were 3.8 & 3.7, respectively. Testers were also asked to share what they liked and disliked about EDUPAL as well as its advantages and disadvantages over their default methods of seeking help. The majority found EDUPAL beneficial and user-friendly. They highlighted fast feedback and constant availability as immediate advantages over other aid methods. The chatbot interface made their experience feel interactive, engaging and personalized. Our recommendations were deemed of good quality but users suggested providing more specific examples for application. Lastly, EDUPAL's anonymous mode provided a "safe zone" for instructors who usually avoid sharing experiences or asking for help.

Based on the favorable feedback received, we conclude that EDUPAL shows a successful pilot system and is worth generalizing to address global knowledge distillation, experience sharing automation, recommendation personalization and

---

[2] Future updates will verify that the user is faculty and that their recommendation is supported by research before incorporating their feedback/rating.

[3] 80% were STEM, 70% were female, experience ranged from 2 to 32 years.

support scalability, in addition to promoting fairness in access to resources, given that EDUPAL is available 24/7 for free. We also recognize the potential of EDU-PAL becoming a screening tool for educational consultations, thus augmenting the impact of instructional designers and making appointments run faster in times of high demand, e.g. during the pandemic.

References

1. Beirne, E., Romanoski, M.P.: Instructional design in higher education: defining an evolving field. An environmental scan of the digital learning landscape, OLC outlook (2018)
2. Burke, R.: Knowledge-based recommender systems. Encyclopedia Libr. Inf. Syst. **69**(Supplement 32), 175–186 (2000)
3. Crawford, J., et al.: Covid-19: 20 countries' higher education intra-period digital pedagogy responses. J. Appl. Learn. Teaching **3**(1), 1–20 (2020)
4. Cutri, R.M., Mena, J.: A critical reconceptualization of faculty readiness for online teaching. Distance Educ. **41**(3), 361–380 (2020)
5. Czerniewicz, L.: What we learnt from "going online" during university shutdowns in South Africa, March 2020. https://philonedtech.com/what-we-learnt-from-going-online-during-university-shutdowns-in-south-africa/
6. García, E., Romero, C., Ventura, S., De Castro, C.: An architecture for making recommendations to courseware authors using association rule mining and collaborative filtering. User Model. User-Adap. Inter. **19**(1–2), 99–132 (2009)
7. Garcia-Martinez, S., Hamou-Lhadj, A.: Educational recommender systems: a pedagogical-focused perspective. Multimedia Services in Intelligent Environments, pp. 113–124 (2013)
8. Han, J., Kamber, M., Pei, J.: 2 - getting to know your data. In: Han, J., Kamber, M., Pei, J. (eds.) Data Mining (Third Edition), pp. 39–82. The Morgan Kaufmann Series in Data Management Systems, 3rd edn. Morgan Kaufmann, Boston (2012). https://doi.org/10.1016/B978-0-12-381479-1.00002-2, https://www.sciencedirect.com/science/article/pii/B9780123814791000022
9. Haugeland, J.: Artificial Intelligence: The Very Idea. MIT press (1989)
10. Hodges, C., et al.: The difference between emergency remote teaching and online learning. Educause Rev. **27**, 1–12 (2020)
11. Kimmons, R., Veletsianos, G., VanLeeuwen, C.: What (some) faculty are saying about the shift to remote teaching and learning, May 2020. https://er.educause.edu/blogs/2020/5/what-some-faculty-are-saying-about-the-shift-to-remote-teaching-and-learning
12. Klašnja-Milićević, A., Ivanović, M., Nanopoulos, A.: Recommender systems in e-learning environments: a survey of the state-of-the-art and possible extensions. Artif. Intell. Rev. **44**(4), 571–604 (2015)
13. Lam, X.N., Vu, T., Le, T.D., Duong, A.D.: Addressing cold-start problem in recommendation systems. In: Proceedings of the 2nd International Conference on Ubiquitous Information Management and Communication, pp. 208–211 (2008)
14. Leung, M., Sharma, Y.: Online classes try to fill education gap during epidemic, February 2020. https://www.universityworldnews.com/post.php?story=2020022108360325
15. Mahmood, T., Ricci, F.: Improving recommender systems with adaptive conversational strategies. In: Proceedings of the 20th ACM Conference on Hypertext and Hypermedia, pp. 73–82 (2009)

16. Means, B., Bakia, M., Murphy, R.: Learning Online: What Research Tells Us About Whether, When and How. Routledge (2014)
17. Moore, M.G.: Three types of interaction (1989)
18. Motala, S., Menon, K.: In search of the 'new normal': reflections on teaching and learning during Covid-19 in a South African University. Southern African Rev. Educ. **26**(1), 80–99 (2020)
19. Ramadoss, B., Balasundaram, S.R.: Management and selection of visual metaphors for courseware development in web based learning. In: 2006 IEEE Conference on Cybernetics and Intelligent Systems, pp. 1–6 (2006)
20. Ricci, F., Rokach, L., Shapira, B.: Introduction to recommender systems handbook. In: Ricci, F., Rokach, L., Shapira, B., Kantor, P.B. (eds.) Recommender Systems Handbook, pp. 1–35. Springer, Boston, MA (2011). https://doi.org/10.1007/978-0-387-85820-3_1
21. Wu, Z.: How a top Chinese university is responding to coronavirus, March 2020. https://www.weforum.org/agenda/2020/03/coronavirus-china-the-challenges-of-online-learning-for-universities/
22. Xie, J., Rice, M.F.: Instructional designers' roles in emergency remote teaching during Covid-19. Distance Educ. 1–18 (2021)

Computer-Supported Human Mentoring for Personalized and Equitable Math Learning

Peter Schaldenbrand(✉) , Nikki G. Lobczowski , J. Elizabeth Richey ,
Shivang Gupta, Elizabeth A. McLaughlin , Adetunji Adeniran ,
and Kenneth R. Koedinger

Carnegie Mellon University, Pittsburgh, PA 15213, USA
pschalde@andrew.cmu.edu

Abstract. Computer tutor data indicate that more learning opportunities yield greater achievement, but also confirm there are gaps in the number and quality of opportunities marginalized students receive that technology alone does not address. Personalized learning with mentors can close this gap in opportunities but is expensive to implement. We introduce a free, web-based application, Personalized Learning^2 (PL^2), designed to improve mentoring efficiency by connecting mentors to intervention and instructional resources. Preliminary findings indicated that PL^2's categorization of students based on math learning software data enabled mentors to focus their efforts, and that mentors found PL^2 resources to positively expand how they taught and mentored.

Keywords: Personalized learning · Mentor augmentation · Motivational resources · Design-based research

1 Introduction

More than 60 years after the Supreme Court's ruling to desegregate schools, American K-12 education remains marred by strikingly inequitable access, opportunities, and learning outcomes across racial groups and income classes. These gaps are especially big in mathematics and they perpetuate inequalities across generations [11]. While these are long-standing problems, researchers have struggled to identify effective solutions. Research undertaken in public schools in high-poverty neighborhoods provides grounds for hope. A large randomized control trial demonstrated that one year of intensive, personalized human tutoring could significantly increase math achievement for minoritized students in high-poverty neighborhoods [2]. Unfortunately, these gains came at a substantial resource cost; with a tutor providing instruction to just two students per class period, the extra costs of nearly $4,000 per student are not feasible in many districts.

AIEd technologies can lower the cost of personalized tutoring and increase student achievement [9,10], but they are not sufficient. An increase in learning

© Springer Nature Switzerland AG 2021
I. Roll et al. (Eds.): AIED 2021, LNAI 12749, pp. 308–313, 2021.
https://doi.org/10.1007/978-3-030-78270-2_55

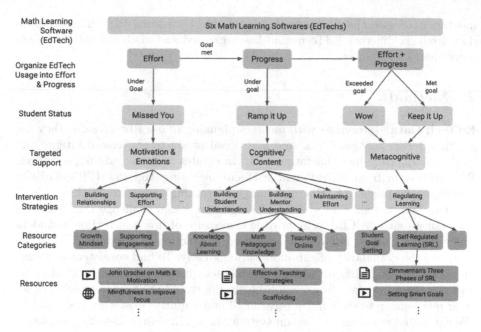

Fig. 1. PL^2 connects EdTech products to mentor and student assisting resources via a system of categorization of student data and resources based on intervention strategies.

opportunities, such as time spent making progress in math tutoring software, leads to an increase in achievement; however, marginalized students experience fewer of these opportunities for learning [5]. Many interventions aim to reduce opportunity gaps in math by addressing non-content related learner variables [1,4]. The conditions of each student and the extent to which their learning is affected is diverse [7,8,13], suggesting that there is no one-size-fits-all solution to the problem. Personalized learning, which is tailored to the social, material, and organizational needs of each child [14], may be the ideal solution, but it is not practical in terms of cost and availability of human resources for every child in the U.S. to receive one-on-one attention from a human tutor [3].

We introduce the Personalized Learning^2 (PL^2) application[1] which is designed to improve mentoring efficiency by recommending instructional resources curated from the Internet to mentors based on their students' usage of educational technology (EdTech) software. Figure 1 depicts how PL^2 connects EdTech data to resources, integrating smoothly into a mentor's workflow. PL^2 provides tools for easily navigating these resources and matching them to students' and mentors' needs. Since PL^2's initial deployment in Summer 2019, 148

---

[1] http://personalizedlearning2.org/.

mentors have used the system with a total of 814 students. PL^2 currently pulls data from six different EdTechs and has organized and made available to users more than 100 resources.

2 Method

EdTech Data. Interviews with mentors planning to use PL^2 revealed they use multiple technology products, and additional products can create fatigue and inefficiencies. To reduce this fatigue, PL^2 integrates mentors' existing softwares. PL^2 pulls data from six different EdTechs including McGraw Hill's ALEKS, Carnegie Learning's MATHia, and Imagine Learning's Imagine Math.

Data from the six EdTechs varies, creating a design challenge for presenting data consistently. All EdTechs provide a measure of how much time a student spent in the system, but not all measured completion of sub-units of curriculum, and the granularity of sub-units varied greatly. To find consistent measurements across EdTechs, we computed abstract quantities in the form of effort and progress using data from each EdTech. Effort and progress were selected for their relationship to students' motivational and cognitive obstacles, respectively. We calculated effort using time on system and curriculum sub-units completed, and progress using accuracy on the sub-units completed.

Resources. Internet resources can be helpful in addressing the opportunity gaps students face by supporting their self-efficacy [12], feelings of belonging [15], growth mindset [16], and utility value of STEM [6]. These resources are scattered across the web and therefore can be difficult to find, and mentors may not know what to search for. PL^2 was designed to help make sense of the unwieldy number of resources available on the internet by selecting, organizing, and summarizing relevant materials according to a three-tiered hierarchy: Strategy $\rightarrow$ Category $\rightarrow$ Resource (see Fig. 1). Strategies are the highest level in the structure for finding appropriate resources for an issue. For example, a mentor may see their student is not putting effort into their work and explore resources within the Supporting Effort strategy. This strategy has categories of resources including Growth Mindset and Supporting Engagement. Within each category there are existing resources (e.g., videos, links to external websites, papers, interactive activities). This structure supports varying degrees of mentor expertise to navigate through resources and allows mentors to create their own resources. Resource strategies were organized according to enabling conditions that were identified through interviews with PL^2 partners as candidate root causes for student success (see Fig. 2).

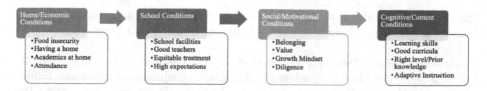

Fig. 2. Themes emerged from interviews indicating many enabling conditions that must be met for students to succeed.

Categorizing EdTech Usage. PL^2 organizes student usage into four categories called "student statuses" (See Fig. 1) that correspond to optimal intervention strategies: Missed You designates a failure to meet effort goals, indicating a motivational or emotional strategy is likely needed; Ramp It Up signifies meeting the effort goals but falling short on progress goals, indicating a potential need for a content or cognitive intervention; Wow represents students exceeding their goals, indicating that the student needs a more challenging goal; and Keep It Up is for students meeting their effort and progress goals, indicating that they are on track. As shown in Fig. 1, there is a hierarchical structure for calculating the student status, which is organized according to progression of the enabling conditions seen in Fig. 2. Motivational needs are prioritized over cognitive needs, and therefore effort is assessed prior to progress.

3 Results

The distribution of statuses in 2020 accounting for 3612 student-weeks in the EdTechs was 28.5% Missed You, 18.2% Ramp it Up, 40.2% Wow, and 13.1% Keep it Up, indicating the categorization strategy can detect variability in students' behaviors. In total, there are 58 resources designed for mentors, 40 for students, and 16 for either students or mentors that have been neatly organized into 3 methods of targeted support, 9 strategies, and 36 resource categories. Thus far, interviews with mentors using PL^2 have led to expressions of PL^2's ability to positively expand the way they teach and mentor by thinking about students in different ways, as illustrated by the following quote from a PL^2 mentor: "I like using the 'Parent Engagement' resource because that is one of the bigger problems I have in my district. It is a great resource that provides me with new/creative ideas on how to engage parents with their child's academics."

4 Conclusions and Future Work

The design of PL^2 engaged a community of mentors, teachers, and students and provided an example of design-based research that is not common in the AIEd community. PL^2 also exhibits a novel attempt at connecting multiple EdTech data streams and methods for comparing and using student data across EdTechs. Future work for the PL^2 project includes empirical studies and validating the efficacy of the application. We also plan to revise our measure of progress to include information about a student advancing through their curriculum.

Acknowledgment. This work is supported by the Chan Zuckerberg Initiative (CZI) Grant # 2018-193694. Any opinions, findings, and conclusions or recommendations expressed in this material are those of the authors and do not necessarily reflect the views of the Chan Zuckerberg Initiative.

References

1. Cohen, G.L., Garcia, J., Apfel, N., Master, A.: Reducing the racial achievement gap: a social-psychological intervention. Science **313**(5791), 1307–1310 (2006). https://doi.org/10.1126/science.1128317, https://science.sciencemag.org/content/313/5791/1307
2. Cook, P., et al.: Not too late: Improving academic outcomes for disadvantaged youth (2015)
3. Donnelly, C.: The use of case based multiple choice questions for assessing large group teaching: Implications on student's learning (2014)
4. Evans, G.W., Rosenbaum, J.: Self-regulation and the income-achievement gap. Early Childhood Res. Q . **23**(4), 504–514 (2008). https://doi.org/10.1016/j.ecresq.2008.07.002, https://www.sciencedirect.com/science/article/pii/S0885200608000549
5. Flores, A.: Examining disparities in mathematics education: achievement gap or opportunity gap? High School J. **91**(1), 29–42 (2007). http://www.jstor.org/stable/40367921
6. Harackiewicz, J.M., Rozek, C.S., Hulleman, C.S., Hyde, J.S.: Helping parents to motivate adolescents in mathematics and science: an experimental test of a utility-value intervention. Psychol. Sci. **23**(8), 899–906 (2012). https://doi.org/10.1177/0956797611435530, pMID: 22760887
7. Hattie, J.: Visible Learning: A Synthesis of Over 800 Meta-Analyses Relating to Achievement. Routledge (2009)
8. Kirschner, P.A., Sweller, J., Clark, R.E.: Why minimal guidance during instruction does not work: an analysis of the failure of constructivist, discovery, problem-based, experiential, and inquiry-based teaching. Educ. Psychol. **41**(2), 75–86 (2006). https://doi.org/10.1207/s15326985ep4102_1
9. Koedinger, K., Anderson, J., Hadley, W., Mark, M.: Intelligent tutoring goes to school in the big city. Int. J. Artif. Intell. Educ. **8**, 30–43 (1997)
10. Pane, J.F., Griffin, B.A., McCaffrey, D.F., Karam, R.: Effectiveness of cognitive tutor algebra i at scale. Educ. Eval. Policy Anal. **36**(2), 127–144 (2014). https://doi.org/10.3102/0162373713507480
11. Rose, H., Betts, J.: The effect of high school courses on earnings. Rev. Econ. Stat. **86**, 497–513 (2004). https://doi.org/10.1162/003465304323031076
12. Siegle, D., McCoach, D.B.: Increasing student mathematics self-efficacy through teacher training. J. Adv. Acad. **18**, 278–312 (2007)
13. Tang, Y., Liang, J., Hare, R., Wang, F.Y.: A personalized learning system for parallel intelligent education. IEEE Trans. Comput. Soc. Syst. **7**(2), 352–361 (2020). https://doi.org/10.1109/TCSS.2020.2965198

14. Vanbecelaere, S., et al.: Technology-mediated personalised learning for younger learners: concepts, design, methods and practice. In: Proceedings of the 2020 ACM Interaction Design and Children Conference: Extended Abstracts, pp. 126–134. IDC 2020, Association for Computing Machinery, New York (2020). https://doi.org/10.1145/3397617.3398059

15. Walton, G.M., Cohen, G.L.: A brief social-belonging intervention improves academic and health outcomes of minority students. Science **331**(6023), 1447–1451 (2011). https://doi.org/10.1126/science.1198364, https://science.sciencemag.org/content/331/6023/1447

16. Yeager, D.S., Dweck, C.S.: Mindsets that promote resilience: when students believe that personal characteristics can be developed. Educ. Psychol. **47**(4), 302–314 (2012). https://doi.org/10.1080/00461520.2012.722805

Internalisation of Situational Motivation in an E-Learning Scenario Using Gamification

Philipp Schaper[✉], Anna Riedmann, and Birgit Lugrin

Human-Computer Interaction, University of Wuerzburg, Würzburg, Germany
philipp.schaper@uni-wuerzburg.de

Abstract. Self-directed learning is of critical importance in adult learning, for example, when taking part in online courses or learning at universities. To work on a challenging topic continuously requires learners to self-motivate. By applying self-determination theory to address the basic psychological needs for competence, autonomy and relatedness, the internalisation of motivation can be fostered. We implemented a learning environment, which addresses these needs using gamification elements to scaffold situational motivation, and compared it with a control version in a user study to investigate the effect of the implemented gamification elements on the internalization of situational motivation. Our results show an internalization of situational motivation with significantly higher internalised and significantly lower extrinsic situational motivation in the gamified version relative to the control condition.

Keywords: E-learning · Motivation · Need satisfaction · Gamification

1 Introduction

Lifelong learning, especially in a self-directed manner is often associated with online courses [6] but this also affects learning in universities [21]. To voluntarily engage in self-directed learning can be challenging, because the learner has to be self-motivated which either requires intrinsic motivation, or having internalised extrinsic motivation, so called identified motivation. Identified motivation is also considered autonomous motivation, in which the person accepts the relevance and importance of the respective behavior [4]. To promote the internalisation of extrinsic motivation in learning, self-determination theory has been considered for several decades [8]. Internalisation occurs if the needs for competence, autonomy and relatedness are satisfied [8], which should in turn promote self-directed learning.

The use of gamification based on self-determination theory allows for a new perspective to address these basic psychological needs [24]. Our work is focused on short-term changes in motivation, i.e. situational motivation, due to adding gamification elements to a learning environment.

© Springer Nature Switzerland AG 2021
I. Roll et al. (Eds.): AIED 2021, LNAI 12749, pp. 314–319, 2021.
https://doi.org/10.1007/978-3-030-78270-2_56

2   Related Work

Self-determination theory is based on the idea that motivation is connected to the basic psychological needs for competence, autonomy and relatedness [8]. The theory suggests a continuum of motivation ranging from extrinsic motivation to intrinsic motivation, with identified and intrinsic motivation being the most self-determined and therefore autonomous motivation. Contexts, which support the fulfillment of the three basic psychological needs, can maintain or even enhance intrinsic and identified motivation, and support internalisation and integration of extrinsic motivation [8]. This indicates that behavior, which is initially motivated externally, is still displayed but now autonomously instead of based on external rewards [7]. To increase autonomous motivation in a learning scenario, it is therefore necessary to convey a sense of fulfillment for competence, autonomy and relatedness [9].

Gamification can be described as the implementation of game-specific elements in non-game contexts [11] to benefit the motivational state of learners [12]. This has already been used in several E-learning environments [5,15,16], especially for the acquisition of a second language [10,22].

However, reviews of gamification research by Nacke et al. [20] as well as Seaborn et al. [24] conclude that gamification in learning environments is often not linked to theoretical constructs and in some instances might even have negative effects [26].

Self-determination theory has explicitly been pointed out to address gamification in learning environments on a theoretical basis [2,23]. The underlying idea is to address psychological needs, via adequate gamification elements: Meaningful decisions for autonomy, competence by positive feedback and relatedness by integrating a narrative.

To the best of our knowledge, there is a research gap in the demonstration of internalisation of situational motivation, based on empirically measured need satisfaction. Studies (e.g. [23]) demonstrated effects of gamification elements on need satisfaction, but not on motivation measures. Other approaches (e.g. [3]) focused on non situational intrinsic motivation, which is clearly relevant but might not be equivalent to introducing internalisation in the short-term for a self-directed learning scenario.

The aim of the present study is to experimentally induce internalisation of situational motivation in an E-learning scenario using gamification. In particular, to lower situational extrinsic motivation while increasing situational identified, and possibly intrinsic motivation. By promoting the internalisation of situational motivation during interaction with a learning environment which satisfies the basic psychological needs, identified or even intrinsic motivation should be enhanced. These forms of motivation are considered to be autonomous motivation, which is a critical prerequisite for self-directed learning, which is focused on the active role of the learner in the learning process [1,13,19]. Therefore, we conducted a user study to empirically test differences in need satisfaction and situational motivation, compared to a control version without gamification.

Addressing the basic psychological needs should result in internalisation of situational motivation. Therefore, we formulate and test the following hypotheses:

- H1a: Higher need satisfaction (competence) in the gamified version
- H1b: Higher need satisfaction (autonomy) in the gamified version
- H1c: Higher need satisfaction (relatedness) in the gamified version
- H2a: Situational intrinsic motivation is higher in the gamified version
- H2b: Situational identified motivation is higher in the gamified version
- H2c: Situational extrinsic motivation is lower in the gamified version.

3 Learning Environment

We implemented a web-based learning environment for learning Spanish, including learning materials, and predefined questions and answers which were completed in a predefined sequence, separated into an introduction, a learning phase and a knowledge test. For the *introduction*, four questions for a simple conversation can be read and heard. The *learning phase* contained three topics: (1) nine leisure activities (2) 13 names of different food items (3) numbers from one to 20. Clicking on the respective word or numeral resulted in the word being read in Spanish and in German. After each topic, a quiz including feedback was used to repeat the vocabulary. The *knowledge test* was separated in the same three topics with four questions for the first two topics (food and numbers) and six for the last topic (activities), all including feedback.

For the gamified version, we selected adequate game elements to address the corresponding psychological needs [23]: A virtual character named 'Sabina' was used to implement a narrative to the learning environment. The questions of the introduction are addressed and answered by Sabina, making her more relatable. Each topic of the learning phase and quiz was linked to Sabina, i.e. finding out about Sabina's hobbies. Feedback in the knowledge test showed a picture of Sabina being happy or sad, depending on the answer. Badges were introduced before the learning phase and awarded after each section was completed. A progress bar indicated the advance in every section.

4 User Study

29 students of the University of Würzburg participated in the 30 min online experiment. Three had to be excluded due to a technical error, resulting in a final sample of $N = 26$ participants (19 female, 6 male) with a mean age of 21.73 ($SD = 2.29$), randomly allocated to both conditions ($N_{gamification} = 13$, $N_{control} = 13$). Students received partial course credit for participation. The participants were instructed not to take part in the study if they had prior knowledge in Spanish.

To assess need satisfaction for competence, autonomy and relatedness we sued the questions by Sheldon and Filak [25]. For situational intrinsic, identified, and extrinsic motivation as well as amotivation, we used the Situational

Motivation Scale questionnaire [14]. The performance in the learning phase and the knowledge test of the learning scenario was tracked for each participant. As additional measures for validation we assessed the acceptance of the learning environment [17] and included the User Experience Questionnaire [18]. Additionally, participants were presented with compulsory text boxes to state positive as well as negative aspects of the learning scenario.

5   Results and Discussion

Due to the small sample size, we used Mann-Whitney U tests with alpha set at .05, see Table 1.

Table 1. Mean values for both conditions. SDs in parentheses. U and p values show results for Mann-Whitney U tests with * indicating significance.

		Control	Gamified	U	p
Need Satisfaction	Relatedness (1–5)	2.26(0.65)	2.51(0.78)	99	.479
	Competence (1–5)	3.62(0.83)	3.92(0.83)	98.50	.479
	Autonomy (1–5)	2.59(0.98)	3.03(0.91)	101.50	.390
Situational Motivation	Intrinsic (1–7)	4.75(1.15)	5.17(0.86)	100.50	.418
	Identified (1–7)	4.27(1.26)	5.23(0.92)	124	.044*
	Extrinsic (1–7)	4.27(1.01)	2.63(0.80)	17.50	<.001*
Additional Measures	Performance learning phase	89%(7%)	92%(7%)	111	.186
	Performance knowledge test	78%(15%)	80%(16%)	87	.920
	Attractiveness (1–7)	4.85(1.50)	5.77(0.82)	115.50	.113
	Perspicuity (1–7)	5.25(1.48)	6.12(0.61)	114.50	.125
	Efficiency (1–7)	4.92(1.31)	5.65(0.62)	116.50	.101
	Dependability (1–7)	4.77(0.59)	4.94(0.51)	94	.650
	Stimulation (1–7)	4.44(1.80)	5.48(0.81)	112.50	.153
	Novelty (1–7)	4.02(1.49)	4.60(1.28)	105.50	.287
	Acceptance E-learning (1–5)	3.67(1.03)	4.22(0.75)	108.50	.223

We did not find a higher need satisfaction in the gamified condition and therefore reject H1a, H1b and H1c. H2a had to be rejected as we found no change in intrinsic motivation. However, we observed that the predicted internalisation for identified motivation and extrinsic motivation were significant, accepting H2b and H2c. All additional measures indicated no potential drawbacks, however there were no significant benefits of the gamified version. A replication study with higher sample size should be conducted to test if significant differences for the measures, which were only descriptively higher in the current study can be found or if the gamification adaptions might have to be further refined. Due to the relevance of short-term changes in self-directed learning, the willingness to continue learning in our experimental setting should also be measured in future settings, as well as learning gain.

6   Conclusion

In this contribution, we investigated the effect on situational motivation by addressing need satisfaction using gamification elements in an E-learning scenario. We successfully demonstrated an effect of our gamified learning environment on the internalisation of situational motivation in respect to extrinsic and identified motivation, however not for intrinsic motivation. This contribution is a first step in our effort in utilising gamification to promote autonomous motivation in adult learning.

Acknowledgments. We would like to thank Johanna Bogner for the preparation of the learning environment.

References

1. Abar, B., Loken, E.: Self-regulated learning and self-directed study in a pre-college sample. Learn. Individ. Differ. **20**(1), 25–29 (2010)
2. Agapito, J.L., Rodrigo, M.M.T.: Investigating the impact of a meaningful gamification-based intervention on novice programmers' achievement. In: Penstein Rosé, C., et al. (eds.) AIED 2018. LNCS (LNAI), vol. 10947, pp. 3–16. Springer, Cham (2018). https://doi.org/10.1007/978-3-319-93843-1_1
3. Barata, G., Gama, S., Jorge, J., Gonçalves, D.: Improving participation and learning with gamification. In: Nacke, L.E., Harrigan, K., Randall, N. (eds.) Proceedings of the First International Conference on Gameful Design, Research, and Applications - Gamification 2013, pp. 10–17. ACM Press, New York (2013)
4. Burton, K.D., Lydon, J.E., D'Alessandro, D.U., Koestner, R.: The differential effects of intrinsic and identified motivation on well-being and performance: prospective, experimental, and implicit approaches to self-determination theory. J. Pers. Soc. Psychol. **91**(4), 750 (2006)
5. Chang, J.W., Wei, H.Y.: Exploring the possibility of using humanoid robots as instructional tools for teaching a second language in primary school. Educ. Technol. Soc. **19**(2), 177–203 (2016)
6. Christensen, G., Steinmetz, A., Alcorn, B., Bennett, A., Woods, D., Emanuel, E.: The MOOC phenomenon: who takes massive open online courses and why? SSRN Electron. J. (2013)
7. Deci, E.L., Ryan, R.M.: Facilitating optimal motivation and psychological well-being across life's domains. Can. Psychol. **49**(1), 14 (2008)
8. Deci, E.L., Ryan, R.M.: Self-Determination Theory, pp. 416–437. Sage Publications, Inc. (2011)
9. Deci, E.L., Schwartz, A.J., Sheinman, L., Ryan, R.M.: An instrument to assess adults' orientations toward control versus autonomy with children: reflections on intrinsic motivation and perceived competence. J. Educ. Psychol. **73**(5), 642–650 (1981). https://doi.org/10.1037/0022-0663.73.5.642
10. Dehghanzadeh, H., Fardanesh, H., Hatami, J., Talaee, E., Noroozi, O.: Using gamification to support learning english as a second language: a systematic review. Comput. Assist. Lang. Learn. 1–24 (2019)

11. Deterding, S., Dixon, D., Khaled, R., Nacke, L.: From game design elements to gamefulness: defining gamification. In: Lugmayr, A., Franssila, H., Safran, C., Hammouda, I. (eds.) Proceedings of the 15th International Academic MindTrek Conference on Envisioning Future Media Environments - MindTrek 2011, pp. 9–15. ACM Press, New York (2011)

12. Dicheva, D., Dichev, C., Agre, G., Angelova, G.: Gamification in education: a systematic mapping study. Educ. Technol. Soc. **18**, 75–88 (2015)

13. Efklides, A.: Interactions of metacognition with motivation and affect in self-regulated learning: the MASRL model. Educ. Psychol. **46**(1), 6–25 (2011)

14. Guay, F., Vallerand, R.J., Blanchard, C.: On the assessment of situational intrinsic and extrinsic motivation: the situational motivation scale (SIMS). Motiv. Emot. **24**(3), 175–213 (2000)

15. Hakulinen, L., Auvinen, T., Korhonen, A.: Empirical study on the effect of achievement badges in trakla2 online learning environment. In: 2013 Learning and Teaching in Computing and Engineering, pp. 47–54. IEEE (2013)

16. Jang, J., Park, J.J.Y., Yi, M.Y.: Gamification of online learning. In: Conati, C., Heffernan, N., Mitrovic, A., Verdejo, M.F. (eds.) AIED 2015. LNCS (LNAI), vol. 9112, pp. 646–649. Springer, Cham (2015). https://doi.org/10.1007/978-3-319-19773-9_82

17. Kreidl, C.: Akzeptanz und Nutzung von E-Learning-Elementen an Hochschulen. Waxmann, Gründe für die Einführung und Kriterien der Anwendung von E-Learning. Münster (2011)

18. Laugwitz, B., Held, T., Schrepp, M.: Construction and evaluation of a user experience questionnaire. In: Holzinger, A. (ed.) USAB 2008. LNCS, vol. 5298, pp. 63–76. Springer, Heidelberg (2008). https://doi.org/10.1007/978-3-540-89350-9_6

19. Mega, C., Ronconi, L., de Beni, R.: What makes a good student? How emotions, self-regulated learning, and motivation contribute to academic achievement. J. Educ. Psychol. **106**(1), 121–131 (2014)

20. Nacke, L.E., Deterding, S.: The maturing of gamification research. Comput. Hum. Behav. **71**, 450–454 (2017)

21. Park, S.Y.: An analysis of the technology acceptance model in understanding university students' behavioral intention to use e-learning. Educ. Technol. Soc. **12**(3), 150–162 (2009)

22. Rawendy, D., Ying, Y., Arifin, Y., Rosalin, K.: Design and development game Chinese language learning with gamification and using mnemonic method. Procedia Comput. Sci. **116**, 61–67 (2017)

23. Sailer, M., Hense, J.U., Mayr, S.K., Mandl, H.: How gamification motivates: an experimental study of the effects of specific game design elements on psychological need satisfaction. Comput. Hum. Behav. **69**, 371–380 (2017)

24. Seaborn, K., Fels, D.I.: Gamification in theory and action: a survey. Int. J. Hum Comput Stud. **74**, 14–31 (2015)

25. Sheldon, K.M., Filak, V.: Manipulating autonomy, competence, and relatedness support in a game-learning context: new evidence that all three needs matter. Br. J. Soc. Psychol. **47**(2), 267–283 (2008)

26. Toda, A.M., Valle, P.H.D., Isotani, S.: The dark side of gamification: an overview of negative effects of gamification in education. In: Cristea, A.I., Bittencourt, I.I., Lima, F. (eds.) HEFA 2017. CCIS, vol. 832, pp. 143–156. Springer, Cham (2018). https://doi.org/10.1007/978-3-319-97934-2_9

Learning Association Between Learning Objectives and Key Concepts to Generate Pedagogically Valuable Questions

Machi Shimmei[✉] ⓘ and Noboru Matsuda ⓘ

North Carolina State University, Raleigh, NC 27695, USA
{mshimme,noboru.matsuda}@ncsu.edu

Abstract. It has been shown that answering questions contributes to students learning effectively. However, generating questions is an expensive task and requires a lot of effort. Although there has been research reported on the automation of question generation in the literature of Natural Language Processing, these technologies do not necessarily generate questions that are useful for educational purposes. To fill this gap, we propose QUADL, a method for generating questions that are aligned with a given learning objective. The learning objective reflects the skill or concept that students need to learn. The QUADL method first identifies a key concept, if any, in a given sentence that has a strong connection with the given learning objective. It then converts the given sentence into a question for which the predicted key concept becomes the answer. The results from the survey using Amazon Mechanical Turk suggest that the QUADL method can be a step towards generating questions that effectively contribute to students' learning.

Keywords: Question generation · MOOCS · Learning engineering

1 Introduction

Creating high-quality questions is important for instructors as valid questions provide insight into their students' learning status, which in turn helps instructors enhance their teaching methods. Answering questions is also an essential part of learning. The benefit of answering questions for learning has been shown in many studies, aka *test-enhanced learning* [1, 2]. On Massive Open Online Course (MOOC), questions are also an influential component that determines the effectiveness of the course. It is reported that students learn better when they practice skills by answering questions than when only watching videos or reading text [3]. However, creating questions that effectively help students' learning requires experience and extensive efforts.

When the question is generated for educational use in particular, with the focus on test-enhanced learning, machine-generated questions should have a pedagogical value in addition to general features such as clarity and fluency. Although there are a number of studies on question generation in the field of AI in education [4, 5], little has been studied about the pedagogical value of the generated questions. To fill this gap, we propose a

© Springer Nature Switzerland AG 2021
I. Roll et al. (Eds.): AIED 2021, LNAI 12749, pp. 320–324, 2021.
https://doi.org/10.1007/978-3-030-78270-2_57

method for generating questions that supposedly ask about the key concepts the students need to learn to attain the learning objectives. There have been no studies that aim to generate questions that align with the learning objectives.

2 Related Work

Recent works on question generation take a data-driven approach using neural networks. Large datasets such as SQuAD [6], NewsQA [7], MSMARCO [8] enabled training a recurrent neural network (RNN) for question generation. The number of studies with the aim of generating questions specifically for educational purposes has been also increasing. The limited number of relevant datasets available is among the primary challenges in educational question generation. Although there are datasets such as SciQ [9], which contains questions from science textbooks, the size of the data is considerably small. Therefore, some studies utilize general question generation datasets to train a model. Wang *et al.* [10] demonstrated that an LSTM-based model, called QG-Net, trained on a SQuAD can be used for generating questions on educational contents.

Another challenge for question generation is how to identify an answer candidate. QG-Net and other models [11–13] require that an input sentence is tagged with a candidate of an answer for the generated question. There are also some models that can find an answer candidate in a given text. For example, Willis *et al.* [14] proposed a key phrase extraction model that outputs an answer candidate from a given paragraph text. QUADL also has the Answer Prediction model that finds an answer candidate (i.e., a target token index). The key difference of our Answer Prediction model from the existing models is that *our proposed Answer Prediction model aims to select target tokens that are aligned with a given learning objective.*

3 Methods

Figure 1 shows an overview of QUADL. Given a pair of a learning objective LO and a sentence S, $<LO, S>$, QUADL generates a question Q that will be suitable to achieve the learning objective LO. The question Q is a verbatim question, which means that the answer can be literally found in the given sentence S. The following is an example of $<LO, S>$ and Q:

Learning objective (LO): Describe metabolic pathways as stepwise chemical transformations either requiring or releasing energy; and recognize conserved themes in these pathways.

Sentence (S): Among the main pathways of the cell are photosynthesis and <u>cellular respiration,</u> although there are a variety of alternative pathways such as fermentation.

Question (Q): Along with photosynthesis, what are the main pathways of the cell?

Answer: Cellular respiration.

Notice that the answer is tagged (underlined in S) in the sentence S. For the sake of explanation, we call the tagged token(s) in the given sentence S as a *target token* hereafter.

QUADL consists of two components: (1) the Answer Prediction model and (2) the Question Conversion model. The Answer Prediction model identifies $<Is, Ie>$, called *token index*, where Is and Ie show the index of the start and end of a target token within a given sentence S relative to the learning objective LO. We adopted BERT, Bidirectional Encoder Representation from Transformers [15] for this Answer Prediction model. The learning objective and sentence were combined as a single input $<LO, S>$ to the model. The final hidden state of the BERT model was fed to the single layer classification model that outputs logit for the start index (Is) and another single layer classification model that outputs logit for the end index (Ie) for each token in the sentence S. The final score was calculated by taking the softmax of the sum of the start logit and end logit for every possible span ($Is < Ie$) in the sentence. The score was also calculated for $< Is = 0, Ie = 0 >$ indicating that the sentence is not suitable to generate a question for the learning objective. The index $<Is, Ie>$ with the largest score became the final prediction. For the rest of the paper, we call sentences that have non-zero indices (i.e., $Is \neq 0$ *and* $Ie \neq 0$) the *target sentences*, whereas others are referred to as the *non-target sentences* (i.e., has the zero token index $<0, 0>$). We created training data for the Answer Prediction model using the text data from existing online courses at Open Learning Initiative[1] (OLI).

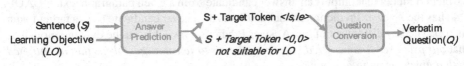

Fig. 1. The QUADL model

Given a sentence with the non-zero target token index, the Question Conversion model generates a question for which the target token becomes the answer. We use an existing bidirectional-LSTM seq2seq model with attention and copy mechanisms, QG-Net [10], for the Question Conversion model. We used an existing *pre-trained* QG-Net model that was trained using SQuAD datasets[2]. We could train the QG-Net using the OLI course data. However, the OLI courses we used for the current study do not contain a sufficient number of verbatim questions—many of the questions are fill-in-the-blank and multiple-choice questions hence not suitable to generate training data for QG-Net.

4   Evaluation

We have the following research questions: **RQ1**:How well does the Answer Prediction model identify target tokens (including zero token indices) in a given sentence relative to a given learning objective? **RQ2**:How well does the pre-trained QG-Net generate questions for a given sentence tagged with the target tokens? To answer the questions,

---

[1] https://oli.cmu.edu.
[2] https://rajpurkar.github.io/SQuAD-explorer/.

we conducted a survey on Amazon Mechanical Turk (AMT). In AMT, for each triplet $<LO, S < Is, Ie>, Q>$ shown, the participants were asked if they agreed or disagreed with the following two statements: (1) To get a question that helps attain the learning objective LO, it is adequate to convert the sentence S into a question whose answer is the token $<Is, Ie>$ highlighted. (2) The question Q is suitable for attaining the learning objectives LO. Each statement corresponds to each research question.

Table 1 summarizes the results for RQ1. The table shows that, for the predictions with a non-zero target index, 70% (123/178) of the predictions *were accepted*. As for the non-target sentence predictions (i.e., the Answer Prediction model output the zero $<0,0>$ index), only 26% (43/164) were accepted. That is, 55% (90/164) of the predicted non-target sentences were considered to be target sentences by participants.

Table 1. The evaluation of the predicted target tokens by the Answer Prediction model. There were 178 sentences that the Answer Prediction model predicted target tokens (non-zero index) and 164 sentences that the model predicted non-target (zero index $<0, 0>$). The table shows how many of them were accepted/not accepted by the majority vote by Amazon Mechanical Turk (AMT) participants.

AMT	Model prediction			
		Non-zero target index $<Is \neq 0, Ie \neq 0>$	Zero-index $<0, 0>$	Total
	Accepted	**123 (70%)**	**43 (26%)**	166 (49%)
	Tie	32 (18%)	25(15%)	57 (17%)
	Not accepted	22 (12%)	90 (55%)	112 (33%)
	Nonsensical	1	6 (4%)	7 (2%)
	Total	178 (100%)	164 (100%)	342 (100%)

As for the RQ2, the results showed that the participants considered that 45% (80/178) of the questions generated by QG-Net (used in QUADL) were appropriate for achieving the associated learning objective. Notice that the result is influenced by the performance of the Answer Prediction model because questions are generated from sentences that the Answer Prediction model predicted target tokens.

5   Conclusion

We proposed QUADL for generating questions that are aligned with the given learning objective. As far as we are aware, there have been no studies that aim to generate questions that are suitable for attaining the learning objectives. The evaluation through Amazon Mechanical Turk revealed that the 70% of the predicted target tokens were considered to be appropriate. The result also showed there is a need for improvement to reduce the false negatives—incorrectly predicting that a given sentence is not suitable to attain the learning objective. The current study utilized a survey on Amazon Mechanical Turk.

Evaluating the effectiveness of generated questions with real students in an authentic context is an important next step to be conducted.

Acknowledgements. The research reported here was supported by National Science Foundation Grant No. 2016966 and No. 1623702 to North Carolina State University.

References

1. Rivers, M.L.: Metacognition about practice testing: a review of learners' beliefs, monitoring, and control of test-enhanced learning. Educ. Psychol. Rev. (2020)
2. Pan, S.C., Rickard, T.C.: Transfer of test-enhanced learning: meta-analytic review and synthesis. Psychol. Bull. **144**(7), 710–756 (2018)
3. Koedinger, K.R., et al.: Learning is not a spectator sport: doing is better than watching for learning from a MOOC. In: Proceedings of the Second (2015) ACM Conference on Learning@ Scale (2015)
4. Kurdi, G., et al.: A systematic review of automatic question generation for educational purposes. Int. J. Artif. Intell. Educ. **30**(1), 121–204 (2020)
5. Pan, L., et al.: Recent advances in neural question generation. arXiv preprint arXiv:1905.08949 (2019)
6. Rajpurkar, P., Jia, R., Liang, P.: Know what you don't know: Unanswerable questions for SQuAD. arXiv preprint arXiv:1806.03822 (2018)
7. Trischler, A., et al.: Newsqa: A machine comprehension dataset. arXiv preprint arXiv:1611.09830 (2016)
8. Bajaj, P., et al.: Ms marco: A human generated machine reading comprehension dataset. arXiv preprint arXiv:1611.09268 (2016)
9. Welbl, J., Liu, N.F., Gardner, M.: Crowdsourcing multiple choice science questions. arXiv preprint arXiv:1707.06209 (2017)
10. Wang, Z., et al.: QG-net: a data-driven question generation model for educational content. In: Proceedings of the Fifth Annual ACM Conference on Learning at Scale (2018)
11. Kim, Y., et al.: Improving neural question generation using answer separation. In: Proceedings of the AAAI Conference on Artificial Intelligence (2019)
12. Nema, P., et al.: Let's Ask Again: Refine Network for Automatic Question Generation. arXiv preprint arXiv:1909.05355 (2019)
13. Yuan, X., et al.: Machine comprehension by text-to-text neural question generation. arXiv preprint arXiv:1705.02012 (2017)
14. Willis, A., et al.: Key phrase extraction for generating educational question-answer pairs. In: Proceedings of the Sixth (2019) ACM Conference on Learning@ Scale (2019)
15. Devlin, J., et al.: Bert: Pre-training of deep bidirectional transformers for language understanding. arXiv preprint arXiv:1810.04805 (2018)

Exploring the Working and Effectiveness of Norm-Model Feedback in Conceptual Modelling – A Preliminary Report

Loek Spitz[1], Marco Kragten[1], and Bert Bredeweg[1,2(✉)] (iD)

[1] Faculty of Education, Amsterdam University of Applied Sciences, Amsterdam, Netherlands
{l.spitz,m.kragten,b.bredeweg}@hva.nl
[2] University of Amsterdam, Informatics Institute, Amsterdam, Netherlands

Abstract. Having learners (K7–10) acquire system thinking skills is challenging. Together with teachers we deploy qualitative representations of complex systems to enable this learning process. Teachers select their own topics for their leaners to work on which makes that lessons vary in content depending on the teacher's preference. Within this setting we face the challenge of adequately coaching learners while they create their knowledge models. For this, we use norm-model based feedback, ignoring learner specific information. Here we report the working and effectiveness of this approach.

Keywords: Engagement · Feedback · Systems thinking · Qualitative reasoning

1 Introduction

Systems thinking is a difficult skill to learn [3, 7, 11, 16, 17]. Learners may easily ignore relevant factors, apply causal relationships incorrectly, fail to see feedback mechanisms and their impact, and not recognize cause-effect patterns across systems (so called transfer). Even transfer to similar systems (near transfer; [12]) is experienced as difficult [6].

In the project 'Denker' (https://denker.nu/) we work towards addressing these challenges by deploying qualitative representations [4] in a workbench for learners to develop their systems thinking skills. Learners use interactive diagrams to create conceptual models and thereby construct their understanding of systems and how they behave.

As modelling is difficult task, adequate feedback is necessary to ensure successful learning [10, 15, 18]. Automated tutoring systems can be valuable instruments for learning [1, 9]. However, these approaches require large amounts of student data and careful design to ensure the quality of the feedback [13, 14].

In our situation, this approach is not possible. We work with multiple teachers from different subject areas who each create their own assignments. The available data is sparse (small groups of learners) and distributed across topics. Moreover, learning domain knowledge is intertwined with acquiring systems thinking skills, and it is not feasible to create an overview of typical errors learners make or misconceptions they have.

© Springer Nature Switzerland AG 2021
I. Roll et al. (Eds.): AIED 2021, LNAI 12749, pp. 325–330, 2021.
https://doi.org/10.1007/978-3-030-78270-2_58

To accommodate this challenge, we have developed a lightweight norm-model based feedback approach. Using a meta-vocabulary, teachers create models of the subject matter that they find relevant. When learners create their models, deviations from this norm are identified using a norm-based mapping approach, ignoring learner specific information. The approach was used and studied in real educational settings.

2   Conceptual Models for System Thinking

Conceptual models are a class of knowledge construction tools that use logical (symbolic, non-numeric) representations for the expression of conceptual knowledge [2, 5, 8]. This logic-base is a crucial asset in facilitating automated feedback to support learners at an individual level in their knowledge construction efforts [10]. Dynalearn is a software that facilitates the creation of conceptual models (https://www.dynalearn.nl). It is organized into a set of distinct levels with increasing complexity [2]. For the work presented in this paper learners worked at level 3. This level allows for cause-effect representations to support reasoning about how changes propagate through a system (Fig. 1). Learners represent the *Entities* (the physical objects and/or abstract concepts) that make up the system, the *Quantities* characterizing each of these entities, and the *Causal* dependencies (+ & –) between the quantities. Quantities have a *Direction change* (∂) and a *Quantity space*. The latter specifics the possible values a quantity can take on. This allows for representing the idea that a system moves through different states (e.g., {0, +, Border, Extreme}).

Additionally, level 3 has ingredients to represent the idea of *Agent* and *Exogenous* quantity behaviour. With this, learners learn to distinguish the 'system' from the 'external factors' affecting it. Finally, the notion of *Correspondence* (C) is used to specify co-occurring vales (e.g., IF Population Size = 0 THEN Natality = 0). When simulating, a *State-graph* appears (sequence of states and transitions) and the *Value history* (overview of values for a sequence of states) can be used to inspect the simulation results.

The automated feedback compares a learner-created model with the norm-model (created by the teacher). After each manipulation executed by the learner in the canvas a new mapping is made using a Monte-Carlo-based heuristic approach. The engine runs for at most 5 s and then returns the best mapping. Next, for each discrepancy the tool provides two options for feedback. *Cueing*: a small red circle is placed around each deviating model ingredient (Q2 in Fig. 1) and a red question mark appears on the right-hand side in the canvas. *Help*: when clicking on the question mark, a message-box appears showing a sentence for each deviation (in Fig. 1: Quantity: Q2: wrong name?).

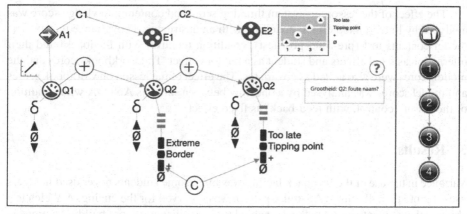

Fig. 1. LHS shows a generic example of a level 3 model. The RHS shows a state-graph consisting of 4 states, each state referring to one of the 4 values in the quantity spaces (0, 0), (+, +), etc.

3 The Model and the Accompanying Lesson

Together with high school teachers we designed a lesson for a grade 9 course. The topic (the Neolithic Age) was decided by the teacher. The lesson was designed over the course of several online meetings between the teacher and the members of the research team. We started by designing a model in Dynalearn describing the shift from hunters and gatherers to agricultural societies during the Neolithic Age. Next, we created a workbook to go along with the model. The goal of the workbook was for the students to be able to work independently during an online session. The workbook contained information about the Neolithic Age, a link to a videoclip, questions to get students thinking about the subject and scaffolds for building the model in Dynalearn.

There was one online session of 2.5 h for each of the two participating classes. In each session the class was split into two subgroups using breakout rooms. For each subgroup a teacher and a member of the research group were available to help the students. The sessions started with the pre-test and finished with the post-test. During the main part of the lesson the students worked independently on the workbook and in Dynalearn. They could use the video session or a text chat to ask questions.

4 Research Method and Data Analysis

Two classes of 46 and 38 students respectively participated in this study. Both worked on the lesson in Dynalearn, one with feedback enabled, the other without. All students had experience with the software; they participated in a similar lesson (at system thinking level 3, but without the feedback facility) earlier in the school year.

A test of six items was developed for measuring students' *system thinking* skills, based on the learning objectives for systems thinking [2]. Three items were added to measure the amount of *content knowledge* that was learned (i.e., the Neolithic Age). A scoring protocol for the *workbook* assignments was created. The quality of student created-*models* were based the number of correct, wrong and missing model-ingredients.

The effect of the lesson on system thinking score and content knowledge score was analysed by linear mixed effect modelling with an unstructured covariance matrix. The model contains test (pre- and post-test), condition (control, with feedback) and their interaction as fixed effects and student as a random effect. The workbook scores and the model scores were also added as covariates. The effect of the lesson on workbook scores and model scores was analysed by a one-way between subjects ANOVA with condition of the lesson (control, with feed-back) as fixed effect.

5   Results

Variance in the use of the feedback facility was high; some students never used it, others uses it more than 40 times. A similar pattern was observed for the cueing. A Welch test showed that students with feedback enabled took significantly more building actions in the model than students without the feedback function ($p < 0.001$). Using ANOVA to compare the means showed that students with feedback enabled completed a significantly larger part of the model ($p = .02$), while there was no significant difference between the groups on percentage of workbook completed ($p = .054$).

There was no significant main or interaction effect of test and condition on system thinking. The mixed effect model with content knowledge as the outcome variable showed a significant main effect on content knowledge score ($\beta = -.26, t = .13. p = .05$) and a significant interaction effect between condition and test ($\beta = .49, t = 2.78. p < .01$). Workbook score had a significant effect on content knowledge score ($\beta = 1.99, t = 4.01, p < .001$), while model score ($\beta = -.03, t = -2.06, p = .04$) had no significant effect on system thinking score. The model explained 46% of content knowledge variability.

There was no significant effect of condition on workbook score, $F(1, 89) = .17, p = 0.68$. There was a significant effect of condition on model score, $F(1, 70) = 10.44, p < 0.01$. Content knowledge score (gain from pre-test to post-test) was not significantly correlated to number of interactions with cueing ($r = .24, p = .12$) and help ($r = .07, p = .65$). Model score was significantly correlated to number of interactions with cueing ($r = .49, p < .001$) and help ($r = .37, p < .01$).

6   Conclusion and Discussion

In this work, investigated how students interact with a lightweight norm-model based feedback approach.

The experimental group was significantly more engaged with the model than the control group. The former took more actions in the model and built it up further, although they did not get further in the workbook. The former can be explained by the fact that they did not get stuck with the model as much, and that they were more aware of the mistakes they made. This is especially relevant during an online lesson, where it is hard for teachers to keep an eye on students' progress and help them with quick hints when they get stuck or make mistakes without noticing. In the control group, many students made mistakes in the model without correcting them, because they did not notice them.

The feedback function seems to be able to partially take over the role of the teacher in this regard.

Having access to the feedback function significantly increased the learning gains regarding content knowledge, but not regarding system thinking. The latter may be explained by the fact that the students had already taken a lesson previously introducing system thinking using Dynalearn. This may have led to a ceiling effect.

Students in the feedback condition had significantly higher model scores at the end of the lesson. Although the lesson goal is to learn, and not to build the best possible model, it is interesting to see that the feedback enabled students create better models.

Although gaining access to the feedback increased learning gains regarding content knowledge, these gains were not correlated with the way students used the cueing and help. This makes sense, because stronger students may achieve high learning gains without needing much feedback, while feedback allows those in need to 'catch up'. Use of help and cueing was correlated with model quality, probably because it happens naturally while building the model and even strong students can use it to correct mistakes.

References

1. van der Bent, R., Jeuring, J., Heeren, B.: The diagnosing behaviour of intelligent tutoring systems. In: Scheffel, M., Broisin, J., Pammer-Schindler, V., Ioannou, A., Schneider, J. (eds.) EC-TEL 2019. LNCS, vol. 11722, pp. 112–126. Springer, Cham (2019). https://doi.org/10.1007/978-3-030-29736-7_9
2. Bredeweg, B., et al.: DynaLearn - an intelligent learning environment for learning conceptual knowledge. AI Mag. **34**(4), 46–65 (2013)
3. Cronin, M., Gonzalez, C., Sterman, J.D.: Why don't well-educated adults understand accumulation? A challenge to researchers, educators, and citizens. Organ. Behav. Hum. Decis. Process. **108**(1), 116–130 (2009)
4. Forbus, K.D.: Qualitative Representations. How People Reason and Learn About the Continuous World. The MIT Press, Cambridge (2018)
5. Forbus, K., Carney, K., Sherin, B., Ureel, L.: VModel: a visual qualitative modeling environment for middle-school students. In: Proceedings of the 16th Innovative Applications of Artificial Intelligence Conference, San Jose (2004)
6. Genter, D., Lowenstein, J., Thompson, L.: Learning and transfer: a general role for analogical encoding. J. Educ. Psychol. **95**(2), 393–408 (2003)
7. Jensen, E., Brehmer, B.: Understanding and control of a simple dynamic system. Syst. Dyn. Rev. **19**(2), 119–137 (2003)
8. Kinnebrew, J.S., Biswas, G.: Modeling and measuring self-regulated learning in teachable agent environments. J. e-learning Knowl. Soc. **7**(2), 19–35 (2011)
9. Kulik, J.A., Fletcher, J.: Effectiveness of intelligent tutoring systems: a meta- analytic review. Rev. Educ. Res. **86**(1), 42–78 (2016)
10. Liem, J.: Supporting conceptual modelling of dynamic systems: a knowledge engineering perspective on qualitative reasoning. University of Amsterdam, Netherlands, Thesis (2013)
11. Moxnes, E.: Misperceptions of basic dynamics: the case of renewable resource management. Syst. Dyn. Rev. **20**(2), 139–162 (2004)
12. Perkins, D.N., Salomon, G.: Transfer of learning. Int. Encyclopedia Educ. **2**, 6452–6457 (1992)
13. Perrotta, C., Selwyn, N.: Deep learning goes to school: toward a relational understanding of AI in education. Learn. Media Technol. **45**(3), 251–269 (2019)

14. Price, T.W., Zhi, R., Dong, Y., Lytle, N., Barnes, T.: The impact of data quantity and source on the quality of data-driven hints for programming. In: Penstein Rosé, C., et al. (eds.) AIED 2018. LNCS (LNAI), vol. 10947, pp. 476–490. Springer, Cham (2018). https://doi.org/10.1007/978-3-319-93843-1_35

15. Sins, P.H.M., Savelsbergh, E.R., van Joolingen, W.R.: The difficult process of scientific modelling: an analysis of novices' reasoning during computer-based modelling. Int. J. Sci. Educ. 27(14), 1695–1721 (2005)

16. Sterman, J.D.: Learning in and about complex systems. Syst. Dyn. Rev. 10(2–3), 291–330 (1994)

17. Sweeney, L.B., Sterman, J.D.: Bathtub dynamics: initial results of a systems thinking inventory. Syst. Dyn. Rev. 16(4), 249–286 (2000)

18. VanLehn, K.: Model construction as a learning activity: a design space and review. Interact. Learn. Environ. 21(4), 371–413 (2013)

A Comparative Study of Learning Outcomes for Online Learning Platforms

Francois St-Hilaire[1], Nathan Burns[1], Robert Belfer[1], Muhammad Shayan[1], Ariella Smofsky[1], Dung Do Vu[1], Antoine Frau[1], Joseph Potochny[1], Farid Faraji[1], Vincent Pavero[1], Neroli Ko[1], Ansona Onyi Ching[1], Sabina Elkins[1], Anush Stepanyan[1], Adela Matajova[1], Laurent Charlin[1,2], Yoshua Bengio[1,2], Iulian Vlad Serban[1], and Ekaterina Kochmar[1,3(✉)]

[1] Korbit Technologies Inc., Montreal, Canada
ekaterina@korbit.ai
[2] Quebec Artificial Intelligence Institute (Mila), Montreal, Canada
[3] University of Bath, Bath, UK

Abstract. Personalization and active learning help educational systems to close the gap between students with varying abilities. We run a comparative head-to-head study of learning outcomes for two popular online platforms: `Platform A`, which delivers content over lecture videos and multiple-choice quizzes, and `Platform B`, which provides interactive problem-solving exercises and personalized feedback. We observe a statistically significant increase in the learning outcomes on `Platform B`. Further, the results of the self-assessment questionnaire suggest that participants using `Platform B` improve their metacognition.

Keywords: Online and distance learning · Models of teaching and learning · Intelligent and interactive technologies · Data science

1 Introduction

We investigate the learning outcomes induced by two popular online learning platforms in a comparative head-to-head study. `Platform A` is a widely-used platform that follows a traditional model, where students learn by watching lecture videos, reading, and testing their knowledge with multiple choice quizzes. In contrast, `Platform B`[1] focuses on personalized, active learning approach with problem-solving exercises [34]. `Platform B` is powered by an AI tutor, which alternates between lecture videos and interactive problem-solving exercises. The AI tutor shows students problem statements and students attempt to solve them. Each incorrect attempt is addressed with personalized pedagogical interventions tailored to student's needs and misconceptions (see Fig. 1).

---

[1] `Platform B` is the Korbit learning platform available at www.korbit.ai.

© Springer Nature Switzerland AG 2021
I. Roll et al. (Eds.): AIED 2021, LNAI 12749, pp. 331–337, 2021.
https://doi.org/10.1007/978-3-030-78270-2_59

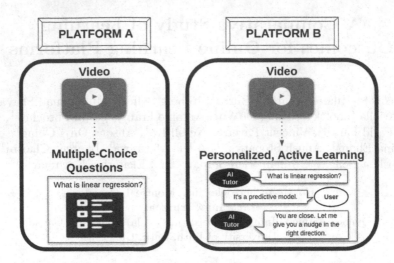

Fig. 1. Platform A follows a traditional learning approach utilizing videos and multiple choice quizzes, while Platform B uses a personalized, active learning approach with problem-solving exercises.

In this study, we formulate and test the following hypothesis:

Hypothesis: Participants studying with Platform B have higher learning gains than those studying with Platform A, because Platform B employs personalized, active learning and problem-based learning and provides a wider and more personalized set of pedagogical elements to its students.

2 Related Work

Online learning platforms have the capability of bridging the gap and addressing inequalities in society caused by uneven access to in-person teaching [13,15,17, 30,39,42]. The current COVID-19 pandemic further exacerbates the need for high quality online education being accessible to a wide variety of students [1,4,28].

Nevertheless, the efficacy of online and distance learning has been challenged by researchers: specifically, it may be hard to address the differences in students' learning needs, styles and aptitudes on such platforms [9,14,37,40]. This calls for approaches that can be adapted and personalized to the needs of each particular student. Studies confirm that personalization is key to successful online learning [26,33], as it can maximize the learning benefits for each individual student [45]. In addition, problem-solving has been shown to be a highly effective approach for learning in various domains [12,18,19,43,44]. Such problem-solving and active learning activities can be addressed by intelligent tutoring systems, which are also capable of giving personalized feedback and explanations and incorporating conversational scaffolding [2,7,8,12,20,21,24,25,27,31,32].

In contrast to previous studies investigating learning outcomes with intelligent tutoring systems, in this study the AI-powered learning platform, Platform

B, is a fully-automated system based on machine learning models [34]. The system is trained from scratch on educational content to generate automated, personalized feedback for students and has the ability to automatically generalize to new subjects and improve as it interacts with new students [35,36].

To evaluate the impact of educational technology and online learning platforms on student learning outcomes, we follow previous research [3,11,16,22,23, 29,38,41]. We adopt the well-established pre-/post-assessment framework, where students are split into intervention groups and their knowledge of the subject is evaluated before and after their assigned intervention. Further, we measure student's metacognition. Students' ability to self-assess and develop self-regulation skills plays a crucial role in online learning [16,25], though studies show that students struggle to evaluate their own knowledge and skills level [5,6,10].

3   Experimental Setup

48 participants were randomly divided between the two platforms, where the first group was asked to study the course from Platform A and the second from Platform B. Each group completed a 3-hour long course on *linear regression*. The majority fall into our target audience of undergraduates (89.6%) studying disciplines not centered around mathematics (e.g. health sciences).

Linear regression was selected as the topic of study since it is one of the most fundamental topics, that is covered early on in any course on machine learning and data science, and the material covering this topic on both platforms is comparable. To ensure a fair comparison, extra care was taken to ensure that the courses and the subtopics they covered were as similar as possible.

The study ran over a 4-day period with strict deadlines and detailed instructions set for the participants. All participants were required to take an assessment quiz on linear regression before the course (*pre-quiz*) and another one after the course (*post-quiz*). The quizzes contained 20 multiple-choice questions each and were equally adapted to both courses, with questions in pre- and post-quizzes isomorphically paired. Using pre- and post-quiz scores, we measure *learning gains* to quantify how effectively each participant has learned. A student's learning gain g is estimated as the difference between their pre-quiz (*pre_score*) and post-quiz (*post_score*) scores. Further, a student's normalized learning gain g_{norm} is calculated by:

$$g_{norm} = \frac{post\_score - pre\_score}{100\% - pre\_score} \qquad (1)$$

4   Results and Discussion

25 participants completed the course on Platform A and 23 on Platform B. Average learning gains are shown in Fig. 2 for the two platforms. The average normalized learning gains for Platform B participants are 49.24% higher than for Platform A participants, with the difference being statistically significant at a 90% confidence level ($p = 0.068$ w.r.t. one-sided t-test). Average raw learning

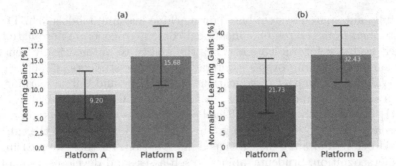

Fig. 2. (a) Average learning gains g with 95% confidence intervals.* (b) Average normalized learning gains g_{norm} with 95% confidence intervals.** Here * and ** indicate a statistically significant difference at 95% and 90% confidence level respectively.

gains for Platform B participants are 70.43% higher than for Platform A participants, with the difference being statistically significant at a 95% confidence level ($p = 0.038$ w.r.t. one-sided t-test). Overall, our hypothesis that learning outcomes are higher for participants on Platform B than on Platform A is confirmed.

We estimate that participants on Platform B spent at least twice as much time doing active learning (problem-solving exercises) compared to participants on Platform A, although the total average study times on the two platforms were equivalent. We further observed that the rate of correct answers on the first try positively correlates with both learning gains ($r = 0.44$) and post-quiz results ($r = 0.46$), and the number of exercises completed positively correlates with the post-quiz score ($r = 0.28$), suggesting that participants who spent more time on active learning performed better and, as a result, obtained higher post-quiz scores and learning gains.

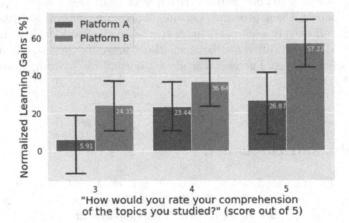

Fig. 3. Normalized learning gains for each self-assessed comprehension rating with 95% confidence intervals. Only 1 participant gave a score lower than 3 (not shown here).

Finally, we evaluated meta-cognitive aspects related to the students' learning experience with the two platforms using a questionnaire. In particular, students were asked the question *"How would you rate your comprehension of the topics you studied?"*. As shown in Fig. 3, it appears that `Platform B` not only induced overall higher learning gains, but also gave participants a more accurate understanding of their knowledge level and helped improve their meta-cognition.

References

1. Adedoyin, O.B., Soykan, E.: Covid-19 pandemic and online learning: the challenges and opportunities. Interactive Learn. Environ. **1–13** (2020)
2. Albacete, P., Jordan, P., Katz, S., Chounta, I.-A., McLaren, B.M.: The impact of student model updates on contingent scaffolding in a natural-language tutoring system. In: Isotani, S., Millán, E., Ogan, A., Hastings, P., McLaren, B., Luckin, R. (eds.) AIED 2019. LNCS (LNAI), vol. 11625, pp. 37–47. Springer, Cham (2019). https://doi.org/10.1007/978-3-030-23204-7_4
3. Barokas, J., Ketterl, M., Brooks, C., Greer, J.: Lecture capture: student perceptions, expectations, and behaviors. In: Sanchez, J., Zhang, K. (eds.) World Conference on E-Learning in Corporate, Government, Healthcare, and Higher Educatio, pp. 424–431 (2010)
4. Basilaia, G., Kvavadze, D.: Transition to online education in schools during a SARS-CoV-2 coronavirus (COVID-19) pandemic in Georgia. Pedagogical Res. **5**(4) (2020)
5. Brown, G.T.L., Harris, L.R.: Student self-assessment. In: McMillan, J.H. (ed.) The SAGE Handbook of Research on Classroom Assessment, pp. 367–393. Sage, Thousand Oaks (2013)
6. Brown, G.T., Andrade, H.L., Chen, F.: Accuracy in student self-assessment: directions and cautions for research. Assess. Educ.: Principles, Policy Practice **22**(4), 444–457 (2015)
7. Büdenbender, J., Frischauf, A., Goguadze, G., Melis, E., Libbrecht, P., Ullrich, C.: Using computer algebra systems as cognitive tools. In: International Conference on Intelligent Tutoring Systems, pp. 802–810 (2002)
8. Chi, M., Koedinger, K., Gordon, G., Jordan, P., Vanlehn, K.: Instructional factors analysis: a cognitive model for multiple instructional interventions. In: EDM 2011 - Proceedings of the 4th International Conference on Educational Data Mining, pp. 61–70 (2011)
9. Coffield, F.J., Moseley, D.V., Hall, E., Ecclestone, K.: Learning Styles for Post 16 Learners: What Do We Know?. Learning and Skills Research Centre/University of Newcastle upon Tyne, London (2004)
10. Crowell, T.L.: Student self grading: perception vs. reality. Am. J. Educ. Res. **3**(4), 450–455 (2015). https://doi.org/10.12691/education-3-4-10
11. Demmans Epp, C., Phirangee, K., Hewitt, J., Perfetti, C.A.: Learning management system and course influences on student actions and learning experiences. Educ. Tech. Research Dev. **68**(6), 3263–3297 (2020). https://doi.org/10.1007/s11423-020-09821-1
12. Fossati, D., Di Eugenio, B., Ohlsson, S., Brown, C., Chen, L.: Data driven automatic feedback generation in the iList intelligent tutoring system. Technol. Instr. Cogn. Learn. **10**(1), 5–26 (2015)

13. Graesser, A., VanLehn, K., Rose, C., Jordan, P., Harter, D.: Intelligent tutoring systems with conversational dialogue. AI Mag. **22**(4), 39–51 (2001)
14. Honey, P., Mumford, A.: The Manual of Learning Styles. Peter Honey Publications, Maidenhead (1992)
15. Hrastinski, S., Stenbom, S., Benjaminsson, S., Jansson, M.: Identifying and exploring the effects of different types of tutor questions in individual online synchronous tutoring in mathematics. Interactive Learn. Environ. **1–13** (2019)
16. Kashihara, A., Hasegawa, S.: A model of meta-learning for web-based navigational learning. Int. J. Adv. Technol. Learn. **2**(4), 198–206 (2005)
17. Koedinger, K., Corbett, A.: Cognitive tutors: technology bringing learning sciences to the classroom. In: Sawyer, R.K. (ed.) The Cambridge Handbook of the Learning Sciences, pp. 61–78. Cambridge University Press, New York (2006)
18. Kolb, D.A.: Experiential Learning Experience as the Source of Learning and Development. Prentice Hall, New Jersey (1984)
19. Kumar, A.N.: Results from the evaluation of the effectiveness of an online tutor on expression evaluation. ACM SIGCSE Bull. **37**(1), 216–220 (2005)
20. Kumar, A.N.: Generation of problems, answers, grade, and feedback–case study of a fully automated tutor. J. Educ. Resourc. Comput. (JERIC) **5**(3), 3-es (2005)
21. Lin, C.F., Yeh, Y.C., Hung, Y.H., Chang, R.I.: Data mining for providing a personalized learning path in creativity: an application of decision trees. Comput. Educ. **68**, 199–210 (2013). https://doi.org/10.1016/j.compedu.2013.05.009
22. Ma, W., Adesope, O.O., Nesbit, J.C., Liu, Q.: Intelligent tutoring systems and learning outcomes: a meta-analysis. J. Educ. Psychol. (2014)
23. Mark, M.A., Greer, J.E.: Evaluation methodologies for intelligent tutoring systems. J. Artif. Intell. Educ. **4**, 129–129 (1993)
24. Melis, E., Siekmann, J.: ActiveMath: an intelligent tutoring system for mathematics. Artif. Intell. Soft Comput. **91–101** (2004)
25. Munshi, A., Biswas, G.: Personalization in OELEs: developing a data-driven framework to model and scaffold SRL processes. In: Isotani, S., Millán, E., Ogan, A., Hastings, P., McLaren, B., Luckin, R. (eds.) AIED 2019. LNCS (LNAI), vol. 11626, pp. 354–358. Springer, Cham (2019). https://doi.org/10.1007/978-3-030-23207-8_65
26. Narciss, S., et al.: Exploring feedback and student characteristics relevant for personalizing feedback strategies. Comput. Educ. **71**, 56–76 (2014)
27. Nye, B.D., Graesser, A.C., Hu, X.: AutoTutor and family: a review of 17 years of natural language tutoring. Int. J. Artif. Intell. Educ. **24**(4), 427–469 (2014)
28. Onyema, E.M., et al.: Impact of Coronavirus pandemic on education. J. Educ. Pract. **11**(13), 108–121 (2020)
29. Penstein, C.R., Moore, J., VanLehn, K., Allbritton, D.: A comparative evaluation of socratic versus didactic tutoring. Ann. Meeting Cogn. Sci. Soc. **23** (2001)
30. Psotka, J., Massey, D., Mutter, S.: Intelligent Tutoring Systems: Lessons Learned. Lawrence Erlbaum Associates, Hillsdale (1988)
31. Rus, V., Stefanescu, D., Baggett, W., Niraula, N., Franceschetti, D., Graesser, A.C.: Macro-adaptation in conversational intelligent tutoring matters. In: Trausan-Matu, S., Boyer, K.E., Crosby, M., Panourgia, K. (eds.) ITS 2014. LNCS, vol. 8474, pp. 242–247. Springer, Cham (2014). https://doi.org/10.1007/978-3-319-07221-0_29
32. Rus, V., Stefanescu, D., Niraula, N., Graesser, A.C.: DeepTutor: towards macro- and micro-adaptive conversational intelligent tutoring at scale. In: Proceedings of the first ACM Conference on Learning@ Scale Conference, pp. 209–210 (2014)

33. Sampson, D., Karagiannidis, C.: Personalised learning: educational, technological and standarisation perspective. Digit. Educ. Rev. **4**, 24–39 (2002)
34. Serban, I.V., et al.: A large-scale, open-domain, mixed-interface dialogue-based ITS for STEM. In: Bittencourt, I.I., Cukurova, M., Muldner, K., Luckin, R., Millán, E. (eds.) AIED 2020. LNCS (LNAI), vol. 12164, pp. 387–392. Springer, Cham (2020). https://doi.org/10.1007/978-3-030-52240-7_70
35. Kochmar, E., Vu, D.D., Belfer, R., Gupta, V., Serban, I.V., Pineau, J.: Automated personalized feedback improves learning gains in an intelligent tutoring system. In: Bittencourt, I.I., Cukurova, M., Muldner, K., Luckin, R., Millán, E. (eds.) AIED 2020. LNCS (LNAI), vol. 12164, pp. 140–146. Springer, Cham (2020). https://doi.org/10.1007/978-3-030-52240-7_26
36. Grenander, M., Belfer, R., Kochmar, E., Serban, I.V., St-Hilaire, F., Cheung, J.C.: Deep discourse analysis for generating personalized feedback in intelligent tutor systems. In: The 11th Symposium on Educational Advances in Artificial Intelligence (2021)
37. Stash, N.V., Cristea, A.I., De Bra, P.M.: Authoring of learning styles in adaptive hypermedia: problems and solutions. In: Proceedings of the 13th International World Wide Web Conference on Alternate Track Papers & Posters, pp. 114–123 (2004)
38. Tan, Y., Quintana, R.M.: What can we learn about learner interaction when one course is hosted on two MOOC platforms? In: Companion Proceedings to the International Conference on Learning Analytics and Knowledge (LAK), pp. 155–156 (2019)
39. Tomkins, S., Ramesh, A., Getoor, L.: Predicting Post-Test Performance from Online Student Behavior: A High School MOOC Case Study. International Educational Data Mining Society (2016)
40. VanLehn, K., Graesser, A.C., Jackson, G., Jordan, P., Olney, A., Rose, C.P.: When are tutorial dialogues more effective than reading? Cogn. Sci. **31**(1), 3–62 (2007)
41. VanLehn, K.: The relative effectiveness of human tutoring, intelligent tutoring systems, and other tutoring systems. Educ. Psychol. (2011)
42. Wang, Y., Paquette, L., Baker, R.: A longitudinal study on learner career advancement in MOOCs. J. Learn. Anal. **1**(3), 203–206 (2014)
43. Wood, D., Wood, H.: Vygotsky, tutoring and learning. Oxf. Rev. Educ. **22**(1), 5–16 (1996)
44. Woolf, B.: Building Intelligent Interactive Tutors. Morgan Kaufmann Publishers, Burlington (2009)
45. Yin, B., Patikorn, T., Botelho, A.F., Heffernan, N.T.: Observing personalizations in learning: identifying heterogeneous treatment effects using causal trees. In: Proceedings of the Fourth (2017) ACM Conference on Learning@ Scale, pp. 299–302 (2017)

Explaining Engagement: Learner Behaviors in a Virtual Coding Camp

Angela E. B. Stewart[1]([✉]) [iD], Jaemarie Solyst[1] [iD], Amanda Buddemeyer[2] [iD],
Leshell Hatley[3] [iD], Sharon Henderson-Singer[4], Kimberly Scott[4] [iD],
Erin Walker[2] [iD], and Amy Ogan[1] [iD]

[1] Carnegie Mellon University, Pittsburgh, PA 15213, USA
angelast@andrew.cmu.edu
[2] University of Pittsburgh, Pittsburgh, PA 15213, USA
[3] Uplift Inc., Washington D.C. 20013, USA
[4] Arizona State University, Tempe, AZ 85281, USA

Abstract. Engagement is critical to learning, yet current research rarely explores its underlying contextual influences, such as differences across modalities and tasks. Accordingly we examine how patterns of behavioral engagement manifest in a diverse group of ten middle school girls participating in a synchronous virtual computer science camp. We form multimodal measures of behavioral engagement from learner chats and speech. We found that the function of modalities varies, and chats are useful for short responses, whereas speech is better for elaboration. We discuss implications of our work for the design of intelligent systems that support online educational experiences.

Keywords: Engagement · Underrepresented learners · Virtual camp

1 Introduction

Empirical research has long confirmed that engagement is essential to learning [2, 6]. Although a precise definition of engagement is elusive, researchers agree that it consists of complexly interwoven behavioral and psychological components [7,15,17]. Given this critical link between engagement and learning, researchers have created innovations to improve outcomes, particularly through AI systems that detect engaged learning behaviors and intervene accordingly. However these works overwhelmingly consider a narrow view of engagement, classifying learners as overall engaged or not. This does not take into account for a given learner how their engagement might vary across interaction modalities (e.g., speaking out loud versus text-based chatting) and tasks. For example, a learner may appear to be disengaged because they are not actively speaking up. However, in a small group setting they might start to talk more as they become more comfortable. These multiple views are important because, as noted by culturally-responsive engagement frameworks, learner behaviors will differ, as their values and cultural norms differ [2]. A better understanding of how engagement manifests across

© Springer Nature Switzerland AG 2021
I. Roll et al. (Eds.): AIED 2021, LNAI 12749, pp. 338–343, 2021.
https://doi.org/10.1007/978-3-030-78270-2_60

these varying contextual factors is crucial to better design of AI systems, as it can inform what is being modeled and the types of interventions that will be most effective.

Our work begins to fill this gap by understanding the behavioral engagement patterns that emerge in a diverse group of middle school girls participating in an online computer science camp. Our focus in this work is behavioral engagement, which refers to a learner's participation and presence in the environment [7,15]. We focus on behavioral engagement because its indicators, such as attendance or participation, are directly measurable [15]. We use chat and speech signals to represent verbal contributions, such as sharing artifacts built in the camp.

Most closely related to our work is that on detecting learners' engaged behaviors [5,8,9,11,13,14,18]. More limited work has sought to explain the types of engaged behaviors that occur in educational environments, such as participation via help-giving behaviors [1,12], self-regulated learning behaviors [4], or on-topicness and frequency of MOOC posts [18]. Recent work has combined signals to form a multimodal understanding of engagement [10,19]. In an environment most similar to ours, Lin et al. [11] studied an online flipped course. They used a combination of log and behavioral data (e.g., punctuality, camera on versus off). They found that students who watched more pre-lecture videos had better conceptual understanding and higher grades, and students who arrived on time with their cameras on interacted more. These works give a more holistic view of engagement by including multiple modalities in their analysis. However, there is still a gap in describing the various contextual factors that influence engagement (e.g., differences across modalities and tasks).

2   Data Collection and Processing

Participants were ten middle school girls (ages 12-14) from diverse racial backgrounds (four white, one Hispanic/Latina and white, one Asian and white, two Hispanic/Latina, one American Indian/Alaskan Native, and one chose not to report). Nine of the ten learners indicated some form of previous programming experience. Learners were monetarily compensated for their participation. The virtual coding camp took place over three days (two to three hours a day), on Zoom, an online videoconferencing platform (https://zoom.us/). Chats and audio were recorded with Zoom's built-in functionality.

We designed the camp to provide a culturally-responsive, introductory computing experience. Culturally-responsive computing aims to address not only technical literacy, but community, culture, and identity [16]. Led by three facilitating instructors, the camp included activities focusing on both computer science concepts and reflections on power and identity (descriptions shown in Table 1). For coding activities, learners used a custom-built, online, block-based programming interface, where the goal was to use code blocks to control a robotic character.

We utilized data from learner chat and speech contributions. We removed data from time periods without relevant activity (e.g., logging into the session).

Table 1. Description of the categories and number of activities are shown.

Category	#	Description
Active Prompt	2	Respond to prompt via chat
Breakout Room Activity	2	Collaborate in small groups to solve a problem
Coding	6	Individual programming assignments following the lessons
Community Building	6	Get to know other learners
Feedback	3	Give facilitators feedback on how to improve the camp
Lesson	9	Learn computer science concepts and how to implement in coding interface
Movement	3	Move around to increase energy
Power and Identity	11	Reflect on culture and representations of power and identity
Presentation	5	Presentations about robots, coding, and notable women of color in computing
Share Out	6	Share coding creations from Coding assignments

We replaced emojis or emoticons (e.g., :)) with the word *emoticon* so they could be included as a single word for analysis. We transcribed the speech data using a third-party service. Two members of the research team quality checked the transcriptions and removed 22 utterances for which the speaker could not be identified. For both modalities, we tokenized words using NLTK [3]. In total, 759 chats and 638 utterances were included in our analyses.

We summarized the signals at the activity level in order to compare behavioral engagement across modalities. To do this, we counted the number of chats or utterances in a given activity, and standardized by the duration (minutes) of the activity. Our final **engagement frequency** metrics were words chatted and words spoken per minute. For each modality, we also calculated a category-level **binary engagement** value as whether the learner engaged at any point during that category (e.g., the learner sent at least one chat).

3 Results and Discussion

The distribution of the proportion of categories in which learners were engaged (binary engagement) is shown in Fig. 1. All learners engaged via chat in at least 50% of the categories, suggesting widespread learner preference for chat. Compared to chats, there was more variability in whether learners spoke aloud. Overall speech was less frequent with 30% of learners speaking in less than half the categories. This finding is unsurprising as there are fewer barriers to chatting than speaking (e.g., no need for a working microphone or quiet space to talk).

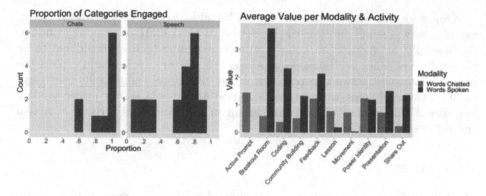

Fig. 1. (Left) For each modality, the distribution of categories in which learners were engaged are shown. (Right) For each activity category, the average words chatted and spoken per minute are shown.

In order to understand how behavioral engagement differs across tasks, we used the engagement frequency metrics to calculate the average words chatted or spoken per minute for each category (Fig. 1). Speech contributions dominate chat contributions for almost every activity, suggesting learners were more verbose when speaking aloud than chatting. We confirmed this finding by calculating the average words per utterance (9.94) compared to the average words per chat (3.58). Taken together with our previous findings, we hypothesize that a frequency-verbosity trade off affects behavioral engagement patterns for chats and speech. As an illustrative example, in a Breakout Room activity, learners collaboratively designed a robot character and provided more in-depth responses aloud than via chat. One learner spoke about hobbies for the robot: "No-no, oddly specific is what makes people actually enjoy... Very specific, quirky things that make you go 'oh, that seems just like what a human would do', is really what brings things together." Another learner added, "Yeah, it gives the robot a personality. It's not just something made in a factory, it has interactions and you can relate to it, in a way." A third learner suggested a robot hobby via chat, "banjo." In this example, the function of the chat was short, quick responses, and learners elaborated aloud. Indeed for the categories where chatting was the dominant contribution modality (Active Prompt, Lesson, Movement), the task at hand required short, quick responses via chat (e.g., an Active Prompt activity was to write conditionals via the chat).

Our findings provide insight into the design of AI in education systems. We show that a one-size-fits all definition of behavioral engagement does not work in practice, as behavioral patterns differed by modality and task. Thus, intelligent systems should consider flexible definitions of engagement that take context into account [7]. Understanding where and why learners are engaging can guide the kinds of interventions that are most appropriate. This is especially important for marginalized learners, whose engaged behaviors might differ [2,7].

Our work has limitations that should be addressed in future research. Our sample size was small, limiting the kinds of statistical analyses we could conduct. Additionally, we focused on behavioral engagement, which is considered to be the product of other psychological processes [15]. Future work should explore the complex interplay between psychological and behavioral components of engagement. That said, this work presents important steps towards understanding behavioral engagement of a diverse group of middle school girls in a virtual computer science camp.

References

1. Ahmed, I., et al.: Investigating help-giving behavior in a cross-platform learning environment. In: Isotani, S., Millán, E., Ogan, A., Hastings, P., McLaren, B., Luckin, R. (eds.) AIED 2019. LNCS (LNAI), vol. 11625, pp. 14–25. Springer, Cham (2019). https://doi.org/10.1007/978-3-030-23204-7_2
2. Bingham, G.E., Okagaki, L.: Ethnicity and student engagement. In: Christenson, S.L., Reschly, A.L., Wylie, C. (eds.) Handbook of Research on Student Engagement, pp. 65–95. Springer, Boston (2012). https://doi.org/10.1007/978-1-4614-2018-7_4
3. Bird, S., Klein, E., Loper, E.: Natural Language Processing with Python: Analyzing Text with the Natural Language Toolkit. O'Reilly Media, Inc. (2009)
4. Cicchinelli, A., et al.: Finding traces of self-regulated learning in activity streams. In: Proceedings of the 8th International Conference on Learning Analytics and Knowledge, pp. 191–200 (2018)
5. Dixson, M.D.: Measuring student engagement in the online course: the online student engagement scale (OSE). Online Learn. **19**(4), n4 (2015)
6. Finn, J.D., Zimmer, K.S.: Student engagement: What is it? Why does it matter? In: Christenson, S.L., Reschly, A.L., Wylie, C. (eds.) Handbook of Research on Student Engagement, pp. 97–131. Springer, Boston (2012). https://doi.org/10.1007/978-1-4614-2018-7_5
7. Fredricks, J.A., Filsecker, M., Lawson, M.A.: Student engagement, context, and adjustment: addressing definitional, measurement, and methodological issues. Learn. Instr. **43**, 1–4 (2016). https://doi.org/10.1016/j.learninstruc.2016.02.002, special Issue: Student engagement and learning: theoretical and methodological advances
8. Hayati, H., Khalidi Idrissi, M., Bennani, S.: Automatic classification for cognitive engagement in online discussion forums: text mining and machine learning approach. In: Bittencourt, I.I., Cukurova, M., Muldner, K., Luckin, R., Millán, E. (eds.) AIED 2020. LNCS (LNAI), vol. 12164, pp. 114–118. Springer, Cham (2020). https://doi.org/10.1007/978-3-030-52240-7_21
9. Henderson, N., Rowe, J., Paquette, L., Baker, R.S., Lester, J.: Improving affect detection in game-based learning with multimodal data fusion. In: Bittencourt, I.I., Cukurova, M., Muldner, K., Luckin, R., Millán, E. (eds.) Artificial Intelligence in Education, pp. 228–239. Springer, Cham (2020). https://doi.org/10.1007/978-3-030-52237-7_19
10. Kim, Y., Butail, S., Tscholl, M., Liu, L., Wang, Y.: An exploratory approach to measuring collaborative engagement in child robot interaction. In: Proceedings of the Tenth International Conference on Learning Analytics & Knowledge, pp. 209–217 (2020)

11. Lin, L.C., Hung, I.C., Chen, N.S., et al.: The impact of student engagement on learning outcomes in a cyber-flipped course. Education Tech. Research Dev. **67**(6), 1573–1591 (2019)
12. Mawasi, A., et al.: Using design-based research to improve peer help-giving in a middle school math classroom. In: Gresalfi, M., Horn, I.S. (eds.) The Interdisciplinarity of the Learning Sciences, 14th International Conference of the Learning Sciences, pp. 1189–1196. International Society of the Learning Sciences (ISLS) (2020)
13. Mota, S., Picard, R.W.: Automated posture analysis for detecting learner's interest level. In: 2003 Conference on Computer Vision and Pattern Recognition Workshop, vol. 5, p. 49. IEEE (2003)
14. Munshi, A., et al.: Modeling the relationships between basic and achievement emotions in computer-based learning environments. In: Bittencourt, I.I., Cukurova, M., Muldner, K., Luckin, R., Millán, E. (eds.) Artificial Intelligence in Education, pp. 411–422. Springer, Cham (2020). https://doi.org/10.1007/978-3-030-52237-7_33
15. Reschly, A.L., Christenson, S.L.: Jingle, jangle, and conceptual haziness: evolution and future directions of the engagement construct. In: Christenson, S.L., Reschly, A.L., Wylie, C. (eds.) Handbook of Research on Student Engagement, pp. 3–19. Springer, Boston (2012). https://doi.org/10.1007/978-1-4614-2018-7_1
16. Scott, K.A., Sheridan, K.M., Clark, K.: Culturally responsive computing: a theory revisited. Learn. Media Technol. **40**(4), 412–436 (2015)
17. Sinatra, G.M., Heddy, B.C., Lombardi, D.: The challenges of defining and measuring student engagement in science. Educ. Psychol. **50**(1), 1–13 (2015). https://doi.org/10.1080/00461520.2014.1002924
18. Yan, W., Dowell, N., Holman, C., Welsh, S.S., Choi, H., Brooks, C.: Exploring learner engagement patterns in teach-outs using topic, sentiment and on-topicness to reflect on pedagogy. In: Proceedings of the 9th International Conference on Learning Analytics & Knowledge, pp. 180–184 (2019)
19. Zhang, Z., Li, Z., Liu, H., Cao, T., Liu, S.: Data-driven online learning engagement detection via facial expression and mouse behavior recognition technology. J. Educ. Comput. Res. **58**(1), 63–86 (2020)

Using AI to Promote Equitable Classroom Discussions: The TalkMoves Application

Abhijit Suresh[1,2($\boxtimes$)], Jennifer Jacobs[1,2], Charis Clevenger[1,2], Vivian Lai[1,2], Chenhao Tan[1,2], James H. Martin[1,2], and Tamara Sumner[1,2]

[1] Institute of Cognitive Science, University of Colorado Boulder, Boulder, USA
{Abhijit.Suresh,Jennifer.Jacobs,Charis.Clevenger,Vivian.Lai,
Chenhao.Tan,James.Martin,Tamara.Sumner}@colorado.edu
[2] Department of Computer Science, University of Colorado Boulder, Boulder, USA

Abstract. Inclusion in mathematics education is strongly tied to discourse rich classrooms, where students ideas play a central role. Talk moves are specific discursive practices that promote inclusive and equitable participation in classroom discussions. This paper describes the development of the TalkMoves application, which provides teachers with detailed feedback on their usage of talk moves based on accountable talk theory. Building on our recent work to automate the classification of teacher talk moves, we have expanded the application to also include feedback on a set of student talk moves. We present results from several deep learning models trained to classify student sentences into student talk moves with performance up to 73% F1. The classroom data used for training these models were collected from multiple sources that were pre-processed and annotated by highly reliable experts. We validated the performance of the model on an out-of-sample dataset which included 166 classroom transcripts collected from teachers piloting the application.

Keywords: Accountable talk · Deep learning · BERT · Equity-focused instruction · AI application · Mathematics · Teacher learning · Classroom discourse

1 Introduction

1.1 Overview of the TalkMoves Application

The TalkMoves application [1] serves as an exemplar for a new type of translational activity enabled by big data: the reification of existing, well-researched theoretical frameworks in deep learning models. In particular, this application

Supported by National Science Foundation under Grant No. 1837986: The TalkBack Application: Automating Analysis and Feedback to Improve Mathematics Teachers' Classroom Discourse.

draws on accountable talk theory [6,9] as a framework for providing fully auto-mated feedback to mathematics teachers on specific discourse moves they and their students used during classroom instruction. The application is an inno-vative AI-driven platform that builds on advances in deep learning for natural language processing and speech recognition to automatically analyze teaching episodes and offer users near real-time feedback.

Using talk moves is an equity-focused endeavor [8]. Forming and sustaining a learning environment based on accountable talk theory can be particularly beneficial for girls and students from home backgrounds, inculturating them into the norms of democratic discourse that can later be realized in wider civic spheres [6]. Shifting away from traditional discourse patterns towards account-able talk makes space for students' contributions, especially for English Lan-guage Learners, through a focus on communicating mathematically and pre-senting arguments rather than acquiring vocabulary and other low-level linguistic skills [4,7]. Furthermore, increased participation by students from non-dominant groups can foster dispositions in which attention is given to competencies and resources rather than deficits and obstacles [3]. The current version of the Talk-Moves application includes feedback on a set of six teacher (keeping everyone together, getting students to relate to another's ideas, restating, pressing for accuracy, revoicing and pressing for reasoning) and four student talk moves (relating to another student, asking for more info, making a claim, and pro-viding evidence/reasoning).

2 Approach

2.1 Data

We collected 461 written transcripts of mathematics teaching episodes from ele-mentary and secondary (K-12) math lessons drawing on multiple sources. In addition, 166 transcripts were available from 21 teachers who piloted the Talk-Moves application during the 2019–2020 academic year. Written transcripts were segmented into sentences using an automated script. Each sentence in the tran-script was manually coded by two annotators for teacher and student talk moves; these coded sentences served as the "ground-truth" training dataset for our mod-els. The annotators established reliability prior to applying the codes and again when they were approximately halfway through coding to ensure that their anno-tations remained accurate and consistent. Reliability, calculated using Cohen's kappa, was high for each talk move at both time periods and ranged from 0.88–1.0.

2.2 Model Development and Performance

The set of 461 transcripts was used for training while the set of 166 transcripts was used as an out-of-sample testing set. The dataset of 461 transcripts is made up of 176,757 sentences including 115,418 teacher utterances and 61,339 student utterances. The data was split into training and validation set according to a 90/10% split. The validation set was stratified to mimic the distribution of the labels in the training set. Similarly, the out of sample test set of 166 transcripts had 49,048 teacher utterances and 13,968 student utterances collected from the pilot study. In this initial study, we chose to train two independent deep learning models to automatically identify the teacher and student talk moves. For the teacher model, starting with LSTM neural networks, we explored different model architectures including recent transformer models such as BERT and RoBERTa [10,11]. The inputs to the model were student-teacher "sentence pairs", which refers to a combination of a teacher sentence concatenated with the immediately prior student sentence. For example, a sentence pair can include a student utterance "I'm pretty sure the numerator would be four" followed by a teacher utterance "Okay, why do you think the numerator would be four?". This sentence pair is a good example of the teacher encouraging a student to reason (pressing for reasoning). The output of the deep learning model is a 7-way sequence classification (softmax) over the six teacher talk moves and "None".

The inputs to the student model were student-student "sentence pairs", which refer to a combination of a student sentence concatenated with the immediately prior student sentence. The output was a 5-way sequence classification (softmax) over the four student talk moves and "None". After hypertuning parameters of the model, we found that RoBERTa base performed the best among the BASE models for both teacher and student sequence classification (see Table 1). We did not find evidence for a significant change in performance on other variants of BERT including XLNet-base and AlBERT-base.

Table 1. Performance of teacher and student models on the out-of-sample test set

	F1 score (in %)	MCC
Teacher Model – BERT-base [2] – RoBERTa- base [5]	76.05 **77.29**	0.7519 **0.7627**
Student Model – BERT-base [2] – RoBERTa-base[5]	71.96 **73.04**	0.6585 **0.6727**

2.3 Confusion Matrix

The confusion matrix for the student talk moves model (see Table 2) indicates that the talk move "relating to another student" performed worse relative to the other student talk moves. We intend to conduct additional experiments to determine whether the performance can be improved by extending sentence pairs to include multiple previous student sentences as context.

Table 2. Confusion matrix of the student talk moves model

RoBERTa-base: 73.04% F1	0	1	2	3	4	Precision	Recall	F1
0 - None	5284	221	139	415	61	0.82	0.86	0.84
1 - Relating to another student	209	407	8	149	84	0.46	0.47	0.47
2 - Asking for more information	59	6	245	17	6	0.56	0.71	0.63
3 - Making a claim	798	209	34	3611	313	0.82	0.73	0.77
4 - Providing Evidence	59	39	11	210	1364	0.74	0.81	0.77

3 Conclusion

The TalkMoves application provides teachers with detailed feedback on their use of research-based discourse practices, in the form of specific instructional talk moves that prior research suggests promote inclusion and equity [6]. The strong performance of both the student and teacher talk moves models illustrates the reliability and robustness of artificial intelligence algorithms applied to noisy real-world classroom data. The student utterances extracted from classroom transcripts are especially challenging to interpret relative to teacher speech. For example, there are numerous instances where the utterances lacked well-formed syntax, including missing words. Despite the limitations of this noisy dataset, BASE models achieved good results in classifying the student talk moves, particularly Roberta BASE which was used for both the teacher and student models. As a next step, we plan to experiment with strategies such that the student model performs on par with the teacher model. In addition, future research is needed to continually improve the model performance of individual talk moves, better understand teachers' perceptions and use of the application, and consider how it can be incorporated into structured professional learning opportunities that promote discourse-rich pedagogy.

References

1. http://talkmoves.com/
2. Devlin, J., Chang, M.W., Lee, K., Toutanova, K.: BERT: pre-training of deep bidirectional transformers for language understanding. arXiv preprint arXiv:1810.04805 (2018)
3. Hand, V.: Seeing culture and power in mathematical learning: toward a model of equitable instruction. Educ. Stud. Math. **80**(1), 233–247 (2012). https://doi.org/10.1007/s10649-012-9387-9
4. Khisty, L.L., Chval, K.B.: Pedagogic discourse and equity in mathematics: when teachers' talk matters. Math. Educ. Res. J. **14**(3), 154–168 (2002). https://doi.org/10.1007/BF03217360
5. Liu, Y., et al.: RoBERTa: a robustly optimized BERT pretraining approach. arXiv preprint arXiv:1907.11692 (2019)

6. Michaels, S., O'Connor, C., Resnick, L.B.: Deliberative discourse idealized and realized: accountable talk in the classroom and in civic life. Stud. Philos. Educ. **27**(4), 283–297 (2008). https://doi.org/10.1007/s11217-007-9071-1
7. Moschkovich, J.: Principles and guidelines for equitable mathematics teaching practices and materials for English language learners. J. Urban Math. Educ. **6**(1), 45–57 (2013)
8. O'Connor, C., Michaels, S.: Supporting teachers in taking up productive talk moves: the long road to professional learning at scale. Int. J. Educ. Res. **97**, 166–175 (2019)
9. O'Connor, C., Michaels, S., Chapin, S.: Scaling down to explore the role of talk in learning: from district intervention to controlled classroom study. In: Socializing Intelligence Through Academic Talk and Dialogue, pp. 111–126 (2015)
10. Suresh, A., Sumner, T., Huang, I., Jacobs, J., Foland, B., Ward, W.: Using deep learning to automatically detect talk moves in teachers' mathematics lessons. In: 2018 IEEE International Conference on Big Data (Big Data), pp. 5445–5447. IEEE (2018)
11. Suresh, A., Sumner, T., Jacobs, J., Foland, B., Ward, W.: Automating analysis and feedback to improve mathematics teachers' classroom discourse. Proc. AAAI Conf. Artif. Intell. **33**, 9721–9728 (2019)

Investigating Effects of Selecting Challenging Goals

Faiza Tahir[(✉)], Antonija Mitrović [iD], and Valerie Sotardi

University of Canterbury, Christchurch, New Zealand
faiza.tahir@pg.canterbury.ac.nz

Abstract. Goal setting is a vital component of self-regulated learning. Numerous studies show that selecting challenging goals has strong positive effects on performance. We investigate the effect of support for goal setting in SQL-Tutor. The experimental group had support for selecting challenging goals, while the control group students could select goals freely. The experimental group achieved the same learning outcomes as the control group, but by attempting and solving significantly fewer, but more complex problems. Causal modelling revealed that the experimental group students who selected more challenging goals were superior in problem solving. We also found a significant improvement in self-reported goal setting skills of the experimental group.

Keywords: Self-regulated learning · Goal setting · Intelligent tutoring system

1 Introduction

Self-Regulated Learning (SRL) is defined as an "active, constructive process whereby learners set goals for their learning and then attempt to monitor, regulate, and control their cognition, motivation and behavior guided and constrained by their goals and the contextual features in the environment" (Zimmerman 2011). The goal-setting theory illustrates that setting difficult goals lead to higher performance (Locke and Latham 1990, 2019). Many studies show the benefits of goal-setting activities (Latham and Yukl 1975; Locke and Latham 2002), the power of self-set goals (Locke 2001), influence of various strategies in goal attainment (Seijts and Latham 2005; Masuda et al. 2015), and the effects of goal commitment (Landers et al. 2017). Zimmerman (2002) reported that students who set precise and actionable goals often reported higher self-awareness and had higher achievements. As mentioned in a meta-review of achievement (Collins 2004), meeting a standard or goal is not enough; one should struggle for excellence. The goal-setting theory discussed the greater effects of task-specific over non-task related goals (Latham and Piccolo 2012) and effects of selecting challenging goals (Latham et al. 2017).

Goal setting has been studied in various learning environments (Melis and Siekmann 2004; Davis et al. 2016; Cicchinelli et al. 2018). In the context of AIED, relevant research connects students' goal-setting behavior with their motivation (Bernacki et al. 2013; Carr et al. 2013; Duffy and Azevedo 2015). Crystal Island (Rowe et al. 2011) asks students

© Springer Nature Switzerland AG 2021
I. Roll et al. (Eds.): AIED 2021, LNAI 12749, pp. 349–354, 2021.
https://doi.org/10.1007/978-3-030-78270-2_62

to solve a mystery by accomplishing eleven goals. Their results reveal that students who achieved more goals significantly improved their learning performance. In Meta-Tutor (Harley et al. 2017), four pedagogical agents support SRL via dialogs with the student. The agents determine the student's previous knowledge, and assist the student in selecting goals. Evaluation of Meta-Tutor revealed that students who collaborated more with agents learnt more. This paper discusses the effects of selecting challenging goals on learning in the context of SQL-Tutor (Mitrovic 2003).

2   Study Design and Procedure

We enhanced SQL-Tutor by adding support for all three phases of the Zimmerman's model (2003), but in this paper we focus on the forethought phase only. SQL-Tutor contains over 300 problems, classified using 38 different problem templates (Mathews 2006). A problem template covers a set of problems, which require the same problem-solving strategy. The 38 problem templates are grouped into eight high-level goals. The student is required to select a goal at the start of each session, and also after achieving a goal. The system always suggests challenging goals. The student is free to select one of the suggested goals, or any other goal.

We use a simple heuristic strategy to select a challenging goal for the student. At the start, students complete a pre-test, with scores ranging from 0 to 9. The initial goal is determined based on the student's pre-test score, while for the subsequent ones the system considers the student's current level (*slevel*). The student level ranges from 1 to 9, and it is determined dynamically, based on the student's success during problem solving (Mitrovic 2003). For example, if the student scored 6 or more on the pre-test (i.e. the median score or higher), the challenging goal should be 8. The goal-setting page shows the number of problems per goal, and the number of problems the student has solved. The previously achieved goals are highlighted. If the student with a low pre-test score selects a very challenging goal, the system would suggest a less challenging one. To achieve a goal, the student needs to complete at least half of the relevant problems, or solve the five most complex problems.

The SRL instrument used in the study was adopted from (Kizilcec et al. 2017). Out of 24 questions, in this paper we only discuss the goal-setting subscale (4 questions). We also added five self-efficacy (SE) questions from the Motivated Strategies for Learning Questionnaire (Pintrich and De Groot 1990). The survey used a five-point Likert scale, ranging from "Not at all true for me" (1) to "Very true for me" (5). The SRL and SE questions were included in Survey 1.

The participants, volunteers from the second-year database course at the University of Canterbury in 2020, were randomly allocated to the experimental (57) and control (42) groups. After providing informed consent, the participants completed the pre-test and Survey 1. The experimental group received support during goal setting, while the control group participants selected goals freely. After selecting a goal, students could choose any problem. The study lasted for four weeks. At the end of the study, students completed the post-test of similar structure and complexity as the pre-test, and completed Survey 2 (which was identical to Survey 1).

We hypothesized that the experimental group would achieve higher learning outcomes (H1). We formed a hypothesis for the experimental group: that selecting challenging goals would affect students' learning positively (H2). Finally, we expected that the support for goal setting would improve students' goal-setting skills (H3).

3 Results

We compared the pre/post-test scores of participants who completed both tests (Table 1). There is no significant difference on pre-test scores of the control (59.88%, sd = 28.82) and experimental groups (55.56%, sd = 29.18). The experimental group improved significantly from pre- to post-test (W = 298, p = .03), but the control group students did not (p = .74). Comparing normalized gains revealed no significant difference. These results partially support hypothesis H1. The control group attempted/completed significantly more problems (Table 1). The experimental group completed significantly more complex problems. These findings show that experimental group achieved higher learning gains by completing fewer but more complex problems.

Table 1. Summary of major statistics: mean (sd)

	Control (42)	Experimental (57)	Significance
Attempted Problems	92.98 (61.86)	57.46 (41.33)	U = 783, p = .003
Completed Problems	91.86 (61.33)	56.44 (41.09)	U = 783, p = .003
Problem Complexity	2.92 (0.96)	3.32 (1.08)	U = 1465.5, p = .057
Time (min)	360.19 (335.33)	296.71 (233.22)	p = .58

Table 2. Summary statistics for the three subgroups: mean (sd)

	SEQ (18)	Mix (25)	SG (14)
Pre-test %	62.97 (27.75)	64.46 (24.01)	61.11 (33.98)
Post-Test %	n = 9, 64.21 (31.81)	n = 12, 77.78 (28.55)	n = 8, 69.45 (21.23)
Time (min)	346.17 (290.49)	283.36 (163.41)	257.0 (263.09)
Attempted goals	6.39 (2.62)	7.04 (1.14)	5.00 (2.18)
Achieved goals	4.72 (2.54)	3.60 (2.43)	1.64 (2.34)
Attempted Problems	78.28 (44.84)	58.96 (38.18)	28.0 (22.34)
Problem Solved	77.50 (44.23)	57.88 (38.02)	26.79 (21.92)
Problem Complexity	2.85 (.74)	3.11 (.86)	4.31 (1.20)

We divided the experimental group post-hoc into three subgroups (Table 2). Fourteen students always accepted the suggested goals (SG), 18 students worked on the goals in the sequential order (SEQ), while the remaining 25 students used a mixed strategy (Mix). We found no significant differences between the subgroups on the pre-/post-test scores and time, but there were statistically significant differences on the number of attempted goals (H = 8.12, p = .017), achieved goals (H = 10.13, p = .006), the number of attempted/solved problems (H = 13.88, p = .001 and H = 14.41, p = .001 respectively), and problem complexity (H = 12.20, p = .002). The post-hoc analyses revealed no significant differences between the SEQ and Mix groups. The SG subgroup attempted significantly fewer goals in comparison to the SEQ (U = 55, p = .006) and Mix groups (U = 94, p = .016), and achieved significantly fewer goals in comparison to the Mix group (U = 77, p = .003). The SG group also attempted/solved significantly fewer problems in comparison to the SEQ (U = 44, p = .002 in both cases) and Mix groups (U = 74.5, p = .003 and U = 71, p = .002 respectively). However, the average problem complexity of solved problems for the SG group was significantly higher in comparison to the SEQ (U = 40.5, p = .001) and Mix (U = 77.5, p = .004) groups.

We analyzed the data using the structural equation model (Fig. 1). We hypothesized that the pretest score and the number of attempted problems will have a positive effect on learning. The variable labelled "Accepted goals" shows how many times students accepted the suggested goals. Because not all students completed the post-test, we use a different measure of learning: the high-

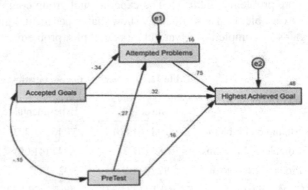

Fig. 1. Multiple mediation model

est achieved goal (HAG). All path coefficients are significant at p < .05 except PreTest -> HAG, and the covariance between Accepted Goals and PreTest. There is a significant negative effect of Accepted goals on Attempted problems. These findings suggest that (1) students who accepted system goals tended to achieve higher goals (the confidence interval [.1345, .7074] does not include zero), and (2) students who accepted suggested goals, despite of attempting fewer problems, achieved higher goals (the confidence interval [−.5903, −.1133]). Students with lower pre-test scores achieved higher goals when they accepted system suggestion. These findings support H2.

To test hypothesis H3, we compared the scores from the two surveys (Table 3). No differences exist at the time of Survey 1 on goal setting and self-efficacy (SE). The goal-setting scores of the experimental group improved significantly (z = −1.93, p = .05), but not in the control group. There is a significant difference (z = −2.97, p < .005) on the goal-setting scores on Survey 2. The SE scores differed both as a function of group and time. At Survey 1, the experimental group had lower SE, but they increased at Survey 2 (z = −1.57, p = .1) whereas the SE scores decreased for the control group (z = −1.86, p = .06). These findings suggest that (a) students who complete the tasks in the absence of the intervention reported lower SE over time; and, (b) the goal-setting intervention may lead to considerable gains in SE, *especially* for students who started with less confidence. Although it is important to investigate further these trends in future research, these findings confirm our Hypothesis 3.

Table 3. Goal setting and self-efficacy scores: mean (sd)

	Goal setting		Self-efficacy	
	Exper. (21)	Control (14)	Exper. (21)	Control (14)
Survey 1	3.56 (0.63)	3.39 (0.64)	3.38 (0.65)	3.5 (0.66)
Survey 2	3.95 (0.65)	3.28 (0.65)	3.74 (0.65)	2.98 (0.67)

Our findings highlight the effects of setting challenging goals under realistic conditions, in a study that lasted four weeks. The limitations of our study are the small sample size and the low completion rates for Survey 2 and post-test. The results are in line with the goal-setting theory. In future work, we will investigate the effects of the intervention on the monitoring and self-reflection SRL phases.

References

Bernacki, M.L., Nokes-Malach, T.J., Aleven, V.: Fine-grained assessment of motivation over long periods of learning with an intelligent tutoring system: methodology, advantages, and preliminary results. In: Azevedo, R., Aleven, V. (eds.) International Handbook of Metacognition and Learning Technologies. SIHE, vol. 28, pp. 629–644. Springer, New York (2013). https://doi.org/10.1007/978-1-4419-5546-3_41

Carr (nee Harris), A., Luckin, R., Yuill, N., Avramides, K.: How mastery and performance goals influence learners' metacognitive help-seeking behaviours when using Ecolab II. In: Azevedo, R., Aleven, V. (eds.) International Handbook of Metacognition and Learning Technologies. SIHE, vol. 28, pp. 659–668. Springer, New York (2013). https://doi.org/10.1007/978-1-4419-5546-3_43

Cicchinelli, A., et al.: Finding traces of self-regulated learning in activity streams. In: Proceedings of the 8th International Conference on Learning Analytics and Knowledge, pp. 191–200 (2018)

Davis, D., Chen, G., Jivet, I., Hauff, C., Houben, G.-J.: Encouraging metacognition & self-regulation in MOOCs through increased learner feedback. In: Learning Analytics and Knowledge Conference, pp. 17–22 (2016)

Duffy, M.C., Azevedo, R.: Motivation matters: interactions between achievement goals and agent scaffolding for self-regulated learning within an intelligent tutoring system. Comput. Hum. Behav. **52**, 338–348 (2015)

Harley, J.M., Taub, M., Azevedo, R., Bouchet, F.: Let's set up some subgoals: understanding human-pedagogical agent collaborations and their implications for learning and prompt and feedback compliance. IEEE Trans. Learn. Technol. **11**(1), 54–66 (2017)

Kizilcec, R.F., Pérez-Sanagustín, M., Maldonado, J.J.: Self-regulated learning strategies predict learner behavior and goal attainment in Massive Open Online Courses. Comput. Educ. **104**, 18–33 (2017)

Landers, R.N., Bauer, K.N., Callan, R.C.: Gamification of task performance with leaderboards: a goal setting experiment. Comput. Hum. Behav. **71**, 508–515 (2017)

Latham, G.P., Piccolo, R.F.: The effect of context-specific versus nonspecific subconscious goals on employee performance. Hum. Resour. Manag. **51**(4), 511–523 (2012)

Latham, G.P., Yukl, G.A.: Assigned versus participative goal setting with educated and uneducated woods workers. J. Appl. Psychol. **60**(3), 299 (1975)

Latham, G.P., Brcic, J., Steinhauer, A.: Toward an integration of goal setting theory and the automaticity model. Appl. Psychol. **66**(1), 25–48 (2017)

Locke, E.A.: Self-set goals and self-efficacy as mediators of incentives and personality. In: Erez, M., Kleinbeck, U., Thierry, H. (eds.) Work Motivation in the Context of a Globalizing Economy, pp. 13–26. Lawrence Erlbaum Associates Publishers, Mahwa (2001)

Locke, E.A., Latham, G.P.: A Theory of Goal Setting & Task Performance: Prentice-Hall, Inc., Hoboken (1990)

Locke, E.A., Latham, G.P.: Building a practically useful theory of goal setting and task motivation: a 35-year odyssey. Am. Psychol. **57**(9), 705 (2002)

Locke, E.A., Latham, G.P.: The development of goal setting theory: a half century retrospective. Motiv. Sci. **5**(2), 93 (2019)

Masuda, A.D., Locke, E.A., Williams, K.J.: The effects of simultaneous learning and performance goals on performance: an inductive exploration. J. Cogn. Psychol. **27**(1), 37–52 (2015)

Mathews, M.: Investigating the effectiveness of problem templates on learning in intelligent tutoring systems. Honours report, Department of Computer Science, University of Canterbury (2006)

Melis, E., Siekmann, J.: Activemath: an intelligent tutoring system for mathematics. In: Rutkowski, L., Siekmann, J.H., Tadeusiewicz, R., Zadeh, L.A. (eds.) ICAISC 2004. LNCS (LNAI), vol. 3070, pp. 91–101. Springer, Heidelberg (2004). https://doi.org/10.1007/978-3-540-24844-6_12

Mitrovic, A.: An intelligent SQL tutor on the web. Artif. Intell. Educ. **13**(2–4), 173–197 (2003)

Pintrich, P.R., De Groot, E.V.: Motivational and self-regulated learning components of classroom academic performance. J. Educ. Psychol. **82**(1), 33 (1990)

Rowe, J.P., Shores, L.R., Mott, B.W., Lester, J.C.: Integrating learning, problem solving, and engagement in narrative-centered learning environments. Artif. Intell. Educ. **21**(1–2), 115–133 (2011)

Seijts, G.H., Latham, G.P.: Learning versus performance goals: when should each be used? Acad. Manag. Perspect. **19**(1), 124–131 (2005)

Zimmerman, B.J.: Becoming a self-regulated learner: an overview. Theory Pract. **41**(2), 64–70 (2002)

Zimmerman, B.J., Schunk, D.H.: Handbook of Self-Regulation of Learning and Performance. Routledge/Taylor & Francis Group, London (2011)

Modeling Frustration Trajectories and Problem-Solving Behaviors in Adaptive Learning Environments for Introductory Computer Science

Xiaoyi Tian[1(✉)], Joseph B. Wiggins[1], Fahmid Morshed Fahid[2],
Andrew Emerson[2], Dolly Bounajim[2], Andy Smith[2], Kristy Elizabeth Boyer[1],
Eric Wiebe[2], Bradford Mott[2], and James Lester[2]

[1] University of Florida, Gainesville, FL, USA
{tianx,jbwiggi3,keboyer}@ufl.edu
[2] North Carolina State University, Raleigh, NC, USA
{ffahid,ajemerso,dbbounaj,pmsmith4,wiebe,bwmott,lester}@ncsu.edu

Abstract. Modeling a learner's frustration in adaptive environments can inform scaffolding. While much work has explored momentary frustration, there is limited research investigating the dynamics of frustration over time and its relationship with problem-solving behaviors. In this paper, we clustered 86 undergraduate students into four frustration trajectories as they worked with an adaptive learning environment for introductory computer science. The results indicate that students who initially report high levels of frustration but then reported lower levels later in their problem solving were more likely to have sought help. These findings provide insight into how frustration trajectory models can guide adaptivity during extended problem-solving episodes.

Keywords: Frustration trajectory · Adaptive learning environments · Problem-solving behavior · Computer science education · Block-based programming

1 Introduction

Affect plays a critical role in human behavior, social interaction, and learning [4, 6]. Learners can benefit from affective states such as engaged concentration, while disengagement and boredom can lead to negative learning outcomes [1,7]. An affective state of particular interest is frustration, which can occur when a learner experiences an impasse or encounters task errors [5]. While repeated frustration can lead to boredom [9] and eventually attrition [8], many studies show that a certain level of frustration can motivate a learner to overcome obstacles during problem solving and can benefit learning [14,16,17].

Modeling student frustration presents significant challenges because frustration is dynamic in nature that has a complex relationship with learner behaviors.

© Springer Nature Switzerland AG 2021
I. Roll et al. (Eds.): AIED 2021, LNAI 12749, pp. 355–360, 2021.
https://doi.org/10.1007/978-3-030-78270-2_63

In computer science learning in particular, students engage in an iterative process of task planning, implementation and testing [5]. Students learning to code may experience more intense and durable frustration than students using highly-scaffolded learning environments [13,18]. Learning environments informed by dynamic changes in frustration could prevent a learner from experiencing prolonged frustration.

This paper investigates the role of learner frustration trajectories in an adaptive, block-based programming environment. Using frustration trajectories generated from learner self-reports, we investigate two research questions: 1) What common trajectories of frustration arise over problem-solving interactions with a block-based programming environment? 2) Do students with different frustration trajectories display different problem-solving behaviors? We identified four distinct frustration trajectories over a series of programming activities, and found that learners' problem-solving behaviors exhibited significant differences across different phases of interactions. These findings can inform the design of adaptive learning environments to promote productive frustration and better optimize individuals' learning experiences.

2 Study

This study utilized a dataset collected from student interactions with a block-based programming environment, PRIME (Fig. 1 (a)), designed for undergraduate introductory computer science to teach basic programming concepts. Students proceed through 20 programming activities over three units. Each unit contains 6 or 7 activities and is designed to be completed in approximately an hour. In the system, students can request hints (top-right panel) for a specific step of a programming problem. More details on the learning environment can be found in our previous work [10,20].

Participants are students in an introductory computing course at a university in the southeastern United States. Students completed a pre-survey about their prior programming experiences, CS attitudes and CS concept assessments (as pre-test) [15]. These student incoming characteristics (pre-test scores, CS attitudes and prior programming experiences) were used as covariates in our following analysis (RQ2). During the learning activities, at the end of each unit, students responded to a seven-item Likert questionnaire [19], which included the question, "I was frustrated while working on this unit." We used student responses to this question collected from each of the end-of-unit surveys as the measurement of frustration. The data used in this paper includes 86 students who attempted at least one activity in each unit and completed all pre-, post-, and end-of-unit surveys, with 67.4% majoring in Computer Science or Computer Engineering. Students attempted a mean of 19.4 programming activities ($SD = 1.8$, $Median = 20$) and completed 15.8 ($SD = 4.8$, $Median = 18$).

The PRIME learning environment logs interaction events. We grouped five frequent interaction events into two problem solving behaviors. *Workspace exploration* involves four events, namely creating blocks, moving blocks, deleting

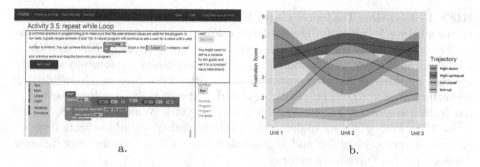

a.   b.

Fig. 1. (a) Block-based programming environment (b) Frustration trajectories

blocks and searching through the toolbox. *Help-seeking* indicates the frequency of students pressing the hint button while solving problems. We calculated all variables as standardized values per PRIME instructional unit with a mean of 0 and standard deviation of 1.

3 Results and Design Implications

RQ1: Frustration Trajectory Clustering. The average frustration reported across all units was 3.18 out of 7 ($SD = 1.97$). Over the course of the programming activities, students' reported frustration increased, with a mean of 2.71 after Unit 1, 3.29 after Unit 2 and 3.52 after Unit 3 ($SD = 1.88, 2.02, 1.93$, respectively). A paired-sample t-test revealed significant change from Unit 1 to Unit 2 ($p = .006$), but no significant difference from Unit 2 to Unit 3 ($p = .259$). To cluster learners' frustration trajectories, we considered both the initial intensity and relative changes of frustration. The clustering vectors were composed of the following features: 1) binary frustration level during Unit 1, split by the median ($Median = 2$); 2) relative frustration changes from Unit 1 to Unit 2; and 3) relative frustration changes from Unit 2 to Unit 3. We performed a k-medoids [11] clustering of learners' frustration and used the distortion elbow [12] to visually determine the optimal number of clusters: four. We refer to the four clusters as *low-equal* (33.7%, 29/86), *low-up* (24.4%, 21/86), *high-up/equal* (27.9%, 24/86) and *high-down* (14%, 12/86). Figure 1 (b) shows the frustration trends of the four clusters, two groups with low frustration and two groups with high frustration at the end of the first unit. Among the two low-starting groups, one group (*low-equal*) remained relatively constant over the entire interaction, $M(SD)_{Unit1,2,3} = 1.28$ (0.45), 1.21 (0.41), 2.40 (1.74), whereas the other group (*low-up*) whose frustration went up dramatically after Unit 2 and moved slightly down after Unit 3, $M(SD)_{Unit1,2,3} = 1.38$ (0.50), 4.29 (1.62), 3.05 (1.83). For students who were highly frustrated after Unit 1, one group (*high-up/equal*) was constantly frustrated over time, $M(SD)_{Unit1,2,3} = 4.42$ (1.43), 5.15 (1.26), 4.75 (1.62), while the other group (*high-down*) became less frustrated in the middle and went up at the end, $M(SD)_{Unit1,2,3} = 5.08$ (0.86), 2.83 (1.40), 4.58 (1.16).

RQ2: Frustration Trajectory and Problem-Solving Behaviors. To investigate whether learners in different frustration trajectories exhibit different frequencies of problem-solving actions, we conducted a one-way MANCOVA to compare the effect of frustration trajectories on *workspace exploration* and *help-seeking* in three units after controlling for student incoming characteristics (described in Sect. 2). The results revealed a statistically significant effect of frustration trends ($F(18, 209.789) = 2.491$, $p =.001$, Wilks' $\Lambda = .578$, partial $\eta^2 = .167$). Post-hoc tests to determine each dependent variable effects revealed that *workspace exploration* and *help-seeking* in Unit 2 are significantly different between the frustration trajectories.

Next, we conducted pairwise comparisons on estimated marginal means of dependent variables in each activity to test between-group differences. The results showed that the *low-equal* group performed a relatively low number of programming actions (*workspace exploration* and *help-seeking*) throughout. The *low-up* students had the highest number of *workspace explorations* ($M(SD) = 0.64(0.99)$) in Unit 2 where their frustration increased. The frequency of *workspace explorations* performed by the *low-up* group was significantly higher than all other groups. Interestingly, the *high-down* group, whose frustration decreased in Unit 2, were frequent help-seekers across all three units ($M(SD)_{Unit1,2,3} = 0.74(1.79)$, $0.81(1.84)$, $0.22(1.84)$). They requested significantly more hints than the two early-low groups in Unit 1, and more than the all three other groups in Unit 2. The results also indicated that while the group with persistent frustration (*high-up/equal* group) conducted an average level of *workspace exploration* in Unit 1 ($M(SD) = 0.26(1.32)$), these students made the lowest number of *workspace exploration* actions in Unit 2 ($M(SD) = -0.55(1.05)$) and Unit 3 ($M(SD)= -0.45(0.87)$) and were significantly lower than *low-up* group in Unit 2 and the *low-equal* group in Unit 3.

Design Implications. A low rate of *workspace exploration* behaviors may significantly predict students' frustration. For previously non-frustrated students (*low-equal*), scarcity of these actions is likely indicative of a smooth progression, and for previously frustrated students (*high-up/equal*), it is more likely a sign of disengagement. This suggests that to accurately detect frustrated learners, it is important to take prior frustration states into account. This finding aligns with prior work indicating frustration could lead to lack of persistence and result in systematic guessing and gaming behaviors [3].

Students with a *high-down* frustration trajectory sought more help by requesting hints. This suggests that providing additional hints and feedback may help close the gap between task difficulty and user knowledge, thus reducing learner's frustration. Another possible solution would be to provide guidance to students on managing their frustration and introducing relevant strategies (e.g., help-seeking). It is important to note that learning environments should not be designed to entirely eliminate frustration, as frustration is inherent in learning, particularly when students encounter challenging problems [2]. Rather, learning environments should be designed to enable students to experience productive levels of frustration by recognizing and regulating it.

Acknowledgements. This work is supported by the National Science Foundation through the grants DUE-1626235 and DUE-1625908. Any opinions, findings, and conclusions or recommendations expressed in this material are those of the author(s) and do not necessarily reflect the views of the National Science Foundation.

References

1. Andres, J.M.A.L., et al.: Affect sequences and learning in Betty's brain. In: Proceedings of the 9th International Conference on Learning Analytics & Knowledge, pp. 383–390 (2019)
2. Baker, R., D'Mello, S., Rodrigo, M., Graesser, A.: Better to be frustrated than bored: the incidence and persistence of affect during interactions with three different computer-based learning environments. Int. J. Hum.-Comput. Stud. **68**(4), 223–241 (2010)
3. Baker, R., Walonoski, J., Heffernan, N., Roll, I., Corbett, A., Koedinger, K.: Why students engage in "gaming the system" behavior in interactive learning environments. J. Interact. Learn. Res. **19**(2), 185–224 (2008)
4. Berry, D.S., Hansen, J.S.: Positive affect, negative affect, and social interaction. J. Pers. Soc. Psychol. **71**(4), 796 (1996)
5. Bosch, N., D'Mello, S.: The affective experience of novice computer programmers. Int. J. Artif. Intell. Educ. **27**(1), 181–206 (2017). https://doi.org/10.1007/s40593-015-0069-5
6. Calvo, R.A., D'Mello, S.: Affect detection: an interdisciplinary review of models, methods, and their applications. IEEE Trans. Affect. Comput. **1**(1), 18–37 (2010)
7. Craig, S., Graesser, A., Sullins, J., Gholson, B.: Affect and learning: an exploratory look into the role of affect in learning with AutoTutor. J. Educ. Media **29**(3), 241–250 (2004)
8. Dillon, J., et al.: Student emotion, co-occurrence, and dropout in a MOOC context. In: Proceedings of the International Conference on Educational Data Mining, pp. 353–357 (2016)
9. D'Mello, S., Graesser, A.: Dynamics of affective states during complex learning. Learn. Instr. **22**(2), 145–157 (2012)
10. Emerson, A., et al.: Predictive student modeling in block-based programming environments with Bayesian hierarchical models. In: Proceedings of the 28th ACM Conference on User Modeling, Adaptation and Personalization, pp. 62–70 (2020)
11. Kaufman, L., Rousseeuw, P.J.: Finding Groups in Data: An Introduction to Cluster Analysis, vol. 344. Wiley, Hoboken (2009)
12. Kodinariya, T.M., Makwana, P.R.: Review on determining number of cluster in k-means clustering. Int. J. Adv. Res. Comput. Sci. Manag. Stud. **1**(6), 90–95 (2013)
13. Liu, Z., Pataranutaporn, V., Ocumpaugh, J., Baker, R.: Sequences of frustration and confusion, and learning. In: Proceedings of the International Conference on Educational Data Mining, pp. 114–119 (2013)
14. Pardos, Z.A., Baker, R.S., San Pedro, M.O., Gowda, S.M., Gowda, S.M.: Affective states and state tests: investigating how affect throughout the school year predicts end of year learning outcomes. In: Proceedings of the Third International Conference on Learning Analytics and Knowledge, pp. 117–124 (2013)
15. Rachmatullah, A., Wiebe, E., Boulden, D., Mott, B., Boyer, K., Lester, J.: Development and validation of the computer science attitudes scale for middle school students (MG-CS attitudes). Comput. Hum. Behav. Rep. **2**, 100018 (2020)

16. Richey, J.E., et al.: More confusion and frustration, better learning: the impact of erroneous examples. Comput. Educ. **139**, 173–190 (2019)
17. Shute, V.J., et al.: Modeling how incoming knowledge, persistence, affective states, and in-game progress influence student learning from an educational game. Comput. Educ. **86**, 224–235 (2015)
18. Stachel, J., Marghitu, D., Brahim, T.B., Sims, R., Reynolds, L., Czelusniak, V.: Managing cognitive load in introductory programming courses: a cognitive aware scaffolding tool. J. Integr. Des. Process Sci. **17**(1), 37–54 (2013)
19. Wiebe, E.N., Lamb, A., Hardy, M., Sharek, D.: Measuring engagement in video game-based environments: investigation of the user engagement scale. Comput. Hum. Behav. **32**, 123–132 (2014)
20. Wiggins, J.B., et al.: Exploring novice programmers' hint requests in an intelligent block-based coding environment. In: Proceedings of the 52nd ACM Technical Symposium on Computer Science Education, pp. 52–58 (2021)

Behavioral Phenotyping for Predictive Model Equity and Interpretability in STEM Education

Marcus Tyler(✉), Alex Liu, and Ravi Srinivasan

The University of Texas at Austin Applied Research Labs, Austin, TX 78758, USA
{mtyler,aliu,rav}@arlut.utexas.edu

Abstract. Predictive models are increasingly being deployed in social and behavioral applications in support of decision making that directly affects people's lives. Given such high stakes, it is important to develop models with interpretable and defensible features, with decisions that are unbiased toward historically marginalized groups. In this work we investigate the use of nonnegative matrix factorization (NMF) for generating interpretable features in an educational setting, combined with a standard bias mitigation algorithm for training predictive models. Our application in this work is predicting enrollment in STEM degrees, and improving fairness of our models through bias mitigation. We perform our experiments on the High School Longitudinal Study of 2009, and evaluate our results using both objective metrics and subjective interpretation of the NMF factors, or behavioral phenotypes. Our empirical results from these experiments suggest that NMF combined with bias mitigation can potentially be used to improve fairness measures while simultaneously aiding in interpretability.

Keywords: Behavioral phenotyping · Matrix factorization · Supervised machine learning · Bias mitigation · High school longitudinal study

1 Introduction

Encouraging equity in Science, Technology, Engineering, and Math (STEM) careers is an important and open problem in the U.S., where many groups are underrepresented. Discrepancies persist in gender representation in STEM education, with men more likely than women to obtain college degrees in STEM fields, even though women comprise nearly half the overall U.S. workforce [2].

Predictive AI models for STEM enrollment are potentially useful tools for understanding why some students choose to enroll in STEM fields, as well as for potentially informing policy and school administration decisions. However, building models that directly impact people's lives requires that the models be both interpretable and equitable. Interpretability of AI models is an active area of research, with matrix and tensor factorization showing promise in medical

© Springer Nature Switzerland AG 2021
I. Roll et al. (Eds.): AIED 2021, LNAI 12749, pp. 361–366, 2021.
https://doi.org/10.1007/978-3-030-78270-2_64

phenotyping [6,7,14] and in education with the Dartmouth StudentLife study [8]. Similarly, there has been growing interest in machine learning research to develop models that are unbiased toward marginalized groups [1], and a variety of bias mitigation methods have been proposed [10].

In this work in progress, we study interpretability and bias mitigation for predicting STEM enrollment on the High School Longitudinal Study of 2009 (HSLS:09), a nationally-representative dataset of 23,503 high school students in the U.S. spanning 2009–2016 [13]. The dataset includes surveys of students, parents, and school administrators, and variables such as coursework, extracurricular activities, academic attitudes, demographics, and income. Prior work on HSLS:09 has endeavored to predict STEM outcomes using conceptual models [9], and machine learning has been employed for dropout prediction [12], but we found no studies that evaluated the effectiveness of bias mitigation measures for machine learning in HSLS:09. In this work we utilize nonnegative matrix factorization (NMF) [4] and bias mitigation to predict STEM degree enrollment with meaningful, interpretable, and equitable features. In particular, we show that factorization discovers meaningful educational phenotypes for STEM enrollment that match previous research studies but in a fully automated manner.

2 Experiments

2.1 Modeling Methods

The HSLS:09 dataset has around 4,000 variables, and we use NMF to obtain a low-rank approximation of the data, which aids in interpretability. We use the NMF function from Python Scikit-learn with l_2 regularization to encourage sparse factorizations [11], which produced more consistent solutions across NMF runs. We fit NMF on a training set, and use the factors from the entire dataset as features in a Random Forest classifier to predict STEM enrollment. We chose Random Forest because it is a baseline method that is not easily interpretable by itself, to demonstrate the flexibility of using NMF features to improve general model interpretability. A comparison of performance with other classifiers is outside the scope of this study, but would be necessary before practical deployment of an AI-based system for predicting STEM enrollment.

To mitigate bias, we implement *Uniform Sampling*, a baseline technique that uses stratified sampling to give each of the four combinations of protected attribute (in this case, gender) and label class equal representation in the training sets [10]. For our experiments we undersampled the three more populous partitions in the training set to be the same size as the smallest partition.

2.2 Experimental Setup

We modeled the subset of 11,559 high school students who went to college, and the subset of variables that took place in high school (2009–2013). Our protected attribute was gender, and our target label was whether the student referenced their first degree as a STEM major, and we removed both of these along with similar highly correlated variables. We performed predictions with 10-fold cross-validation with varying NMF rank, and evaluated predictive accuracy using area under the curve (AUC). To evaluate fairness we used Disparate Impact and Theil Index from the AI Fairness 360 project [3]. Disparate Impact measures group fairness with the ratio of the rate of favorable predicted outcomes for women versus men, and ideally equals 1. Theil Index measures the entropy of favorable outcomes for individuals, and ideally is 0.

2.3   Results

In Fig. 1 we plot the AUC and Disparate Impact by rank, and compare with predictions on the raw (full-rank) features, which show that NMF with bias mitigation improves fairness without significantly sacrificing AUC. We see a slight tradeoff between accuracy and fairness by rank, which suggests that low-rank NMF could potentially be used for bias mitigation independent of uniform sampling. For subsequent analysis we focus on the case where rank = 60 with bias mitigation. This case had a Theil Index = 0.137, improving on the raw features without mitigation case with Theil Index = 0.201.

To evaluate interpretability, we look at the ranking of variables within each factor. In Table 1 we show the top three most important factors for predicting STEM enrollment, and their top variables. The top factor focuses on grades, while factors 2 and 3 focus on student identity in science and math. The identity

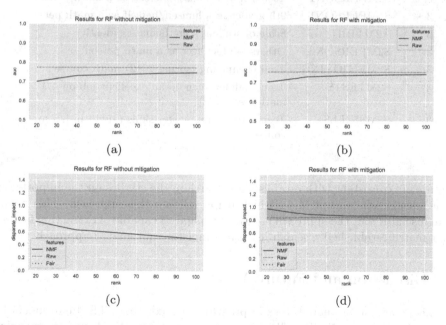

Fig. 1. Evaluation metrics by rank, with and without mitigation. The green region represents target fairness range for Disparate Impact.

Table 1. Top 3 factors for predicting STEM enrollment (rank = 60, mitigated)

Factor	Variable	Description
1	X3TGPAWGT	Overall GPA computed, honors-weighted
1	X3TAFGPATOT	GPA for all academic courses, failed courses excluded
1	X3TAGPAWGT	GPA for all academic courses, honors weighted
1	X3TGPATOT	Overall GPA computed
1	X3TCREDACAD	Credits earned in academic courses
1	X3TGPAACAD	GPA for all academic courses
1	X3TSTATYR09	School year 2009/10 transcript availability
2	S2SATTENTION	How often paid attention to spring 2012 science teacher
2	S2SPERSON1	Teenager sees himself/herself as a science person
2	S2STCHINTRST	[...] science teacher makes science interesting
2	S2STCHEASY	[...] science teacher makes science easy to understand
2	S2SUSEJOB	Teenager thinks science is useful for future career
2	S2SENJOYING	9th grader is enjoying fall 2009 science course very much
2	X2SCIINT	Scale of student's interest in fall 2009 science course
3	S2MPERSON1	Teenager sees himself/herself as a math person
3	S1MPERSON1	9th grader sees himself/herself as a math person
3	X2MTHID	Scale of student's mathematics identity
3	S2MPERSON2	Others see teenager as a math person
3	S2MENJOYING	Teen is enjoying (spring 2012) math course
3	S2MTESTS	Teen confident can do an excellent job on [...] math tests
3	S2MSKILLS	Teen certain can master skills taught in [...] math course

factors are consistent with established research in the studies on the importance of academic self-concept [4,5], which helps validate this approach. This result demonstrates that the approach used in this paper is capable of extracting factors that match established research using a completely automated approach.

3 Discussion and Future Work

In this paper, we presented work in progress on predicting STEM enrollment in HSLS:09, a large and nationally-representative dataset. We show how a combination of standard machine learning methods can automatically reveal insights in educational data which are consistent with established research while simultaneously addressing issues of bias. This suggests that such a combined approach can be used to provide new insights into datasets and protected attributes that are less understood in the literature.

In future work we plan to study other protected attributes, such as race and the intersectionality of race and gender. For example, we intend to explore whether racial disparities in STEM education present different underlying factors as gender disparities. Our results also suggest that low-rank NMF might potentially be used for bias mitigation on its own, and we intend to study fairness in factorization further and compare with established mitigation methods.

Acknowledgements. This work was supported by the U.S. Office of Naval Research (ONR N00014-19-1-2625).

References

1. Barocas, S., Selbst, A.D.: Big data's disparate impact. Calif. L. Rev. **104**, 671 (2016)
2. Beede, D.N., Julian, T.A., Langdon, D., McKittrick, G., Khan, B., Doms, M.E.: Women in STEM: a gender gap to innovation. Economics and Statistics Administration Issue Brief (04–11) (2011)
3. Bellamy, R.K.E., et al.: AI Fairness 360: an extensible toolkit for detecting, understanding, and mitigating unwanted algorithmic bias, October 2018. https://arxiv.org/abs/1810.01943
4. Berry, M.W., Browne, M., Langville, A.N., Pauca, V.P., Plemmons, R.J.: Algorithms and applications for approximate nonnegative matrix factorization. Comput. Stat. Data Anal. **52**(1), 155–173 (2007)
5. Ertl, B., Luttenberger, S., Paechter, M.: The impact of gender stereotypes on the self-concept of female students in stem subjects with an under-representation of females. Front. Psychol. **8**, 703 (2017)
6. Ho, J.C., et al.: Limestone: high-throughput candidate phenotype generation via tensor factorization. J. Biomed. Inform. **52**, 199–211 (2014)
7. Ho, J.C., Ghosh, J., Sun, J.: Extracting phenotypes from patient claim records using nonnegative tensor factorization. In: Ślezak, D., Tan, A.-H., Peters, J.F., Schwabe, L. (eds.) BIH 2014. LNCS (LNAI), vol. 8609, pp. 142–151. Springer, Cham (2014). https://doi.org/10.1007/978-3-319-09891-3_14
8. Hosseinmardi, H., Kao, H.T., Lerman, K., Ferrara, E.: Discovering hidden structure in high dimensional human behavioral data via tensor factorization. arXiv preprint arXiv:1905.08846 (2019)
9. Jiang, S., Simpkins, S.D., Eccles, J.S.: Individuals' math and science motivation and their subsequent stem choices and achievement in high school and college: a longitudinal study of gender and college generation status differences. Dev. Psychol. **56**(11), 2137 (2020)
10. Kamiran, F., Calders, T.: Data preprocessing techniques for classification without discrimination. Knowl. Inf. Syst. **33**(1), 1–33 (2012). https://doi.org/10.1007/s10115-011-0463-8
11. Pedregosa, F., et al.: Scikit-learn: machine learning in Python. J. Mach. Learn. Res. **12**, 2825–2830 (2011)

12. Sansone, D.: Beyond early warning indicators: high school dropout and machine learning. Oxford Bull. Econ. Stat. **81**(2), 456–485 (2019)
13. United States Department of Education. Institute of Education Sciences. National Center for Education Statistics: High School Longitudinal Study, 2009–2013 [United States] (2016). https://doi.org/10.3886/ICPSR36423.v1
14. Wang, Y., et al.: Rubik: knowledge guided tensor factorization and completion for health data analytics. In: Proceedings of the 21th ACM SIGKDD International Conference on Knowledge Discovery and Data Mining, pp. 1265–1274 (2015)

Teaching Underachieving Algebra Students to Construct Models Using a Simple Intelligent Tutoring System

Kurt VanLehn[✉] [iD], Fabio Milner[iD], Chandrani Banerjee, and Jon Wetzel[iD]

Arizona State University, Tempe, AZ, USA
{kurt.vanLehn,milner,cbanerj1,jwetzel4}@asu.edu

Abstract. An algebraic model uses a set of algebraic equations to describe a situation. Constructing such models is a fundamental skill, but many students still lack the skill, even after taking several algebra courses in high school and college. For underachieving college students, we developed a tutoring system that taught students to decompose the to-be-modelled situation into schema applications, where a schema represents a simple relationship such as distance-rate-time or part-whole. However, when a model consists of multiple schema applications, it needs some connection among them, usually represented by letting the same variable appear in the slots of two or more schemas. Students in our studies seemed to have more trouble identifying connections among schemas than identifying the schema applications themselves. This paper describes a newly designed tutoring system that emphasizes such connections. An evaluation was conducted using a regression discontinuity design. It produced a marginally reliable positive effect of moderate size (d = 0.4).

Keywords: Intelligent tutoring system · Algebraic model construction · Algebra story problem solving · Algebra word problem solving

1 The Research Problem and Prior Work on It

Constructing models is a fundamental and important skill. According to the Next Generation Science Standards [1], "developing and using models" is one of 8 key scientific practices. According to the Common Core State Standards for Mathematics (CCSSM) [2], "modeling with mathematics" is one of its 8 key mathematical practices.

Students are introduced to model construction with arithmetic story problems in primary school, and then algebraic story problems in secondary school. Both are notoriously difficult. Many methods for teaching model construction have been investigated [see 3 for review].

Several researchers have applied Kintch's theory of text comprehension to model construction [4–12]. The theory posits that students construct equations by matching schemas against their understanding of the story. Each match of a schema fills slots of the schema and produces an equation. Nathan et al. [6] observed that some relationships were obvious to students and some were not. For example, given this story:

© Springer Nature Switzerland AG 2021
I. Roll et al. (Eds.): AIED 2021, LNAI 12749, pp. 367–371, 2021.
https://doi.org/10.1007/978-3-030-78270-2_65

Six seconds after an F-35 fighter jet passes over some militants, they fire an FIM-92 Stinger missile at the plane. The plane flies at full speed, 537 m/s. The missile flies at its full speed, 750 m/s. How long will it take the missile to catch up with the plane? the equations below illustrate a model that conforms to the theory.

- *Dplane = 537 * Tplane*; obvious application of the Motion schema
- *Dmissile = 740 * Tmissile*; obvious application of the Motion schema
- *Dplane = Dmissile*; nonobvious application of the Overtake schema
- *Tplane = Tmissile + 6*; nonobvious application of the Comparison schema

In a series of design-based research and quantitative studies, we converged on an instructional design that taught schemas explicitly using an example-based intelligent tutoring system [13]. The results were positive but not statistically reliable.

In order to understand the remaining impediments to learning, we tutored students individually. Students seemed to have more trouble identifying connections among schemas than identifying the schema applications themselves. We redesigned the instruction to replace the traditional concept of a variable denoting a quantity with variables as connections between slots. The next section describes the tutoring system.

2   The OMRaaT Tutoring System

Figure 1 shows a solved problem in the new tutoring system, which is named OMRaaT: an acronym for One Mathematical Relationship at a Time. Each row is a schema application. The boxes are slots. When students select the name of a schema, a new row is added to the table with empty slots labelled by the text above them. The first few slots describe the schema application. The Motion schema (first two rows of Fig. 1) has 4 description slots; the Equality schema (third row) has 2; the Addition schema (last row) has 2. Students fill a description slot by selecting from a menu including all possible description slot fillers.

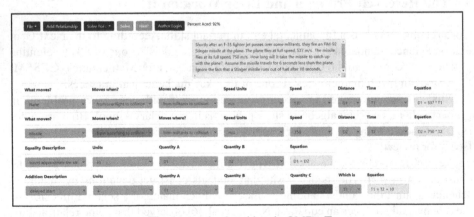

Fig. 1. A solved problem in OMRaaT

When the student finishes filling all the description slots of a schema application, the slots turn red (incorrect) or green (correct). On the third incorrect attempt, the slot turns yellow and shows the correct entry. The yellow coloring and the delayed feedback are intended to discourage guessing. Also, the percentage of slots filled correctly on the first attempt is displayed at the top of the window (e.g., "Percent Aced: 92%").

After all the description slots of a schema application have been filled correctly, the student fills in the remaining "quantity" slots. For the Motion schema, there are 3 quantity slots, for distance, rate and time. To fill a quantity slot of the Motion schema or other obvious schemas, students select from a menu that has numbers mentioned in the story (e.g., 537, 750 and 6 for the example above) and "Unknown." If they select Unknown, the system invents a variable name that is unique to the slot. Thus, the variables denote slots. When all the quantity slots have been filled, the student gets red/green feedback.

When students fill the two quantity slots of an Equality schema application, they select from a menu that has only the variables defined in the obvious schema applications. When they fill the quantity slots in an Addition schema application, they select from a menu with Unknown, all the variables of obvious schema applications and all the numbers in the story. The Addition schema has students first enter 3 quantities involved in an addition relationship, then select one of them to be the sum.

When students have finished correctly entering all the schema applications required, they click the "Solve for:" button to select one of the variables, then click the Solve button, which pops up a window with the numerical value of the solved-for variable.

The overall instruction was implemented as a module of 60 pages in the university's LMS, Canvas. Most pages were OMRaaT problems. A few pages were text with examples. A few more pages were problems to be done on the assessment system, called the Solver. To solve a problem on the Solver, the student types in equations using only numbers that appear in the problem statement. When the student presses the Solve button, it solves the equations or displays an error message. Every Solver-problem page was followed by a page with a video of the instructor solving the same Solver problem using OMRaaT-style reasoning.

3 An Evaluation

To evaluate the OMRaaT module, we used a regression discontinuity design in the context of an undergraduate class on modeling. The class historically has many mathematically challenged students. Students above the cutoff on the Solver-based pretest (the no-treatment group) were prevented from taking the OMRaaT modules. Students below the cutoff (the treatment group) were required to take the module. They could be considered underachieving students with regards to algebraic modeling skill.

To be fair to the students in both groups, algebraic modeling was taught twice bracketed by tests. The sequence was: (1) pretest, (2) OMRaaT module for the treatment group only, (3) mid-test, (4) Solver-based instruction on algebraic modelling for all students, and (5) post-test. Students' scores on the mid-test and post-test counted towards their grade, whereas the score on the pre-test was only used for placement in the treatment or no-treatment condition.

Of the 53 students who consented to have their data used for this evaluation and took both the pre-test and mid-test, 29 were in the treatment group and 23 were in the no-treatment group. One student whose pre-test score was below the cutoff did only 4 of the assigned OMRaaT problems, and thus was excluded from the treatment group. All the other treatment students did most of the problems.

Figure 2 shows the regression plot for the two groups. The solid line is the diagonal, where mid-test scores equals pre-test scores. The good news is that all but one of the treatment students gained: their mid-test scores were larger than their pre-test scores, often by large amounts. The majority of the no-treatment students also gained, which could be due to a test-retest effect or gaining familiarity with the Solver. However, a significant proportion (approximately one-third) of the no-treatment

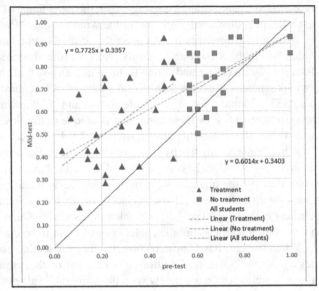

Fig. 2. Results of the OMRaaT module evaluation

students actually decreased their scores from pre-test to mid-test. The normalized gain scores of the treatment group were higher than those of the no-treatment group, but not reliably so ($p = 0.08$, $d = 0.42$).

However, the success of a regression discontinuity design hinges on whether the regression lines of the two groups (shown in blue and red) fit the data better than a regression line for the union of the two groups (shown in green). The OMRaaT module raised the regression line of the treatment group by about 0.09 above the no-treatment group. As the standard deviation of the post-test scores was 0.21, an 0.09 increase corresponds to an effect size of 0.41. However, the double regression line model was not reliability different from the single regression line model ($p = 0.27$). Our interpretation is that there may be a positive benefit, but it is too small to show up reliably given the large scatter in the data and the small number of data points.

The bottom line is that there appears to be a moderately large positive effect ($d = 0.4$), but its existence is doubtful due to the large variance in scores and the small sample. The fact that the OMRaaT module improved the scores of underachieving students is remarkable and welcome, but more data are needed to be sure that the positive apparent benefit actually exists.

References

1. NGSS, Next Generation Science Standards: For States, By States. The National Academies (2013)

2. Common Core State Standards for Mathematics (2011)
3. VanLehn, K.: Model construction as a learning activity: a design space and review. Interact. Learn. Environ. **21**(4), 371–413 (2013)
4. Kintsch, W., Greeno, J.G.: Understanding and solving word arithmetic problems. Psychol. Rev. **92**, 109–129 (1985)
5. Cummins, D.D., Kintsch, W., Reusser, K., Weimer, R.: The role of understanding in solving word problems. Cogn. Psychol. **20**, 405–438 (1988)
6. Nathan, M.J., Kintsch, W., Young, E.: A theory of algebra-word-problem comprehension and its implications for the design of learning environments. Cogn. Instr. **9**(4), 329–389 (1992)
7. Fuchs, L., Fuchs, D., Seethaler, P.M., Barnes, M.A.: Addressing the role of working memory in mathematical word-problem solving when designing intervention for struggling learners. ZDM Math. Educ. **52**(1), 87–96 (2019). https://doi.org/10.1007/s11858-019-01070-8
8. Riley, M.S., Greeno, J.G.: Developmental analysis of understanding language about quantities and of solving problems. Cogn. Instr. **5**(1), 49–101 (1988)
9. Marshall, S.P.: Schemas in Problem Solving. Cambridge University Press, Cambridge (1995)
10. Fuchs, L.S., et al.: The effects of schema-broadening instruction on second grader's word-problem performance and their ability to represent word problems with algebraic equations: a randomized control study. Elem. Sch. J. **110**(4), 440–463 (2010)
11. Jitendra, A.K., et al.: Effects of a research-based intervention to improve seventh-grade students' proportional problem solving: a cluster randomized trial. J. Educ. Psychol. **107**(4), 1019–1034 (2015)
12. Xin, Y.P., Jitendra, A.K., Deatline-Buchman, A.: Effects of mathematical word problem-solving instruction on middle school students with learning problems. J. Spec. Educ. **39**(3), 181–192 (2005)
13. Vanlehn, K., Banerjee, C., Milner, F., Wetzel, J.: Teaching algebraic model construction: a tutoring system, lessons learned and an evaluation. Int. J. Artif. Intell. Educ. **30**(3), 459–480 (2020)

Charisma and Learning: Designing Charismatic Behaviors for Virtual Human Tutors

Ning Wang[✉], Aditya Jajodia, Abhilash Karpurapu, and Chirag Merchant

Institute for Creative Technologies, University of Southern California,
Los Angeles, USA
nwang@ict.usc.edu

Abstract. Charisma is a powerful device of communication. Research on charisma on a specific type of leader in a specific type of organization – teachers in the classroom - has indicated the positive influence of a teacher's charismatic behaviors, often referred to as immediacy behaviors, on student learning. How do we realize such behaviors in a virtual tutor? How do such behaviors impact student learning? In this paper, we discuss the design of a charismatic virtual human tutor. We developed verbal and nonverbal (with the focus on voice) charismatic strategies and realized such strategies through scripted tutorial dialogues and pre-recorded voices. A study with the virtual human tutor has shown an intriguing impact of charismatic behaviors on student learning.

Keywords: Charisma · Pedagogical agents · Virtual human

1 Introduction

Charisma is one the oldest and most effective forms of leadership. Decades of research have established the strong connection between charismatic leadership and positive outcomes for organizations [4–6,12]. In the past decade, a great number of researchers have focused the investigation of charisma on a specific type of leader in a specific type of organizations – teachers in the classrooms. This body of research has indicated the positive influence of a teacher's charismatic behavior on the learning environment [7], including student engagement [10], effort [16], on-task behavior, motivation, and achievement of better learning outcomes [2]. In education, charismatic skills are often studied as *immediacy* behaviors, which are communication behaviors that reduce social and psychological distance between people [13,15]. In a classroom, nonverbal immediacy behaviors include eye contact, smiling, movement around the classroom, and relaxed body posture [1,17], and verbal immediacy behaviors include the use of personal examples, humor, inviting input, providing personalized feedback, and even simple gestures such as addressing and being addressed by students by name [8]. How to create immediacy and harness the efficacy of teacher immediacy in virtual learning environments such as intelligent tutoring systems? In

© Springer Nature Switzerland AG 2021
I. Roll et al. (Eds.): AIED 2021, LNAI 12749, pp. 372–377, 2021.
https://doi.org/10.1007/978-3-030-78270-2_66

this paper, we discuss our preliminary work in designing charismatic behaviors for a virtual human tutor. We conducted a study to evaluate the impact of the virtual human tutor on student learning. The results not only show interesting interactions of charisma expressed through different modalities, but also shed light on how to automatically generate charismatic speech, voices, and gestures for virtual characters.

2   The ALIVE! Testbed

To design and study the charismatic behaviors of a virtual human tutor, we developed a testbed called ALIVE!, where a virtual human gives a lecture on the human circulatory system. The lecture includes 83 utterances [3] and several pop-up quizzes. For experimental purposes, we developed a "baseline" of the lecture, where no verbal charismatic strategies were used, and a "charismatic" version with utterances in the "baseline" lecture modified using verbal charismatic strategies. The voice-over of the lecture was recorded in a sound studio by a male staff member unrelated to the project. The recording instructions focused on speaking generally in a flat and monotone voice for non-charismatic voices, while using an animated voice for charismatic ones. We conducted a study with ALIVE!, where participants were randomly assigned to one of the six conditions: Non-Charismatic Text Only, Charismatic Text Only, Non-Charismatic Text with Non-Charismatic Voice, Non-Charismatic Text with Charismatic Voice, Charismatic Text with Non-Charismatic Voice and Charismatic Text with Charismatic Voice. In the study, the participants filled out a Background Survey and took a Pre-Test about the human circulatory system. Then, the participants went through the lecture in ALIVE!. After that, the participants filled out a Post-Interaction Survey and took a Post-Test on the human circulatory system. Each study session was designed to last one hour. Details of the verbal and nonverbal behaviors of the virtual tutor, and the measures are described in [18].

3   Results

We recruited 122 participants from Amazon Mechanical Turk for the study. A preliminary analysis on perceived charisma is discussed in [18]. In this paper, we focus on the analysis of the impact of charismatic behavior on learning.

Perceived Charisma. We conducted a manipulation check to see if the use of charismatic strategies in text and voice had impacted perceived charisma (measured via item "The virtual human is charismatic"). A one-way ANOVA comparing the two text-only conditions indicates a significant difference between the charismatic text (CT) and non-charismatic text (NT) conditions ($p = .012, F = .62, M_{CT} = 2.65, M_{NT} = 3.67$). This suggests that the manipulation of verbal charismatic strategies is successful on perceived charisma. A two-way ANOVA test comparing the four conditions where the virtual human tutor gave the lecture in voice indicates a significant main effect of the text ($p = .013, F = 6.504$)

but not the voice ($p = .187, F = 1.772$) on perceived charisma. The interaction between charismatic text and voice is not statistically significant ($p = .466, F = .536, M_{NTNV} = 2.68, M_{NTCV} = 2.83, M_{CTNV} = 3.12, M_{CTCV} = 3.66$, NV: non-charismatic voice, CV: charismatic voice). This indicates that the manipulation of tutorial text was successful. But the manipulation of voice did not result in a significant difference in perceived charisma.

Domain Knowledge. To measure the learning of domain knowledge, we graded the 47 multiple-choice questions on the post-test (score ranges from 0 to 100). A one-way ANOVA test comparing the two text-only conditions shows that there was no significant difference between the charismatic and non-charismatic text conditions ($p = .443, F = 1.796, M_{CT} = 34.1, M_{NT} = 32.1$). A two-way ANOVA test comparing the four conditions where the virtual human tutor gave the lecture in voice indicates that there is no significant main effect of the text ($p = .266, F = 1.258$) or the voice ($p = .935, F = .007$). There is, however, a significant interaction between these two variable on post-test scores ($p = .046, F = 4.125, M_{NTNV} = 27.65, M_{NTCV} = 29.78, M_{CTNV} = 33.29, M_{CTCV} = 29.58$). Means from the four conditions show that when non-charismatic tutorial text was used, the addition of charismatic voice resulted in better learning performance. However, when charismatic text was used, the addition of the charismatic voice resulted in decreased learning performance.

Self Efficacy. We measured self-efficacy with the same item in the pre- and post-survey. We first built a repeated-measure general linear model (GLM) to compare the changes in self-efficacy between the two text-only conditions. Overall, there is a significant within-subject effect: the participants' self-efficacy increased from pre-interaction to post-interaction ($p < .001, F = 60.199$). There is no significant main effect on the type of tutorial text ($p = .859, F = .032$). This indicate the change in self-efficacy did not differ between the charismatic text and non-charismatic text condition. We then conducted a second repeated-measure GLM to compare the four conditions where the virtual human tutor gave the lecture in voice. There is a significant within-subject effect that indicates the self-efficacy increased from pre- to post-interaction ($p < .001, F = 80.445$). The main effect of the type of tutorial text is not significant ($p = .178, F = 1.851$). Neither is the main effect of the type of voice ($p = .991, F < .001$).

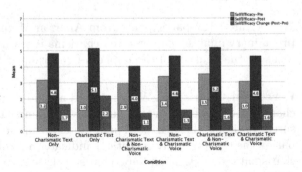

Fig. 1. The means of the pre-, post- and change in self-efficacy for six experimental conditions.

The interaction between the two variables on changes in self-efficacy is statistically significant ($p = .039, F = 4.406$). Means shown in Fig. 1 indicate a similar trend in the interaction of these two variables on post-test scores: when non-charismatic tutorial text was used, the addition of charismatic voice resulted in a larger increase of self-efficacy. However, when charismatic text was used, the addition of the charismatic voice resulted in the same level of change in self-efficacy.

4 Discussion

In this paper, we discussed the design and a study of charismatic verbal and non-verbal behavior for a virtual human tutor. Results indicates that the charismatic verbal strategies alone (i.e., without voice) did not make a significant impact on learning. When the charismatic tutorial is spoken through charismatic or non-charismatic voice, the voice alone did not make a significant impact on learning either. There is, however, a significant interaction between the two on learning outcomes. The results on the interaction effect indicates that the impact of charismatic strategies may depend on how they are expressed through different modalities. For example, the charismatic verbal strategies may depend on the way such strategies where spoken. Intriguingly, when the virtual human tutor gave a lecture written with charismatic strategies, using charismatic voices, it did not result in a "best" or "doubled" impact on student learning, compared to when no verbal strategies were used, or no charismatic voices were used. In fact, its impact on student learning is more similar to when non-charismatic voices were used and when the tutorial dialogue is stripped of any charismatic verbal strategies. On the other hand, the two conditions where the use of charismatic strategies were "mismatched" in verbal and nonverbal communications, e.g., charismatic tutorial dialogue spoken in a non-charismatic voice, resulted in a similar impact on student learning. This suggests a "congruency effect" of charismatic verbal and nonverbal strategies on learning. The congruency and conflict between verbal and nonverbal communications have been well studied on its impact on negotiation [11], trust [14], parent-child interaction [9], etc. While research in the communication literature suggests that congruency of verbal and nonverbal messages is important for accurate and persuasive communication, research in educational psychology indicates the opposite [11]. The results here suggest that when the verbal and nonverbal behavior (e.g., voice) are incongruent with each other, it can potentially result in a better learning outcome, compared to when they are congruent with each other.

In this study, the use of an animated voice did not result in higher ratings of perceived charisma, compared to when a monotone voice was used. In the post-survey, we asked the participants how much they agree with the statement that the virtual human tutor "Used an animated voice". A two-way ANOVA test show that there is no significant main effect of voice on the participants' rating ($p = .095, F = 2.853$). The main effect of voice is not significant ($p = .175, F = 1.876$) and the interaction between voice and text is not significant either $(p = .204, F = 1.640)$. Additionally, Pearson's correlation shows that ratings on this item are significantly correlated with the perceived charisma $(r = .463, p < .001s)$. The two results combined suggest that, in the future, making the voice more animated can potentially improve the virtual human tutor's perceived charisma.

Acknowledgement. This research was supported by the National Science Foundation (NSF) under Grant #1816966. Any opinions expressed in this material are those of the authors and do not necessarily reflect the views of the NSF.

References

1. Andersen, J.F.: Teacher immediacy as a predictor of teaching effectiveness. Ann. Int. Commun. Assoc. **3**(1), 543–559 (1979)
2. Bolkan, S., Goodboy, A.K.: Transformational leadership in the classroom: fostering student learning, student participation, and teacher credibility. J. Instruct. Psychol. **36**(4) (2009)
3. Chi, M.T., Siler, S.A., Jeong, H., Yamauchi, T., Hausmann, R.G.: Learning from human tutoring. Cogn. Sci. **25**(4), 471–533 (2001)
4. DeGroot, T., Aime, F., Johnson, S.G., Kluemper, D.: Does talking the talk help walking the walk? An examination of the effect of vocal attractiveness in leader effectiveness. Leadersh. Q. **22**(4), 680–689 (2011)
5. Dumdum, U.R., Lowe, K.B., Avolio, B.J.: A meta-analysis of transformational and transactional leadership correlates of effectiveness and satisfaction: an update and extension. In: Transformational and Charismatic Leadership: The Road Ahead 10th Anniversary Edition, pp. 39–70. Emerald Group Publishing Limited (2013)
6. Gasper, J.M.: Transformational leadership: An integrative review of the literature (1992)
7. Goodboy, A.K., Bolkan, S.: Leadership in the college classroom: the use of charismatic leadership as a deterrent to student resistance strategies. J. Classroom Interaction 4–10 (2011)
8. Gorham, J.: The relationship between verbal teacher immediacy behaviors and student learning. Commun. Educ. **37**(1), 40–53 (1988)
9. Grebelsky-Lichtman, T.: Children's verbal and nonverbal congruent and incongruent communication during parent-child interactions. Hum. Commun. Res. **40**(4), 415–441 (2014)
10. Harvey, S., Royal, M., Stout, D.: Instructor's transformational leadership: university student attitudes and ratings. Psychol. Rep. **92**(2), 395–402 (2003)
11. Johnson, D.W., McCarty, K., Allen, T.: Congruent and contradictory verbal and nonverbal communications of cooperativeness and competitiveness in negotiations. Commun. Res. **3**(3), 275–292 (1976)
12. Judge, T.A., Piccolo, R.F.: Transformational and transactional leadership: a meta-analytic test of their relative validity. J. Appl. Psychol. **89**(5), 755 (2004)
13. Mehrabian, A., et al.: Silent Messages, vol. 8. Wadsworth Belmont, CA (1971)
14. Morioka, S., et al.: Incongruence between verbal and non-verbal information enhances the late positive potential. PLoS ONE **11**(10) (2016)
15. Myers, S.A., Zhong, M., Guan, S.: Instructor immediacy in the Chinese college classroom. Commun. Stud. **49**(3), 240–254 (1998)

16. Pounder, J.S.: Full-range classroom leadership: implications for the cross-organizational and cross-cultural applicability of the transformational-transactional paradigm. Leadership 4(2), 115–135 (2008)
17. Richmond, V.P., McCroskey, J.C., Kearney, P., Plax, T.G.: Power in the classroom vii: linking behavior alteration techniques to cognitive learning. Commun. Educ. **36**(1), 1–12 (1987)
18. Wang, N., Pacheco, L., Merchant, C., Skistad, K., Jethwani, A.: The design of charismatic behaviors for virtual humans. In: Proceedings of the 20th ACM International Conference on Intelligent Virtual Agents, pp. 1–8 (2020)

AI-Powered Teaching Behavior Analysis by Using 3D-MobileNet and Statistical Optimization

Ruhan Wang[1], Jiahao Lyu[1], Qingyun Xiong[1], and Junqi Guo[1,2(✉)]

[1] School of Artificial Intelligence, Beijing Normal University, Beijing 100875, China
[2] Engineering Research Center of Intelligent Technology and Education Application, Ministry of Education, Beijing 100875, China

Abstract. Artificial intelligent technology can realize multi-angle analysis and feedback of teaching process. This paper provides an innovative auxiliary for classroom teaching evaluation and fills in the lack of teacher behavior analysis in AI application. Firstly, a 3D-MobileNet framework is proposed for behavior recognition, which can process time-domain information for the video through layered training. Next, we design a comprehensive model by using both the analytic hierarchy process and entropy weight method (AHP-EW) to output the quantitative results of the teaching evaluation in three dimensions. This model combines the subjective and objective weights through a statistical optimization strategy to improve the credibility. Finally, we test our model on a 45-min teaching video, and compare it with the existing model in various aspects, proving that our method is highly feasible and competitive.

Keywords: Smart education · Teaching behavior · 3D-MobileNet · AHP-EW

1 Introduction

In the process of classroom teaching, different teaching attitude and styles play important reference roles for the teaching evaluation. The traditional teaching evaluation mainly uses the information recorded by scaling manually [1]. In recent years, information [2–7] and artificial intelligence [8–10] technology has been widely used in the field of classroom teaching evaluation. However, these researches often focus on reflecting teachers' effects through the performance of students [11], and has a strong subjectivity to the evaluation methods and indicators. To solve these, we propose a teaching analysis method in this paper, which achieves accurate recognition of teaching behavior based on computer vision technology and then realizes a exact mapping from behavioral data to evaluation indicators through a statistical optimization algorithm.

2 Methodology

For an original teaching video, we first input it into the behavior recognition model to classify it into eight kinds of teaching behaviors. Next, we use the behavior analysis

I. Roll et al. (Eds.): AIED 2021, LNAI 12749, pp. 378–383, 2021.
https://doi.org/10.1007/978-3-030-78270-2_67

model to quantify each dimension of the teaching process based on the behavior s' frequency, and finally obtain the quantitative evaluation (Fig. 1).

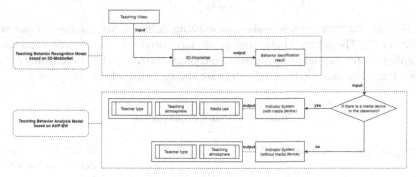

Fig. 1. Overall design flow chart

2.1 A 3D-MobileNet Based Teaching Behavior Recognition Model

Convolutional neural network has been widely used in the field of computer vision due to its advantages in feature extraction and processing [12–19]. We choose MobileNet [20] and it is optimized into a three-dimensional input network through hierarchical training inspired by P3D [21] and I3D [22]. Parameters of the convolutional layer are partially expanded before the fully connection layer, and then expanded into the full connection layer after further reducing the size of the feature map (Fig. 2).

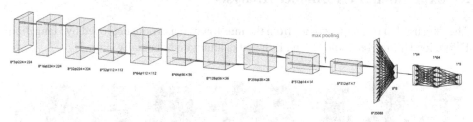

Fig. 2. 3D-MobileNet network structure diagram

2.2 An AHP-EW Based Teaching Behaviors Analysis Model

Inspired by the work of AHP-EW [23, 24], we use the analytic hierarchy process (AHP) [25, 26] and entropy weight method (EW) [27, 28] to get the subjective weight w_S and objective weight w_O respectively. Then get the comprehensive weight w through the optimization model below:

$$min\sum\nolimits_{i=1}^{n}(w_i - w_{oi})^2$$

$$s.t. \begin{cases} w_i \geq w_j (i < j) \\ min\{w_{si}, w_{oi}\} \leq w_i \leq max\{w_{si}, w_{oi}\} \\ \sum_{i=1}^{n} w_i = 1 \end{cases}$$

It is worth mentioning that finding an existing behavior set suitable for our work is difficult [29]. The selected behaviors should have obvious characteristics and cannot have ambiguity for the identification work, and should be relevant to the teaching process analysis later. The final analysis indicators are shown below (Table 1).

Table 1. Indicator display chart

Teaching analysis A	Teacher type B1	Lecture C1	Teaching textbook D1
		Book guide C2	Pointing at blackboard D2
		Blackboard-writing C3	No gesturing D3
		Multimedia-type C4	Hand-Stroking D4
		Interactive C5	Raising hand D5
	Teaching atmosphere B2	Inactive C6	Bending over desktop D6
		Active C7	Walking around D7
	Media use B3	Frequency C8	Writing on the board D8

3   Experiment Performance Analysis

The data used in this paper comes from the intelligent classroom deployed by Hangzhou Hikvision Digital Technology Co., Ltd. and is annotated manually (Fig. 3).

Fig. 3. Data set schematic diagram (A: writing on the board; B: bending over the desktop; C: hand stroking; D: no gesture; E: pointing at blackboard (projection); F: raising hand; G: teaching textbook exercises; H: walking around)

3.1   Teaching Behavior Recognition Model

To evaluate the effectiveness of the 3D-MoblieNet, we select 3D CNN [30, 31] and ResNet CRNN [32] for comparing. 800 videos which each contains 8 frames are used to train. The accuracy in behavior recognition and the test results on a 45-min (standard class length) middle school mathematics teaching video are shown in the following table, where MobileNet_1 and MobileNet_2 respectively refer to the two layers of 3D-MobileNet. The hierarchical training significantly improves the accuracy of MobileNet, which can reach 94.63% on our data set (Table 2).

Table 2. Model accuracy and identification time comparison diagram

	3D CNN	ResNet CRNN	MobileNet_1	MobileNet_2
Accuracy	79.29%	84.88%	83.73%	94.63%
Recognizing time	489.0255162 s	552.5230791 s	423.5016813 s	

3.2 Teaching Behaviors Analysis Model

Through the method in Sect. 2.2, the comprehensive weight for analysis as Table 3.

Table 3. The comprehensive weight for teaching analysis by AHP-EW

Comprehensive weight in traditional class						
Indicator	Lecture	Book guide	Board-writing	Interactive	Inactive	Active
D1	0.1995	0.3783	0.0287	0.0465	0.2812	0.0293
D2	0.0238	0.0289	0.3511	0.0351	0.125	0.1555
D3	0.2873	0.1537	0.0974	0.0718	0.25	0.0524
D4	0.1995	0.1188	0.0706	0.2594	0.0625	0.208
D5	0.0665	0.0769	0.0506	0.3394	0.0312	0.243
D7	0.1995	0.2146	0.0506	0.2201	0.125	0.1555
D8	0.0238	0.0289	0.3511	0.0277	0.125	0.1555

To evaluate the advantages of AHP-EW on actual teaching, we find 990 students and 10 experts to evaluate the same teaching video, collecting their evaluation to obtain the normalized recognized result. After using 3D-MobileNet to get the recognition results, we obtain the evaluation results by using only AHP, only EW and AHP-EW for the same video. We solve the mean square error between the results obtained by the three evaluation methods and the expert scoring, finding that AHP-EW is the closest to the people's evaluation.

4 Conclusion

In this paper, we first propose a 3D-MobileNet architecture for video behavior recognizing, then combine AHP and EW by an optimization method to ensure the rich objective theoretical basis and the subjectivity of the evaluation work itself. After that, we evaluate the performance of the proposed method, which verifies the superiority of our method. Our method will be helpful for further teaching evaluation and adjustment, providing powerful support for teaching quality inspection.

Acknowledgements. This research is sponsored by National Natural Science Foundation of China (No.61977006), "Educational Big Data R&D and its Application"—Major Big Data Engineering Project of National Development and Reform Commission 2017, and Beijing Advanced Innovation Center for Future Education (BJAICFE2016IR-004).

References

1. Hiebert, J.: Teaching Mathematics in Seven Countries: Results from the TIMSS 1999 Video Study. DIaNe Publishing (2003)
2. Flanders, N.A.: Intent, action and feedback: a preparation for teaching. J. Teach. Educ. **14**(3), 251–260 (1963)
3. Amidon, E.J., Hough, J.J.: Interaction analysis: theory, research and application (1967)
4. Wang, D., Han, H., Liu, H.: Analysis of instructional interaction behaviors based on OOTIAS in smart learning environment. In: 2019 Eighth International Conference on Educational Innovation through Technology (EITT), pp. 147–152. IEEE, October 2019
5. Li, B.H., Zhang, X.Y.: Classroom teaching video analysis software design and realization based on ITIAS. Comput. Eng. Softw. **01**, 46–50 (2019)
6. Fu: Educational informationalization and educational information disposal. Doctoral dissertation (2002)
7. Prieto, L.P., Sharma, K., Kidzinski, Ł, Rodríguez-Triana, M.J., Dillenbourg, P.: Multimodal teaching analytics: automated extraction of orchestration graphs from wearable sensor data. J. Comput. Assist. Learn. **34**(2), 193–203 (2018)
8. Holstein, K., McLaren, B.M., Aleven, V.: Co-designing a real-time classroom orchestration tool to support teacher–AI complementarity. J. Learn. Anal. **6**(2), 27–52 (2019)
9. Long, Y., Aleven, V.: Supporting students' self-regulated learning with an open learner model in a linear equation tutor. In: Lane, H.C., Yacef, K., Mostow, J., Pavlik, P. (eds.) AIED 2013. LNCS (LNAI), vol. 7926, pp. 219–228. Springer, Heidelberg (2013). https://doi.org/10.1007/978-3-642-39112-5_23
10. Wu, S.: Rain classroom, an intelligent teaching tool under the background of mobile internet and big data. Collection **17**, 26–32 (2018)
11. Kasparova, A., Celiktutan, O., Cukurova, M.: Inferring student engagement in collaborative problem solving from visual cues. In: Companion Publication of the 2020 International Conference on Multimodal Interaction, pp. 177–181, October 2020
12. Gu, J., et al.: Recent advances in convolutional neural networks. Pattern Recogn. **77**, 354–377 (2018)
13. Krizhevsky, A., Sutskever, I., Hinton, G.E.: Imagenet classification with deep convolutional neural networks. In: Advances in Neural Information Processing Systems, vol. 25, pp. 1097–1105 (2012)
14. LeCun, Y., et al.: Backpropagation applied to handwritten zip code recognition. Neural Comput. **1**(4), 541–551 (1989)
15. Sermanet, P., Eigen, D., Zhang, X., Mathieu, M., Fergus, R., LeCun, Y.: OverFeat: integrated recognition, localization and detection using convolutional networks. arXiv preprint arXiv:1312.6229 (2013)

16. Zeiler, M.D., Fergus, R.: Visualizing and understanding convolutional networks. In: Fleet, D., Pajdla, T., Schiele, B., Tuytelaars, T. (eds.) ECCV 2014. LNCS, vol. 8689, pp. 818–833. Springer, Cham (2014). https://doi.org/10.1007/978-3-319-10590-1_53
17. Simonyan, K., Zisserman, A.: Very deep convolutional networks for large-scale image recognition. arXiv preprint arXiv:1409.1556 (2014)
18. He, K., Sun, J.: Convolutional neural networks at constrained time cost. In: Proceedings of the IEEE Conference on Computer Vision and Pattern Recognition, pp. 5353–5360 (2015)
19. Srivastava, R.K., Greff, K., Schmidhuber, J.: Highway networks. arXiv preprint arXiv:1505.00387 (2015)
20. Howard, A.G., et al.: MobileNets: efficient convolutional neural networks for mobile vision applications. arXiv preprint arXiv:1704.04861 (2017)
21. Qiu, Z., Yao, T., Mei, T.: Learning spatio-temporal representation with pseudo-3D residual networks. In: Proceedings of the IEEE International Conference on Computer Vision, pp. 5533–5541 (2017)
22. Carreira, J., Zisserman, A.: Quo vadis, action recognition? A new model and the kinetics dataset. In: Proceedings of the IEEE Conference on Computer Vision and Pattern Recognition, pp. 6299–6308 (2017)
23. Xu, S., Xu, D., Liu, L.: Construction of regional informatization ecological environment based on the entropy weight modified AHP hierarchy model. Sustain. Comput.: Inf. Syst. **22**, 26–31 (2019)
24. Gang, L.I., Jianping, L.I., Xiaolei, S.U.N., Meng, Z.: Research on a combined method of subjective-objective weighing and the its rationality. Manage. Rev. **29**(12), 17 (2017)
25. Teknomo, K.: Analytic hierarchy process (AHP) tutorial. Revoledu. com 1–20 (2006)
26. Forman, E.H., Gass, S.I.: The analytic hierarchy process—an exposition. Oper. Res. **49**(4), 469–486 (2001)
27. Gray, R.M.: Entropy and Information Theory. Springer, Heidelberg (2011). https://doi.org/10.1007/978-1-4419-7970-4
28. Qiyue, C.: Structure entropy weight method to confirm the weight of evaluating index. Syst. Eng. Theory Pract. **30**(7), 1225–1228 (2010)
29. Liu, Q., et al.: Classroom teaching behavior analysis method based on artificial intelligence and its application. China Video Educ. **9**, 13–21 (2019)
30. Tran, D., Bourdev, L., Fergus, R., Torresani, L., Paluri, M.: Learning spatiotemporal features with 3D convolutional networks. In: Proceedings of the IEEE International Conference on computer Vision, pp. 4489–4497 (2015)
31. Ji, S., Xu, W., Yang, M., Yu, K.: 3D convolutional neural networks for human action recognition. IEEE Trans. Pattern Anal. Mach. Intell. **35**(1), 221–231 (2012)
32. Song, Y., Tan, L., Zhou, L., Lv, X., Ma, Z.: Video action recognition based on hybrid convolutional network. In: Sun, X., Wang, J., Bertino, E. (eds.) Artificial Intelligence and Security. ICAIS 2020. LNCS, vol. 12240, pp. 451–462. Springer, Cham (2020). https://doi.org/10.1007/978-3-030-57881-7_40

Assessment2Vec: Learning Distributed Representations of Assessments to Reduce Marking Workload

Shuang Wang[1]([✉]), Amin Beheshti[1]([✉]), Yufei Wang[1], Jianchao Lu[1],
Quan Z. Sheng[1], Stephen Elbourn[2], Hamid Alinejad-Rokny[3],
and Elizabeth Galanis[4]

[1] Macquarie University, Sydney, Australia
{shuang.wang,amin.beheshti,michael.sheng}@mq.edu.au,
{yufei.wang,jianchao.lu}@students.mq.edu.au
[2] ITIC Pty LTD., Sydney, Australia
stephen.elbourn@itic.com.au
[3] BML Lab, UNSW Sydney, Sydney, Australia
h.alinejad@unsw.edu.au
[4] Deakin University, Melbourne, Australia
egalanis@deakin.edu.au

Abstract. Reducing instructors workload in online and large-scale learning environments could be one of the most important factors in educational systems. To address this challenge, techniques such as Artificial Intelligence has been considered in tutoring systems and automatic essay scoring tasks. In this paper, we construct a novel model to enable learning distributed representations of assessments namely Assessment2Vec and mark assessments automatically with Supervised Contrastive Learning loss which will effectively reduce instructors' workload in marking large number of assessments. The experimental results based on the real-world datasets show the effectiveness of the proposed approach.

Keywords: Assessment2Vec · Assessment marking · Natural language processing · Supervised contrastive learning · AI-enabled education

1 Introduction

Over the past years, different solutions have been proposed to reduce instructors workload and facilitate teachers' and students interactions. An early example of such a technique is one-to-one interaction models, which is well suited for courses with a small number of students. However, in a large classroom or in a massive open online course with hundreds of enrolled students, assessment marking can be a very challenging task. Accordingly, reducing instructors workload in online and large-scale learning environments is one of the most important factors in educational systems. To address this challenge, Artificial Intelligence (AI) has been used in intelligent tutoring systems and automatic essay scoring tasks. Language

© Springer Nature Switzerland AG 2021
I. Roll et al. (Eds.): AIED 2021, LNAI 12749, pp. 384–389, 2021.
https://doi.org/10.1007/978-3-030-78270-2_68

model pre-training is used to represent assessment to vectors since it has been shown to be effective to improve natural language processing (NLP) tasks [1–4]. To mark essay assessments automatically, it is difficult to transform essay assessment answers to vectors since approaches such as word representation [5] may lose information in a sentence and approaches such as sentence2vec [6] may lose information in an assessment. Contrastive learning has recently become a dominant component in self-supervised learning for NLP which aims to embed augmented versions of the same sample close to each other while trying to distinguish embeddings from different samples [7,8]. How to use Supervised Contrastive Learning (SCL), however, can be challenging to mark assessments. Since physical resources (e.g. GPU) are limited [9], it is difficult to construct an efficient model with less resources and better performance. To address these challenges, we propose a novel representation model, Assessment2Vec, to transform assessments to vectors where assessments are marked in terms of SCL loss. The major contributions of the paper are summarized as: (i) A new general representation model is constructed which represents assessments to vectors. (ii) A comprehensive experimental evaluation on four models with eight real groups of data has been performed. The results show the effectiveness of proposed models in making assessments.

2   Related Work

To reduce instructors' workload, artificial intelligence techniques have been used in education [10–16]. Automated scoring systems have been studied, which typically utilize machine learning techniques to automatically grade essays [11,13–16] and short answers [10,12]. Deep neural network and artificial intelligence have been used in different applications, which has motivated us to use them in assessment marking problems. In real scenarios, computing resource is usually limited, which was not considered in [11–16]. NLP has been used widely in different applications. The sentence similar function was studied based on word2vector similar elements [5] while word2vector only represented static word information. Sentence2vector considers information in a sentence while it ignores information in assessments. The assessment level representation [11,13–16] ignores word or sentences information. In this paper, BERT (Bidirectional Encoder Representations from Transformers) [15] and long short-term memory (LSTM) [17] network are combined to represent assessments to a vector which considers information among words and sentences. Contrastive learning has recently become a dominant component in self-supervised learning for NLP. It aims to embed augmented versions of the same sample close to each other while trying to distinguish embeddings from different samples [7,8]. In this paper, SCL is firstly used on marking assessments which has never been considered in existing studies [11–16].

3   Assessment Representation and the Models

Problem Description: In this paper we aim to mark assessments automatically. Formally, given an answer of an exam question, an output score is obtained.

For a specific exam, there are many different questions which correspond to various assessments. Separated models are constructed for different questions. The pre-trained BERT [18] is used to represent words and sentences in the given answers. Bi-directional LSTM model [17] is explored for learning word-level and sentence-level information when we have little resources to fine tune the whole BERT model. To mark assessments efficiently, four different question models are constructed to represent assessments which includes frozen BERT-Sentence bi-LSTM (B-S-LSTM), frozen BERT-Word bi-LSTM (B-W-LSTM), frozen B (BERT) and fine tuning B. **Frozen B-S-LSTM:** For sentence level representation, an assessment is embedded by word vectors and is split into sentences by sentence split tool. Sentences are processed by frozen BERT and bi-LSTM. **Frozen B-W-LSTM:** For word level representation, assessment is processed by frozen BERT. All word embedding vectors are processed by bi-LSTM with sentence level representation. **Frozen B and Fine-tuning B:** For answer level representation, frozen B and fine-tuning B are constructed. Each word is embedded by word representation. Parameters are not fine tuned in frozen B but are in fine tuning B.

Supervised Contrastive Learning. After assessments are represented according to different levels of representation models, a sigmoid activation function is used to mark assessments as $s(X) = sigmoid(WX + b)$ where W and b are the metric parameters. The loss for SCL term is indicated to capture similarities between assessments of the same class and contrast them with assessments from other classes. A SCL term is used for fine-tuning pre-trained language models [19]. For a multi-class classification problem with S classes where S is the full score on assessments, a batch of training assessments of size N, $(H_{outt}, s_t)(t \in \{1, \cdots, N\})$ are studied where s_t is the score for assessment h_t. L_{MSE} is denoted as mean square error (MSE) loss function. For assessments, the difference of scores for the assessments is lower which implies that assessments are similar. According to the supervised information, the loss function L_S is defined based on SCL as $L_S = -\sum_{t=1}^{N} \sum_{j=1}^{|P|} \frac{1}{|P|} \log \frac{\exp(H_{outt} \cdot H_{outj}/\tau)}{\sum_{k=1(k \neq t)}^{N} \exp(H_{outk} \cdot H_{outt}/\tau)}$ where P is the set for H_{outt} which has the same predicted score and τ is the temperature parameter. The loss function L is defined in terms of L_{MSE} and L_{SCL} as $L_{SCL} = \alpha L_{MSE} + \beta L_S$ where α and β are scalar weighting hyper parameters that can be tuned for each downstream assessment.

4   Experiments

Our experimental study uses the automated student assessment prize (ASAP) Dataset, a widely used benchmark for essay assessment. Five-fold cross-validation is evaluated in the training split of the ASAP Dataset by quadratic weighted Kappa (QWK). To analyze the performance of the proposed model, experiments are conducted for different models. The MSE, and SCL loss are used respectively to measure the difference between the predicted values and ground truth for these models. Different models are implemented in the Python

Table 1. 5-fold experiment results in the test split of the ASAP dataset.

Models	Loss	Prompt								
		1	2	3	4	5	6	7	8	Avg.
Fr B-S-LSTM	MSE	0.756	0.669	0.646	0.730	0.778	0.754	0.784	0.618	0.722
	SCL	0.752	**0.678**	0.649	0.742	0.777	0.754	0.786	0.615	0.724
Fr B-W-LSTM	MSE	0.791	0.627	0.654	0.761	0.783	0.774	0.795	**0.671**	0.733
	SCL	0.790	0.642	0.659	0.777	0.785	0.763	0.799	**0.671**	0.742
Fr B	MSE	0.631	0.486	0.367	0.490	0.600	0.457	0.602	0.372	0.507
Ft B	MSE	0.793	0.672	0.664	0.802	**0.797**	**0.804**	0.799	0.630	0.753
	SCL	**0.807**	0.671	**0.674**	**0.819**	**0.797**	0.795	**0.825**	0.657	**0.764**
Ft B [15]	MSE	0.829	0.391	0.762	0.886	0.876	0.584	0.818	0.540	0.711
CNN+LSTM [13]	MSE	0.821	0.688	0.694	0.805	0.807	0.819	0.808	0.644	0.761

programming language with the Pytorch library on a single NVIDIA P100 GPU with 16G memory. In this paper, we use *BERT-Base* [18] as the pre-trained language model. With parameter calibration, we empirically set the batch size for BERT and bi-LSTM models to 8, the maximum epochs to 30, the learning rate to 5×10^{-5}, the max length to 512, and the drop out probabilities to 0.5.

Experimental Result Analysis. After parameter calibration, four different models are evaluated by eight essay groups. To compare different models, the results for QWK are shown in Table 1 where Ft is represented as fine tuning and Fr is as frozen. From the result in Table 1, the performance is better for Ft B with SCL than MSE. When there are many questions for assessments, Ft B consumes significant computing resources while Fr B consumes limited computing resources. Fine tuning BERT is suitable for one question each time while Fr B is used for different questions. The performance improves significantly from 0.507 (Fr B) to 0.733 (Fr B-W-LSTM). According to Table 1, the values (0.733, 0.742) of Fr B-W-LSTM is better than (0.722, 0.724) of Fr B-S-LSTM. SCL also has improvement on Fr B from 0.733 to 0.742 by Fr B-W-LSTM and from 0.722 to 0.724 by Fr B-S-LSTM. According to Table 1, the average value of MSE for Ft B is 0.753 higher than 0.711 obtained in [15] with the same datasets. The average values of SCL for Ft B is 0.764, slightly higher than 0.761 obtained in [13]. Therefore, the proposed models are more efficient than existing models.

5 Conclusion

Large classes present serious challenges in assessment marking. In this paper, we analyzed the assessment marking problems. A general model is constructed for the problems with limited computing resources. SCL is used to improve the performance on the proposed models. We conducted a series of experiments to prove that even with limited computing resources, the proposed models can still get better performance. As an ongoing and future work, we are leveraging storytelling approaches [20,21] to facilitate summarizing, analysing, and presenting similar assessments.

Acknowledgement. We Acknowledge the AI-enabled Processes (AIP) Research Centre and ITIC Pty Ltd for funding this research.

References

1. Dai, A.M., Le, Q.V.: Semi-supervised sequence learning. In: Advances in Neural Information Processing Systems, vol. 28, pp. 3079–3087 (2015)
2. Peters, M.E., et al.: Deep contextualized word representations. arXiv preprint arXiv:1802.05365 (2018)
3. Radford, A., Narasimhan, K., Salimans, T., Sutskever, I.: Improving language understanding with unsupervised learning. Technical report, OpenAI (2018)
4. Beheshti, A., Benatallah, B., Sheng, Q.Z., Schiliro, F.: Intelligent knowledge lakes: the age of artificial intelligence and big data. In: U, L.H., Yang, J., Cai, Y., Karlapalem, K., Liu, A., Huang, X. (eds.) WISE 2020. CCIS, vol. 1155, pp. 24–34. Springer, Singapore (2020). https://doi.org/10.1007/978-981-15-3281-8_3
5. Yuan, X., Wang, S., Wan, L., Zhang, C.: SSF: sentence similar function based on word2vector similar elements. J. Inf. Process. Syst. **15**(6), 1503–1516 (2019)
6. Zhou, T., Zhang, Y., Lu, v; Classifying computer science papers. In: Proceedings of the 25th International Joint Conference on Artificial Intelligence, pp. 9–15 (2016)
7. Jaiswal, A., Babu, A.R., Zadeh, M.Z., Banerjee, D., Makedon, F.: A survey on contrastive self-supervised learning. Technologies **9**(1), 2 (2021)
8. Fang, H., Xie, P.: CERT: contrastive self-supervised learning for language understanding. arXiv preprint arXiv:2005.12766 (2020)
9. Strubell, E., Ganesh, A., McCallum, A.: Energy and policy considerations for deep learning in NLP. arXiv preprint arXiv:1906.02243 (2019)
10. Burrows, S., Gurevych, I., Stein, B.: The eras and trends of automatic short answer grading. Int. J. Artif. Intell. Educ. **25**(1), 60–117 (2015)
11. Dasgupta, T., Naskar, A., Dey, L., Saha, R.: Augmenting textual qualitative features in deep convolution recurrent neural network for automatic essay scoring. In: Proceedings of the 5th Workshop on Natural Language Processing Techniques for Educational Applications, pp. 93–102 (2018)
12. Sung, C., Dhamecha, T.I., Mukhi, N.: Improving short answer grading using transformer-based pre-training. In: Isotani, S., Millán, E., Ogan, A., Hastings, P., McLaren, B., Luckin, R. (eds.) AIED 2019. LNCS (LNAI), vol. 11625, pp. 469–481. Springer, Cham (2019). https://doi.org/10.1007/978-3-030-23204-7_39
13. Taghipour, K., Ng, H.T.: A neural approach to automated essay scoring. In; Proceedings of the 2016 Conference on Empirical Methods in Natural Language Processing, pp. 1882–1891 (2016)
14. Uto, M., Okano, M.: Robust neural automated essay scoring using item response theory. In: Bittencourt, I.I., Cukurova, M., Muldner, K., Luckin, R., Millán, E. (eds.) AIED 2020. LNCS (LNAI), vol. 12163, pp. 549–561. Springer, Cham (2020). https://doi.org/10.1007/978-3-030-52237-7_44
15. Uto, M., Xie, Y., Ueno., M.: Neural automated essay scoring incorporating handcrafted features. In: Proceedings of the 28th International Conference on Computational Linguistics, pp. 6077–6088, Barcelona, Spain (Online). International Committee on Computational Linguistics, December 2020

16. Wang, Y., Wei, Z., Zhou, Y., Huang, X.-J.: Automatic essay scoring incorporating rating schema via reinforcement learning. In: Proceedings of the 2018 Conference on Empirical Methods in Natural Language Processing, pp. 791–797 (2018)
17. Hochreiter, S., Schmidhuber, J.: Long short-term memory. Neural Comput. 9(8), 1735–1780 (1997)
18. Devlin, J., Chang, M.-W., Lee, K., Toutanova, K.: BERT: pre-training of deep bidirectional transformers for language understanding. arXiv preprint arXiv:1810.04805 (2018)
19. Gunel, B., Du, J., Conneau, A., Stoyanov, V.: Supervised contrastive learning for pre-trained language model fine-tuning. arXiv preprint arXiv:2011.01403 (2020)
20. Tabebordbar, A., Beheshti, A., Benatallah, B.: ConceptMap: a conceptual approach for formulating user preferences in large information spaces. In: Cheng, R., Mamoulis, N., Sun, Y., Huang, X. (eds.) WISE 2020. LNCS, vol. 11881, pp. 779–794. Springer, Cham (2019). https://doi.org/10.1007/978-3-030-34223-4_49
21. Beheshti, A., Tabebordbar, A., Benatallah, B.: iStory: intelligent storytelling with social data. In: Companion of The 2020 Web Conference 2020, Taipei, Taiwan, 20–24 April 2020, pp. 253–256. ACM/IW3C2 (2020)

Toward Stable Asymptotic Learning with Simulated Learners

Daniel Weitekamp[✉], Erik Harpstead, and Kenneth Koedinger

Carnegie Mellon University, Pittsburgh, PA 15213, USA
weitekamp@cmu.edu

Abstract. Simulations of human learning have shown potential for supporting ITS authoring and testing, in addition to other use cases. To date, simulated learner technologies have often failed to robustly achieve perfect performance with considerable training. In this work we identify an impediment to producing perfect asymptotic learning performance in simulated learners and introduce one significant improvement to the Apprentice Learner Framework to this end.

Keywords: Simulated learners · Cognitive modeling · Authoring tools

1 Introduction

Simulated learners are simulations of human learning that learn to perform tasks through an interactive process of demonstrations and feedback provided either by a human tutor or an Intelligent Tutoring System (ITS). Simulated learners have the potential to revolutionize learning technologies on a number of fronts, Matsuda demonstrated that students can learn by teaching a simulated learner called SimStudent [11], and Li showed the potential of SimStudent for cognitive model discovery [5]. Additionally, Matsuda, Maclellan, and Weitekamp [7,10, 13] have demonstrated the use of simulated learners as a potential means of authoring ITSs [12], such as cognitive tutors [4], more efficiently than comparable methods [2] that do not employ simulated learners.

For the purposes of using simulated learners as authoring tools it is desirable that the performance of the simulated learner asymptotically tends toward zero error. This capability ensures that the ultimate tutoring system behavior learned by the agent does not mark correct student responses as incorrect or incorrect responses as correct. We explore the asymptotic performance of simulated learners using the Apprentice Learner (AL), a modular software library for creating simulated learners instantiating different mechanistic theories of learning [6]. In this work, we identify a new source of learning failure that prevent AL agents from achieving zero training error and demonstrate an adjustment to the AL framework that allows it to recover from this failure mode. More broadly this work identifies and remedies an issue unexplored by prior inductive task learning literature, how an agent can recover from an incorrect induction made early in training to asymptotically acquire a knowledge state functionally equivalent to a set of ground-truth procedural knowledge.

© Springer Nature Switzerland AG 2021
I. Roll et al. (Eds.): AIED 2021, LNAI 12749, pp. 390–394, 2021.
https://doi.org/10.1007/978-3-030-78270-2_69

2   Training Test Domain: Multi-column Addition ITS

To demonstrate issues in aysmptotic training behavior, we use mutli-column addition as a simple prototypical example. We train our agents on an ITS implemented with CTAT's [1] nools [9] model tracer [3] that supports practice on what is often called the "standard" or "traditional" algorithm for adding large numbers, whereby the digits of the numbers to be summed are aligned in columns, summed column by column, with an extra "carry" row that is used for carrying the tens' digit from one column to the next.

```
          □  □  1
          5  3  9
       +  4  2  1
          ─────────
       □  9  6  0
```

3   A Brief Overview of the Apprentice Learner Framework

Apprentice Learner agents learn a set of skills sufficient to apply target tasks by learning in an interactive process with an ITS or human author. Each skills that an AL agent learns has at least four parts *how*, *where*, *when*, and *which*, that are each learned by different learning mechanisms. *How-learning* learns the *how-part* a composition of domain general functions that produces an action. For example, after trying a number of operations in different combinations *how-learning* might learn the *how-part* Mod10(Add(?.v, ?.v)) which takes two interface elements (the ?s) as arguments, sums their values (the .v's) and takes their one's digit (i.e. the modulus of 10).

A found RHS can work for a particular example but fail to work in general, for example another explanation an AL agent may come up with from the previous example is Copy(?.v) or just copy the value of the second value in a column into the carry slot. For 539+421 this would work for the first carry step, but would fail in general. In the AL agents used in this study a skill is identified by its RHS, so if the RHS happens to be wrong a new one must be induced, and the old one must be overridden or discarded.

Where-learning learns the *where-part*—a set of rules that pick out a set of arguments for a RHS, for example all of the numbers above the line in a column, and a "selection", the field into which the evaluation of the RHS will be placed. *When-learning* learns the *when-part* conditions, over the whole state, under which a skill should fire given a proposed *where-part* binding. Finally *which-learning* learns a policy for picking which potential application of a skill should be applied if multiple pass the *where-* and *when-part* rules. Given space constraints the reader should refer to prior work for further details about these learning mechanisms [6,14].

4   Addressing Lingering Weak and Overgeneral Skills

AL agents may need to observe several examples of taking particular problem steps to induce the correct *how-part* for the true skill associated with that kind of step. In the meantime a weak (i.e. not correct in all situations) *how-part* can be induced. Skills with weak *how-parts* will tend to be buried by building up a low *which-part* utility through repeated negative feedback, whereas correct skills may accumulate some negative feedback as their *when-part* conditions are refined and fewer later on. Consequently correct skills tend to override weak skills by accumulating a higher *which* utility. Prior work with simulated learners has shown that overriding via *which* utility works well in many domains [8], but we have identified some circumstances that necessitate a revaluation of this method.

For instance, it is possible for *how-parts* to be misattributed to the wrong skill. Consider for example, the case of an untrained simulated learner seeing $215 + 846$, and asking for examples of how to do the first few steps. Adding the 5 and 6 produces 11, creating an opportunity for the same skill to be induced and attributed to the first two actions (which should utilize seperate skills)—placing a 1 below and carry a 1 to the next column. Since the interface elements on which the two actions act are different, the *where-learning* mechanism for that skill will over-gencralize the conditions constraining legal bindings of the selection field causing the agent to apply the skill in a number of absurd ways.

4.1   Two Methods for Addressing Overgeneralization Errors

To address overgeneralization issues in AL agents we present two possible implementation changes and evaluate each independently. First, we implement a means for faulty skills to be removed including those that have overgeneralized. Second, we implement a *where-learning* mechanism that is capable of undoing generalization errors. A key observation in both proposed implementation changes is that faulty skills, either those with incorrect RHSs or overgeneralized where-part rules, will tend to make more errors than non-faulty ones, especially late in the training process. From a cognitive standpoint these can be thought of as persistent weak hypotheses of the true procedure. But these weak hypotheses should not persist indefinitely in the face of negative reinforcement, and should eventually be given up on.

Our first implementation change is to add a new removal utility, a number between 0 and 1 that when lowered below a threshold of .2 signals that a skill should be removed from an agent. We try three different functions of "p" and "n" (the numbers of instances of positive and negative feedback) for this utility: 1) the proportion correct $p/(p+n)$ (same as the *which* utility) 2) double counted negatives $p/(p+2n)$, and 3) nonlinearly counted negatives $p/p+n+1/4n^2)$. Nonlinearly counted negatives implements the intuition that skills that persistently produce errors after considerable training are more likely to be faulty than skills that only produce errors initially while a skills *when-part* rules are still being refined.

Our second implementation change introduces a fourth condition called "recovering where" that enables overgeneralized *where-part* conditions to return to a more specific state. Each newly generalized set of *where-part* conditions has its own removal utility that is updated, when applicable, with positive or negative feedback, and is removed when the utility calculated on the counts of positive and negative feedback falls below a threshold of .5.

4.2 Results of Implementation Changes

For all tested variations of the two proposed implementation changes we ran 100 agents on 100 3 × 3 multi-column addition problems. The first problem is always fixed to $215 + 846$ to ensure that a large number of the agents exhibit the *where-part* overgeneralization issue, and the remaining 99 problems are sampled randomly.

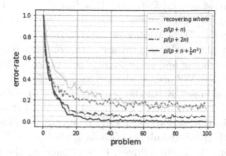

Fig. 1. Comparison of four recovery methods from *where-part* overgeneralization

Among the implementations of skill removal utility, nonlinearly counted negatives (i.e. "$p/(p + n + 1/4n^2)$") reliably removed overgeneralized and persistent weak skills, while the other methods still exhibited persistant error. We suspect that the 'recovering where' condition was ineffective because each competing *where-part* generalization shares a *when-learning* mechanism, meaning that even if bad generalizations are eliminated, eventually a considerable number of unusual training instances centered around irrelevant selection fields will remain in the *when* training history, making it far more challenging to establish a set of consistent *when-part* conditions. Whole skill removal by contrast is a consistently more effective method of over-generalization removal since removing an entire skill allows for a new skill to be induced in its place, giving *when-learning* a clean slate to work with.

5 Conclusion

In this work we have identified challenges to achieving asymptotic performance with simulated learners and remediated sources of persistent asymptotic error in simulated learners implemented with the Apprentice Learner Framework.

References

1. Aleven, V., McLaren, B.M., Sewall, J., Koedinger, K.R.: The cognitive tutor authoring tools (CTAT): preliminary evaluation of efficiency gains. In: Ikeda, M., Ashley, K.D., Chan, T.-W. (eds.) ITS 2006. LNCS, vol. 4053, pp. 61–70. Springer, Heidelberg (2006). https://doi.org/10.1007/11774303_7
2. Aleven, V., Sewall, J., McLaren, B.M., Koedinger, K.R.: Rapid authoring of intelligent tutors for real-world and experimental use. In: Sixth IEEE International Conference on Advanced Learning Technologies (ICALT 2006), pp. 847–851. IEEE (2006)
3. Blessing, S.B., Gilbert, S.B., Ourada, S., Ritter, S.: Authoring model-tracing cognitive tutors. Int. J. Artif. Intell. Educ **19**(2), 189–210 (2009)
4. Koedinger, K.R., Anderson, J.R., Hadley, W.H., Mark, M.A.: Intelligent tutoring goes to school in the big city (1997)
5. Li, N., Cohen, W.W., Koedinger, K.R., Matsuda, N.: A machine learning approach for automatic student model discovery. In: Edm, pp. 31–40. ERIC (2011)
6. Maclellan, C.J., Harpstead, E., Patel, R., Koedinger, K.R.: The apprentice learner architecture: closing the loop between learning theory and educational data. In: International Education Data Mining Society (2016)
7. MacLellan, C.J., et al.: Authoring tutors with complex solutions: a comparative analysis of example tracing and simstudent. In: The 2nd AIED Workshop on Simulated Learners. CEUR-WS.org, Madrid, Spain (2015)
8. MacLellan, C.J., Koedinger, K.R.: Domain-general tutor authoring with apprentice learner models. Int. J. Artif. Intell. Educ. 1–42 (2020). https://link.springer.com/journal/40593/topicalCollection/AC_d47d1642e3402a9a8134c35afb7851c8
9. Martin, D.: Nools. https://github.com/noolsjs/nools
10. Matsuda, N., Cohen, W.W., Koedinger, K.R.: Teaching the teacher: tutoring simstudent leads to more effective cognitive tutor authoring. Int. J. Artif. Intell. Educ. **25**(1), 1–34 (2015)
11. Matsuda, N., Keiser, V., Raizada, R., Stylianides, G., Cohen, W.W., Koedinger, K.R.: Learning by teaching simstudent – interactive event. In: Biswas, G., Bull, S., Kay, J., Mitrovic, A. (eds.) AIED 2011. LNCS (LNAI), vol. 6738, p. 623. Springer, Heidelberg (2011). https://doi.org/10.1007/978-3-642-21869-9_124
12. VanLehn, K.: The relative effectiveness of human tutoring, intelligent tutoring systems, and other tutoring systems. Educ. Psychol. **46**(4), 197–221 (2011)
13. Weitekamp, D., Harpstead, E., Koedinger, K.: An interaction design for machine teaching to develop AI tutors. In: CHI (2020). in press
14. Weitekamp, D., Ye, Z., Rachatasumrit, N., Harpstead, E., Koedinger, K.: Investigating differential error types between human and simulated learners. In: Bittencourt, I.I., Cukurova, M., Muldner, K., Luckin, R., Millán, E. (eds.) AIED 2020. LNCS (LNAI), vol. 12163, pp. 586–597. Springer, Cham (2020). https://doi.org/10.1007/978-3-030-52237-7_47

A Word Embeddings Based Clustering Approach for Collaborative Learning Group Formation

Yongchao Wu[✉], Jalal Nouri, Xiu Li, Rebecka Weegar, Muhammad Afzaal,
and Aayesha Zia

Stockholm University, Stockholm, Sweden
{yongchao.wu,jalal,xiu.li,rebeckaw,muhammad.afzaal}@dsv.su.se

Abstract. Today, collaborative learning has become quite central as a method for learning, and over the past decades, a large number of studies have demonstrated the benefits from various theoretical and methodological perspectives. This study proposes a novel approach that utilises Natural Language Processing(NLP) methods, particularly pre-trained word embeddings, to automatically create homogeneous or heterogeneous groups of students in terms of knowledge and knowledge gaps expressed in assessments. The two different ways of creating groups serve two different pedagogical purposes: (1) homogeneous group formation based on students' knowledge can support and make teachers' pedagogical activities such as feedback provision more time efficient, and (2) the heterogeneous groups can support and enhance collaborative learning. We evaluate the performance of the proposed approach through experiments with a dataset from a university course in programming didactics.

Keywords: Collaborative learning · Artificial intelligence · Natural language processing · Word embeddings · AI · NLP

1 Introduction

Over the past decades, there is a large consensus which asserts the higher achievement effects of collaborative learning on individual cognitive development as compared to individualistic learning and traditional instructional methods [1–3]. It is pointed out in [4,5] that a better interaction could be triggered by student knowledge levels and interests in the group. In this study, we attempt to automatically create collaborative learning groups based on student knowledge levels expressed in assignments. Group formation is one of the most important steps in collaborative learning. Many studies have investigated computer-supported approaches to optimise group formation. Researchers have explored the effectiveness of ontology engineering [6], Genetic algorithms (GA) [7,8], and multi-objective GA [9,10] to establish optimised groups. Machine

© Springer Nature Switzerland AG 2021
I. Roll et al. (Eds.): AIED 2021, LNAI 12749, pp. 395–400, 2021.
https://doi.org/10.1007/978-3-030-78270-2_70

learning based clustering approaches have also been explored in [11–13]. We found that most of the group criteria are naturally equipped with semantic information. For example, a programming course might cover topics of *loop, abstraction, data structure*. Nevertheless, the current machine learning-based approaches only regard group criteria as numerics, namely one-hot embeddings, which fail to catch semantic information. In the world of artificial intelligence, word embeddings have shown great power at catching rich semantic relationships between words [14]; a classic example of word embeddings calculation is $vec(Madrid) - vec(Spain) + vec(France) \approx vec(Paris)$. We argue that semantic information from the text-formatted group criteria would be helpful for group formation. Thus, we propose and present a group formation approach that forms homogeneous and heterogeneous groups based on students' knowledge levels that can catch rich semantic information of group criteria.

2   Method Formulation

In this study, we develop a model that can automatically compose homogeneous and heterogeneous collaborative learning groups G_{ho} and G_{he}, based on student knowledge levels in a course. Each student's knowledge level is determined by which topics the student has mastered or not mastered on the course. For G_{ho}, the group members should share similar knowledge levels and intend to master similar topics. On the other hand, for G_{ho}, the group members intend to master different topics.

A group G_i consists of a list of students $\{s_1, s_2, \dots\}$. Each student has a mastered topic list M and an unmastered topic list L. Each topic in the L and M is in one-word or multiple-word text form. Considering that L and M are super small text data, and the word order of L and M does not affect the list's general topical information, we use average embedding of all topics as the list representation [15]. We deploy fastText[1] word embeddings, which is effective at catching rich linguistic information [16], to map each topic in L and M to a vector $v_{t_i} \in \mathbb{R}^{1 \times 300}$. Thus, the corresponding student s_i could be mapped to a student vector $v_{s_i} \in \mathbb{R}^{1 \times 300} = \frac{\sum_{i=1}^{N_s} v_{t_i}}{N_s}$, where, N_s is the number of topics in the topic list. We measure the cosine similarity and dissimilarity between v_{s_i} and v_{s_j} to tell the similarity or difference of student knowledge levels: $sim(v_{s_i}, v_{s_j}) = \frac{v_{s_i} \cdot v_{s_j}}{\|v_{s_i}\| \times \|v_{s_j}\|}$; $dis(v_{s_i}, v_{s_j}) = 1 - sim(v_{s_i}, v_{s_j})$. For all students $s_1, s_2, \dots, s_n$, we could establish two matrices M_{sim} and M_{dis} which consists of similarity or dissimilarity between any two student vectors. Then we feed M_{sim} or M_{dis} to an unsupervised spectral clustering algorithm [17] to get clusters as initial candidates G_{ini}. The main reason to use spectral clustering is that it can cluster both similar and different objects together though designed object distance matrices like M_{sim} and M_{dis}. For the spectral clustering algorithm, the number of desired clusters is determined by the user. In this study, for n

---

[1] https://fasttext.cc/.

students, we choose cluster size for G_{ini} as $\sqrt{\frac{n}{2}}$ recommended by the work of [18]. Note that in a collaborative learning group, the group size boundary is no more than eight [19]. Thus, we optimise larger group candidates in G_{ini} which consists of over eight students into smaller groups. After the optimisation step, G_{ho} or G_{he} could be achieved, and group sizes vary from two to eight.

3 Experiment

The dataset we used in the experiment is a digital quiz collected in a programming didactics course at Stockholm University that was given in 2020. The quiz contains 16 questions, and the course instructor has labeled each question with a programming-related topic, such as *loop, condition, parallelism, operators*. Overall, the quiz contains the performance of 121 students. Students' private information is removed to anonymise the data completely.

We measure the group homogeneity Q_{ho_k} and heterogeneity Q_{he_k} as group validation metrics. To measure the Q_{ho_k} and Q_{he_k} we consider evaluating the intra-group similarity and topic diversity. For a group in G_{ho}, the intra-group similarity $\frac{2\sum_{i=1}^{l-1}\sum_{j=i+1}^{l-1} sim(v_{t_i},v_{t_j})}{l\times(l-1)}$ should be higher because students share similar knowledge levels. While for a group in G_{he}, the group members should lack proficiency in diverse topics, which results in a bigger number of unmastered topics per person $\frac{n_{L_k}}{n_{S_k}}$. We could get the average homogeneity and heterogeneity as: $Q_{ho} = \frac{\sum_{k=1}^{N} Q_{ho_k}}{N}$ and $Q_{he} = \frac{\sum_{k=1}^{N} Q_{he_k}}{N}$, where N is the final cluster size for G_{ho} or G_{he}. Inspired by the work [20], Chi-Square χ^2 and Log-likelihood G^2 are used in our work to measure the inter-group distance as triangulate metrics. Suppose we regard unmastered topics in the group as features, a higher χ^2 and G^2 scores reflect that inter-group distance is larger, and each group has more significant patterns of topic features, which result in homogeneous groups. In contrast, lower χ^2 and G^2 scores signify that inter-group distance is smaller, and topic features are similarly distributed among groups, which reflect characteristics of heterogeneous groups. To evaluate the robustness of our approach, we add random grouping and one-hot embeddings-based approach as comparison sets. Student groups with sizes between 2 and 8 are formulated from a student sample size of 10, 20, 30, ..., 90, 100. For each student size, we perform random sampling from the overall 121 students 10 times, and calculate average metrics Q_{ho}, Q_{he}, χ^2 and G^2.

3.1 Experiment Results and Discussion

The experiment results for group validation measurement regarding homogeneity Q_{ho}, heterogeneity Q_{he}, Chi-Square χ^2, and Log-likelihood G^2 for different sets could be referred to in Fig. 1. Across experiment sets with different student sizes, G_{ho} leads the highest scores on chi-square and log-likelihood, reflecting that in the homogeneous collaborative learning groups created by our approach, inter-group distance is large and each group has significant patterns. G_{ho} also achieves

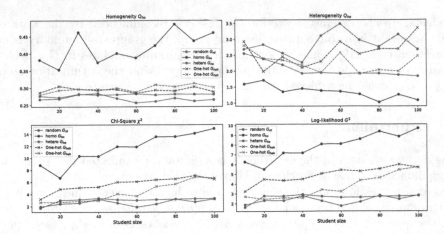

Fig. 1. The experiment results for group validation measurement for comparing our approach with random and one-hot based approach

the highest homogeneity score, which means students share similar knowledge levels in each group. G_{ho} gets the lowest score on heterogeneity, which is consistent with the fact that students intend to lack less diverse topics in a homogeneous collaborative learning group.

In a heterogeneous collaborative learning group, students are very different in terms of knowledge levels. The intra-group diverse knowledge levels will contribute to a small inter-group distance. G_{he} gets the lowest score on homogeneity, Chi-Square, and log-likelihood. However, its scores are only slightly lower than G_{rd}'s because the random sampling-based approach intends to create heterogeneous collaborative learning groups. G_{he} gets the highest heterogeneity score, which indicates students could learn more topics from each other in a heterogeneous collaborative learning group created by our approach. Compared to the one-hot embedding-based approach, when building G_{ho}, our approach achieved lower scores on heterogeneity and higher scores on homogeneity, Chi-Square, and log-likelihood. On the other hand, our approach achieved higher scores on heterogeneity and lower scores on homogeneity, Chi-Square, and log-likelihood when forming G_{he}. Our approach outperforms the one-hot embedding-based approach through the experiments when creating both homogeneous and heterogeneous groups. G_{ho} and G_{he} could support teachers' feedback provision practices as well as effective collaborative learning.

4   Conclusions and Future Work

We have presented a novel word embeddings-based approach to form collaborative learning groups based on student knowledge levels by catching group criteria semantic information. Experiments show that our approach outperforms the traditional one-hot embedding-based machine learning clustering approach. High-validation homogeneous and heterogeneous groups could be generated through our approach. An interesting direction for future work would be to apply word embeddings with other student characteristics, such as cognitive and personality traits. We intend to study the effect on the actual learning and collaborating process through experimental studies by deploying it in a formal learning environment.

References

1. Johnson, D., Johnson, R.: An educational psychology success story: social interdependence theory and cooperative learning. Educ. Res. **38**(5), 365–376 (2009).https://doi.org/10.3102/0013189X09339057
2. Johnson, D.W., Johnson, R.T., Buckman, L.A., Richards, P.S.: The effect of prolonged implementation of cooperative learning on social support within the classroom. J. Psychol. Interdisc. Appl. **119**(5), 405–11 (1985). https://doi.org/10.1080/00223980.1985.10542911
3. Slavin, R.E.: Research on cooperative learning and achievement: what we know, what we need to know. Contemp. Educ. Psychol. **21**(1), 43–69 (1996). https://doi.org/10.1006/ceps.1996.0004
4. Yang, S.: Context aware ubiquitous learning environments for peer-to-peer collaborative learning. Educ. Technol. Soc. **9**, 188–201 (2006)
5. Michaelsen, L. K., Knight, A. B., Fink, L. D.: Team-based learning: a transformative use of small groups (2002)
6. Isotani, S., Inaba, A., Ikeda, M., Mizoguchi, R.: An ontology engineering approach to the realization of theory-driven group formation. Int. J. Comput. Support. Coop. Learn. **4**, 445–478 (2009)
7. Moreno, J., Ovalle, D.A., Vicari, R.M.: A genetic algorithm approach for group formation in collaborative learning considering multiple student characteristics. Comput. Educ. **58**(1), 560–69 (2012). https://doi.org/10.1016/j.compedu.2011.09.011
8. Chen, C.M., Kuo, C.H.: An optimized group formation scheme to promote collaborative problem-based learning. Comput. Educ. **133**, 94–115 (2019). https://doi.org/10.1016/j.compedu.2019.01.011
9. Garshasbi, S., Mohammadi, Y., Graf, S., Garshasbi, S., Shen, J.: Optimal learning group formation: a multi-objective heuristic search strategy for enhancing intergroup homogeneity and intra-group heterogeneity. Expert Syst. Appl. **118**, 506–521 (2019)
10. Miranda, P.B., Mello, R.F., Nascimento, A.C.: A multi-objective optimization approach for the group formation problem. Expert Syst. Appl. **162**, 113828 (2020)
11. Maina, E.M., Oboko, R.O., Waiganjo, P.W.: Using machine learning techniques to support group formation in an online collaborative learning environment. Int. J. Intell. Syst. Appl. **9**(3), 26–33 (2017). https://doi.org/10.5815/ijisa.2017.03.04
12. Chen, L., Yang, Q.H.: A group division method based on collaborative learning elements. In: The 26th Chinese Control and Decision Conference (2014 CCDC), pp. 1701-1705 (2014). https://doi.org/10.1109/CCDC.2014.6852443
13. Bourkoukou O., Bachari E.E., Boustani A.E.: Building effective collaborative groups in e-learning environment. In: Ezziyyani M. (eds) Advanced Intelligent Systems for Sustainable Development (AI2SD 2019). AI2SD 2019. Advances in Intelligent Systems and Computing, vol. 1102. Springer, Cham (2020). https://doi.org/10.1007/9783030366537_11

14. Mikolov, T., Sutskever, I., Chen, K., Corrado, G., Dean, J.: Distributed represen-
    tations of words and phrases and their compositionality. In: Proceedings of the
    26th International Conference on Neural Information Processing Systems - vol. 2,
    pp. 3111–19. NIPS 2013. Red Hook, NY, USA: Curran Associates Inc. (2013)
15. Wu, L., et al.: Word Mover's Embedding: from Word2Vec to document embedding.
    In: Proceedings of the 2018 Conference on Empirical Methods in Natural Language
    Processing. Belgium, Brussels, October 2018
16. Bojanowski, P., Grave, E., Joulin, A., Mikolov, T.: Enriching word vectors with
    subword information. Trans. Assoc. Comput. Linguist. **5**, 135–146 (2017)
17. Ng, A., Jordan, M.I., Weiss, Y.: On spectral clustering: analysis and an algorithm.
    In: NIPS (2001)
18. Han, J., Kamber, M., Pei, J.: 10 - Cluster analysis: basic concepts and methods. In:
    Han, J., Kamber, M., Pei, J. (eds) Data Mining (Third Edition). Boston: Morgan
    Kaufmann (The Morgan Kaufmann Series in Data Management Systems), pp.
    443–495 (2012). https://doi.org/10.1016/B978-0-12-381479-1.00010-1
19. Hooper, S., Hannafin, M.J.: Cooperative CBI: the effects of heterogeneous versus
    homogeneous grouping on the learning of progressively complex concepts. J. Educ.
    Comput. Res. **4**(4), 413–424 (1988). https://doi.org/10.2190/T26C-3FTH-RNYP-
    TV30
20. Cavaglià, G.: Measuring corpus homogeneity using a range of measures for inter-
    document distance. In: Proceedings of the Third International Conference on Lan-
    guage Resources and Evaluation (LREC 2002). Las Palmas, Canary Islands - Spain:
    European Language Resources Association (ELRA) (2002)

Intelligent Agents Influx in Schools: Teacher Cultures, Anxiety Levels and Predictable Variations

R. Yamamoto Ravenor[✉] [iD]

Ochanomizu University, 2-1-1 Otsuka, Bunkyo-ku 112-8610, Tokyo, Japan
yamamoto.ravenor@ocha.ac.jp

Abstract. Artificially intelligent robots entered Japanese schools in 2017 in the same impetuous way in which computers flooded education worldwide during the 80s. Unlike computers, which became indispensable to school culture, AI robots have yet to find a place in the classroom. This paper presents a pilot study aimed at finding clusters of common teacher attitude to better plan the deployment of such AI agents in the future. The results of teacher surveys, first, indicated that the most influential factor in teacher adoption of such technology is coding literacy rather than age or study major; and secondly, served to train machine learning algorithms and develop a "culture stress system" that predicts teacher anxiety and recommends an optimum number of AI agents that targeted teachers in a school can plausibly accommodate.

Keywords: Teacher-AI interaction · Teacher cultures · AI school population

1 Introduction

In a variety of forms, artificially intelligent (AI) agents/tools are entering education, particularly AI humanoid robots are being seen in many Japanese schools. The increase in this profoundly different element of a school "population", in Japan alone having reached the order of thousands, could metaphorically be considered a form of immigration, as an influx of "cerebral" entities into a population introduces new cultures, languages and ways of doing basic things. Integration of AI agents in schools is somewhat different from that of more limited forms of technology, which touch less on human sensitivities. The arrival of a non-human group in human cultures of students and teachers may represent a stress upon these cultures. Students are the usual focus of related studies, yet teachers are being largely ignored.

This paper leverages on the sense that the ideal of AI "immigration" to schools is not alone to enhance students' learning experience, but primarily to create a better teacher. The questions before us are: What impacts the likelihood of teacher anxiety towards working with AI teaching tools? And how many such tools should a particular school take in at a time? The whole inquiry runs into the underlying theory that cooperative interaction "between the human and the electronic members of the partnership" enhances

I. Roll et al. (Eds.): AIED 2021, LNAI 12749, pp. 401–405, 2021.
https://doi.org/10.1007/978-3-030-78270-2_71

human capability [1]. The study attempts to answer the first question by seeking statistical relationships between a variety of demographic factors and teacher attitude towards working with the AI tools. Then, using the results of the analysis, the study designs a "culture stress system" (CSS) in an attempt to predict teacher anxiety and recommend the number of AI agents likely to be successfully adopted.

In contrast to existing literature, this study suggests that age is not the key factor for determining teacher comfort in dealing with AI agents. Results show that coding literacy is more important. Moreover, by including the length of teachers' coding experience as a feature in the dataset used to train the machine learning (ML) algorithm selected to develop the CSS, the model achieved better accuracy.

2   Softbank Pilot Project

In 2017 Softbank Robotics mobilized humanoid resources with the declared purpose of making a social contribution to education. Around 2,000 robots (called "Pepper"), equipped with artificial intelligence and based on an open and programmable platform, were lent to 282 public elementary and junior high schools across Japan, giving programming education experiences to approximately 91,000 people, according to the company data. The project yielded inconsistent results. It publicised success stories of some students winning programming competitions with Pepper and praised the creativity of some teachers using Pepper as programming teaching tools, but the returns of Pepper, for instance, signaled substantial failure. Based on data collected (for the purpose of this study) from 23 city education boards responsible with the distribution of Pepper in schools, it is estimated that, presently, the number of robots decreased to less than half the initial population (363 out of 795); no other technology of comparable versatility being known to have taken its place to equivalent scale. Had Softbank anticipated potential discomfort on the part of teacher users, many of these resources may have been put to better use, while the challenges to teacher readiness may have been strategically addressed.

In trying to understand the factors that influenced the adoption of Pepper in Japanese schools, the study followed the classic work of Waller [2], who first explored the small cultural world of school and the entanglements of interrelations in it. This research looks for such relations at a micro-level view of school cultures, i.e. the level of the individual teacher, and uses the plural to denote that many teacher cultures exist [3] and can influence the decision to adopt an innovation [4]. In this paper, the specific concept of "teacher cultures" refers to habits and/or attitudes of teachers, characteristic of an age, academic discipline and/or expertise, as well as literacies, which AI agents may throw stresses upon, some teachers thus finding it hard to accommodate them. Many studies related to the use of technology for teaching attribute teacher adoption patterns to cultural factors [5–8] and some find significant relationship between training and improvement in teacher attitudes towards technology [9]. Despite recent evidence that age is a factor related to the teachers' use of technology [10], older studies contend the opposite [11, 12]. The literature related to the topic presented here is copious, however, findings in studies focused on conventional computers may not all prove instructive in the context of AI, which may call for new types of scholarship. Interaction between teachers and AI agents in schools is rare and understudied.

This research is divided into two parts, according to the objectives. Part one gives a survey of the cultures of 134 teachers (who received at least one robot each to use freely in programming class) based on four factors: one attitudinal — anxiety, and three demographic — age, academic major and coding literacy. Teacher anxiety towards Pepper was determined plainly using a three-choice answer format for the question "Do you mind working with Pepper in the classroom?". The ambiguity of the question in English disappears when translated into Japanese, yet retains the indirect communication style reported to be preferred by the Japanese [13], while the answer choices "Mind a lot", "Mind a little" and "Do not mind" help avoid neutral answers that Japanese respondents are found to often choose [14]. Part two focused on addressing the demand for quantification when faced with anxiety caused by changes in population composition. The study created the CSS, a Python program that prompts the user to enter the number of targeted teachers and their respective data, and, based on the conclusions derived from preceding analysis, predicts teacher culture stress (anxiety) levels and recommends an optimum number of Pepper robots for each teacher. Six well-known ML models were trained for the purpose and the selection was based on accuracy.

3 Analysis of Teacher Adoption of AI Robots

Partly confirming prior work and the obvious, the results of teacher survey conducted this year indicate that older teachers and those specialized in fields other than STEM are indeed more likely to feel anxiety working with Pepper. Approximately 82% of teachers aged 40 and younger, and 80% of teachers with STEM-related degrees do not mind Pepper (culture stress is absent) (see Fig. 1 and Fig. 2). No group is over-represented in the survey sample. However, the fact that, among teachers who mind (culture stress is present), on the one hand, 22% are aged 40 and younger (see Fig. 1) and, on the other hand, 92% of the STEM-specialized are older than 40 (see Fig. 2), suggests that the factor of age may actually be a marker for experience with computer coding, as younger teachers were more likely to have code learning opportunities. In total, 40% of respondents mind Pepper, 89% of whom declared to have no coding experience. 86% of all code-literate teachers do not mind working with Pepper. 96% of all teachers who mind Pepper a lot (acute culture stress) do not have experience with coding (see Fig. 3). Coding literacy clearly determines the likelihood of teacher anxiety about Pepper.

Based on the relationships previously found, the following ML models were trained via the Scikit-Learn Python library: Naïve Bayes, logistic regression, K nearest neighbor classifier, decision tree, random forest and support vector machine. Random forest algorithm outperformed the others in terms of accuracy and was used for further development of the system. It was trained with hyperparameter max_depth $= 5$ and achieved an accuracy of 80.59% with coding literacy, age and study major as model features, by including the length of coding experience, more age ranges and majors that, for simplification, are not illustrated in the figures above. The maximum number of AI tools that the CSS assigns a teacher is 3, assumed from the rough estimates which showed that the initial average of 7 Peppers per school halved. Although there is generally one targeted teacher per school, the CSS can also handle more (charts data, detailed analysis and code are available from the author upon request).

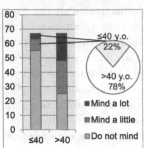

Fig. 1. Culture stress by age.

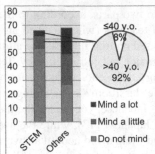

Fig. 2. Culture stress by major.

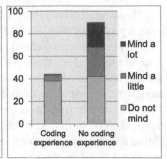

Fig. 3. Culture stress by coding literacy.

4 Conclusion

Unlike other studies which identified age as a factor that encourages or discourages the teacher adoption of technology, the evidence presented here identifies coding literacy as a prerequisite for direct cooperative endeavors between a teacher and an AI agent (regardless of its complexity). The deployment of Pepper robots in classrooms notwithstanding teacher readiness to integrate them exerted stress on some teacher cultures, which partly explains the poor uptake of this technology by many Japanese schools. The ML *culture stress system* was created to assist programmes such as Softbank's in reducing the risk of human and/or non-human resources being misallocated. Larger training datasets are likely to improve the model's performance. What other factors would limit teachers' adoption of such technology and where else does the discomfort come from, are some future research directions to be considered in due course.

Typically, schools teach neither students nor teachers how to use computers, tablets, smartphones etc. on the grounds that this is simply learnt by osmosis. The same cannot be said of, for instance, open-source technologies, such as those on which Pepper and many tools for AI, ML, big data and cloud are built. Teachers do not all in the same way make use of their teaching tools (blackboard, projector etc.), and the same may be expected from the way in which they would use their AI tools. Unless an AI is designed to standardize teaching practices, the limits to which it will aid a teacher will coincide not only with the limits to which it is developed, but also with the limits of that teacher's will and ability to perform more effectively with it than alone. In conclusion, predicting these limits with a model such as the one presented here is the *sine qua non* for rational placement of AI in a school.

Acknowledgements. This work is funded by the JSPS KAKENHI Grant-in-Aid for Early-Career Scientists, under grant # JP20K13900. All opinions expressed in this material are solely the author's and do not necessarily reflect the views of the author's organization, JSPS or MEXT. Grateful thanks are due to Prof. Dr. David Alan Grier for his continual guidance and acute observations from which the study profited, and to Muhammad Haseeb UR Rehman Khan for the assistance with ML work.

References

1. Licklider, J.C.: Man-computer symbiosis. IRE Trans. Hum. Factors Electron. **1**, 4–11 (1960)
2. Waller, W.: The Sociology of Teaching. Wiley, New York (1932)
3. Sachs, J., Smith, R.: Constructing teacher culture. Br. J. Sociol. Educ. **9**(6), 421–439 (1988)
4. Rogers, E.M.: Diffusion of Innovations, 5th edn. Free Press, New York (2003)
5. Anderson, T., Varnhagen, S., Campbell, K.: Faculty adoption of teaching and learning technologies: contrasting earlier adopters and mainstream faculty. Can. J. High. Educ. **28**(23), 71–78 (1998)
6. Geoghegan, W.H.: Whatever happened to instructional technology? Paper Presented at the 22nd Annual Conference of the International Business Schools Computing Association, Baltimore (1994)
7. Ertmer, P.A., Ottenbreit-Leftwich, A.T.: Teacher technology change. J. Res. Technol. Educ. **42**(3), 255–284 (2010)
8. Demetriadis, S., et al.: "Cultures in negotiation": teachers' acceptance/resistance attitudes considering the infusion of technology into schools. Comput. Educ. **41**(1), 19–37 (2003)
9. Yildirim, S.: Effects of an educational computing course on pre-service and in-service teachers: a discussion and analysis of attitudes and use. J. Res. Comput. Educ. **32**(4), 479–496 (2000)
10. Fraillon, J., Ainley, J., Schulz, W., Friedman, T., Gebhardt, E.: Preparing for Life in a Digital Age. The IEA International Computer and Information Literacy Study International Report, pp. 180–209. Springer, Heidelberg (2014). https://doi.org/10.1007/978-3-319-14222-7
11. Hayden, M.A.: The structure and correlates of technological efficacy. In: Petry, J.R. (ed.) Proceedings of the Annual Meeting of the Mid-South Educational Research Association, pp. 1–29. Meridian, Mississippi (ERIC Document Reproduction Service no. ED 394 998) (1995)
12. Honeyman, D.S., White, W.J.: Computer anxiety in educators learning to use the computer: a preliminary report. J. Res. Comput. Educ. **20**(2), 129–138 (1987)
13. Pan, Y., Landreth, A., Park, H., Hinsdale-Shouse, M., Schoua-Glusberg, A.: Cognitive interviewing in non-English languages: a cross-cultural perspective. In: Harkness, J.A., et al. (eds.) Survey Methods in Multinational, Multiregional, and Multicultural Contexts, pp. 91–113. Wiley, Hoboken (2010)
14. Yang, Y., Harkness, J.A., Chin, T.-Y., Villar, A.: Response style and culture. In: Harkness, J.A., et al. (eds.) Survey Methods in Multinational, Multiregional, and Multicultural Contexts, pp. 203–223. Wiley, Hoboken (2010)

WikiMorph: Learning to Decompose Words into Morphological Structures

Jeffrey T. Yarbro[(✉)] and Andrew M. Olney[(✉)]

University of Memphis, Memphis, TN 38152, USA
{jyarbro2,aolney}@memphis.edu

Abstract. This paper presents WikiMorph, a tool that automatically breaks down words into morphemes, etymological compounds (morphemes from root languages), and generates contextual definitions for each component. It comes in two flavors: a dataset and a deep-learning-based model. The dataset was extracted from Wiktionary and contains over 450k entries. We then used this dataset to train a GPT-2 model to generalize and decompose any word into morphemes and their definitions. We find that the model accurately generates complex breakdowns when given a high-quality initial definition.

Keywords: Morphemes · GPT-2 · Wiktionary · Etymological compounds

1 Introduction

The ability to recognize the morphological structure of words and the meaning of the morphemes within that structure positively correlates with vocabulary development and reading comprehension [4, 7, 8, 27]. Unfortunately, there are not many tools designed to increase morphological awareness. While morpheme segmentation tools and datasets are available [2, 6, 18, 20, 22], these lack critical elements for learning, such as definitions for each morpheme or a sense of its etymology. We attempt to fill this void by introducing WikiMorph: a dataset and deep-learning model. The dataset was collected by extracting user-inputted morphological data from a December 2020 version of the English Wiktionary XML dump file. This dataset contains morphemes (from English and root languages), PoS tags, and contextual definitions for each morpheme. Since Wiktionary lacks morpheme entries for some words, we also train a GPT-2 model on this dataset to generalize and break down any English word.

The model receives two inputs: a word and, optionally, its definition. If a definition is not received, the model will attempt to generate a definition for the input word. From there, it autoregressively generates a word breakdown which includes morphemes and contextualized definitions. See Sect. 3 for results.

2 WikiMorph: Dataset and Model

Wiktionary is an online, multilingual dictionary sponsored by the Wikimedia Foundation that contains a wide variety of information useful for NLP tasks. For this paper,

I. Roll et al. (Eds.): AIED 2021, LNAI 12749, pp. 406–411, 2021.
https://doi.org/10.1007/978-3-030-78270-2_72

we are primarily interested in the definition and etymology sections of Wiktionary. The etymology section is of particular importance since it often contains annotated morphological segmentations for words. These segmentations can either be in English or from root languages such as Latin or Ancient Greek. We will refer to morphemes from root languages as etymological compounds throughout this paper. These compounds are useful since they give additional insights into English words and often allow further morpheme segmentation within the root language.

Extracting data from Wiktionary comes with many challenges. Most notably, standardization. Wiktionary was primarily designed to allow for flexible formatting to make it easy for authors across the web to contribute. This flexibility makes it essential first to regularize the formatting of Wiktionary. We do this by looping through the XML file and applying many regular expressions. These regular expressions aim to remove markup codes and allow our morpheme extraction algorithm to grab all relevant data.

Wiktionary does not require authors to input morpheme segmentations when a word falls under a common rule. Meaning that some affixes are regularly void of morphological entries, and therefore, unacceptable for this work. Most of these missing affixes are suffixes that change the grammatical context of the word. (e.g., making dog plural by adding -s). To combat this, we created a list of common suffixes that did not have regular entries in Wiktionary and used a series of heuristics to find the root morpheme. We then check a word corpus to see whether the root morpheme is an actual word and use DistillBERT word embeddings to see whether it is similar to the base word.

To extract morphemes, we deploy a recursive methodology. This methodology first attempts to find English morphemes within the Etymology section of Wiktionary. If found, we proceed to search Witkionary's entry for each of these found morphemes to see whether they too contain annotated morpheme segmentations. We repeat this process until the word cannot be broken down further in English. We then perform a similar lookup in root languages we deemed as "good" for each English morpheme. With "good" in this context meaning that we found examples where the language gave additional insights not seen in the English breakdown alone. If multiple etymological breakdowns were found, we chose only one with two criteria in mind. (1) Does the compound have a complete Wiktionary entry? (2) How insightful is the root language for English words? While rankings varied based on criteria 1, the system typically prefers Latin and Ancient Greek compounds since they are well-represented in Wiktionary and many morphologically complex words are derived from them.

Words often have different meanings depending upon the context. The same is true for morphemes within a word. We account for this by choosing the best definition entry for each morpheme using word embeddings from two models: DistillBERT and Spacy's Core model [10, 21]. We perform two operations for each morpheme definition. (1) Definition Similarity: Cosine similarity between the base word and morpheme's definition. (2) Addition Similarity: Adds word vectors from other morphemes within the base word to the current morpheme's definition vector, then takes the cosine similarity between the new vector and the base word's definition vector. We then perform a weighted average operation over the values and choose the definition with the highest average.

Since Wiktionary does not guarantee that word entries have a complete morpheme breakdown, it is necessary to filter out any of our extractions containing incomplete breakdowns. We do this by looping through the extracted morphemes and using a series of heuristics on each root morpheme to ensure completeness. These heuristics consist of the following checks: (1) Checks the number of syllables within the word [1]. (2) Checks the word frequency [19]. (3) Checks the number of etymological compounds. (4) Checks to see whether there are any common affixes.

We then train the WikiMorph model by extending the large variant of GPT-2 made available by Hugging Face [24]. GPT-2 is an autoregressive model that uses the decoding blocks of the transformer architecture [11, 23]. It contains 36 decoding blocks with 774M parameters. We use 16-bit precision for lesser memory requirements and greater training speed. We then fine-tune the pretrained GPT-2 model for three epochs.

To assess the model, we removed 1500 samples from the dataset prior to training. We then perform an ablation test on these samples to see how the model performs when it receives an input definition vs. when it does not. For both conditions, the aim is to test how well the model segments morphemes and its ability to generate contextualized definitions. To test its segmentation ability, we use accuracy and character-level ROUGE1 as a sanity check to ensure that the model did not produce wildly different morphemes [26]. To evaluate how well the model generates contextualized definitions, we use word-level ROUGE1 (with stemming) and cosine similarity between the generated definition and ground-truth definition using RoBERTa word embeddings [15]. For definition evaluation, the metrics are only performed when both the generated and ground-truth sample have an instance of the same morpheme to ensure alignment.

3   Results and Discussion

The results in Table 1 show that the model performed well at segmenting morphemes with over 85% accuracy for both English and etymology. For English morphemes, the model also did well at matching the characters within the ground-truth's segmentation, demonstrating that it is unlikely to give wildly different results even when the segmentations are different. The only notable differences in morpheme characters came in examples such as the ones shown in Fig. 1. Here the model adds characters to properly form the root morpheme "perceive".

Table 1. Results showing model performance and differences between when the model receives an input definition (+) vs. when it does not (−).

	Metrics	English morphemes		Etymology morphemes	
		+ Def	− Def	+ Def	− Def
Morpheme segmentation	Accuracy	0.925	0.887	0.890	0.854
	Character ROUGE1	0.992	0.985	0.572	0.559
Definition generation	ROUGE1 Score	0.808	0.528	0.945	0.931
	RoBERTa Sim.	0.754	0.421	0.933	0.913

It is important to note that some differences between the sample and generated breakdowns are not errors. As stated in Sect. 2, Wiktionary does not guarantee a full morphological breakdown for word entries, leading to some entries within our dataset not having a complete segmentation. However, since morphemes are repeatable in words (e.g., a- appears in arise and amoral), there is a probability that the morpheme is correctly segmented in other entries of the dataset—allowing the model to pick up on this probability and act as a denoiser, as shown in Fig. 1B.

```
"Word": "perceptor",                  A) Dataset Sample    "Word": "perceptor",                                              B) Generated Sample
"PoS": "Noun",                                             "Definition": "That which perceives ",
"Definition": "That which perceives.",                     "Morphemes": [
"Derivational Morphemes": [                                    {
    {                                                             "Morpheme": "perceive",
        "Affix": "perceptor",                                     "Definition": "To become aware of, through the physical senses or by thinking;
        "Language": "en",                                         to see; to understand.",
        "PoS": "Noun",                                            "Etymology Compounds": [{
        "Meaning": "That which perceives.",                           "Etymology Compound": "per",
        "Etymology Compounds": [                                      "Compound Meaning": "by, through"
            {                                                     },{
                "Affix": "perceptor",                                 "Etymology Compound": "capiō",
                "Language": "la",                                     "Compound Meaning": "to take"
                "PoS": "Noun",                                    }]
                "Meaning": "a receiver, imbiber"          },{
            }                                                 "Morpheme": "-or",
        ]                                                     "Definition": "Creates an agent noun, indicating a person who does something."
    }                                                         "Etymology Compounds": [{
}                                                                 "Etymology Compound": null,
                                                                  "Compound Meaning": null
                                                              }]
```

Fig. 1. A) Real sample from WikiMorph dataset. B) Generated output when given a definition.

The model showed considerable improvements while generating English definitions when given an input definition. This result is not surprising. When the model receives a definition, it has a good initialization—allowing it to pay attention and reference that good definition during the generation of each morpheme's definition. If the model does not receive an input definition, it has no context about the word aside from what it might have learned during training. Without this context, it can hallucinate while generating the initial definition, cascading additional errors across subsequent morphemes.

Interestingly, while the English contextualized definitions were significantly worse when the model did not receive a definition, the generated definitions for etymological compounds only saw a slight degradation. We speculate the reasons for this are due to three reasons. (1) The definitions are often much shorter for root languages than in English, thereby decreasing the probability that the model makes an error on an early token leading to subsequent errors. (2) There are fewer definition entries for each etymological compound. (3) These affixes frequently appear throughout many different words in our dataset, giving the model many opportunities to memorize the result.

4 Conclusion

This paper presents WikiMorph, a novel dataset and GPT-2-based model designed to help students learn morphology. The dataset extracted is one of the largest morpheme datasets to date and the only large-scale dataset containing contextualized definitions and etymological compounds. The trained WikiMorph model displayed an impressive ability to generate word breakdowns; however, further evaluation is required to determine its effectiveness in learning environments.

Acknowledgments. This material is based upon work supported by the National Science Foundation (1918751, 1934745) the Institute of Education Sciences (R305A190448).

References

1. Ash, S.: Jg2p (2018). https://github.com/steveash/jg2p
2. Balota, D.A., et al.: The English lexicon project. Behav. Res. Methods **39**(3), 445–459 (2007). https://doi.org/10.3758/BF03193014
3. Blais, C., Fiset, D., Arguin, M., Jolicoeur, P., Bub, D., Gosselin, F.: Reading between eye saccades. PLoS One **4**(7), (2009). https://doi.org/10.1371/journal.pone.0006448
4. Bowers, P.N., Kirby, J.R., Hélène Deacon, S.: The effects of morphological instruction on literacy skills: a systematic review of the literature. Rev. Educ. Res. **80**(2), 144–179 (2010). https://doi.org/10.3102/0034654309359353
5. Burani, C., Marcolini, S., De Luca, M., Zoccolotti, P.: Morpheme-based reading aloud: evidence from dyslexic and skilled Italian readers. Cognition **108**(1), 243–262 (2008). https://doi.org/10.1016/j.cognition.2007.12.010
6. Creutz, M., Lagus, K.: Unsupervised discovery of morphemes. In: Proceedings of the ACL-2002 Workshop on Morphological and Phonological Learning, pp. 21–30 (2002). https://doi.org/10.3115/1118647.1118650
7. Duncan, L.G.: Language and reading: the role of morpheme and phoneme awareness. Curr. Dev. Disord. Rep. **5**(4), 226–234 (2018). https://doi.org/10.1007/s40474-018-0153-2
8. Goodwin, A.P., Ahn, S.: A meta-analysis of morphological interventions: effects on literacy achievement of children with literacy difficulties. Ann. Dyslexia **60**(2), 183–208 (2010). https://doi.org/10.1007/s11881-010-0041-x
9. Gwilliams, L.: How the brain composes morphemes into meaning. Philos. Trans. R. Soc. B Biol. Sci. **375**(1791), (2020). https://doi.org/10.1098/rstb.2019.0311
10. Honnibal, M., Montani, I., Van Landeghem, S., Boyd, A.: SpaCy: industrial-strength natural language processing in Python. Zenodo (2021). https://doi.org/10.5281/zenodo.1212303
11. Hoppe, S., Toussaint, M.: Qgraph-bounded Q-learning: stabilizing model-free off-policy deep reinforcement learning. ArXiv (2020)
12. Kirov, C., Sylak-Glassman, J., Que, R., Yarowsky, D.: Very-large scale parsing and normalization of wiktionary morphological paradigms. In: Proceedings of the 10th International Conference on Language Resources and Evaluation, LREC 2016, pp. 3121–3126 (2016)
13. Krizhanovsky, A.A.: Transformation of wiktionary entry structure into tables and relations in a relational database schema, 10 (2010). http://arxiv.org/abs/1011.1368.
14. Bensoussan, M., Laufer, B.: Lexical guessing in context in EFL reading comprehension. J. Res. Reading **7**(1), 15–31 (1984). https://doi.org/10.1111/j.1467-9817.1984.tb00252.x
15. Liu, Y., et al.: RoBERTa: a robustly optimized BERT pretraining approach. ArXiv, no. 1 (2019)
16. Luong, M., Manning, C.D.: Better word representations with recursive neural networks for morphology. In: CoNLL-2013, pp. 104–113 (2003)
17. Metheniti, E., Neumann, G., van Genabith, J.: Linguistically inspired morphological inflection with a sequence-to-sequence model (2020). http://arxiv.org/abs/2009.02073
18. Metheniti, E., Neumann, G.: Wikinflection: massive semi-supervised generation of multilingual inflectional corpus from Wiktionary. In: TLT 2018, pp. 147–161 (2018)
19. Speer, R., Chin, J., Lin, A., Jewett, S., Nathan, L.: LuminosoInsight/Wordfreq (2018). https://doi.org/10.5281/zenodo.1443582

20. Sánchez-Gutiérrez, C.H., Mailhot, H., Deacon, S.H., Wilson, M.A.: MorphoLex: a derivational morphological database for 70,000 English words. Behav. Res. Methods **50**(4), 1568–1580 (2017). https://doi.org/10.3758/s13428-017-0981-8
21. Sanh, V., Debut, L., Chaumond, J., Wolf, T.: DistilBERT, a distilled version of BERT: smaller, faster, cheaper and lighter, pp. 2–6. ArXiv (2019)
22. Smit, P., Virpioja, S., Grönroos, S.-A., Kurimo, M.: Morfessor 2.0: toolkit for statistical morphological segmentation, pp. 21–24 (2015). https://doi.org/10.3115/v1/e14-2006
23. Vaswani, A., et al.: Attention is all you need. In: Advances in Neural Information Processing Systems (NIPS), December 2017, pp. 5999–6009 (2017)
24. Wolf, T., et al.: Transformers: state-of-the-art natural language processing. ArXiv (2019). https://doi.org/10.18653/v1/2020.emnlp-demos.6
25. Zhu, Y., Vuli, I., Korhonen, A.: For learning word representations (2013)
26. Lin, C.-Y.: ROUGE: a package for automatic evaluation of summaries. In: Proceedings of the ACL Workshop: Text Summarization Braches Out 2004, p. 10 (2004)
27. Hayashi, Y., Murphy, V.: An investigation of morphological awareness in Japanese learners of English. Lang. Learn. J. **39**(1), 105–120 (2011). https://doi.org/10.1080/0957173100363614

Individualization of Bayesian Knowledge Tracing Through Elo-infusion

Michael Yudelson$^{(\boxtimes)}$ 📖

Yudelson Consulting LLC, Pittsburgh, PA 15217, USA
http://www.yudelson.info

Abstract. For as long as the Bayesian Knowledge Tracing (BKT) approach is known, so are the attempts to account for not only skill-level but individual student factors. A lot of computational methods to implement individualization in BKT were proposed over the past 25 years as BKT existed. To this day, virtually all individualization approaches were not suited for easy implementation. Either they were purely analytical (only fit for post-hoc analyses) or required significant computational effort to realize (e.g., calibrating individual factors as students cleared units of content).

In this work, we discuss implementing the individualization of BKT using a mechanism of an Elo rating schema. Elo has been established in the educational domain for some time and offers tangible theoretical and practical benefits. We show that infusing BKT even with an Elo component using a single parameter to track student-specific factors results in significant quantitative and qualitative improvements to modeling student learning. This approach is easy to implement in a system already featuring BKT.

Keywords: Bayesian Knowledge Tracing · Elo rating schema

1 Introduction

Bayesian Knowledge Tracing (BKT) [4] is one of the most researched approaches to tracking student learning in computer-assisted problem-solving applications. One of the popular thrusts of BKT-centric research is accounting for student-specific factors. Standard BKT contains only skill-specific parameters, although, admittedly, these parameters are population parameters. The issue of individualization first raised in the original BKT paper targets the variability in how students learn and perform above and beyond what is captured by skill-specific components of the model. Despite a significant volume of publications on the topic of individualizing BKT, the resulting approaches are largely analytical due to computational, implementational, and other considerations. This is why individualization of BKT was largely an analytical topic and, to the best of our knowledge, was never deployed in a setting other than experimental.

In this work, we are proposing an approach that, while remaining in the stream of iBKT research, is first and foremost operationalizable. Our suggestion

© Springer Nature Switzerland AG 2021
I. Roll et al. (Eds.): AIED 2021, LNAI 12749, pp. 412–416, 2021.
https://doi.org/10.1007/978-3-030-78270-2_73

is easy to implement in any new product suitable to be equipped with a standard BKT and easy to be added to a product already using standard BKT as its core student modeling method. We call our approach individualization through Elo-infusion or simply Elo-infusion. Elo is a family of rating schema methods used in multiple contexts from rating players in competitive sports to online dating and education. We are borrowing Elo's mechanism of tracking, in this case, student proficiency while solving problems and infusing a standard BKT with it.

2 Bayesian Knowledge Tracing and Its Individualization

There are four types of model parameters used in Bayesian Knowledge Tracing: an initial probability of knowing the skill a priori – $p(L_0)$ (or p-init), probability of student's knowledge of a skill transitioning from *not known* to a *known* state after an opportunity to apply it – pT (or p-transit), the probability to make a mistake when applying a known skill – pS (or p-slip), and the probability of correctly applying a not-known skill – pG (or p-guess).

One notable individualization approach proposes to split BKT parameters to per-skill and to-per student components. Student and skill components are added in log-odds space and transformed back to probability space: see Eq. 1a. Here, w is one of the BKT parameters being individualized, w_k is the per-skill component of the BKT parameter, and w_i is the per-student component. This iBKT model is fit using a coordinate gradient descent method [6].

$$w = \sigma(logit(w_k) + logit(w_i)) \tag{1a}$$
$$\sigma(x) = 1/(1 + e^{-x}) \tag{1b}$$
$$logit(y) = ln(y/(1-y)) \tag{1c}$$

3 Elo Rating Schema

Elo rating schema was proposed by Arpad Elo [1] and was originally used to rate chess players. Recently, it's been successfully used for tracking learners performance. Pelánek et al. use Elo to track knowledge of Geography [2]. Math Garden [3] deployed in K-12 setting in the Netherlands is based on Elo too.

Elo capturing students and items is shown in Eq. 2a. Here, p_{ij} – is the probability student answers the item correctly, m_{ij} – is the log-odds value of that probability, s_i – is student's unidimensional ability (initially 0), and b_j – is question/problem unidimensional difficulty (initially 0). Tracked Elo values are updated as new data points are observed according to Eqs. 2b–2c. K is a sensitivity parameter controlling the magnitude of the update.

$$p_{ij} = Pr(X_{ij} = 1) = \sigma(m_{ij}) = 1/(1 + e^{-m_{ij}}) = 1\Big/\big(1 + e^{-(s_i - b_j)}\big) \tag{2a}$$
$$s_i = s_i + K \cdot (X_{ij} - p_{ij}) \tag{2b}$$
$$b_j = b_j - K \cdot (X_{ij} - p_{ij}) \tag{2c}$$

4 Elo-infusion

We have devised an individualized Elo-infused BKT by combining per-student and per-skill components like in an iBKT approach featured in [6] and shown in Eq. 1a. The per-skill parts of parameters (w_k) remain the same, while the per-student part (s_i) is taken from Elo (rf. Eq. 3). The $sgn(w)$ function maps every individualized BKT parameter to $\{-1, 1\}$. For the pL_0 (prior mastery), $(1 - pF) = 1$ (not forgetting), pT (learning), $1 - pS$ (not slipping), pG (guessing), $sgn(w) = 1$. One could commonly call this group *positive effects on performance*. For the remaining vector-matrix parameters $sgn(w) = -1$. Running values of the parameters are updated according to standard BKT or Elo rules. As a result, we have four BKT parameters per skill and one parameter for updating all student ratings. Just like in the iBKT approach by Yudelson and colleagues [6], different subsets of BKT parameters could be the target of infusion. One could infuse all – priors (p-init), transitions (p-learn), and emissions (p-slip and p-guess), or any combination of the three groups.

$$w = \sigma(sgn(w) \cdot s_i + logit(w_k)) \tag{3}$$

5 Data

We used the datasets from the KDD Cup 2010[1]. The data was contributed by Carnegie Learning Inc., a publisher of math curricula and a producer of intelligent tutoring systems for middle school and high school. There are two datasets, Algebra I, and Bridge to Algebra, both collected in the 2008–2009 school year. We removed the rows that had no skill tagging. Just like in the original Cognitive Tutor, skills are treated as unique within sections of math content. We obtained original BKT parameters shipped with the Cognitive Tutor product from Carnegie Learning, Inc.

6 Computational Experiments

We implemented an Elo-infused BKT model described above based on the hmm-scalable [6]. To test the new approach, we used the datasets described above to compare it to the standard BKT using parameters shipped with the original Cognitive Tutor as well as to standard BKT and iBKT models fit to the data.

The task of the gradient-based search for sensitivity parameter K was simplified by enumerating candidate sensitivities from 0.0001 to 1.0 using a factor of 5 and 2 (yielding values ending in 5 and 1). The fitting of the BKT part of the model remains computationally correct even after introducing the Elo-infusion. Different parameter scopes were targeted. Namely, just p-init (Pi), p-init and

---

[1] KDD Cup 2010 Educational Data mining Challenge http://pslcdatashop.web.cmu.edu/KDDCup.

Table 1. Best infused models, their respective scopes, and Elo sensitivities K compared to reference models.

Dataset	Model	Infusion scope	Sensitivity	Accuracy	RMSE
Algebra I	Shipped BKT			0.7557	0.4367
Algebra I	Fit BKT			0.8304	0.3566
Algebra I	Elo-infused fit BKT	Pi, A, B	0.030	0.8325	0.3532
Algebra I	Elo-infused shipped BKT	Pi, A, B	0.300	0.8200	0.3731
B. to Algebra	Shipped BKT			0.7994	0.3840
B. to Algebra	Fit BKT			0.8333	0.3516
B. to Algebra	Elo-infused fit BKT	Pi, A	0.010	0.8351	0.3494
B. to Algebra	Elo-infused shipped BKT	Pi, A, B	0.300	0.8234	0.3624

p-learn (Pi, A), just p-learn (A), and all parameters p-init, p-learn, p-slip, and p-guess (Pi, A, B).

To draw comparisons between the alternative models, we used a combined 5 times 2-fold student-stratified cross-validation F-test to compare models [5]. This approach was validated on multiple datasets and shown reliable model ranking results. The use of 2-fold cross-validation defends against increased overlap of the training sets when the number of folds is 3 or more. This approach has a low Type I error rate.

In terms of model performance metrics, we used accuracy and root mean squared error (RMSE). Accuracy lets us know how often the model predicts the right answer outcome (right or wrong). RMSE tells us how far numerically from the correct outcome our prediction was. For each metric, we computed an aggregated mean value across 10 training-prediction rounds. Whenever necessary, we applied the 5x2-fold F-test to obtain a significance value (p-value) for every pair of the models compared.

7   Results

The computational cross-validation experiments are summarized in Table 1. There, we give averages over 10 prediction tasks (5 random runs of 2-fold validation). For each of the considered datasets the lowest performing reference point is the shipped model. The best performing model is the Elo-infused fit BKT. It is worth noting, that the Elo-infused shipped BKT (fixed BKT skill parameters but) is a significant improvement over the shipped model. Elo-infused shipped BKT is about half-way between the reference shipped BKT and the best Elo-infused fit BKT in both datasets. In terms of pairwise comparisons of accuracy/RMSE model performances – all are statistically significantly different even if correction to account for multiple comparisons are made.

References

1. Elo, A.E.: The Rating of Chessplayers, Past and Present. Arco Publishers, Milton (1978)
2. Nižnan, J., Pelánek, R., Rihák, J.: Student models for prior knowledge estimation. In: Proceedings of the 8th International Conference on Educational Data Mining, New York, NY, USA, pp. 109–116. ACM (2015)
3. Hofman, A., Jansen, B., de Mooij, S., Stevenson, C., van der Maas, H.: A solution to the measurement problem in the idiographic approach using computer adaptive practicing. J. Intell. **6**(1), 14 (2018)
4. Corbett, A.T., Anderson, J.R.: Knowledge tracing: modeling the acquisition of procedural knowledge. User Model. User-Adap. Inter. **4**(4), 253–278 (1995)
5. Alpaydm, E.: Combined 5x2 cv F-test for comparing supervised classification learning algorithms. Neural Comput. **11**(8), 1885–1892 (1999)
6. Yudelson, M.V., Koedinger, K.R., Gordon, G.J.: Individualized Bayesian knowledge tracing models. In: Lane, H.C., Yacef, K., Mostow, J., Pavlik, P. (eds.) AIED 2013. LNCS (LNAI), vol. 7926, pp. 171–180. Springer, Heidelberg (2013). https://doi.org/10.1007/978-3-642-39112-5_18

Self-paced Graph Memory Network for Student GPA Prediction and Abnormal Student Detection

Yue Yun[1], Huan Dai[1], Ruoqi Cao[2], Yupei Zhang[1](✉) (iD), and Xuequn Shang[1](✉)

[1] School of Computer Science, Northwestern Polytechnical University,
Xi'an 710129, Shaanxi, China
{yundayue,daihuan}@mail.nwpu.edu.cn, {ypzhaang,shang}@nwpu.edu.cn
[2] National Research University Higher School of Economics, 20 Myasnitskaya Ulitsa,
Moscow 101000, Russia
rcao@hse.ru

Abstract. Student learning performance prediction (SLPP) is a crucial step in high school education. However, traditional methods fail to consider abnormal students. In this study, we organized every student's learning data as a graph to use the schema of graph memory networks (GMNs). To distinguish the students and make GMNs learn robustly, we proposed to train GMNs in an "easy-to-hard" process, leading to self-paced graph memory network (SPGMN). SPGMN chooses the low-difficult samples as a batch to tune the model parameters in each training iteration. This approach not only improves the robustness but also rearranges the student sample from normal to abnormal. The experiment results show that SPGMN achieves a higher prediction accuracy and more robustness in comparison with traditional methods. The resulted student sequence reveals the abnormal student has a different pattern in course selection to normal students.

Keywords: Student learning performance prediction · Self-paced learning · Graph memory networks · Abnormal student detection

1 Introduction

A crucial demand in higher education is SLPP [1,2]. Unfortunately, almost all of related researches [3–6] mainly focus on GPA prediction. They treated GPA as the only indicator for evaluating students and failed to consider one abnormal situation: there are some students whose learning patterns are clearly different from others, e.g., promising students or gifted students. Note, they are less but important, and GPA is not enough.

For identifying them, we propose an effective method with strong interpretability in educational sense, and we obtain the inspiration from regression models: students from mentioned abnormal situation are equal to outliers of

© Springer Nature Switzerland AG 2021
I. Roll et al. (Eds.): AIED 2021, LNAI 12749, pp. 417–421, 2021.
https://doi.org/10.1007/978-3-030-78270-2_74

training samples in regression models. Then, we aim to design a robust prediction model for the task of SLPP, and identify students in mentioned abnormal situation, i.e., ASD task. In addition, for mining underlying information, we proposed graph memory networks (GMN) model [7] as our basic model. Then, to improve the robustness of GMNs and enhance the precision of training loss, we combine self-paced learning [8] (SPL) with GMNs, and proposed SPGMN algorithm. Specially, our two tasks are combined with SPGMN for better interpretability: we train a SPGMN model for SLPP task on training samples, then, we apply the trained model to handle the ASD task.

Definition 1. *Abnormal student is the student whose learning pattern is clearly different from others. And, ASD is the abbreviation of* **abnormal student decision**. *Specially, this definition is with respect to the data aspect.*

Summarily, the contributions of our work are: 1) SPGMN can enhance the robustness of GMNs model in SLPP task; 2) improve the precision and interpretablity of ASD task; 3) Experimental results indicate that SPGMN can really enhance the robustness of GMNs. The ASD experiment proofs that GPA is not enough, and we propose learning pattern as an supplement.

2   Proposed Methods

Self-paced Graph Memory Network. SPL incorporates a self-paced function and a pacing parameter into the learning objective of GMN to optimize the order of samples and model parameters. Each sample is assigned to a weight to reflect the easiness of the sample. Then, these weighted samples are gradually involved into training from easy to complex, as shown as follows:

$$\min_{\mathbf{w},\mathbf{v}} \mathbb{E}(\mathbf{w},\mathbf{v}) = \sum_{i=1}^{n} \mathbf{v}_i \cdot \ell_i(y_i, g(\mathbf{x}_i, \mathbf{w})) + f(\mathbf{v}, \lambda^k) \tag{1}$$

where y_i is the ground-truth label of the i-th graph, $\ell(\cdot)$ is the loss function of GMN, $\mathbf{x}_i$ is the i-th trained graph, $\mathbf{w}$ is the parameters, $f(\mathbf{v}; k)$ is a dynamic self-paced function with respect to $v \& k$ (e.g., $f(v, \lambda^k) = \lambda^k(\frac{1}{2}\mathbf{v^2} - \mathbf{v})$), and λ is the age of the SPGMN to control the learning pace, k means k-th iteration, while $\mathbf{v} = [v_1, v_2, ..., v_n]$ selects which samples involved into training data.

Abnormal Student Detection Based on SPGMN. Here, we introduce our ASD method based on SPGMN algorithm, i.e., SPGMN-ASD. We first train a SPGMN model for SLPP task on the training data. After enough iterations, we obtain a trained SPGMN model and the training loss of each student. Obviously, the set of losses for abnormal students (outliers) has clearly difference with the set of losses for normal students (majority of training samples). Based on this consideration, we first sort α values (the loss of target student divide the sum of all losses). Then we randomly set one break point from the list of non-zero α values and obtain two series of α. Finally, if these two series cannot satisfy t-test, the corresponding students of series with larger α values are detected as abnormal students.

3   Experiments

Here, we design our experiments based on real educational dataset from X-university, GPA-data. GPA-data consists of the registration information of 600 students, exam scores for courses they had enrolled (Each student totally enrolls about 70 to 80 courses.) and the background information of all courses. Here, data is constructed in the form of graphs (one graph represents a student). The nodes are courses that students had ever enrolled, and top-5 similar nodes are connected with non-weighted edges. Specially, 600 students are classified into 4 categories by GPAs, and GPA $\in [1, 2, 3, 4]$.

Student Learning Performance Prediction. To compare the performance of SPGMN and some compared baseline/SLPP methods, i.e., GMN [7], convolutional neural network (CNN) [4], K-Nearest Neighbor (KNN) [3], decision trees (DTs) [9], naïve Bayes (NBs) [10], we follow the experimental protocol in [11] and perform 10-fold cross-validation and report the mean accuracy over all folds.

Table 1. SLPP results. Note, GPA-k means the data of first k terms of GPA-data.

Methods	Datasets						
	GPA-1	GPA-2	GPA-3	GPA-4	GPA-5	GPA-6	GPA-7
KNN	44.83	42.83	40.17	45.00	40.67	36.83	37.50
DTs	49.83	50.50	59.50	59.83	60.67	61.00	60.33
NBs	35.17	40.83	55.83	54.17	51.00	50.50	50.33
CNN	69.83	68.83	70.00	72.33	74.17	77.83	79.67
GMN	76.00	79.50	89.17	91.17	92.67	93.84	94.50
SPGMN	**77.00**	**81.00**	**90.17**	**93.20**	**94.33**	**96.00**	**97.50**

We predict the final GPAs based on the data of the first k term ($k \in [1, 7]$), i.e., GPA-k. Table 1 shows that: (1) in all semesters, our SPGMN model consistently achieves better results w.r.t compared methods; (2) as semesters go, the advantage of SPGMN is amplified, i.e., SPGMN has strong robustness.

Abnormal Students Detection. Here, we test our ASD method on the training dataset of GPA-7. Specially, we randomly repeat SPGMN-ASD 400 times and consider the k student with highest frequency appeared as abnormal students. Note, k is the average number of results of 400 experiments.

Figure 1(a) suggests that: 1) the abnormal series and the normal series has a noticeable gap at 511-th point; 2) these two series come from two different distribution. Then, Fig. 1(b) indicates that GPA is not enough to evaluating students: 1) 100% of promising students were abnormal, i.e., they need more attention from teachers; 2) For the students whose GPA is 4, 56% of them

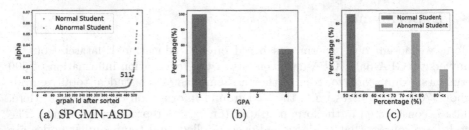

Fig. 1. (a): Results of ASD. (b): y-axis is the percent of all students who abnormal. (c): x is the percentage of major-related courses cut of all courses. y-axis is the percentage of student with different learning pattern cut of all students. (Color figure online)

were abnormal, i.e., those gifted students need more challenged works while the remaining 44% may not need.

By studying the 30 outliers, we found that they are not errors. Instead, it represents a noticeable event: the strategy of course selection (i.e., learning pattern) of abnormal students is different from normal students[1]. As shown in Fig. 1(c): 1) normal students (blue bars) are more likely to select non-major-related courses, e.g., *Basic of Finance* for students of materials science and technology; 2) abnormal students (orange bars) are more likely to choose major-related courses, e.g., *Database* for students of computer science; 3) when students enrolled more major-related courses, their learning patterns are more likely abnormal, i.e., learning patterns can be considered as a supplement of GPAs.

4   Conclusion

This paper combines SLPP task and ASD task for strong interpretability of ASD task. And we combine GMNs model and SPL for enhancing the robustness of GMNs for SLPP task and improving the precision of ASD task. Then, experiments verify the mentioned advancements. Finally, we propose the learning pattern as an supplement of GPA by analyzing the results of ASD experiment.

Acknowledgement. This research is funded by the National Natural Science Foundation of China (Grants No. 61802313, U1811262, 61772426), the Fundamental Research Funds for Central Universities (Grant No. G2018KY0301) and the education and teaching reform research project of Northwestern Polytechnical University (Grant No. 2020JGY23).

References

1. Kuh, G.D., Kinzie, J., Buckley, J.A., Bridges, B.K., Hayek, J.C.: Piecing Together the Student Success Puzzle: Research, Propositions, and Recommendations: ASHE Higher Education Report, vol. 116. Wiley, Hoboken (2011)

---

[1] Here, we do not consider these required courses which students must enrolled.

2. Zhang, Y., Dai, H., Yun, Y., Liu, S., Lan, A., Shang, X.: Meta-knowledge dictionary learning on 1-bit response data for student knowledge diagnosis. Knowl. Based Syst. **205**, 106290 (2020)
3. Al-Shehri, H., et al.: Student performance prediction using support vector machine and k-nearest neighbor. In 2017 IEEE 30th Canadian Conference on Electrical and Computer Engineering (CCECE), pp. 1–4. IEEE (2017)
4. Ma, Y., Zong, J., Cui, C., Zhang, C., Yang, Q., Yin, Y.: Dual path convolutional neural network for student performance prediction. In: Cheng, R., Mamoulis, N., Sun, Y., Huang, X. (eds.) WISE 2020. LNCS, vol. 11881, pp. 133–146. Springer, Cham (2019). https://doi.org/10.1007/978-3-030-34223-4_9
5. Zhang, Y., Yun, Y., Dai, H., Cui, J., Shang, X.: Graphs regularized robust matrix factorization and its application on student grade prediction. Appl. Sci. **10**(5), 1755 (2020)
6. Zhang, Y., An, R., Cui, J., Shang, X.: Undergraduate grade prediction in Chinese higher education using convolutional neural networks. In: 11th International Learning Analytics and Knowledge Conference, LAK 2021, pp. 462–468 (2021)
7. Khasahmadi, A.H., Hassani, K., Moradi, P., Lee, L., Morris, Q.: Memory-based graph networks (2020). arXiv preprint arXiv:2002.09518
8. Meng, D., Zhao, Q., Jiang, L.: A theoretical understanding of self-paced learning. Inf. Sci. **414**, 319–328 (2017)
9. Kabakchieva, D.: Student performance prediction by using data mining classification algorithms. Int. J. Comput. Sci. Manag. Res. **1**(4), 686–690 (2012)
10. Amra, I.A.A., Maghari, A.Y.: Students performance prediction using KNN and Naïve Bayesian. In: 2017 8th International Conference on Information Technology (ICIT), pp. 909–913. IEEE (2017)
11. Ying, Z., You, J., Morris, C., Ren, X., Hamilton, W., Leskovec, J.: Hierarchical graph representation learning with differentiable pooling. In: Advances in Neural Information Processing Systems, pp. 4800–4810 (2018)

Using Adaptive Experiments to Rapidly Help Students

Angela Zavaleta-Bernuy[1]([✉]), Qi Yin Zheng[1], Hammad Shaikh[1], Jacob Nogas[1], Anna Rafferty[2], Andrew Petersen[1], and Joseph Jay Williams[1]

[1] University of Toronto, Toronto, Canada
angelazb@cs.toronto.edu
[2] Carleton College, Northfield, USA

Abstract. Adaptive experiments can increase the chance that current students obtain better outcomes from a field experiment of an instructional intervention. In such experiments, the probability of assigning students to conditions changes while more data is being collected, so students can be assigned to interventions that are likely to perform better. Digital educational environments lower the barrier to conducting such adaptive experiments, but they are rarely applied in education. One reason might be that researchers have access to few real-world case studies that illustrate the advantages and disadvantages of these experiments in a specific context. We evaluate the effect of homework email reminders in students by conducting an adaptive experiment using the Thompson Sampling algorithm and compare it to a traditional uniform random experiment. We present this as a case study on how to conduct such experiments, and we raise a range of open questions about the conditions under which adaptive randomized experiments may be more or less useful.

Keywords: Reinforcement learning · Randomized experiments · Multi-Armed bandits · A/B testing · Field deployment

1 Introduction

Instructors frequently look for ways to support their students and improve their performance. With access to online learning environments, instructors can gather feedback in a larger scale setting. Since optimal instructional designs and scaffolds may not be known ahead of time, multiple possibilities can be tested using Uniform Random (UR) A/B experiments, also known as randomized control trials. In a UR A/B experiment, students are uniformly assigned to the different conditions that an instructor or researcher would like to test to learn about their relative effectiveness. One challenge of this approach is how to use data more rapidly to help current students. To mitigate this, we can aim to maximize total learning by having most students being subject to the more effective conditions as they become known.

© Springer Nature Switzerland AG 2021
I. Roll et al. (Eds.): AIED 2021, LNAI 12749, pp. 422–426, 2021.
https://doi.org/10.1007/978-3-030-78270-2_75

Adaptive randomization is an effective strategy for assigning more students to the current most optimal condition, while retaining the ability to test other conditions. We use a Multi-Armed Bandit (MAB) algorithm that uses machine learning to increase the number of students assigned to the current most effective condition (or *arm*) [1,7]. MAB are commonly used for rapid use of data in different areas such as marketing to optimize the benefits of the users and balance exploration vs. exploitation [1,3]. For this study, we used Thompson Sampling (TS), a probability matching algorithm, where the probability of assignment is proportional to the probability that the arm is optimal [1].

In this paper, we present a real-world experiment to illustrate the benefits and limitations of using UR A/B experiments and TS in educational settings. First, we use UR A/B experiments to evaluate different versions of emails in a homework reminder intervention to determine if a more effective version can be identified. We then compare the results of UR A/B experiments with the TS results to study its performance and benefits. Our experimental design allows us to compare classical balanced A/B comparisons side-by-side with a TS adaptive experiment to evaluate the trade-offs of using each of these methods.

2   Multi-Armed Bandit (MAB) Algorithms

To optimize the experience of students, we use the TS algorithm designed to solve MAB problems, useful for adaptively assigning participants to conditions. [7].

The stochastic MAB problem is the problem of sequentially choosing from a discrete set of actions to maximize cumulative reward, where a reward is some measure of the effectiveness of the chosen action (arm). In this paper, we focus on the MAB problem with binary rewards. More precisely, we choose between K versions (arms), and we denote the choice of action at step t of the experiment by x_t. Assuming we choose the $k-th$ arm, where $k \in \{1, 2, \ldots, K\}$, then $x_t = k$, and we receive a reward r_t with probability p_k.

TS shows strong empirical performance in maximizing the cumulative reward [5]. TS is a Bayesian algorithm that maintains a posterior distribution over each reward p_k. In our case, we use a Beta prior with parameters α_k and β_k. Arms are chosen by sampling values from the posteriors over each arm, and choosing the arm corresponding to the highest sample drawn. The posterior distribution is then updated based on the chosen action $x_t = k$ and observed reward r_t. We use a uniform prior for each arm, $\alpha_k = \beta_k = 1$, for all $k \in \{1, 2, ..., K\}$.

$$(\alpha_k, \beta_k) \leftarrow \begin{cases} (\alpha_k, \beta_k) & \text{if } x_t \neq k \\ (\alpha_k, \beta_k) + (r_t, 1 - r_t) & \text{if } x_t = k \end{cases}$$

3   Methods: Traditional and Adaptive Experimentation

For three consecutive weeks, we tested four different versions of the emails to investigate which might be more effective in leading students to click on the homework link appended in the email. To evaluate how TS adapts the distribution of students to each email version, we sent the messages in four different batches on consecutive days of the week (Tuesday to Friday).

Using UR, each of the four email reminders has the same probability of being assigned to a student. For the TS algorithm, the probability of assignment of the email reminders version is proportional to the *reward* (in our case, click rate) in all the previous batches, which is updated after each batch.

For our interventions, we used two different variations of the TS algorithm. For Weeks 1 and 2, we used a UR-TS Hybrid where the TS updating of the probability that an arm has the highest click rate used data from the UR participants too. This hybrid is called $\epsilon[0.5]$-TS, because with epsilon (epsilon $= 0.5$) probability, arms are assigned using UR. This takes inspiration from past algorithms like epsilon-greedy [6] and top-two TS [4]. This is interesting to investigate because scientists may want to get the benefits of obtaining data under UR (in case TS introduces biases [2]) for analysis, while also then using that data to *improve* the performance of the TS algorithm. For Week 3, we applied traditional TS that did not use the data from the UR assignment.

4   Analysis and Results

Using a panel regression with week fixed effects (i.e., include indicators for Week2-$\epsilon[0.5]$-TS and Week3-TS), we can aggregate the uniform portion of the experiment for the three weeks to evaluate the effects of the four arms on student click rate. We find that the average click rate in our sample across the three weeks is around 19%. All four arms had click rates are within 2% of each other and their difference is not statistically significant regardless of time and participant effects. The lack of an optimal arm suggests that all students were assigned to fairly similar treatments. These results are robust to also including student fixed effects within the panel regression model.

For both of the later weeks, $\epsilon[0.5]$-TS behaves similarly, favouring one particular arm. Some arms are assigned a substantially higher number of students and the others far fewer, but the seemingly favoured arm varies across the two weeks and is not consistent in Week2-$\epsilon[0.5]$-TS. In Table 1, we show the cumulative reward per arm and per batch, which influenced the probability of assignment. Even though the probability of assignment aligns with the click rates from the previous batch, the algorithm drastically shifts to the arm with the highest reward and leaves minimal room for exploration in batch 3. As the click rates were not consistent across the different batches, we are unable to identify the presence of the most efficient arm, which is reflected on the cumulative rewards in Table 1, and could be a result of there being minimal differences between arms.

5   Discussion and Limitations

We present a case study of adaptive experimentation, using the Thompson Sampling MAB algorithm, when a difference between arms is not observed (i.e., the

Table 1. The Table presents the cumulative click rate (CCR) and the probability of assignment (PA) for each arm in the next batch for the three weeks of interventions. The arms represent a version of the message. Each batch represents a separate day of message deployments: Monday (Batch 1), Tuesday (Batch 2), Wednesday (Batch 3), and Thursday (Batch 4). The cumulative click rate of each arm represents the percentage of people who received the message that clicked accumulated up to and including that batch. The probability of assignment is the probability that a particular arm will be chosen in the next batch calculated using 1000000 Monte Carlo simulations. Bold values highlight the arm with the highest probability of assignment for each batch on every week.

		Arm 1		Arm 2		Arm 3		Arm 4	
		CCR	PA	CCR	PA	CCR	PA	CCR	PA
Week1-ϵ[0.5]-TS	Batch 1	0.200	0.117	0.277	**0.659**	0.212	0.177	0.167	0.047
	Batch 2	0.219	**0.466**	0.22	0.443	0.149	0.040	0.152	0.052
	Batch 3	0.206	0.434	0.209	**0.452**	0.137	0.017	0.163	0.097
	Batch 4	0.163	0.077	0.205	**0.697**	0.161	0.104	0.162	0.122
Week2-ϵ[0.5]-TS	Batch 1	0.213	0.116	0.086	0.000	0.246	0.225	0.300	**0.659**
	Batch 2	0.198	0.082	0.14	0.004	0.244	0.311	0.266	**0.603**
	Batch 3	0.197	0.065	0.186	0.041	0.261	**0.666**	0.235	0.229
	Batch 4	0.194	0.052	0.209	0.109	0.249	**0.505**	0.240	0.334
Week3-TS	Batch 1	0.231	**0.477**	0.153	0.056	0.22	0.389	0.157	0.078
	Batch 2	0.233	**0.926**	0.137	0.030	0.135	0.033	0.120	0.011
	Batch 3	0.183	**0.510**	0.129	0.039	0.133	0.070	0.174	0.381
	Batch 4	0.181	0.405	0.152	0.086	0.144	0.076	0.181	**0.433**

arms/conditions are equally effective), according to a traditional Uniform Random experiment. We illustrate how TS may randomly favour an arm, even when giving this arm more frequently has no consequences for participants. The TS algorithm minimizes regret—it aims to keep participants from being assigned to sub-optimal arms—so in the case where arms are equivalent, one could argue that any of them could be presented. However, this can be problematic for scientific inference and statistical analysis [2].

One limitation is that there might have been unobserved confounding variables that caused the click rates to change in one particular batch (e.g., students had an assignment due or a test), which will also affect the algorithm – this is one concern to keep in mind in applying MAB algorithms for adaptive experimentation.

6   Conclusion and Future Work

This paper provides an example of a real-world intervention using adaptive experiments, which can help instructors and researchers to use the results from experiments to more rapidly benefit students. We illustrate an instance of conducting such experiments using the TS algorithm, where the results suggest there is no difference between arms/conditions. We hope that this paper provides a first step for instructors and researchers to investigate adaptive experimentation in education. Future work can explore how the algorithm behaves in a wider variety of scenarios, such as different batch sizes and structure, and alternative differences between the arm/conditions.

References

1. Lomas, J.D., et al.: Interface design optimization as a multi-armed bandit problem. In: Proceedings of the 2016 CHI Conference on Human Factors in Computing Systems, pp. 4142–4153 (2016)
2. Rafferty, A., Ying, H., Williams, J.: Statistical consequences of using multi-armed bandits to conduct adaptive educational experiments. JEDM J. Educ. Data Min. **11**(1), 47–79 (2019)
3. Rafferty, A.N., Ying, H., Williams, J.J.: Bandit assignment for educational experiments: benefits to students versus statistical power. In: Penstein Rosé, C., et al. (eds.) AIED 2018. LNCS (LNAI), vol. 10948, pp. 286–290. Springer, Cham (2018). https://doi.org/10.1007/978-3-319-93846-2_53
4. Russo, D.: Simple Bayesian algorithms for best arm identification. In: Conference on Learning Theory, pp. 1417–1418 (2016)
5. Sutton, R.S., Barto, A.G.: Reinforcement Learning: An Introduction. MIT Press, Cambridge (2018)
6. Watkins, C.J.C.H.: Learning from delayed rewards (1989)
7. Williams, J.J., Rafferty, A.N., Tingley, D., Ang, A., Lasecki, W.S., Kim, J.: Enhancing online problems through instructor-centered tools for randomized experiments. In: Proceedings of the 2018 CHI Conference on Human Factors in Computing Systems, pp. 1–12 (2018)

A Comparison of Hints vs. Scaffolding in a MOOC with Adult Learners

Yiqiu Zhou[1](✉), Juan Miguel Andres-Bray[1], Stephen Hutt[1], Korinn Ostrow[2], and Ryan S. Baker[1]

[1] University of Pennsylvania, Philadelphia, PA 19104, USA
{zyq,andresju,huts,rybaker}@upenn.edu
[2] Worcester Polytechnic Institute, Worcester, MA 01609, USA
ksostrow@wpi.edu

Abstract. Scaffolding and providing feedback on problem-solving activities during online learning has consistently been shown to improve performance in younger learners. However, less is known about the impacts of feedback strategies on adult learners. This paper investigates how two computer-based support strategies, *hints* and *required scaffolding questions*, contribute to performance and behavior in an edX MOOC with integrated assignments from ASSISTments, a web-based platform that implements diverse student supports. Results from a sample of 188 adult learners indicated that those given scaffolds benefited less from ASSISTments support and were more likely to request the correct answer from the system.

Keywords: Feedback strategies · Hints · Scaffolding · MOOC · ASSISTments

1 Introduction

Studies have consistently demonstrated the potential of computer-based scaffolding in promoting learning gains during online learning [1–3]. A recent meta-analysis found a moderate effect in problem-based learning in STEM education across various learning contexts [4]. However, the implementation of tutoring strategies varies a great deal (e.g., by types of feedback, the number of levels, and timing) [5–9], resulting in questions about how well results generalize to new platforms and populations.

In this paper, we investigate the effectiveness of two types of tutoring strategies in the context of adult learners: hints and required scaffolds (see Sect. 2) – replicating the methods originally used by Razzaq and Heffernan [2]. Although we use the same platform as [2] (ASSISTments), our experiment differs from the prior study in two ways. First, our study focuses on adult learners, a comparatively underexplored population [3, 4]. Second, we explore how scaffolding strategies influence learners' interactions within a more open learning environment (a MOOC). As MOOCs become an increasingly complex form of content delivery, we sought to understand how feedback strategies influence adult learner's performance and self-regulation.

© Springer Nature Switzerland AG 2021
I. Roll et al. (Eds.): AIED 2021, LNAI 12749, pp. 427–432, 2021.
https://doi.org/10.1007/978-3-030-78270-2_76

2 Method

This work leverages data collected from students enrolled in the edX MOOC Big Data and Education (BDEMOOC) [11]. The course provided eight weeks of content and utilized ASSISTments to deliver assignments each week.

Integration between edX and ASSISTments was made possible by Learning Tools Interoperability (LTI) standards [12, 13]. In each week of BDEMOOC, learners were given an assignment via ASSISTments including 10–11 problems. For each problem, learners could make multiple attempts and request multiple hints. In general, there were three to six levels of hints per problem, followed by the option to request the correct answer to the problem. Students received full credit for completed assignments regardless of the number of attempts or hints requested.

This paper focuses on Week 2 of the course, in which learners were randomly assigned to receive either *hints* or *scaffolding*. Problem content was the same across conditions. Learners in the *hint* condition could request hints on-demand for each problem, the same as all other weeks of the course. Learners in the *scaffold* condition received the same assignment but with scaffolding questions instead of hints. These learners could request to break the problem down before attempting to answer, (similar to requesting a hint). Alternatively, the sequence automatically started if their first answer was incorrect. Once the scaffold sequence was initiated in either case, learners were *required* to complete the entire sequence to proceed to the next problem.

Our dataset was comprised of 188 learners who completed the Week 1 assignment and at least started the Week 2 assignment. To analyze learning gains, we also considered a subset of this data: students that completed *both* weeks and received at least one hint/scaffold in Week 2 (see Table 1).

Table 1. Descriptive statistics of participants

	N started Week 2	N completed Week 2
Learners	188	144
Scaffold condition	110	81
Hint condition	78	63

2.1 Measures

From ASSISTments data [14], we derived prior knowledge (operationalized as the percentage of correct first attempts in the week 1 assignment) and two measures of learning performance: the percentage of correct first attempts and the number of times the student requested the correct answer. *Correct Answer Requests* was operationalized as the proportion of questions for which learners requested the correct answer (referred to as *bottom-out hints* in prior work [16]). It should be noted that these two measures have opposite implications: higher correct answer requests implies that the student gave up on a larger proportion of questions, whereas more correct first attempts indicate less need for assistance and thus better learning.

We also collected each learner's interaction and clickstream data from within the edX platform [15]. Based on prior work [16], we derived two measures: 1) time spent interacting with discussion forums, and 2) time spent watching video lectures. Both values (measured in seconds) were calculated from clickstream data by calculating the time between clicks. These durations were then summed per resource per learner. Click-events with durations of an hour or longer were treated as disengaged and were excluded from the sums.

3   Results

ASSISTments Data. We first considered if condition (*hints* or *scaffolding*) impacted assignment completion. An ANOVA test indicated no main effect of condition on assignment completion ($F(1, 186) = 1.61, p = 0.21$). However, we observed a significant interaction between prior knowledge ($M = 0.80, SD = 0.58$) and condition, ($\beta = 0.59, p = 0.01, df = 184$), indicating that students with lower prior knowledge were significantly less likely to complete the assignment if they were in the *scaffold* condition.

The remainder of our reported analysis considers only students who completed both the Week 1 and Week 2 assignments and received at least one hint/scaffold in Week 2 (see Table 1). We first examined if prior knowledge was different between the groups. A t-test found no significant difference in prior knowledge by condition, $t(136.72) = -0.36, p = 0.72$. Table 2 provides an overview of regressing two performance measures onto condition with prior knowledge as a covariate.

Table 2. Regression analysis of Week 2 performance measures: first attempt (or the percentage of correct first attempts) and correct answer requests.

Predictors	First attempt		Correct answer requests	
	std. β	p	std. β	P
(Intercept)	0.07	**<0.001**	−0.15	**<0.001**
Condition [Scaffold]	−0.13	0.388	0.28	**0.053**
Prior	0.35	**0.003**	−0.38	**0.001**
Condition [Scaffold] * Prior	0.20	0.175	−0.25	0.083

No significant effects of condition were observed for correct first attempts. However, when predicting correct answer requests, our analyses showed main effects for both condition and prior knowledge (Table 2). Simple slopes analysis showed that less knowledgeable learners (1 SD below the mean) in the *scaffolding* condition tended to ask for the correct answer more frequently ($p < 0.01$), as did average (at the mean) learners ($p < 0.05$).

We next considered how the computer-based tutoring strategies impacted learners' interactions with two MOOC resources: lecture videos and the discussion forum. We regressed time spent on each resource during Weeks 2 to 8 onto condition (*hints vs. scaffolds*), including the respective time spent in Week 1 as a covariate to account for individual differences (Table 3 and Table 4). No effects were observed beyond Week 5, so the regression results for these weeks are omitted from the tables.

We note no main effect of condition for use of either resource. We did, however, observe interactions between prior usage and condition when predicting future usage. Learners who previously spent more than average time viewing videos were less likely to do so in the future if assigned to the scaffolding condition. For forum use, learners that had previously high forum use were more likely to continue to have high forum use if in the scaffolding condition.

Table 3. Results from the regression analysis conducted on time spent (TS) on lecture video use from Weeks 2 to 5 of the MOOC.

Predictors	TS Videos Week 2		TS Videos Week 3		TS Videos Week 4		TS Videos Week 5	
	std. β	p	std. β	p	std. β	p	std. β	p
(Intercept)	0.10	0.212	0.09	0.867	0.19	0.816	−0.02	0.171
Condition [Scaffold]	−0.16	0.643	−0.13	0.529	−0.30	0.489	0.02	0.472
TS_Videos_Wk1	0.77	**<0.001**	0.86	**<0.001**	0.85	**<0.001**	0.41	**0.001**
Condition [Scaffold] * TS_Videos_Wk1	−0.08	0.526	−0.23	0.063	−0.41	**0.004**	0.16	0.286

Table 4. Results from the regression analysis conducted on time spent (TS) on forum use from Weeks 2 to 5 of the MOOC.

Predictors	TS Forum Week 2		TS Forum Week 3		TS Forum Week 4		TS Forum Week 5	
	std. β	p	std. β	p	std. β	p	std. β	p
(Intercept)	−0.04	**<0.001**	−0.13	**0.001**	0.05	**0.02**	−0.16	**0.030**
Condition [Scaffold]	0.06	0.377	0.22	0.482	−0.08	0.879	0.27	0.873
TS_Forum_Wk1	0.16	0.174	0.03	0.765	0.17	0.186	0.04	0.763
Condition [Scaffold] * TS_Forum_Wk1	0.36	**0.020**	0.55	**<0.001**	−0.08	0.632	0.48	**0.003**

4 Discussion and Conclusions

This study detailed how feedback strategies (*hints* and *required scaffolding* after errors) impacted adult learners' performance and interactions within a MOOC. Our results revealed that *scaffolding* was associated with poorer performance and that this influence was mediated by prior knowledge. Less knowledgeable learners in the *scaffolding* condition requested significantly more correct answers, indicating that they benefited less from scaffolds and failed to solve later problems. This is contrary to [19], which showed that middle schoolers with low prior knowledge benefited more from scaffolding.

One potential explanation might be the difference in learner groups. Scaffolding may hinder instead of support MOOC learners as it breaks the expected balance between external and internal regulation [20], especially for learners who may expect greater agency. For MOOC learners (typically adults) who value autonomy in regulating the learning process [21, 22], requiring them to complete scaffolds may negatively impact performance and future learning behaviors. Future work should investigate purely on-demand *scaffolding* (i.e., learners are not required to complete full sequences) to examine the learning differences that additional agency may afford.

As such, it will be important for future research to consider how and when feedback is delivered to adult learners. With increasing use of learning technologies by adult populations, it is important to consider what K-12 research generalizes to older populations with different learning demands. Although the implementation of *scaffolding* differs across learning systems, this work serves as an initial step towards developing effective feedback standards for adult online learners.

References

1. Belland, B.R., Walker, A.E., Kim, N.J., Lefler, M.: Synthesizing results from empirical research on computer-based scaffolding in STEM education: a meta-analysis. Rev. Educ. Res. **87**, 309–344 (2017). https://doi.org/10.3102/0034654316670999
2. Razzaq, L., Heffernan, N.T.: Scaffolding vs. hints in the assistment system. In: Ikeda, M., Ashley, K.D., Chan, T.W. (eds.) International Conference on Intelligent Tutoring Systems, vol 4053, pp 635–644. Springer, Heidelberg (2006). https://doi.org/10.1007/11774303_63
3. Ma, W., Adesope, O.O., Nesbit, J.C., Liu, Q.: Intelligent tutoring systems and learning outcomes: a meta-analysis. J. Educ. Psychol. **106**, 901–918 (2014). https://doi.org/10.1037/a0037123
4. Kim, N.J., Belland, B.R., Walker, A.E.: Effectiveness of computer-based scaffolding in the context of problem-based learning for stem education: Bayesian meta-analysis. Educ. Psychol. Rev. **30**(2), 397–429 (2017). https://doi.org/10.1007/s10648-017-9419-1
5. Narciss, S.: Designing and evaluating tutoring feedback strategies for digital learning. Digit. Educ. Rev. **23**, 7–26 (2013)
6. Proske, A., Narciss, S.: Supporting prewriting activities in academic writing by computer-based scaffolds: is more support more meaningful? In: Zumbach, J., Schwartz, N., Seufert, T., Kester, L. (eds.) Beyond Knowledge: The Legacy of Competence, pp. 275–284. Springer, Dordrecht (2008). https://doi.org/10.1007/978-1-4020-8827-8_36
7. Hattie, J., Timperley, H.: The power of feedback. Rev. Educ. Res. **77**, 81–112 (2007). https://doi.org/10.3102/003465430298487

8. Shute, V.J.: Focus on formative feedback. Rev. Educ. Res. (2008). https://doi.org/10.3102/0034654307313795
9. Mory, E.H.: Feedback research revisited. In: Handbook of Research on Educational Communications and Technology. Lawrence Erlbaum Associates Publishers, pp. 745–783 (1996)
10. Devolder, A., van Braak, J., Tondeur, J.: Supporting self-regulated learning in computer-based learning environments: systematic review of effects of scaffolding in the domain of science education. J. Comput. Assist. Learn. **28**, 557–573 (2012). https://doi.org/10.1111/j.1365-2729.2011.00476.x
11. Baker, R.S.: Big data in education. University of Pennsylvania (2020). https://www.edx.org/course/big-data-and-education
12. Severance, C., Hanss, T., Hardin, J.: IMS learning tools interoperability: enabling a mash-up approach to teaching and learning tools. Technol. Instr. Cogn. Learn. **7**, 245–262 (2010)
13. IMS Learning Tools Interoperability 2.0. http://www.imsglobal.org/lti-v2-introduction
14. Heffernan, N.T., Heffernan, C.L.: The ASSISTments ecosystem: building a platform that brings scientists and teachers together for minimally invasive research on human learning and teaching. Int. J. Artif. Intell. Educ. **24**(4), 470–497 (2014). https://doi.org/10.1007/s40593-014-0024-x
15. EdX: EdX Research Guide (2020). https://edx.readthedocs.io/projects/devdata/en/latest/using/package.html
16. Moreno-Marcos, P.M., Alario-Hoyos, C., Munoz-Merino, P.J., et al.: Prediction in MOOCs: a review and future research directions. IEEE Trans. Learn. Technol. **12**, 384–401 (2019). https://doi.org/10.1109/TLT.2018.2856808
17. Wang, Y., Baker, R.S., Paquette, L.: Behavioral predictors of MOOC post-course development. In: CEUR Workshop Proceedings, pp. 100–111 (2017)
18. Roll, I., Baker, R.S.J.d., Aleven, V., Koedinger, K.R.: On the benefits of seeking (and avoiding) help in online problem-solving environments. J. Learn. Sci. 23, 537–560 (2014). https://doi.org/10.1080/10508406.2014.883977
19. Razzaq, L., Heffernan, N.T., Lindeman, R.W.: What level of tutor interaction is best? In: Artificial Intelligence in Education, pp. 222–229 (2007)
20. Ifenthaler, D.: Determining the effectiveness of prompts for self-regulated learning in problem-solving scenarios. Educ. Technol. Soc. **15**, 38–52 (2012)
21. Tiedeman, D.V., Knowles, M.: The adult learner: a neglected species. Educ. Res. **8**, 20–22 (1979). https://doi.org/10.2307/1174362
22. Sitzmann, T., Ely, K.: A meta-analysis of self-regulated learning in work-related training and educational attainment: what we know and where we need to go. Psychol. Bull. **137**, 421–442 (2011). https://doi.org/10.1037/a0022777

An Ensemble Approach for Question-Level Knowledge Tracing

Aayesha Zia[✉], Jalal Nouri, Muhammad Afzaal, Yongchao Wu, Xiu Li,
and Rebecka Weegar

Stockholm University, Stockholm, Sweden
{aayesha,jalal,muhammad.afzaal,yongchao.wu,xiu.li,rebeckaw}@dsv.su.se

Abstract. Knowledge tracing—where a machine models the students'
knowledge as they interact with coursework—is a well-established area in
the field of Artificial Intelligence in Education. In this paper, an ensemble approach is proposed that addresses existing limitations in question-
centric knowledge tracing and achieves the goal of predicting future question correctness. The proposed approach consists of two models; one is
Light Gradient Boosting Machine (LightGBM) built by incorporating
all relevant key features engineered from the data. The second model is
a Multiheaded-Self-Attention Knowledge Tracing model (MSAKT) that
extracts historical student knowledge of future question by calculating
their contextual similarity with previously attempted questions. The proposed model's effectiveness is evaluated by conducting experiments on a
big Kaggle dataset achieving an Area Under ROC Curve (AUC) score of
0.84 with 84% accuracy using 10fold cross-validation.

Keywords: Adaptive learning · Knowledge tracing · Question-level
prediction · Artificial Intelligence · Intelligent education

1 Introduction

The advancements in learning analytics and artificial intelligence have shown
potential to transform traditional modalities of education. One such advancement relates to the use of educational data to track students' knowledge state
[1]. In the case of question-level assessment, knowledge tracing provides an interpretation of the learner's current knowledge level and models their mastery of
the knowledge component to which future questions are related [2].

Historically, Bayesian Knowledge Tracing (BKT) has been the most popular
knowledge tracing method [3]. Due to different reasons, improved extensions of
BKT were introduced such as individualised BKT [4,5]. However for question
level knowledge tracing these approaches showed no or limited ability to take
into account the underlying knowledge concepts that were common for multiple
questions, which resulted in poor prediction performance. Lately, Deep Knowledge Tracing (DKT) [6–9] models have been proposed that utilised Recurrent
Neural Networks (RNN) such as LSTM for knowledge tracing. Although these

© Springer Nature Switzerland AG 2021
I. Roll et al. (Eds.): AIED 2021, LNAI 12749, pp. 433–437, 2021.
https://doi.org/10.1007/978-3-030-78270-2_77

models have provided state-of-the-art performance, they have lacked to convey a good explanation for the estimated knowledge levels, as they have overlooked the context-based relationship in a sequence of questions. Furthermore, current approaches for predicting if students will answer the next question correctly have also overlooked valuable features such as question content, type, and difficulty level when modelling knowledge levels. In this study we address these limitations by answering the following research question:

Q1: What learner attributes, beyond performance on previous questions, support the modeling of students' knowledge levels in order to better predict performance on the next question?

In order to answer this research question we employ an ensemble approach that is based on two models; LightGBM [10] and MSAKT [11]. LightGBM contributes by providing a list of significant learner features that generate better prediction regarding the learner's current knowledge level. MSAKT finds contextual similarities and differences among question contents in order to predict learners' performance on the next question in an assessment.

2   Proposed Model

In this section, we present our ensemble approach that consists of four main phases, 1) data collection, 2) pre-processing, 3) models training, and 4) ensemble building.

In the data collection phase, students' question data is collected which contains information of previously attempted questions and their correctness (whether they are correct or not). The dataset used in this work is about TOEIC (Test of English International Communication) taken from the Kaggle competition of Riiid Answer Correctness Prediction [12]. It consists of 101,230,332 records of 393,656 users and keeps information about 13,523 unique questions.

The pre-processing phase handles the missing values in the collected data by averaging feature values. Afterwards, the more useful features were derived from the preprocessed data such as degree of question concept correctness (Score of same concept questions/Count of same concept questions), and correctness or understanding level of a user (Cumulative user score/Cumulative questions count). This phase outputs a set of 25 significant features including 10 basic and 15 engineered features.

The third phase of the proposed approach is the main phase where two models are trained separately in which one is the boosting model (LightGBM) and the other is the knowledge tracing model (MSAKT). LightGBM is based on a popular boosting machine learning algorithm Gradient Boosting Decision Tree (GBDT) [13]. It is trained on pre-processed engineered features and the boosting rounds are set to 4000 with 40 early stopping rounds. On the other hand, in MSAKT model the historic student performance on knowledge concepts given in the questions is utilised for training. It identifies contextual relation among questions to track the student mastery for the concept of next question.

In the last fourth phase, weighting ensemble strategy was employed on built models to get the final prediction. Firstly, the result of each model was given the weight of the actual output prediction value according to their predictive accuracy. For instance, in this work, LightGBM was assigned weight of 0.3 while MSAKT was allocated a weight of 0.7 due to the relatively higher performance prediction accuracy of MSAKT as compared to LightGBM. Secondly, the weighted outputs of both models were added to get the new prediction output. Mathematically, this is defined by Eq. 1.

$$z = 0.30(x) + 0.70(y) \tag{1}$$

where

$$x = prediction(LightGBM) \tag{2}$$

$$y = prediction(MSAKT) \tag{3}$$

$$z = prediction(Proposed\ Ensemble) \tag{4}$$

The resultant ensemble output z decided the final prediction.

3 Experiments and Results

In this section, the performance of the proposed model is evaluated on the Kaggle dataset by train-test data split (70-30 split) and 10-fold cross validation. The performance of the proposed ensemble model is compared with the individual participating models (LightGBM, SAKT) of the ensemble. The obtained results in terms of accuracy and AUC for data split and cross validation schemes are reported in Table 1. According to the results, the proposed approach outperformed LightGBM and MSAKT in all experiments and the highest results with 84.20% accuracy and 0.84 score of AUC were achieved by a 10-fold cross validation. By comparing LightGBM with MSAKT, it was observed that LightGBM exceeded in percentage accuracy and obtained accuracy of 72.70% and 72.76% by a 70-30 split and 10-fold cross validation respectively. On the other hand, MSAKT showed improved performance in the AUC metric with maximum score of 0.76 for both 70-30 split and 10-fold cross validation.

Table 1. Performance comparison of proposed ensemble with participating models.

Models	70-30 split		10-fold CV	
	Accuracy (%)	AUC	Accuracy (%)	AUC
LightGBM	72.70	0.74	72.76	0.76
Multiheaded-SAKT	72.51	0.76	72.46	0.76
Proposed work	77.66	0.79	84.20	0.84

3.1 Performance Comparison of Weighted Ensembles

In this section, the proposed approach is empirically analysed by assigning ten different weight distributions to the models being ensembled. The rationale behind this analysis is to get the best weighted ensemble for the proposed approach. Table 2 demonstrates the performance comparison for different weight combinations of models by 70-30 data split. The obtained results as percentage accuracy and AUC score highlighted that the prediction output of the MSAKT model in the proposed ensemble has more significance than LightGBM. It is also evident by the results that as weight to the prediction of MSAKT model increases, then accuracy also increases until the best weighted ensemble with 77.66% accuracy and 0.79 AUC score is achieved. After analysing the performance of all weight distributions, one weighted ensemble with the distribution of 0.70 for the MSAKT model and 0.30 to LightGBMmodel proved to be the best.

Table 2. Comparative analysis of different weight distributions to proposed ensemble.

Model weights		Accuracy (%)	AUC
LightGBM weight	Multiheaded-SAKT weight		
0.90	0.10	72.72	0.65
0.80	0.20	73.57	0.67
0.70	0.30	74.23	0.68
0.60	0.40	75.06	0.71
0.50	0.50	76.27	0.73
0.40	0.60	77.52	0.77
0.30	0.70	77.66	0.79
0.20	0.80	76.66	0.79
0.10	0.90	75.11	0.79

4 Conclusion

In this paper, we presented an ensemble approach for predicting students' performance on the next question in an exam, for instance. Firstly, in this approach, a rich set of features were engineered to build an efficient boosting model Light-GBM. Secondly, students' performance records for the previous questions were traced by the MSAKT model. Thirdly, outputs of MSAKT and LightGBM were assigned weights of 0.70 and 0.30 respectively which were later summed to get a final prediction about correctness in response to the next question. More importantly, the evaluation was performed on the Kaggle dataset that resulted in an

AUC score of 0.84 and 84% accuracy. In terms of limitations, parameter tuning in both the boosting model (LightGBM) and the knowledge tracing model (MSAKT) was not focused which might affect the hyperparameter setting for the ensemble model negatively. A future task would be to emphasise on the parameter fine-tuning and test the proposed approach on students' data taken from various academic courses.

References

1. Zhao, J., Bhatt, S., Thille, C., Zimmaro, D., Gattani, N.: Interpretable personalized knowledge tracing and next learning activity recommendation. In: Proceedings of the Seventh ACM Conference on Learning@ Scale, pp. 325–328 (2020)
2. Sonkar, S., Waters, A.E., Lan, A.S., Grimaldi, P.J., Baraniuk, R.G.: qDKT: question-centric deep knowledge tracing. arXiv preprint arXiv:2005.12442 (2020)
3. Pardos, Z.A., Heffernan, N.T.: Modeling individualization in a Bayesian networks implementation of knowledge tracing. In: De Bra, P., Kobsa, A., Chin, D. (eds.) UMAP 2010. LNCS, vol. 6075, pp. 255–266. Springer, Heidelberg (2010). https://doi.org/10.1007/978-3-642-13470-8_24
4. Lee, J.I., Brunskill, E.: The impact on individualizing student models on necessary practice opportunities. International Educational Data Mining Society (2012)
5. Khajah, M., Wing, M., Lindsey, R., Mozer, M.: Integrating latent-factor and knowledge-tracing models to predict individual differences in learning. In: Educational Data Mining 2014. Citeseer (2014)
6. Piech, C., et al.: Deep knowledge tracing. In: Advances in Neural Information Processing Systems, pp. 505–513 (2015)
7. Wang, Z., Feng, X., Tang, J., Huang, G.Y., Liu, Z.: Deep knowledge tracing with side information. In: Isotani, S., Millán, E., Ogan, A., Hastings, P., McLaren, B., Luckin, R. (eds.) AIED 2019. LNCS (LNAI), vol. 11626, pp. 303–308. Springer, Cham (2019). https://doi.org/10.1007/978-3-030-23207-8_56
8. Su, Y., et al.: Exercise-enhanced sequential modeling for student performance prediction. In: Proceedings of the Thirty-Second AAAI Conference on Artificial Intelligence, vol. 32, no. 1 (2018)
9. Zhang, J., Shi, X., King, I., Yeung, D.Y.: Dynamic key-value memory networks for knowledge tracing. In: Proceedings of the 26 International Conference on World Wide Web, pp. 765–774 (2017)
10. Ke, G., et al.: LightGBM: a highly efficient gradient boosting decision tree. Adv. Neural Inf. Process. Syst. **30**, 3146–3154 (2017)
11. Pandey, S., Karypis, G.: A self-attentive model for knowledge tracing. arXiv preprint arXiv:1907.06837 (2019)
12. Kaggle Homepage. https://www.kaggle.com/c/riiid-test-answer-prediction/data/. Accessed 25 Nov 2020
13. Friedman, J.H.: Greedy function approximation: a gradient boosting machine. Ann. Stat. **29**, 1189–1232 (2001)

Industry and Innovation

Scaffolds and Nudges: A Case Study in Learning Engineering Design Improvements

Stephen E. Fancsali(✉), Martina Pavelko, Josh Fisher, Leslie Wheeler, and Steven Ritter

Carnegie Learning, Inc., Pittsburgh, PA 15219, USA
{sfancsali,mpavelko,jfisher,lwheeler,
sritter}@carnegielearning.com

Abstract. We present a brief case study of a multi-year learning engineering effort to iteratively redesign the problem-solving experience of students using the "Solving Quadratic Equations" workspace in Carnegie Learning's MATHia intelligent tutoring system. We consider two design changes, one involving additional scaffolds for the problem-solving task and the next involving a "nudge" for learners to more rapidly and readily engage with these scaffolds and discuss resulting changes in the relative proportion of students who fail to master skills associated with this workspace over the course of two school years.

Keywords: Learning engineering · Intelligent tutoring system · Instructional design

1 Introduction

Carnegie Learning instructional designers, developers, and learning engineers continuously seek to identify areas for instructional and user experience improvements in MATHia, an intelligent tutoring system (ITS) formerly known as Cognitive Tutor [1]. An evolving set of prioritized topics (or workspaces) are tracked via an internal learning engineering [2, 3] dashboard, with priorities for improvement efforts set based on a number of metrics [4], including the proportion of students who fail to master each workspace's fine-grained knowledge components (KCs; [5]) and an "attention metric" index that combines information about failures to reach KC mastery with information about the number of users that encounter particular content, the amount of time it takes students to complete the topic, and other practical elements of the learner experience.

MATHia workspace improvement efforts take a variety of forms, most of which roughly align with steps for "design-loop" adaptivity [6] described in recent literature [7, 8]. While improvement can take the form of relatively sophisticated changes to KC models (and task redesign to reflect these changes), parameters for KC models, problem selection algorithms, among other changes, in what follows, we present a case study focusing on two relatively simple task-design changes within problems in a workspace called "Solving Quadratic Equations" and the relative impact of these changes on the

© Springer Nature Switzerland AG 2021
I. Roll et al. (Eds.): AIED 2021, LNAI 12749, pp. 441–445, 2021.
https://doi.org/10.1007/978-3-030-78270-2_78

442   S. E. Fancsali et al.

proportion of students who fail to master KCs in this workspace over large-scale deployments of MATHia over two school years (SYs). One change introduced additional, optional scaffolding to the task of solving a quadratic equation while the other merely represented a "nudge" to encourage students to more rapidly engage with this optional scaffolding. The scaffolding, by itself, had little impact on learner KC mastery, while the subsequent "nudge" encouraging the use of such scaffolding does appears to have substantially increased the proportion of students completing the workspace successfully. We illustrate the changes made and promising recent data indicating that small changes like these "nudges" may have a large impact, before pointing to future work.

2 "Solving Quadratic Equations" MATHia Workspace

Solving quadratic equations is a hallmark of Algebra I curricula. One Algebra I workspace in MATHia focuses on using its menu-based equation solver tool to apply the quadratic formula to solve quadratic equations. First, the student transforms a given equation into the form $ax^2 + bx + c = 0$, using transformations available in a menu. Next, the learner is expected to select "Apply Quadratic Formula" from the equation solver menu. In the problem illustrated in Fig. 1, the student started with $x^2 - 4x = -1$ and has added 1 to both sides. The student then selected "Apply Quadratic Formula". Figure 1's screenshot presents the result of this choice in the 2018–19 and 2019–20 SY releases of MATHia. Applying the quadratic formula involves several cognitive steps, including identifying the a, b and c terms, substituting those terms into the quadratic formula, and simplifying. Students have previously performed these steps on simpler expressions, and some are comfortable performing these steps for the quadratic formula, while others require or prefer more guidance. MATHia offers the student an optional "scratchpad" tool, which provides scaffolding at the student's request. Figure 2 shows the scratchpad "expanded" in the 2018–19 SY MATHia release. The scratchpad presents the student with the quadratic formula and scaffolding to input the values of coefficients a, b, and c for the formula.

Fig. 1. Problem-solving in "Solving Quadratic Equations" after the student has selected "Apply Quadratic Formula" from the equation solving menu in MATHia in 2018–19 and 2019–20 SYs.

Despite this optional scaffolding, in the 2018-19 SY, 32.1% of 6,698 students who completed this workspace failed to reach mastery of the six KCs tracked by this workspace's "skillometer" using Bayesian Knowledge Tracing [9] before reaching the maximum number of problems set for this workspace by its designers. These students moved on to the next topic in their curriculum sequence, and their teacher was alerted via MATHia's reports and the LiveLab teacher orchestration tool. This high rate of students failing to reach mastery made the workspace a target for data-driven improvement via Carnegie Learning's interdisciplinary learning engineering efforts.

Fig. 2. Problem-solving in 2018–19 SY MATHia with scaffolding "scratchpad" opened by the student. Compare to Fig. 1 where the scratchpad is unopened (by default).

3 Iterative Redesign

For the 2019–20 SY MATHia release, additional scaffolding was added to the optional scratchpad (see Fig. 3) to help with frequent arithmetic errors observed in student data. In addition to providing scaffolds for coefficients a, b, and c, the redesigned scratchpad scaffolded calculating the quadratic formula sub-terms: $-b$, $4ac$, b^2, and $2a$. Despite these scaffolds, the proportion of students failing to reach mastery only declined by 0.2% points in 2019–20 compared to 2018-19; the median and average time to completion decreased by approximately ten minutes (see Table 1).

With failures to reach mastery still at this level, for the 2020–21 SY, instructional designers chose to, by default, expand the scratchpad for students after they select "Apply Quadratic Formula" from the solving menu. The screenshot of Fig. 3 represents the state of the MATHia interface after the student selects "Apply Quadratic Formula" by default; the student no longer needs to expand the (optional) scratchpad scaffolds.

So far in the 2020–21 SY (through March 1, 2021), with the additional scratchpad scaffolding displayed by default, student failures to reach mastery have decreased by approximately 30% (from 34.4% to 24.1%) compared to the prior SY through March 1 (of 2020) (see Table 1).

444   S. E. Fancsali et al.

4   Discussion and Future Work

The space of data-driven improvements and redesigns in ITSs like MATHia is vast, but sometimes simple changes can have substantial impact. We highlight here a particular workspace where two relatively simple design changes were made over the course of two SYs to illustrate improvement compared to a baseline SY. Optional, additional scaffolding alone does not appear to have had the intended impact, but a "nudge" to engage with this scaffolding appears likely to be having an impact. Evidence presented is far from definitive, and more can be done to decrease the rate at which students fail to master the workspace's KCs. Several more sophisticated changes are also often made to an individual workspace in a given SY release of MATHia. We intend to increase the number and frequency of large-scale A/B tests of instructional improvements and redesigns using the UpGrade open-source architecture [10] in real classrooms.

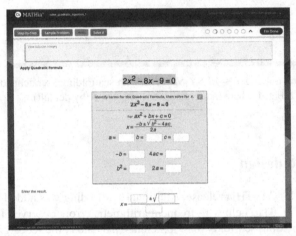

Fig. 3. Problem-solving in 2019–20 & 2020–21 SY MATHia with additional scaffolding provided by the opened "scratchpad" (opened by default in 2020–21 SY).

Table 1. Usage and performance metrics for "Solving Quadratic Equations" for two complete school years (SY) and for the present SY through March 1. Metrics for 2019–20 through March 1, 2020 are provided for comparison to the present (2020–21) SY (through March 1, 2021).

| | Complete SY | | Up to March 1 | |
	2018–19	2019–20	2019–20	2020–21
Completions	6,698	6,565	2,203	2,081
Mastery failures	2,151	2,093	758	503
% Mastery failures	32.1%	31.9%	34.4%	24.1%
Average time (min)	43.6	33.3	31.0	29.7
Median time (min)	35.6	25.7	25.0	21.5

Acknowledgements. This work is sponsored in part by the National Science Foundation under award The Learner Data Institute (Award #1934745). The opinions, findings, and results are solely the authors' and do not reflect those of the funding agency.

References

1. Ritter, S., Anderson, J.R., Koedinger, K.R., Corbett, A.T.: Cognitive tutor: applied research in mathematics education. Psychon. Bull. Rev. **14**, 249–255 (2007)
2. Simon, H.A.: The job of a college president. Educ. Rec. **48**(Winter), 68–78 (1967)
3. Rosé, C.P., McLaughlin, E.A., Liu, R., Koedinger, K.R.: Explanatory learner models: why machine learning (alone) is not the answer. Br. J. Educ. Technol. **50**, 2943–2958 (2019)
4. Fancsali, S.E., Liu, H., Sandbothe, M., Ritter, S.: Targeting design-loop adaptivity. In: Proceedings of the Fourteenth International Conference on Educational Data Mining (2021)
5. Koedinger, K.R., Corbett, A.T., Perfetti, C.: The knowledge-learning-instruction framework: bridging the science-practice chasm to enhance robust student learning. Cogn. Sci. **36**(5), 757–798 (2012)
6. Aleven, V., McLaughlin, E.A., Glenn, R.A., Koedinger, K.R.: Instruction based on adaptive learning technologies. In: Handbook of Research on Learning and Instruction, 2nd edn. pp. 522–560. Routledge, New York (2017)
7. Huang, Y., Aleven, V., McLaughlin, E., Koedinger, K.: A general multi-method approach to design-loop adaptivity in intelligent tutoring systems. In: Bittencourt, I.I., Cukurova, M., Muldner, K., Luckin, R., Millán, E. (eds.) AIED 2020. LNCS (LNAI), vol. 12164, pp. 124–129. Springer, Cham (2020). https://doi.org/10.1007/978-3-030-52240-7_23
8. Huang, Y., et al.: A general multi-method approach to data-driven redesign of tutoring systems. In: LAK21: 11th International Learning Analytics and Knowledge Conference (LAK21), pp. 161–172. ACM, New York (2021)
9. Corbett, A.T., Anderson, J.R.: Knowledge tracing: modeling the acquisition of procedural knowledge. User Model. User-Adapt. Interact. **4**, 253–278 (1994)
10. Ritter, S., Murphy, A., Fancsali, S.E., Fitkariwala, V., Patel, N., Lomas, J.D.: UpGrade: an open source tool to support A/B testing in educational software. In: Proceedings of the First Workshop on Educational A/B Testing at Scale (at Learning @ Scale 2020) (2020)

Condensed Discriminative Question Set for Reliable Exam Score Prediction

Jung Hoon Kim, Jineon Baek, Chanyou Hwang, Chan Bae,
and Juneyoung Park[✉]

Riiid! AI Research, Seoul, Republic of Korea
{junghoon.kim,jineon.baek,cy.hwang,chan.bae,juneyoung.park}@riiid.co

Abstract. The inevitable shift towards online learning due to the emergence of the COVID-19 pandemic triggered a strong need to assess students using shorter exams whilst ensuring reliability. This study explores a data-centric approach that utilizes feature importance to select a discriminative subset of questions from the original exam. Furthermore, the discriminative question subset's ability to approximate the students exam scores is evaluated by measuring the prediction accuracy and by quantifying the error interval of the prediction. The approach was evaluated using two real-world exam datasets of the Scholastic Aptitude Test (SAT) and Exame Nacional do Ensino Médio (ENEM) exams, which consist of student response data and the corresponding the exam scores. The evaluation was conducted against randomized question subsets of sizes 10, 20, 30 and 50. The results show that our method estimates the full scores more accurately than a baseline model in most question sizes while maintaining a reasonable error interval. The encouraging evidence found in this paper provides support for the strong potential of the on-going study to provide a data-centric approach for exam size reduction.

Keywords: Machine learning · Score prediction · Predictive uncertainty

1   Introduction

Using machine learning to reliably assess student grades is of great interest to the AI in Education (AIED) community [2,3,6]. However, without securing reliability of the models, tasks such as grade prediction can become problematic for companies [7]. Having a set of measurable standards for confidence in the model's predictions will allow for industry stakeholders to validate the model's reliability. An application of particular interest is shortened online exams, which became more prominent in continuous assessment for the distance education caused by the COVID-19 outbreak. These shorter online exams are required to not only be shorter in length but also to maintain the level of assessment accuracy shown by full-length in-person exams [10,11].

One approach that allows for enhancing reliability to a machine learning model is through the measurement of prediction uncertainty. A loss function

© Springer Nature Switzerland AG 2021
I. Roll et al. (Eds.): AIED 2021, LNAI 12749, pp. 446–450, 2021.
https://doi.org/10.1007/978-3-030-78270-2_79

presented in the works of Lakshminarayanan allows for the interpretation of the model's confidence in its predictions [5]. This provides insight into the uncertainty in the model outputs which is crucial information for various educational stakeholders.

This paper aims to present an ongoing-study to implement a reliable methodology for producing a discriminative subset of question. Specifically, a Random Forest model is utilized to create a smaller question subset through feature importance analysis and an exam score prediction model is utilized to predict the final grade with an individualized prediction error interval for each grade prediction. The procedure is evaluated using two real-world datasets with exam questions and exam scores. The preliminary results demonstrate encouraging findings on the potential of the current research for real-world applications.

2   Methodology

2.1   Score Prediction Using Smaller Exams

The task of exam size reduction can be defined as finding a shorter version of the original exam by using a reduced subset Q' of the full question set Q. Each question and the user's response was used as the feature. An efficient method that could be employed is using feature importance metrics from a Random Forest regressor [1,4]. After calculating the feature importance of each question, the questions can be sorted by order of importance and the top n questions will be used as an appropriately reduced discriminative set Q' that represents the full question set Q.

In order to more accurately predict exam scores using a discriminative question set, a Transformer-based deep learning model known as Assessment Modeling (AM) is employed. It employs a BERT-like pre-training and downstream task methodology which shows superior performance in score prediction tasks [2]. In the setting of this study, the pre-training stage is conducted on the full question set Q to learn the representations of all questions and student responses that capture the educational context of response data.

2.2   Error Analysis

One of the key trade-offs from shortening the exam via selecting a discriminative question set is the decrease of prediction accuracy. Therefore it becomes critical to be able to assess the acceptable level of error that can possibly arise from predicting exam scores with the discriminative question set. This study employs the method suggested by [5] to implement a loss function that trains the predicted deviation σ for individual predicted score.

The value of finding σ for predictions is that it gives insight into the reliability of the model. In the context of examinations, the reliability of the predicted score provides important information for the student, adding to the educational value of the exam. By providing a σ value, a student can more accurately assess their learning.

To calculate σ values for each predicted score, the score prediction model mentioned in the previous section is slightly modified. An additional layer that takes the outputs of the Transformer model is appended to the network. Unlike the score prediction model which trains the data by updating the losses of the scores, this model is trained by updating the losses in maximum likelihood estimation.

Assuming a normal distribution of the scores, the log-likelihood function with predicted score $\hat{s}$, predicted variance σ^2 and actual score s,

$$\ell(\hat{s}, \sigma) = \ln(\mathcal{L}(\hat{s}, \sigma^2|x)) = -\frac{\ln(\sigma^2)}{2} - \frac{(s - \hat{s})^2}{2\sigma^2} + (\text{constant})$$

shows how likely parameters $\hat{s}$ and σ match the distribution of the data. In other words, if $\hat{s}$ and σ maximizes the likelihood function, the parameter values will be a good representation of the actual data distribution.

3   Results

3.1   Prediction Error of Reduced Question Set

We evaluated the prediction error of selected n questions leveraging the method described in Sect. 2 (Q'_{RF}) and random selection (Q'_{RD}). Table 1 shows the MAE (Mean Absolute Error) of the predicted score of Q'_{RD} and Q'_{RF} for the different values of the question size $|Q'| = n$ in SAT and ENEM data. The experiment was conducted on question sizes $n = 10, 20, 30, 50$. The result shows that Q'_{RF} performs better even when compared to the best performing random question set on both the dataset. The gain in performance is more prominent as n gets smaller. Also, it is observed that predicted deviation σ also decreases as the size n increases.

Table 1. MAE and σ_{avg} of question sets Q'_{RD} and Q'_{RF} for size n in SAT (left) and ENEM (right).

n	Best Q'_{RD}	Worst Q'_{RD}	Mean Q'_{RD}	Q'_{RF}	n	Best Q'_{RD}	Worst Q'_{RD}	Mean Q'_{RD}	Q'_{RF}
10 MAE	100.722	120.922	110.349	88.131	10 MAE	136.254	181.839	160.197	102.274
σ_{avg}	140.248	131.440	134.462	122.178	σ_{avg}	167.751	228.123	174.384	118.681
20 MAE	73.319	81.739	78.362	67.908	20 MAE	110.810	149.814	126.276	79.455
σ_{avg}	89.288	99.471	92.432	90.529	σ_{avg}	100.189	131.349	127.014	95.217
30 MAE	61.199	69.341	63.651	58.681	30 MAE	87.596	128.102	103.789	68.406
σ_{avg}	71.795	75.296	73.706	69.132	σ_{avg}	100.840	139.452	100.260	74.175
50 MAE	44.175	49.297	47.060	42.475	50 MAE	74.396	90.524	80.716	57.408
σ_{avg}	59.026	60.682	54.977	56.049	σ_{avg}	74.396	82.977	78.675	66.267

3.2   Reliability of the Error Estimation σ

We evaluated the reliability of our model's error estimation σ. Figure 1 presents the confidence intervals predicted by the score model on Q'_{RF} ($n = 10$). The x

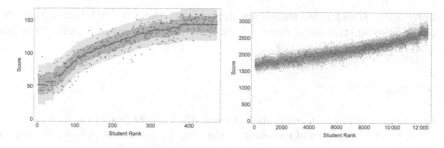

Fig. 1. Prediction interval of our model on question set Q'_{RF} ($n = 10$) in SAT (left) and ENEM dataset (right).

axis value i of the figure denotes the i'th student in the testing set. The students are ordered in increasing order of predicted score $\hat{s}_i$, which is depicted as the orange curve. The actual score s_i is depicted by the blue dots, and the dark band area (resp. light band area) is the confidence interval $\hat{s}_i \pm \sigma_i$ (resp. $\hat{s}_i \pm 2\sigma_i$) of the ith student's score estimated by the model.

The empirical rule states that for a normal distribution, the interval $\hat{s}_i \pm \sigma_i$ (resp. $\hat{s}_i \pm 2\sigma_i$) should have a confidence of 68% (resp. 95%). Indeed, for the SAT data, 72.0% of the actual scores lie in the dark band $\hat{s}_i \pm \sigma_i$ and 96.2% of the actual scores lie in the light band $\hat{s}_i \pm 2\sigma_i$. For the ENEM data, 62.4% of the actual scores lie in the dark band $\hat{s}_i \pm \sigma_i$ and 92.5% of the actual scores lie in the light band $\hat{s}_i \pm 2\sigma_i$. This assures the quality of the predicted scores s_i and the estimated error σ_i of our model on the shortened exam Q'_{RF}.

The results suggest that a set of $n = 10$ questions in Q'_{RF} is capable of comparing a student's performance in our setting. For example, a student with a predicted score 2345.9 ± 105.6 of ENEM (the 75th percentile in predicted score) is highly likely to outperform another student with a predicted score 2143.1 ± 112.0 of ENEM (the 50th percentile in predicted score), as the assumption of normality on both students' score gives a probability of 82.4% ($Z = (2345.9 - 2143.1)/(105.6 + 112.0)$).

4   Conclusion

In this paper, we evaluated a data-centric approach to identify discriminative question sets and for approximating student exam scores. The experiments results conducted on two real-world exams (e.g. SAT & ENEM) show that a properly selected question set can approximate a student's exam score with reasonable error intervals. In our experiments, question sets selected via feature importance produced by a Random Forest model has outperformed randomized question sets of all sizes. The findings from this study is encouraging for both education and machine learning. While the power of feature importance have been deemed important in educational research [9], we empirically validated the potential of such. Overconfident prediction of machine learning models, even accurate ones, could have significant impact on the involved individuals [5]. Therefore, demonstrating a careful approach not only ensures reliability but can also ensure trust [8].

References

1. Breiman, L.: Random forests. Mach. Learn. **45**(1), 5–32 (2001)
2. Choi, Y., et al: Assessment modeling: fundamental pre-training tasks for interactive educational systems. arXiv preprint arXiv:2002.05505 (2020)
3. Hellas, A., et al.: Predicting academic performance: a systematic literature review. In: Proceedings Companion of the 23rd Annual ACM Conference on Innovation and Technology in Computer Science Education, pp. 175–199 (2018)
4. James, G., Witten, D., Hastie, T., Tibshirani, R.: An Introduction to Statistical Learning, vol. 112. Springer, Heidelberg (2013). https://doi.org/10.1007/978-1-4614-7138-7
5. Lakshminarayanan, B., Pritzel, A., Blundell, C.: Simple and scalable predictive uncertainty estimation using deep ensembles. In: Advances in Neural Information Processing Systems, pp. 6402–6413 (2017)
6. Meier, Y., Xu, J., Atan, O., Van der Schaar, M.: Predicting grades. IEEE Trans. Signal Process. **64**(4), 959–972 (2015)
7. Mouta, A., Sánchez, E.T., Llorente, A.P.: Blending machines, learning, and ethics. In: Proceedings of the Seventh International Conference on Technological Ecosystems for Enhancing Multiculturality, pp. 993–998 (2019)
8. Sani, S.M., Bichi, A.B., Ayuba, S.: Artificial intelligence approaches in student modeling: half decade review (2010–2015). IJCSN-Int. J. Comput. Scie. Netw. **5**(5) (2016)
9. Sweeney, M., Rangwala, H., Lester, J., Johri, A.: Next-term student performance prediction: a recommender systems approach. arXiv preprint arXiv:1604.01840 (2016)
10. Vie, J.J., Popineau, F., Bruillard, É., Bourda, Y.: A review of recent advances in adaptive assessment. In: Peña-Ayala, A. (ed.) Learning Analytics: Fundaments, Applications, and Trends, vol. 94, pp. 113–142. Springer, Cham (2017). https://doi.org/10.1007/978-3-319-52977-6_4
11. Zhang, S., Chang, H.H.: From smart testing to smart learning: how testing technology can assist the new generation of education. Int. J. Smart Technol. Learn. **1**(1), 67–92 (2016)

Evaluating the Impact of Research-Based Updates to an Adaptive Learning System

Jeffrey Matayoshi$^{(\boxtimes)}$ (iD), Eric Cosyn, and Hasan Uzun

McGraw Hill ALEKS, Irvine, CA, USA
{jeffrey.matayoshi,eric.cosyn,hasan.uzun}@aleks.com

Abstract. ALEKS is an adaptive learning system covering subjects such as math, statistics, and chemistry. Several recent studies have looked in detail at various aspects of student knowledge retention and forgetting within the system. Based on these studies, various enhancements were recently made to the ALEKS system with the underlying goal of helping students learn more and advance further. In this work, we describe how the enhancements were informed by these previous research studies, as well as the process of turning the research findings into practical updates to the system. We conclude by analyzing the potential impact of these changes; in particular, after controlling for several variables, we estimate that students using the updated system learned 9% more on average.

Keywords: Adaptive assessment · Forgetting · Retrieval practice

1 Introduction

Since the time of Ebbinghaus and his work on the now famous forgetting curve [2,5], the study of memory and retention has long been a significant focus of educational research. Over the years, numerous techniques and methods—important examples of which include *spaced practice* [8] and *retrieval practice* [17]—have been shown to help with the long-term retention of knowledge. Within the artificial intelligence in education (AIED) field specifically, learning systems have benefited greatly both by modeling forgetting [4,16,25] and using personalized review schedules [10,15,21,22,27].

The particular system at the center of this work is ALEKS ("**A**ssessment and **LE**arning in **K**nowledge **S**paces") [14], an adaptive learning system covering subjects such as math, statistics, and chemistry. A key feature of ALEKS is its recurring *progress test*, an assessment that is given to a student after a certain amount of learning has occurred. The progress test focuses on the ALEKS problem types a student has recently learned and, among other things, functions as a mechanism for both spaced practice and retrieval practice—as such, it plays a critical role in ensuring students retain their newly acquired knowledge [13]. While the benefits of spaced practice [8,26] and retrieval practice [3,9,17–19] are well-documented, user feedback has shown that students working in the ALEKS system prefer to spend their time learning new material, rather than

© Springer Nature Switzerland AG 2021
I. Roll et al. (Eds.): AIED 2021, LNAI 12749, pp. 451–456, 2021.
https://doi.org/10.1007/978-3-030-78270-2_80

being assessed by a progress test. Based on these considerations, we began a project with the goal of updating the ALEKS progress test to be shorter and more efficient, thereby giving students additional time to learn and advance in the system—importantly, however, we wanted to do this in such a way as to retain the core benefits of the progress test on knowledge retention.

To that end, we conducted a series of studies [11–13] in an attempt to (a) more completely understand how the retention of knowledge works within the ALEKS system and (b) identify the specific factors that affect this retention. In this work, we discuss how the results from these previous studies were used to make research-based enhancements to ALEKS and the progress test; along the way, we also describe some of the challenges we encountered while implementing these updates. Finally, we conclude with an analysis of the performance of the system after the changes were deployed to production.

2   Previous Research on Forgetting in ALEKS

Our previous research unveiled several findings that informed our updates to the ALEKS system. Using forgetting curves to model knowledge retention in ALEKS, we observed that content on the "edge" of a student's current knowledge decays at a faster rate than content that is "deeper" in the student's knowledge [11]. Building on this result, subsequent work identified other factors affecting knowledge retention in ALEKS—arguably the most significant finding was that the specific characteristics of the problem types have the largest impact on this retention [12]. We also found that a neural network model could effectively use this information to predict the retention of individual problem types [12].

Other results from these studies provided further useful information. For example, we found that students experience a sort of "assessment fatigue" and are less likely to answer a question later in an ALEKS test [11], further highlighting the need to shorten the duration of the progress test. Next, we found evidence that, as a mechanism for retrieval practice, the progress test is more effective when a longer delay exists between the initial learning of a problem type and its appearance on the test [13]. Finally, we observed that further learning of related material in ALEKS can function as a type of retrieval practice. In particular, this act of learning was associated with higher rates of retention compared to the retrieval practice that occurs with a progress test [13], suggesting that learning more and advancing further in the system could be linked to better retention.

3   From Research to Development

Based on these insights, we decided to use the neural network model from [12] to target the problem types students are likely to forget, as these stand to benefit the most from being asked in a progress test. While the previous iteration of the test covered all topics the students recently learned, this targeted approach allowed us to substantially reduce the length of the progress test. Further efficiency was also gained by focusing on problem types learned less recently; as mentioned previously, our research indicated a benefit to delaying the retrieval practice that occurs in progress tests. Finally, other smaller enhancements were made, with the overall goal of helping students learn problem types more efficiently.

The next challenge was to implement these changes within the constraints of a production environment—as these updates would be used by millions of students, the computations needed to (a) run efficiently at scale and (b) be easy to monitor. To address the former concern, we optimized the computational efficiency of the neural network model from [12]. As one example, the original model used a recurrent neural network (RNN) to process the sequential data from students learning in ALEKS. To reduce the resulting computational burden, the RNN was replaced with a set of (non-sequential) features that captured similar information and gave comparable performance. Next, to facilitate the monitoring of the neural network computations, a dedicated database was designed to capture the outputs from the model, along with the relevant input features, making it easy to validate its performance. Additionally, as a side benefit such data would then be readily accessible for any further retraining of the model.

4   Analysis

In this section we use a quasi-experimental design to analyze the impact of the updates on student learning within ALEKS. As the enhancements were pushed to the production servers in July of 2020, we focus on students who started working in the system on or after August 1st, 2020; for these students, we gather data from their activities in ALEKS through the end of the year. To obtain our control group, we find students who started working in the system on August 1st, 2019 or later, and we gather their data through the end of 2019. Additionally, we restrict our search to a selection of seven different math courses that had no version upgrades or changes to their content over this combined time period, as such changes could confound the comparison. These include courses starting as low as fifth grade math and as high as college-level precalculus. Lastly, we require that each student has completed enough work for at least one progress test to be assigned by the system—importantly, the mechanism that assigns this first progress test has not been affected by any of the enhancements to the system.

Table 1. Comparison of the 2019 and 2020 student populations.

Year	Students	Average		
		Length	Hours	Number
2019	166,635	19.6	2.3	3.5
2020	154,098	13.3	1.7	3.5

To compare the behavior of the progress test for these populations, in Table 1 we show the following averages: number of questions on each progress test (length); cumulative hours spent in progress tests per student (hours); and number of progress tests taken per student (number). While the average number of tests is similar between the two populations, the average number of questions and cumulative hours decrease by about 32% and 26%, respectively. The slightly smaller decrease in average hours is likely due to the focus of the updated progress test on problem types students are more likely to forget—these problem types are more challenging and typically take longer for students to answer.

Next, to measure the amount of learning for each student, we compute the difference between the number of problem types the student knows at the beginning of the course—as measured by the *initial test* given in ALEKS—compared to what they know at the end of each study period; we refer to this as the *learning gain* of the student. While we have taken care to try and equalize the student populations from 2019 and 2020, in light of the COVID-19 pandemic it seems unlikely these populations are completely equivalent. Furthermore, from past experience we expect a dependence—or correlation—in the data for students within the same math course, as these students tend to have more similarities.

To address these issues, we fit a multilevel model with the Linear Mixed Effects (LME) class from the `statsmodels` [20] Python library, using the learning gain as our dependent variable and a separate random intercept for each math course. We focus our analysis on an indicator variable that is 1 for students using the updated progress test and 0 otherwise; the coefficient of the variable, β_1, estimates the change in learning gain between the student populations. We also introduce additional independent variables to adjust for differences in the underlying characteristics of the students; these include the time spent learning in the system, amount of learning activity, score on the initial test (as a "ceiling effect" occurs with students who start with more knowledge), and number of problem types in the course (as another ceiling effect occurs with smaller courses). As some of these variables are measured *post-treatment*—i.e., after students have interacted with the progress test—we use the two-step regression procedure outlined in [1] to adjust for post-treatment bias. Specifically, the first step of the procedure is used to adjust for the post-treatment variables, which then allows us to make an unbiased estimate of β_1 in the second step [1,6,7,23,24]. The resulting estimate for β_1 is 9.8 with a 95% confidence interval of (9.6, 10.0). Thus, holding the other variables constant, students using the updated progress test have an estimated learning gain that is higher by 9.8 problem types on average. As the mean learning gain for students using the original progress test is 104.4, this represents an estimated improvement of approximately 9%.

5   Conclusion

In this work we described a set of research-based enhancements to the ALEKS adaptive learning system, with these enhancements being made to help students learn more and advance further in the system. After adjusting for several variables, a comparison of before and after data indicated that, on average, students using the updated system learned 9% more. In the context of the ongoing COVID-19 pandemic, we find this last result to be encouraging. Given that the pandemic has compounded existing inequities within education, we hope that the improvements made to the ALEKS system can, at least in some part, help students who may otherwise be struggling with their learning.

References

1. Acharya, A., Blackwell, M., Sen, M.: Explaining causal findings without bias: detecting and assessing direct effects. Am. Polit. Sci. Rev. **110**(3), 512 (2016)
2. Averell, L., Heathcote, A.: The form of the forgetting curve and the fate of memories. J. Math. Psychol. **55**, 25–35 (2011)
3. Bae, C.L., Therriault, D.J., Redifer, J.L.: Investigating the testing effect: retrieval as a characteristic of effective study strategies. Learn. Instr. **60**, 206–214 (2019)
4. Choffin, B., Popineau, F., Bourda, Y., Vie, J.J.: DAS3H: modeling student learning and forgetting for optimally scheduling distributed practice of skills. In: Proceedings of the 12th International Conference on Educational Data Mining, pp. 29–38 (2019)
5. Ebbinghaus, H.: Memory: A Contribution to Experimental Psychology. Originally published by Teachers College, Columbia University, New York (1885). Translated by Henry A Ruger and Clara E Bussenius (1913)
6. Goetgeluk, S., Vansteelandt, S., Goetghebeur, E.: Estimation of controlled direct effects. J. Roy. Stat. Soc.: Ser. B (Stat. Methodol.) **70**(5), 1049–1066 (2008)
7. Joffe, M.M., Greene, T.: Related causal frameworks for surrogate outcomes. Biometrics **65**(2), 530–538 (2009)
8. Kang, S.H.: Spaced repetition promotes efficient and effective learning: policy implications for instruction. Policy Insights Behav. Brain Sci. **3**(1), 12–19 (2016)
9. Karpicke, J.D., Roediger, H.L.: The critical importance of retrieval for learning. Science **319**(5865), 966–968 (2008)
10. Lindsey, R.V., Shroyer, J.D., Pashler, H., Mozer, M.C.: Improving students long-term knowledge retention through personalized review. Psychol. Sci. **25**(3), 639–647 (2014)
11. Matayoshi, J., Granziol, U., Doble, C., Uzun, H., Cosyn, E.: Forgetting curves and testing effect in an adaptive learning and assessment system. In: Proceedings of the 11th International Conference on Educational Data Mining, pp. 607–612 (2018)
12. Matayoshi, J., Uzun, H., Cosyn, E.: Deep (un)learning: using neural networks to model retention and forgetting in an adaptive learning system. In: Isotani, S., Millán, E., Ogan, A., Hastings, P., McLaren, B., Luckin, R. (eds.) AIED 2019. LNCS (LNAI), vol. 11625, pp. 258–269. Springer, Cham (2019). https://doi.org/10.1007/978-3-030-23204-7_22
13. Matayoshi, J., Uzun, H., Cosyn, E.: Studying retrieval practice in an intelligent tutoring system. In: Proceedings of the Seventh ACM Conference on Learning @ Scale, pp. 51–62 (2020)
14. McGraw Hill ALEKS: What is ALEKS? (2021). https://www.aleks.com/about_aleks
15. Pavlik, P.I., Anderson, J.R.: Using a model to compute the optimal schedule of practice. J. Exp. Psychol. Appl. **14**(2), 101 (2008)

16. Qiu, Y., Qi, Y., Lu, H., Pardos, Z.A., Heffernan, N.T.: Does time matter? Modeling the effect of time with Bayesian knowledge tracing. In: Proceedings of the 4th International Conference on Educational Data Mining, pp. 139–148 (2011)
17. Roediger III, H.L., Butler, A.C.: The critical role of retrieval practice in long-term retention. Trends Cogn. Sci. **15**, 20–27 (2011)
18. Roediger III, H.L., Karpicke, J.D.: The power of testing memory: basic research and implications for educational practice. Perspect. Psychol. Sci. **1**(3), 181–210 (2006)
19. Roediger III, H.L., Karpicke, J.D.: Test-enhanced learning: taking memory tests improves long-term retention. Psychol. Sci. **17**(3), 249–255 (2006)
20. Seabold, S., Perktold, J.: Statsmodels: econometric and statistical modeling with Python. In: 9th Python in Science Conference (2010)
21. Settles, B., Meeder, B.: A trainable spaced repetition model for language learning. In: Proceedings of the 54th Annual Meeting of the Association for Computational Linguistics (Volume 1: Long Papers), pp. 1848–1858 (2016)
22. Tabibian, B., Upadhyay, U., De, A., Zarezade, A., Schölkopf, B., Gomez-Rodriguez, M.: Enhancing human learning via spaced repetition optimization. Proc. Natl. Acad. Sci. **116**(10), 3988–3993 (2019)
23. Vansteelandt, S.: Estimating direct effects in cohort and case-control studies. Epidemiology 851–860 (2009)
24. Vansteelandt, S., et al.: On the adjustment for covariates in genetic association analysis: a novel, simple principle to infer direct causal effects. Genet. Epidemiol.: Off. Publ. Int. Genet. Epidemiol. Soc. **33**(5), 394–405 (2009)
25. Wang, Y., Heffernan, N.T.: Towards modeling forgetting and relearning in ITS: preliminary analysis of ARRS data. In: Proceedings of the 4th International Conference on Educational Data Mining, pp. 351–352 (2011)
26. Weinstein, Y., Madan, C.R., Sumeracki, M.A.: Teaching the science of learning. Cogn. Res.: Principles Implicat. **3**(1), 1–17 (2018). https://doi.org/10.1186/s41235-017-0087-y
27. Xiong, X., Wang, Y., Beck, J.B.: Improving students' long-term retention performance: a study on personalized retention schedules. In: Proceedings of the Fifth International Conference on Learning Analytics And Knowledge, pp. 325–329. ACM (2015)

Back to the Origin: An Intelligent System for Learning Chinese Characters

Jinglei Yu, Jiachen Song, Yu Lu$^{(\boxtimes)}$, and Shengquan Yu

Advanced Innovation Center for Future Education, Faculty of Education,
Beijing Normal University, Beijing, China
luyu@bnu.edu.cn

Abstract. Learning Chinese characters is a challenging task for both native and foreign beginners. One major reason is that most Chinese characters in writing are distinct from each other and lack of directly phonetic clues. Fortunately, many Chinese characters' original forms have iconicity that indicates their meanings. By leveraging on these characteristics and the latest computer vision (CV) techniques, we design and build an intelligent system that could automatically retrieve the iconic and original forms of Chinese characters. Furthermore, the system could provide learners with different styles of the character in a chronological order to bridge the original form and the most commonly used one. Specifically, we adopt the SE-Resnet-50 classification model for both character recognition and style recognition tasks, and design a dedicated retrieval mechanism to properly select the representative characters in different styles for learners. A specific user interface is designed for beginners to upload, recognize, remember, and understand the Chinese characters.

Keywords: Language learning · Character recognition · Computer vision

1 Introduction

Chinese character recognition is difficult for non-native and even native beginners due to several reasons. First, the structure of individual character normally depicts an object or represents some abstract notions rather than based on alphabet. Hence, different Chinese characters in writing are distinct from each other and the number of commonly used characters exceeds 3,000. Second, each Chinese character can be written in five major chronologically formed scripts that are still utilized today, namely seal script, clerical script, cursive script, running script, and regular script. Many characters' current commonly used forms in regular script cannot reflect any visual meaning that can be easily perceived by beginners, while their original forms in seal script can. In addition, Chinese characters are not directly related to their pronunciations, and thus the learners have to recognize the characters apart from the pronunciations.

© Springer Nature Switzerland AG 2021
I. Roll et al. (Eds.): AIED 2021, LNAI 12749, pp. 457–461, 2021.
https://doi.org/10.1007/978-3-030-78270-2_81

Fig. 1. The chronological evolving process of the character "Fire" in five scripts.

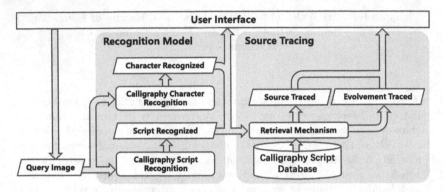

Fig. 2. The simplified block diagram of the system.

The previous studies have shown that iconicity, i.e., structural similarity between the character and its referent, is effective for children in reading [5]. Fortunately, many Chinese characters' original forms are iconic that can be easily recognized. Taking character "Fire" as example, its original form in seal script is a simplified picture of fire. Figure 1 illustrates the chronological evolving process of it in the five scripts. Hence, we design and implement an intelligent system that could automatically recognize and trace back to the original form of the given Chinese character. Beside the original form, the system could also show the evolvement in the five scripts of the given character. The system specifically utilizes the calligraphy collections with high art value, which demonstrates the elegant forms of different scripts and easily arouses the learners' interests.

2 System Design

The system consists of two modules, namely *recognition model* module and *source tracing* module. Both modules connect to the upper user interface and support the interaction with the learners. Figure 2 illustrates the block diagram of the built system. Briefly speaking, the system first receives the input image of an individual character through the user interface. Since structural similarities of the same character among five scripts are generally higher than them among different characters, we treat character recognition and script recognition tasks separately for more accurate classification. In the recognition model module, two convolutional neural network (CNN) [3,4] models are utilized to accomplish the two tasks. The recognition results would be delivered to the source tracing

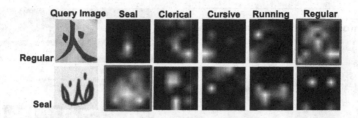

Fig. 3. Important region visualization of two query images on five classes. The brighter the regions are, the more attention the model pays to for the particular class. The correct recognition results are marked in red and illustrated on the left.

module. The source tracing module receives the recognized character and then retrieves different scripts of it in calligraphy script database. The database stores 35,563 images of the Chinese characters in the five scripts, and the representative images would be selected based on the input image. The selection criteria are based on the latent features extracted by the CNN model for calligraphy script recognition. Finally, the system displays the selected images in the chronological order to demonstrate the evolving process of the character, particularly highlighting the seal script as its original form. We will elaborate the two modules in the following parts.

2.1 Recognition Model Module

In recognition model module, both the character recognition and script recognition tasks can be regarded as the image classification task in CV. Specifically, we adopt SE-ResNet-50 [2] to recognize both the character and its script. ResNet [1,8] is a widely used CNN structure, which utilizes residual learning to train deep neural network. We adopt 50 layers ResNet as a based backbone with 49 convolutional layers and one fully connected layer. Meanwhile, squeeze-and-excitation (SE) blocks work as the self-attention function [7] on channels to improve classification accuracy.

For the character recognition model, we set each character as a class and first select 150 characters having five scripts. In total, 35,563 images of 150 characters are used for training and validation. Using 5-fold cross validation, the F1 score of the built model achieves 0.90. For the script recognition model, we treat each script as a class and five in total. In this way, the model is trained to learn the latent feature of each calligraphy script. We utilize gradient-weighted class activation mapping (Grad-CAM) [6] method to visualize the localization of the important regions, where the model attempts to discriminate different classes separately. As shown in Fig. 3, the model correctly responses to the ground-truth class, meanwhile responses to several local regions that partially reflect the script features in other classes. A total of 70,111 images of 2,177 characters are collected and, using 5-fold cross validation, the F1 score of the built model achieves 0.88.

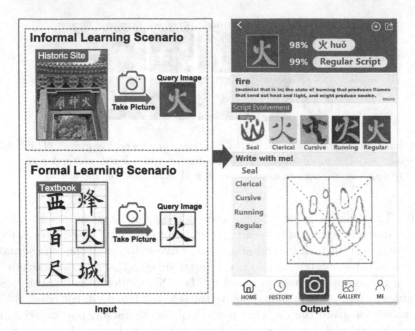

Fig. 4. The system usage scenarios and its user interface.

2.2 Source Tracing Module

Source tracing module selects different scripts of the recognized character in calligraphy script database using the retrieval mechanism. Specifically, the mechanism takes advantage of the features learned by the script recognition model. We extract features before last fully connected layer in float array with a size of 512, then measure the Euclidean distance of features between the query and script dataset images. The selections are the five representative characters in each script that are most similar to the input image.

2.3 User Interface

Learners could easily use the built system to learn the individual Chinese character in either formal or informal learning environment, such as taking pictures and uploading one photo from a historic site or a textbook. As shown in Fig. 4, the recognized character and its script would be shown at the top of the user interface, including the pronunciation and the models' confidences. The user interface also provides translation and explanation in English. Besides, the evolvement of the recognized character is displayed from the seal script to the regular script in the chronological order. More importantly, the original form of the recognized character is shown in the center of the user interface, learners could also counterdraw the character in different scripts to consolidate memory during writing.

3 Conclusion

By leveraging on the latest CV techniques in artificial intelligence, we design and implement an intelligent system to trace back the origin and evolvement of the Chinese characters. Empowered by the system, learning Chinese characters could be more intriguing, meaningful, and effective. We are deploying the system in multiple local schools and meanwhile improving the system to cover more basic Chinese characters.

Acknowledgement. This research is supported by the National Natural Science Foundation of China (No. 62077006), the Fundamental Research Funds for the Central Universities.

References

1. He, K., Zhang, X., Ren, S., Sun, J.: Deep residual learning for image recognition. In: Proceedings of the IEEE Conference on Computer Vision and Pattern Recognition, pp. 770–778 (2016)
2. Hu, J., Shen, L., Sun, G.: Squeeze-and-excitation networks. In: Proceedings of the IEEE Conference on Computer Vision and Pattern Recognition, pp. 7132–7141 (2018)
3. Krizhevsky, A., Sutskever, I., Hinton, G.E.: ImageNet classification with deep convolutional neural networks. Adv. Neural. Inf. Process. Syst. **25**, 1097–1105 (2012)
4. LeCun, Y., et al.: Backpropagation applied to handwritten zip code recognition. Neural Comput. **1**(4), 541–551 (1989)
5. Luk, G., Bialystok, E.: How iconic are Chinese characters? Bilingualism **8**(1), 79 (2005)
6. Selvaraju, R.R., Cogswell, M., Das, A., Vedantam, R., Parikh, D., Batra, D.: Grad-CAM: visual explanations from deep networks via gradient-based localization. In: Proceedings of the IEEE International Conference on Computer Vision, pp. 618–626 (2017)
7. Vaswani, A., et al.: Attention is all you need. arXiv preprint arXiv:1706.03762 (2017)
8. Xie, S., Girshick, R., Dollár, P., Tu, Z., He, K.: Aggregated residual transformations for deep neural networks. In: Proceedings of the IEEE Conference on Computer Vision and Pattern Recognition, pp. 1492–1500 (2017)

Doctoral Consortium

Automated Assessment of Quality and Coverage of Ideas in Students' Source-Based Writing

Yanjun Gao[(⊠)] and Rebecca J. Passonneau

Penn State University, University Park, PA 16802, USA
{yug125,rjp49}@psu.edu

Abstract. Source-based writing is an important academic skill in higher education, as it helps instructors evaluate students' understanding of subject matter. To assess the potential for supporting instructors' grading, we design an automated assessment tool for students' source-based summaries with natural language processing techniques. It includes a special-purpose parser that decomposes the sentences into clauses, a pre-trained semantic representation method, a novel algorithm that allocates ideas into weighted content units and another algorithm for scoring students' writing. We present results on three sets of student writing in higher education: two sets of STEM student writing samples and a set of reasoning sections of case briefs from a law school preparatory course. We show that this tool achieves promising results by correlating well with reliable human rubrics, and by helping instructors identify issues in grades they assign. We then discuss limitations and two improvements: a neural model that learns to decompose complex sentences into simple sentences, and a distinct model that learns a latent representation.

Keywords: Natural language processing · Content analysis · Rubric-based writing assessment

1 Introduction

Source-based writing is an important academic skill for students in higher education. In source-based writing, students engage in a series of learning tasks including identifying and understanding the important information from subject matter and composing text, therefore it helps the instructors evaluate students' learning progress [4,6,16]. Rubric-based writing assessment provides informative feedback to students by indicating the quality and coverage of knowledge the students mention in their writing [9,13]. To assist instructors in rubric-based grading, automated assessment for source-based writing using natural language processing (NLP) techniques has attracted attention in recent years [8,14]. In this work, we study how well performance of a system developed with NLP techniques aligns with instructional rubrics for summarization, and discuss two enhancements using neural models. The content alignment between rubrics and

© Springer Nature Switzerland AG 2021
I. Roll et al. (Eds.): AIED 2021, LNAI 12749, pp. 465–470, 2021.
https://doi.org/10.1007/978-3-030-78270-2_82

students text identifies important content that is missing in students' summaries, thus could help students and instructors diagnose issues to work on.

Summarization is a process of content selection, abstraction and integration with background knowledge [7]. Previous work has used a wise-crowd content evaluation method that constructs a content model representing overlapping ideas from a small set of reference summaries [12]. In manual wise-crowd evaluation, similar ideas from reference summaries are first allocated into content units in a content model, and then used for scoring target summaries. The content model represents the important ideas and their distribution (i.e. number of occurrences) identified in reference summaries, and could be regarded as an automated rubric. [11,12] demonstrated high correlation with manual rubrics in assigning scores to students summaries. Here we develop a fully automated pipeline for wise-crowd content evaluation, PyrEval [3]. PyrEval constructs a content model from reference summaries, and outputs scores for the quality and coverage of students' content, with score justifications in the form of specific student ideas that match the model content units. It contains the following modules: a decomposition parser that extracts clausal units from sentences through pre-defined syntax rules, a WTMF method that converts clause meaning into distributional semantic vectors [5], a novel algorithm that allocates idea units based on semantic similarity of clauses, and a WMIN algorithm [15] that matches student writing to the content model.

To test PyrEval correlation with human rubrics, our work investigate three sets of student writing samples with instructional rubrics collected under different disciplines and classroom settings: computer science (CS) students' post-workshop writing, CS students' source-based summaries, and students' case briefs from an intervention for underserved law students. In the CS student's post-workshop writing dataset, a large class of CS students summarized their participation in a workshop about *Critical Thinking*. Their writing was graded by a professor with a rubric about the important content in the workshop [1]. For CS student's source-based writing, students were asked to write a summary and an argumentative essay from reading three articles given the topics of *Autonomous Vehicle* (AV) and *CryptoCurrency* (CC). The assignments were graded by TAs using a multi-dimensional rubrics that we designed [2], then later by raters. In the case brief assignment, law school students wrote briefs from reading court cases materials as part of an educational intervention to improve their legal writing.

The rubrics for post-workshop dataset and legal briefs dataset are not evaluated with reliability studies, which we did not have the resources to conduct. For the CS students' source-based writing, we collaborated with the instructor to design the assignment and rubric. After the course ended, we trained two raters to re-grade the assignment using the rubric and test the reliability. We found that the trained raters were able to apply the rubric in a reliable way, and that the TAs grades did not correlate with the reliable raters' grades.

2   Results

In this section, we examine the question of whether PyrEval could effectively assess student writing. We show initial results from running PyrEval on the three sets of writing and correlating the scores with human rubrics. We show that the detailed matches between students' writing and the content model output by PyrEval could be used as feedback for instructors, and we analyze the limitation of current components of PyrEval.

2.1   Experiments on CS Students' Post-workshop Writing

There are 135 student submissions in the students' post-workshop writing sample. The professor designed a 10-point rubric with each item representing an important idea. We include 13 samples that are scored full or 9 points together with a sample written by the professor as the reference set, and run PyrEval to construct three content models from three sets of six randomly selected reference samples. Then we run PyrEval to score the target 122 samples with the three content models and output three sets of scores. We apply Pearson and Spearman rank correlation among pairs of automated scores generated from different sets of reference samples, and pairs of automated scores and human scores. The three pairs of automated scores achieved high correlation with each other, reporting Pearson correlation from 0.68 to 0.76, and Spearman correlation from 0.69 to 0.77. When correlating with human score, PyrEval achieves Pearson correlation from 0.46 to 0.49, and Spearman Rank correlation from 0.44 to 0.46. Recall that there is no study of rubric reliability, thus the instructor's grades are not necessarily reliable, analyzed in next section.

   We also find that PyrEval helps the instructor with finding issues in the scoring. In the analysis where we categorize student scores as below average, average or above average, PyrEval and the instructor agree on samples scored within average ($N = 50$). There are 8 agreements on below average and 10 agreements on above average. No student received above average from PyrEval and below average from human, but three submissions are categorized into below average from PyrEval and above average from instructor. After reviewing PyrEval's fine-grained scoring output, the instructor corrected the scores to below average, because these three samples had no overlapping ideas with the rubric. The instructor was lenient with these samples given that the text were very fluent.[1]

2.2   Experiments on CS Students' Source-Based Summaries

The assignment was part of a class for computer science freshmen. Students could choose one of the two topics (three readings per topic) to summarize. At the end, we received 42 submissions for AV and 37 for CC. After the class ended,

---

[1] We direct readers to [1,3] for detailed output from PyrEval that shows content alignments between reference summaries and students summaries. PyrEval is downloadable at https://github.com/serenayj/PyrEval.

we gave two raters training on applying rubric in scoring. Details of rubric design and rater agreement could be found in [2].

We apply PyrEval on students' summaries of the assignments and correlate PyrEval scores to two trained raters' scores. PyrEval reports 0.66 and 0.72 Pearson correlation on AV and CC, respectively, indicating a high correlation with human rubric. Error analysis shows that the decomposition parser in PyrEval has limited performance on complex sentences due to the syntax-based approaches and inaccurate parsing. We believe that replacing the current decomposition parser of PyrEval with our recently developed neural model could potentially enhance PyrEval performance, given training data with adequate coverage.

2.3   Experiments on Students' Case Briefs

There are 42 student briefs to be scored automatically, and 5 reference briefs written by assistant instructors to construct a content model. The rubric uses a 5-point scale, including criteria emphasized important contents appeared in court cases and reasoning process. Similar to the workshop writing set, there is no reliability study of the rubric. We run PyrEval on the 5 reference briefs to construct a content model and score the 42 students briefs. PyrEval achieves 0.53 Pearson correlation with human rubric. After identifying two outliers of student briefs that include unexpected uses of quotes and written in extremely terse way, the Pearson correlation goes up to 0.70, showing a strong correlation with human scores. From our observation, PyrEval is good at capturing definitional statements attributed to its semantic representation method. However, it has limitations such as name entities, coreferences, and domain knowledge. Furthermore, PyrEval tends to generate propositions that are lengthier and incomplete, resulting in less accurate semantic vectors and performance downgrade in content model construction and scoring.

3   Improvements and Conclusions

Recall that PyrEval is limited by the performance of decomposition parser and semantic representation method. To enhance PyrEval and ultimately apply it on essays where the text is more complex, we propose two improvements with the emphasis of linguistic structure and discourse modeling. The first improvement is a new method to decompose sentences into simple sentences derived from tensed clauses. We formulate the problem as new graph edit task, and train a neural parser to predict when to break the sentences, and insert or drop words to output complete and fluent sentences. The second improvement is a new sentence representation method that encodes word sequences and their syntax relations using graph neural network. Such method introduces richer representation compared to WTMF and sequence modeling. For evaluation, we collect a large essay dataset from a large social science class and annotate over 39K complex sentences using Amazon Mechanical Turk. The annotation involves splitting and rephrasing complex sentences into simple and complete sentences. We also prepare data for connective prediction model, a task relevant to essay coherence modeling and discourse analysis [10]. We plan to evaluate both new models using this dataset and improve PyrEval with new models.

We present PyrEval, a software tool with NLP techniques, on student summarization evaluation. PyrEval correlates well with human rubrics and show its potential in providing informative feedback to instructors, as the scores are justified through content alignment with the model. We also identify two enhancement on PyrEval components to provide more accurate representation for contents in complex sentences.

References

1. Gao, Y., Davies, P.M., Passonneau, R.J.: Automated content analysis: a case study of computer science student summaries. In: Proceedings of the Thirteenth Workshop on Innovative Use of NLP for Building Educational Applications, pp. 264–272 (2018)
2. Gao, Y., et al.: Rubric reliability and annotation of content and argument in source-based argument essays. In: Proceedings of the Fourteenth Workshop on Innovative Use of NLP for Building Educational Applications, pp. 507–518 (2019)
3. Gao, Y., Sun, C., Passonneau, R.J.: Automated pyramid summarization evaluation. In: Proceedings of the 23rd Conference on Computational Natural Language Learning (CoNLL), pp. 404–418 (2019)
4. Graham, S., et al.: Teaching secondary students to write effectively. Educator's practice guide. What works clearinghouse.^{TM} ncee 2017–4002. What Works Clearinghouse (2016)
5. Guo, W., Diab, M.: Modeling sentences in the latent space. In: Proceedings of the 50th Annual Meeting of the Association for Computational Linguistics (Volume 1: Long Papers), pp. 864–872 (2012)
6. Hirvela, A., Du, Q.: "Why am i paraphrasing?": undergraduate esl writers' engagement with source-based academic writing and reading. J. English Acad. Purposes 12(2), 87–98 (2013)
7. Kintsch, W., Van Dijk, T.A.: Toward a model of text comprehension and production. Psychol. Rev. 85(5), 363 (1978)
8. Klein, R., Kyrilov, A., Tokman, M.: Automated assessment of short free-text responses in computer science using latent semantic analysis. In: Proceedings of the 16th Annual Joint Conference on Innovation and Technology in Computer Science Education, pp. 158–162 (2011)
9. Lundstrom, K., Diekema, A.R., Leary, H., Haderlie, S., Holliday, W.: Teaching and learning information synthesis: an intervention and rubric based assessment. Commun. Inf. Lit. 9(1), 4 (2015)
10. Nadeem, F., Nguyen, H., Liu, Y., Ostendorf, M.: Automated essay scoring with discourse-aware neural models. In: Proceedings of the Fourteenth Workshop on Innovative Use of NLP for Building Educational Applications, pp. 484–493 (2019)
11. Passonneau, R.J., Chen, E., Guo, W., Perin, D.: Automated pyramid scoring of summaries using distributional semantics. In: Proceedings of the 51st Annual Meeting of the Association for Computational Linguistics (Volume 2: Short Papers), pp. 143–147 (2013)

12. Passonneau, R.J., Poddar, A., Gite, G., Krivokapic, A., Yang, Q., Perin, D.: Wise crowd content assessment and educational rubrics. Int. J. Artif. Intell. Educ. **28**(1), 29–55 (2018)
13. Rakedzon, T., Baram-Tsabari, A.: To make a long story short: a rubric for assessing graduate students' academic and popular science writing skills. Assess. Writ. **32**, 28–42 (2017)
14. Ranalli, J., Link, S., Chukharev-Hudilainen, E.: Automated writing evaluation for formative assessment of second language writing: investigating the accuracy and usefulness of feedback as part of argument-based validation. Educ. Psychol. **37**(1), 8–25 (2017)
15. Sakai, S., Togasaki, M., Yamazaki, K.: A note on greedy algorithms for the maximum weighted independent set problem. Discret. Appl. Math. **126**(2–3), 313–322 (2003)
16. Sampson, V., Enderle, P., Grooms, J., Witte, S.: Writing to learn by learning to write during the school science laboratory: helping middle and high school students develop argumentative writing skills as they learn core ideas. Sci. Educ. **97**(5), 643–670 (2013)

Impact of Intelligent Tutoring System (ITS) on Mathematics Achievement Using ALEKS

Rashmi Khazanchi[✉]

3072 Wendlock Drive, Marietta, GA 30062, USA
rashmi_khazanchi@mitchell.k12.ga.us

Abstract. This doctoral research will explore the Impact of the Intelligent Tutoring System (ITS), such as the Assessment and LEarning in Knowledge Spaces (ALEKS), on students' mathematics achievement, affective engagement, and cognitive engagement on 9th-grade students from two different school districts with somewhat similar demographics. This proposed quasi-experimental study will compare ALEKS-led (while teachers were present) versus teacher-led instructions to provide additional support to struggling students for fifty minutes per week for six weeks. This research will also explore the challenges posed by using ITS in the classroom.

Keywords: Artificial intelligence in education · Intelligent tutoring systems · Personalized learning · Adaptive education · Assessment and LEarning in knowledge spaces (ALEKS) · Student engagement · Student motivation · Knowledge construction

1 Introduction

1.1 Overview of the Need

Mathematical knowledge and skills are vital for academic achievement and influence the life quality of life of individuals [10]. Historically, students in the United States lag behind students in other developed countries, as reported by the Trends in Mathematics and Science Study (TIMSS) conducted by the National Center for Educational Statistics. In 2019 on TIMSS mathematical scale, the U.S. 8th-graders ranked 11th with an average mathematics score of 515 [15]. The National Assessment of Educational Progress (NAEP) reported the mathematics achievement scores of the U.S. eighth-grade students, indicating that only 34% of students perform at or above proficiency level, which is not significantly different from 2017 scores [14].

1.2 Assessment and LEarning in Knowledge Spaces (ALEKS)

ALEKS is a web-based learning system that incorporates AI. ALEKS uses Knowledge Space Theory (KST) for student assessments and learning based on a combinatorial and probabilistic model [4, 8]. ALEKS Assessment (pretest), comprise of 20–30 free

© Springer Nature Switzerland AG 2021
I. Roll et al. (Eds.): AIED 2021, LNAI 12749, pp. 471–475, 2021.
https://doi.org/10.1007/978-3-030-78270-2_83

response questions to determine initial knowledge (approximately 45 min). ALEKS produces a solved example when an explanation is requested for the current problem, and a similar new problem is generated for the student to solve. All the topics that students are ready to learn are presented in the individualized pie chart along with their learning progress [18].

For 9th grade, the learning concepts/subject matter in math is divided into 494 topics. Instead of providing scores for mastery of the content, it describes precisely where the student is in terms of knowledge and what they need to learn. To ensure retention of the learned topic, ALEKS reassess the student periodically by providing mixed questions and adjust the student's knowledge based on their responses.

2   Related Studies

A meta-analytic study on ALEKS's effectiveness on students' learning indicated that students' learning outcomes were like traditional classroom teaching. [9]. Another study conducted using the hybrid teaching approach, where ALEKS was incorporated to teach a graduate-level introductory statistics course. The results indicated no significant difference between hybrid and face-to-face instructional approaches [19]. In one study, a randomized experimental design was implemented with two groups ALEKS versus Teacher-led, to improve struggling students' mathematical skills. ALEKS was implemented in the presence of teachers in the after-school setting. Students performed at a similar level in both groups. However, students using ALEKS required significantly less assistance from the teachers with daily assignment completion [3].

3   Theoretical Framework

ITS, such as ALEKS, provides feedback based on responses. Students can use step-based hints if they face difficulty in solving a problem. For students to understand these hints requires considerable effort, supported by educational psychology and cognitive science [17]. ALEKS is based on a theoretical framework of KST that uses students' assessments and learning based on a combinatorial and probabilistic model [4, 8]. The KST represents the domain as a knowledge map, which further consists of a massive number of knowledge states (KS). Hence, KST can precisely predict students' current knowledge state and what a student is ready to learn.

Research Questions: The following research questions have been formulated for this proposed research:

RQ1) Does the time spent on ALEKS show statistically significant improvement in mathematics achievement scores than traditional teacher-led instructions, as measured by Pretests and Post-Tests?

RQ 2) Is there a statistically significant variation in students' affective engagement between students receiving instruction from the time spent on ALEKS versus students receiving traditional teacher-led instructions using SEI?

RQ3) Is there a statistically significant variation in students' cognitive engagement between students receiving instruction from the time spent on ALEKS versus students receiving traditional teacher-led instructions using SEI?

Variables: The Independent Variable (IV) is the time spent on ALEKS, and the dependent variables (DVs) are mathematics achievement, student's affective, and cognitive engagement in mathematics (Algebra l). ALEKS (Assessment and LEarning in Knowledge Spaces) is a web-based Intelligent Tutoring System that uses AI for learning and assessment. The DV, mathematics achievement, is defined as the understanding/ knowledge of mathematical concepts and skills as measured by their performance on assessments. The DV, student engagement, is an observable manifestation of the student's energy and effort in action, evidenced through a range of indicators such as cognitive, behavioral, and emotional engagement [1, 6, 13, 16].

3.1 Design and Methods

This proposed study will use a quasi-experimental design and investigate mathematics achievement (ITS-led versus teacher-led) and student engagement. ALEKS was chosen for this study as it was adopted recently by the school district. The ALEKS curriculum is aligned with state standards and has shown a positive association between the assessment performance and the state test scores. This study will be conducted on approximately one hundred 9th grade students from two schools with similar demographics. 100% of the students in both districts belong to an economically disadvantaged population, with approximately 9.2% of students are proficient in mathematics in School 1, and 12% of students are proficient in school 2 [5, 11]. Students in school 1 will be in the experimental group, and students from school 2 will be in the control group (nonrandomized). For the experimental group, teachers will use the McGraw curriculum "Reveal" and ALEKS as a supplemental tool. ALEKS will be used in the math support class as an instructional tool. This research study will be implemented for approximately five periods/55 min each per week for six weeks. The control group will only use teacher-led instructions for the same amount of time as the experimental group, and they do not have access to ALEKS. However, students in the math support class are taught by a different teacher. Students will take a pretest to generate a baseline about what students know. Pretest (Quiz) generated by ALEKS will determine students' prior knowledge in mathematics for the given topic. For this proposed study, both the experimental and control groups' overall performance will be compared with 81 topics out of 494, which falls in Unit 2: Reasoning with Linear Equations and Inequalities as per State standards. The content and the state standards will remain the same for both the experimental and the control groups. A Student Engagement Instrument (SEI), a brief -35 item self-reporting survey that measures students' cognitive (Control and Relevance of School Work, Future Aspirations and Goals, and Intrinsic Motivation) and affective engagement (Teacher-Student Relationships, Peer Support at School, Family Support for Learning) will be used to measure student engagement in the schools [1]. SEI will be deployed after the post-test, which will be administered after the completion of the unit. The SEI is a 5-point Likert scale varying from strongly agree, agree, neither agree nor disagree, disagree, to strongly disagree. The low scores indicate a high level of student engagement. The reliability of this scale in terms of internal consistency ranged from .76–.88 [2], and the test-retest interrater is .60–.62. The criterion-related validity reported positive correlations between engagement and indicators of academic performance.

4 Data Analyses

This proposed quasi-experimental study involves various descriptive and inferential statistics regarding ALEKS- led versus teacher-led instructions and student engagement. Independent sample t-test, ANCOVA will be used to identify whether mean differences between ALEKS versus teacher-led are statistically significantly different from one another.

5 Impact/Importance of Research

This proposed study will evaluate the effectiveness of intelligent tutoring systems, ALEKS, in the mathematics achievement of 9th-grade students (Algebra I) and shed light on students' cognitive and affective engagement using an intelligent tutoring system ALEKS. The findings from this proposed study will explore ALEKS effectiveness as a supplemental tool, identify challenges faced by students and teachers, offer suggestions for future implementation of ITS, and help school stakeholders determine the adequate integration of the ITS to improve students' learning outcomes.

6 Limitations/Future Research

ALEKS utilizes prerequisite hierarchies to represent the connection between items and formulas based on knowledge space theory [8]. The approaches used by ITS are mainly based on simple heuristics to access student mastery, such as whether the students get three correct in a row [12], which is based on the ASSIStments system, one of the most widely used ITS.

ALEKS analyzes diagnostic assessment and progress reports of students to ensure appropriate placement in ALEKS courses. After a diagnostic assessment, ALEKS generates a pie chart that indicates what a student already knows and ready to learn. If the diagnostic assessment results reveal that the student has shown insufficient progress or less than 15% mastery, then ALEKS recommends moving students to a lower-level course. Teachers may decline the suggestion to move students to the lower-level course to take state-level assessments. ALEKS has problem-solving tools, such as a calculator within the program, and is available only for certain assignments. The program assumes that students only use the calculator provided by ALEKS. ALEKS uses algorithms to provide personalized instructions and generates printable assignments for offline work based on the students' present knowledge.

Some future recommendations for this research study will be to conduct a similar study with different adaptive intelligent tutoring systems in mathematics and compare their efficacy with the ALEKS. This study can be replicated with different student participants from various grade levels.

References

1. Appleton, J.J., Christenson, S.L., Furlong, M.J.: Student engagement with school: critical conceptual and methodological issues of the construct. Psychol. Sch. **45**(5), 369–386 (2008). https://doi.org/10.1002/pits.20303

2. Appleton, J.J., Christenson, S.L., Kim, D., Reschly, A.L.: Measuring cognitive and psychological engagement: validation of the student engagement instrument. J. Sch. Psychol. **44**(5), 427–445 (2006). https://doi.org/10.1016/j.jsp.2006.04.002

3. Craig, S.D., et al.: The impact of a technology-based mathematics after-school program using ALEKS on student's knowledge and behaviors. Comput. Educ. **68**, 495–504 (2013). https://doi.org/10.1016/j.compedu.2013.06.010

4. Doble, C., Matayoshi, J., Cosyn, E., Uzun, H., Karami, A.: A data-based simulation study of reliability for an adaptive assessment based on knowledge space theory. Int. J. Artif. Intell. Educ. **29**(2), 258–282 (2019). https://doi.org/10.1007/s40593-019-00176-0

5. Downloadable data: The Governer's Office of Student Achievement, 17 February 2020. https://gosa.georgia.gov/report-card-dashboards-data/downloadable-data

6. Eccles, J., Wang, M., Christenson, S.L., Reschly, A.L., Wylie, C.: Handbook of Research on Student Engagement (2012). https://doi.org/10.1007/978-1-4614-2018-7

7. Falmagne, J.C., Albert, D., Doble, C., Eppstein, D., Hu, X. (eds.): Knowledge Spaces: Applications in Education. Springer, Heidelberg (2013). https://doi.org/10.1007/978-3-642-35329-1

8. Falmagne, J.C., Koppen, M., Villano, M., Doignon, J.P., Johannesen, L.: Introduction to knowledge spaces: How to build, test, and search them. Psychol. Rev. **97**(2), 201 (1990). https://doi.org/10.1037/0033-295X.97.2.201

9. Fang, Y., Ren, Z., Hu, X., Graesser, A.C.: A meta-analysis of the effectiveness of ALEKS on learning. Educ. Psychol. **39**(10), 1278–1292 (2019). https://doi.org/10.1080/01443410.2018.1495829

10. Fung, F., Tan, C.Y., Chen, G.: Student engagement and mathematics achievement: Unraveling main and interactive effects. Psychol. Sch. **55**(7), 815–831 (2018). https://doi.org/10.1002/pits.22139

11. GADOE: GADOE CCRPI reporting system. CCRPI09 (2019). https://ccrpi.gadoe.org/Reports/Views/Shared/_Layout.html

12. Heffernan, N.T., Heffernan, C.L.: The ASSISTments ecosystem: building a platform that brings scientists and teachers together for minimally invasive research on human learning and teaching. Int. J. Artif. Intell. Educ. **24**(4), 470–497 (2014). https://doi.org/10.1007/s40593-014-0024-x

13. Kuh, G.D.: What student affairs professionals need to know about student engagement. J. Coll. Stud. Dev. **50**(6), 683–706 (2009). https://doi.org/10.1353/csd.0.0099

14. NAEP: The nation's report card (2019). https://www.nationsreportcard.gov/highlights/mathematics/2019/

15. NCES: National center for education statistics (2019). https://nces.ed.gov/timss/results19/index.asp#/math/intlcompare

16. Skinner, E., Pitzer, J.R.: Developmental dynamics of student engagement, coping, and everyday resilience. In: Christenson, S.L., Reschly, A.L., Wylie, C. (eds.) Handbook of Research on Student Engagement, pp. 21–44. Springer, Boston (2012). https://doi.org/10.1007/978-1-4614-2018-7_2

17. Wittwer, J., Renkl, A.: Why instructional explanations often do not work: a framework for understanding the effectiveness of instructional explanations. Educ. Psychol. **43**(1), 49–64 (2008). https://doi.org/10.1080/00461520701756420

18. Xie, J., Essa, A., Mojarad, S., Baker, R.S., Shubeck, K., Hu, X.: Student learning strategies and behaviors to predict success in an online adaptive mathematics tutoring system. In: Proceedings of the International Conference on Educational Data Mining, pp. 460–465 (2017)

19. Xu, Y.J., Meyer, K.A., Morgan, D.D.: A mixed-methods assessment of using an online commercial tutoring system to teach introductory statistics. J. Stat. Educ. **17**(2) (2009). https://doi.org/10.1080/10691898.2009.11889524

Designing and Testing Assessments and Scaffolds for Mathematics Practices in Science Inquiry

Joe Olsen$^{(\boxtimes)}$ and Janice Gobert

Rutgers University, New Brunswick, NJ 08901, USA
joseph.olsen@rutgers.edu

Abstract. This project seeks to operationalize the mathematics practices that are needed for mathematically-integrated science inquiry, as outlined in the NGSS, and examine the effects of scaffolding these on competencies on the mathematics practices and on content acquisition. Using Evidence-Centered Design, I will design and pilot an assessment that can assess and scaffold competency in mathematics practices integral to virtual scientific inquiry. The assessment and scaffolds will be piloted with students and tested as to their efficacy at improving students' competencies on the practices of interest. After scaffolding on the practices (versus control), students will investigate two new phenomena. Statistical analyses will test if competencies on the practices are predictive of content acquisition in the new topics. The fine-grained operationalization of the practices into sub-practices and the empirical link of scaffolding these on content acquisition have implications for both research as well as science assessment and instruction.

Keywords: Mathematics practices · Science inquiry · Intelligent tutoring systems

1 Introduction

The Next Generation Science Standards (NGSS) identifies using mathematics (NGSS practice 5, we exclude computational thinking here) as an essential practice in science inquiry that students must master during high school [1]. Evidence supports that math is important for understanding science. First, practices like quantitative reasoning (akin to mathematics use in the NGSS), is related to physics content acquisition [3, 11]; and on the other hand, mathematics is a barrier to achievement in STEM [19–21] for a myriad of reasons, one of which is that students do not interpret mathematical symbols and procedures in a semantically-meaningful way, thereby limiting deep understanding of science phenomena [22, 24]. Given the emphasis of the NGSS on the integration of mathematics with science, what is now needed are rigorous operational definitions of the cognitive skills/processes used in mathematics-integrated science inquiry that can support rich domain content learning [c.f., 3, 22, 23, 25].

This dissertation will help fill the gaps in the literature and extend the assessments and scaffolds within Inq-ITS, an online inquiry environment that automatically assesses and scaffolds student inquiry practices using patented algorithms [4, 5]. Inq-ITS has recently

© Springer Nature Switzerland AG 2021
I. Roll et al. (Eds.): AIED 2021, LNAI 12749, pp. 476–481, 2021.
https://doi.org/10.1007/978-3-030-78270-2_84

been extended to include making predictions as one of the *using mathematics* practices [26, 27], but Inq-ITS does not currently assess or scaffold mathematics practices more fully, which could, in turn, support the development of robust, mathematically-integrated understandings of science phenomena (i.e., content). Below, I propose three studies as a part of a larger dissertation project that will expand Inq-ITS's functionalities.

2 Study 1: Developing an Assessment

Automated assessments and interventions can provide feedback to students in real time when it is most beneficial for learning [6]. In the first of the three studies in this project, I will develop a performance assessment that can measure the NGSS mathematics practices and its sub-practice competencies in science inquiry. This assessment will be developed using the Evidence-Centered Design (ECD) framework [7–9]. Important to both assessment and scaffold development, the result will be an operationalization of the important sub-practices of interest here, which in turn will be implemented into Inq-ITS so that that students can be assessed in real time while they conduct scientific inquiry. This study will also result in a set of scaffolds that are driven by the real-time assessment and target potential student difficulties.

2.1 Methods

The Evidence-Centered Design Process. The ECD framework decomposes an assessment into a set of models that structure assessment tasks. Together, the models and assessment tasks can elicit student competencies and provide clear and convincing evidence for claims about student competencies [7–9]. For a more complete treatment of the ECD process, see [7].

The final assessment tasks, after necessary iterations, can then be implemented in Inq-ITS, where knowledge-engineered algorithms will be used to both evaluate student competencies on the practices of interest and scaffold students on these practices [14–16]. These algorithms will be used for assessment and to drive the scaffolds on the sub-practices on which students need support.

Participants and Materials. I will engage approximately 10 high school students, chosen from a diverse sample, in inquiry with the new assessment tasks. Students will be asked to think aloud [12] as they conduct inquiry in order to understand what kind of knowledge and cognitive processes are elicited by the assessment tasks and what difficulties students have on the mathematics practices of interest. The initial versions of the assessment tasks will target practices identified in the literature. Student think-aloud data will be used to iterate on the design of the assessment tasks in order to identify sub-practices of the competencies of interest and ensure the tasks are eliciting evidence of students' competencies [c.f., 13].

Analyses. Student performance will be analyzed via their think aloud tasks data [17], which will be coded for students' competencies (none, partial, full) on the sub-practices identified in Study 1. From this analysis, I will determine if the assessment tasks are eliciting evidence of students' competencies on mathematical practices of interest and if there are unexpected sub-practices that were not included in the original design of the assessment. Hence, think-aloud data can inform iterations to the design of the assessment to account for unanticipated sub-practices. The data will also inform the development of scaffolds that can provide feedback on student performance.

3   Study 2: Piloting the Assessment

The second study to be conducted with a larger sample of students will affirm which sub-practices from the first study are challenging to students, as well as to determine if the automated scaffolds from Study 1 can successfully improve student performance on the new assessment tasks, as evidenced by improved performance on the respective sub-components of the inquiry practice of interest. The second study will also prepare students for two new investigations in Study 3, which will determine if scaffolding of the practices identified in Study 1 is predictive of content learning.

3.1   Methods

Participants and Materials. High school students will complete a pre-test of mathematics practices in the context of science inquiry that will focus on the practices and sub-practices defined in Study 1. All students will complete one mathematics-enhanced Inq-ITS lab that includes the new assessment tasks. The investigation of each question advances in discrete stages, including asking questions; carrying out investigations; graphing, modeling, analyzing and interpreting data; and explaining findings. The new assessment tasks developed in study 1 will be incorporated into the analyzing and interpreting data phase of inquiry. Students will be randomly assigned to a scaffolding vs. control condition in which the system will deliver meta-tutoring [25] at the sub-practice level when it detects that students are not demonstrating competence on the sub-components underlying the mathematics practices.

Measures and Analyses. Student performance on tasks in Inq-ITS will be captured in log files [16, 17] and analyzed using the knowledge-engineered algorithms from Study 1. A pre/post test of mathematics practices targeting the sub-practice competencies will also be developed alongside the assessment tasks to be included in Inq-ITS in order to obtain a system-independent measure of practice competence. Questions may include asking students to coordinate changes in quantities represented on a graph [c.f., 3].

Student performance scores will be determined in two ways: 1) on the practice and sub-practice level by knowledge-engineered algorithms, scoring students as either 0 or 1 on each sub-practice, as is typical in ITS work [18], and 2) the pre/post test scores. I will look for evidence of learning at a fine-grained level through a hierarchical logistic regression model that will determine if the amount of scaffolding received by students is predictive of improved performance on future opportunities to engage in each sub-practice.

4   Study 3: Transfer Effects of Mathematics Practices

As indicated in the introduction, I hypothesize that scaffolding students on the sub-practices of mathematizing during science inquiry will enable students to deeply learn physics content from their inquiry. This study will test this hypothesis by examining physics content learning gains from Inq-ITS investigations that are new to the students to determine if mathematics practice competencies predict content acquisition.

4.1   Methods

Participants and Materials. The same high school students from study 2 will participate in one of two Inq-ITS labs with the new assessment activities. No students will be scaffolded on the new mathematics practices. In one lab, students will investigate Newton's Universal Law of Gravitation, while in another lab students will investigate objects in free-fall. Each lab will consist of answering three questions about the phenomena by engaging in mathematically-integrated science inquiry. Students will complete pre/post assessments that focus on content understandings that are relevant to the new investigations in order to obtain an additional measure of content learning.

Measures and Analyses. As in study 2, student performance will be measured on the practice and sub-practice level by knowledge-engineered algorithms, with overall practice scores being determined by averaging across binary sub-practice scores. In addition, a pre/post test will measure student knowledge of physics content, either Newton's Universal Law of Gravitation or free-fall.

Regression analyses will determine if mathematics practice scores during inquiry in Inq-ITS are predictive of posttest scores measuring content understandings while controlling for pretest scores. Students who score higher on mathematics practices (as determined by Inq-ITS algorithms and scores on the mathematics practices post-test from study 2) are predicted to also score higher on the content post-test during the transfer tasks in study 3.

5   Discussion

This project will contribute a set of assessment tasks that target operationalized NGSS-aligned mathematics practices for science inquiry; these can guide future research on the teaching and learning inquiry. Additionally, these assessments can be auto-scored and their practices auto-scaffolded at scale. Lastly, the empirical link to content learning provides evidence that instructors and researchers should focus on deep integration of mathematics and science content to support science content learning in inquiry.

References

1. Next Generation Science Standards Lead States: Next generation science standards: for states, by states. National Academies Press, Washington (2013)

2. Pruitt, S.L.: The next generation science standards: the features and challenges. J. Sci. Teacher Educ. **25**(2), 145–156 (2014)
3. Panorkou, N., Germia, E.F.: Integrating math and science content through covariational reasoning: the case of gravity. Math. Thinking Learn. 1–26 (2020)
4. Gobert, J.D., Sao Pedro, M., Raziuddin, J., Baker, R.S.: From log files to assessment metrics: measuring students' science inquiry skills using educational data mining. J. Learn. Sci. **22**, 521–563 (2013)
5. Gobert, J., Moussavi, R., Li, H., Sao Pedro, M., Dickler, R.: Scaffolding students' on-line data interpretation during inquiry with Inq-ITS. In: Cyber-physical laboratories in engineering and science education. Springer (2018)
6. Koedinger, K.R., Corbett, A.: Cognitive tutors: technology bringing learning sciences to the classroom. In: Sawyer, R.K. (ed.) The Cambridge Handbook of the Learning Sciences, pp. 61–77. Cambridge University Press, Cambridge (2005)
7. Mislevy, R.J., Haertel, G., Riconscente, M., Rutstein, D.W., Ziker, C.: Evidence-centered assessment design. In: Mislevy, R.J., Haertel, G., Riconscente, M., Rutstein, D.W., Ziker, C. (eds.) Assessing Model-Based Reasoning Using Evidence-Centered Design, pp. 19–24. Springer, Cham (2017). https://doi.org/10.1007/978-3-319-52246-3_3
8. Hendrickson, A., Ewing, M., Kaliski, P.: Evidence-centered design: recommendations for implementation and practice. J. Appl. Test. Technol. **1**(1), 1–27 (2013)
9. Mislevy, R.J., Almond, R.G., Lukas, J.F.: A brief introduction to evidence-centered design. ETS Res. Rep. Ser. **2003**(1), i–29 (2003)
10. Gobert, J.D., Moussavi, R., Li, H., Sao Pedro, M., Dickler, R.: Real-time scaffolding of students' online data interpretation during inquiry with Inq-ITS using educational data mining. In: Auer, M., Azad, A., Edwards, A., de Jong, T. (eds.) Cyber-Physical Laboratories in Engineering and Science Education, pp. 191–217. Springer, Cham (2018)
11. Castillo-Garsow, C.: Continuous quantitative reasoning. Quant. Reason. Math. Model: Driver STEM Integr. Educ. Teach Context **2**, 55–73 (2017)
12. Ericsson, K.A., Simon, H.A.: Verbal reports as data. Psychol. Rev. **87**(3), 215 (1980)
13. Gobert, J.D., Sao Pedro, M.A.: Digital assessment environments for scientific inquiry practices. Wiley Handb. Cogn. Assess. Frameworks Methodol. Appl. 508–534 (2017)
14. Studer, R., Benjamins, V.R., Fensel, D.: Knowledge engineering: principles and methods. Data Knowl. Eng. **25**(1–2), 161–197 (1998)
15. Gobert, J.D., Sao Pedro, M.A., Baker, R.S., Toto, E., Montalvo, O.: Leveraging educational data mining for real-time performance assessment of scientific inquiry skills within microworlds. J. Educ. Data Min. **4**(1), 104–143 (2012)
16. Chi, M.T.: Quantifying qualitative analyses of verbal data: a practical guide. J. Learn. Sci. **6**(3), 271–315 (1997)
17. Carlson, M., Jacobs, S., Coe, E., Larsen, S., Hsu, E.: Applying covariational reasoning while modeling dynamic events: a framework and a study. J. Res. Math. Educ. **33**(5), 352–378 (2002)
18. Anderson, J.R., Corbett, A.T., Koedinger, K.R., Pelletier, R.: Cognitive tutors: lessons learned. J. Learn. Sci. **4**(2), 167–207 (1995)
19. Basson, I.: Physics and mathematics as interrelated fields of thought development using acceleration as an example. Int. J. Math. Educ. Sci. Technol. **33**(5), 679–690 (2002)
20. Sadler, P.M., Tai, R.H.: Success in introductory college physics: the role of high school preparation. Sci. Educ. **85**(2), 111–136 (2001)
21. Ellis, J., Fosdick, B.K., Rasmussen, C.: Women 1.5 times more likely to leave STEM pipeline after calculus compared to men: lack of mathematical confidence a potential culprit. PloS One **11**(7), e0157447 (2016)
22. Redish, E.F., Kuo, E.: Language of physics, language of math: disciplinary culture and dynamic epistemology. Sci. Educ. **24**(5), 561–590 (2015)

23. Bradshaw, G.F., Langley, P.W., Simon, H.A.: Studying scientific discovery by computer simulation. Science **222**(4627), 971–975 (1983)
24. Nixon, R.S., Godfrey, T.J., Mayhew, N.T., Wiegert, C.C.: Undergraduate student construction and interpretation of graphs in physics lab activities. Phys. Rev. Phys. Educ. Res. **12**(1), 010104 (2016)
25. VanLehn, K.: Model construction as a learning activity: a design space and review. Interact. Learn. Environ. **21**(4), 371–413 (2013)
26. Sao Pedro, M.: Real-time formative assessment of NGSS mathematics practices for high school physical science [Grant] (2018)
27. Dickler, R.: An intelligent tutoring system and teacher dashboard to support mathematizing during science inquiry. In: Isotani, S., Millán, E., Ogan, A., Hastings, P., McLaren, B., Luckin, R. (eds.) AIED 2019. LNCS (LNAI), vol. 11626, pp. 332–338. Springer, Cham (2019). https://doi.org/10.1007/978-3-030-23207-8_61

Contextual Safeguarding in Education: Bayesian Network Risk Analysis for Decision Support

Matthew Woodruff[1](✉) ⓘ and Graham Feek[2]

[1] Department of Computer Science, University of Surrey, Guildford, Surrey GU2 7XH, UK
m.woodruff@surrey.ac.uk
[2] Greenwood Academy Trust, Nottingham NG4 2JY, UK

Abstract. A Multi Academy Trust in the UK operates thirty-five academies educating 17,000 students across seven local authority areas. Significant societal problems are increasing risk to young people, including exploitation and violent crime associated with gang culture and drugs. A predictive analytics system is being implemented to support the delivery of contextual safeguarding, where the interplay of the school, peer, family and community environments determine the safeguarding risk. Due to the intense level of human activity required by safeguarding teams to identify and intervene with those at risk, Bayesian network risk modelling is being integrated with traditional analytics to extend and augment human capacity. The participants are keenly aware of the potential for harm from this data; in its collation, appropriateness of methods, accuracy and validity of output, and the human interpretation and resulting actions and impact on young people.

Keywords: Contextual safeguarding · Bayesian network · Predictive analytics · Ethics · Ethical framework · Risk assessment

1 A Landscape of Rising Safeguarding Concern in Schools

Over the past eight years, knife crime levels in the UK have increased by 42% - with 20% of these perpetrated by those under the age of 18. Over 47,000 incidents were reported by the UK Home Office in 2019 alone [1]. On the 31st January 2019 the Home Secretary announced new measures, including Knife Crime Prevention Orders [2].

A Multi Academy Trust in the UK currently has 35 academies educating approximately 17,000 pupils across seven local authority areas. Recently this Trust has sought expertise from both multinational technology vendors and local data science and AI companies to address data and analytics to help identify young people at risk, and together have built a consortium to include a prominent safeguarding system in the UK K-12 market, and is now fostering wider dialogue with the local authority, other government agencies, and academic and other bodies such as the Institute for Ethical AI in Education.

Multi agency approaches to prevention have been identified as essential [3], where police, education, health and social sectors share information and work together. All too often, barriers prevent the timely sharing of information not only between partners, but within single agencies themselves. Agencies, including schools, the police, LAs and community partnerships need to get better at sharing information about gang networks in order to safeguard these children and other pupils [4].

© Springer Nature Switzerland AG 2021
I. Roll et al. (Eds.): AIED 2021, LNAI 12749, pp. 482–486, 2021.
https://doi.org/10.1007/978-3-030-78270-2_85

2   Theoretical Framework and Methodology

2.1   Anticipated Solution

The solution aggregates data across all schools, to identify patterns in over all students. Firstly, the schools Management Information System, providing daily attendance patterns, fixed and permanent exclusions, school moves, conduct/behaviour, achievement, academic performance and contextual information. Second, the safeguarding system used for day to day recording and tracking of safeguarding concerns, and a secure documentation store; and finally open data such as from the Department of Education, and the Police.

Data is kept up to date daily, and provides the source for integrated analytics that provides secure, and role based appropriate, views of the information to the practitioners that need it. Vitally, it combines these multiple sources of data into a single view – for the first-time allowing comparison and correlation between all systems.

2.2   Risk Classification

Risk classification leverages data associated with different aspects of the student context, as proposed by the contextual safeguarding framework [3] (Fig. 1).

Fig. 1. Contextual safeguarding framework used to define classification algorithms.

Ultimately the system will learn from pattern matching contextual factors with the identification of where risk has translated into the actuality of the outcome (e.g. an instance of violent crime or exploitation). Training for these models is being provided by expert human practitioner input with historical data, and the creation of specific rules to capture current human processes and some that are not currently practical to be applied by a human.

2.3   Theoretical Model

Initial findings from this work already show that the human processes on the classification of those at risk is highly algorithmic in nature, and the initial model worked up is described visually in Fig. 2 – as the 'Spheres of influence'. A sample of the rules in each sphere is provided in Table 1.

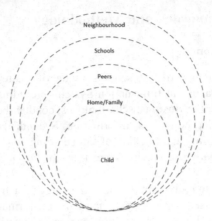

Fig. 2. Contextual safeguarding framework used to define classification algorithms.

Table 1. Classification rules.

Context	Classification rule
Young person	3 month moving-average safeguarding incidents logged
Family	Linked accounts incident activity, contact and address changes
Peers	Patterns of linked account school attendance, esp. morning late
School	Rapid change to patterns of attainment, behavior or exclusion
Neighbourhood	Crime and hospital statistics for home postcodes, Income Deprivation Affecting Children Index (IDACI) measures

Within each sphere, the safeguarding lead performs a number of different actions required to obtain as complete as possible picture of the context, in order for it to inform an overall level or risk and therefore the actions that are required to mitigate that risk and/or directly intervene. The initial dataset covers 1000 secondary phase students and this pilot shall extend in an ethically controlled manner to 17,000. Student records contain approximately 20 features (such as Date of Birth, Level of Special Educational Need, measures of deprivation, prior attainment, ethnicity). Further to this, features shall be built from other available facts, such as changing patterns in attendance, achievement, behavior, and online engagement.

Interviews to capture and classify these observations are underway. They include gathering data from various school online systems such as the count of safeguarding incidents recorded in the last 30 days and trends over time. This contextual safeguarding risk (R_{CS}) could therefore be defined as:

$$R_{CS} = f_1(C, H, P, S, N)$$

where C is the risk coefficient of the Sphere of Influence of the Child defined by the function

$$C = f_2(c_1, \ldots, c_n)$$

and H is the risk coefficient of the Sphere of Influence of the Home defined by the function

$$H = f_3(h_1, \ldots, h_n)$$

and P is the risk coefficient of the Sphere of Influence of the Peers defined by the function

$$P = f_4(p_1, \ldots, p_n)$$

and S is the risk coefficient of the Sphere of Influence of the School defined by the function

$$S = f_5(s_1, \ldots, s_n)$$

and N is the risk coefficient of the Sphere of Influence of the Neighbourhood defined by the function

$$N = f_6(n_1, \ldots, n_n)$$

3 Expected Contribution to AIED

It is paramount that full consideration is given in implementation to the ethical considerations of the use of data for the purposes of tailoring safeguarding interventions. To understand both the context of the use of the technology – not because we 'can' but because we 'should' – and the impact on human lives.

Research and policy on the ethics of AI in general is developing [5] and in public sector and education is emergent [6]. There are also the beginnings of work on self-assessment tools [7]. However, a concrete policy framework does not yet exist that is specific to Ethical AI in K-12 education, or indeed specifically one of the most sensitive aspects: safeguarding young people. Two specific contributions to AIED would be the application of differential privacy to the datasets in order to support the scaling and reuse of the models (generalizability) and also a concept of human practitioner controlled 'fairness' – adjustment of the bias of the model which can be set to different level depending on the sensitivity of the context.

In addition to these two important ethical constraints to the model, this research also seeks to identify if a Bayesian network approach is the most appropriate methodology (due to inherent advantages over neural network and other 'black box' approaches through dealing with both uncertainty and explainability), or if there are other machine learning methods that might be likewise suitable (and effective) in this context, such as generalized multiple linear regression, or causal modeling techniques.

References

1. Allen, G., Audickas, L.: Knife Crime in England and Wales, House of Commons Library Briefing Paper (2019). https://researchbriefings.parliament.uk/ResearchBriefing/Summary/SN04304. Accessed 2 Feb 2021
2. Home Office. Home Secretary announces new police powers to deal with knife crime, 31 January 2019. https://www.gov.uk/government/news/home-secretary-announces-new-police-powers-to-deal-with-knife-crime. Accessed 2 Feb 2021
3. Firmin, C.: Contextual safeguarding, November 2017. https://contextualsafeguarding.org.uk/assets/documents/Contextual-Safeguarding-Briefing.pdf. Accessed 2 Feb 2021
4. The Annual Report of Her Majesty's Chief Inspector of Education, Children's Services and Skills 2017/18, 4 December 2018. https://assets.publishing.service.gov.uk/government/uploads/system/uploads/attachment_data/file/761606/29523_Ofsted_Annual_Report_2017-18_041218.pdf. Accessed 2 Feb 2021
5. High-level expert group on Artificial Intelligence, European Commission. Ethics Guidelines for Trustworthy AI (2019). https://ec.europa.eu/digital-single-market/en/news/ethics-guidelines-trustworthy-ai. Accessed 5 Nov 2021
6. Leslie, D.: Understanding artificial intelligence ethics and safety: a guide for the responsible design and implementation of AI systems in the public sector. The Alan Turing Institute (2019). https://doi.org/10.5281/zenodo.3240529
7. The Institute for Ethical AI in Education: Provisional guidance for educational institutions deploying artificial intelligence: a self-assessment tool (2020). https://www.buckingham.ac.uk/wp-content/uploads/2020/02/Provisional-Guidance-for-Educational-Institutions_-a-self-assessment-tool.pdf. Accessed 15 Feb 2021
8. Woolf, B.: Ethics in AIED – Who cares? AIED (2019)
9. Author, F.: Article title. Journal 2(5), 99–110 (2016)
10. Author, F., Author, S.: Title of a proceedings paper. In: Editor, F., Editor, S. (eds.) CONFERENCE 2016. LNCS, vol. 9999, pp. 1–13. Springer, Heidelberg (2016)
11. Author, F., Author, S., Author, T.: Book title. 2nd edn. Publisher, Location (1999)
12. Author, F.: Contribution title. In: 9th International Proceedings on Proceedings, pp. 1–2. Publisher, Location (2010)
13. LNCS Homepage. http://www.springer.com/lncs. Accessed 21 Nov 2016

Author Index

Printed in the United States
by Baker & Taylor Publisher Services